CONSTRUCTION PRINT READING

CONSTRUCTION PRINT READING

Leonard Koel

CADD Illustrations by Christopher Alexander

Delmar Publishers

an *International Thomson Publishing company* I(T)P®

Albany • Bonn • Boston • Cincinnati • Detroit • London • Madrid
Melbourne • Mexico City • New York • Pacific Grove • Paris • San Francisco
Singapore • Tokyo • Toronto • Washington

NOTICE TO THE READER

Cover Design by Nicole Reamer
Cover Photo: Benjamin Shearn/FPG International LLC

Delmar Staff:
Publisher: Alar Elken
Aquisitions Editor: Tom Schin
Production Manager: Larry Main
Art Director: Nicole Reamer
Senior Project Editor: Christopher Chien
Editorial Assistant: Fionnuala McAvey

Online Services

Delmar Online
To access a wide variety of Delmar products and services on the World Wide Web, point your browser to:
http://www.delmar.com
or email: info@delmar.com

A service of I(T)P

COPYRIGHT © 2000
By Delmar Publishers

The ITP logo is a trademark under license

Printed in the United States of America

For more information, contact:

Delmar Publishers
3 Columbia Circle, Box 15015
Albany, New York, 12212-5015

International Thomson Publishing Europe
Berkshire House
168-173 High Holborn
London WC1V7AA
United Kingdom

Nelson ITP, Australia
102 Dodds Street
South Melbourne,
Victoria, 3205 Australia

Nelson Canada
1120 Birchmount Road
Scarborough, Ontario
M1K 5G4, Canada

International Thomson Publishing France
Tour Maine-Montparnasse
33 Avenue du Maine
75755 Paris Cedex 15, France

International Thomson Editores
Seneca 53
Colonia Polanco
11560 Mexico D. F. Mexico

International Thomson Publishing GmbH
Königswinterer Strasße 418
53227 Bonn
          Germany

International Thomson Publishing Asia
60 Albert Street
#15-01 Albert Complex
Singapore 189969

International Thomson Publishing Japan
Hirakawa-cho Kyowa Building, 3F
2-2-1 Hirakawa-cho, Chiyoda-ku,
Tokyo 102, Japan

ITE Spain/Paraninfo
Calle Magallanes, 25
28015-Madrid, España

  2 3 4 5 6 7 8 9 10   XXX 03 02 01 00 99
Library of Congress Card Number:

ISBN: 07668-08394

Library of Congress Cataloging-in-Publication Data

Koel, Leonard.
     Construction print reading / Leonard Koel ; CAD illustrations by
Chris Alexander.
          p.   cm.
     ISBN 07668-08394
     1. Building—Details—Drawings. 2. Blueprints. 3. Computer-aided
design. I. Title.
TH431.K64 1999
692'.1—dc21
                                        99-27345
                                        CIP

Contents

PREFACE

Construction Print Reading is intended for beginning to intermediate students. Developing the ability to read plans (also referred to as prints) is a skill that can be acquired by most persons who are willing to apply themselves. At management level, a clear understanding of a set of building plans is necessary to organize and supervise the entire construction process. General contractors and subconstractors must carefully study the plans to estimate the cost of a project before bidding on the work. An understanding of building plans is a requisite for the more highly skilled building trade workers on the job. These are the men and women who apply their knowledge and skill to transform the plans from a drawing to the actual construction. *Construction Print Reading* can also be a valuable beginning text for those persons planning a career in architecture and building design.

A complete set of building plans normally includes information about the general construction, electrical, and plumbing work, as well as heating, ventilation, and air conditioning. Also included is information applying to tile setting, painting, masonry, and other areas of work. A well-organized building project requires that the supervisors and trade persons of the many crafts work together. Therefore, understanding each others' information included in the plans is important.

Construction Print Reading is broken down into eight major sections. The sections are divided into units. Review questions are placed at the end of each unit, and answers are found in the appendix of the book. One or more section tests are provided at the end of each section, and answers to these tests are found in the *Instructor's Guide.* A glossary of construction terms used in the text is provided for the further knowledge and convenience of the reader.

Section 1, Construction Drawings, describes the basic language of building plans. This language includes the many lines, symbols, and abbreviations found on construction drawings. In addition, Section 1 compares orthographic and pictorial drawings.

Section 2, Fundamentals of Construction Plans, presents a detailed analysis of all the drawings in a set of house plans. A pictorial comparison is provided for each orthographic drawing on the plans. At the same time, the major components in the plan drawings are discussed in the text. Each unit in Section 2 also includes some general construction information related to the drawings.

Section 3, Electrical, Plumbing, HVAC Plans, discusses in greater detail more complex electrical, plumbing, and heating-ventilating-air conditioning drawings for larger residential and commercial projects.

Section 4, Plan Development, discusses some major factors that influence the design of a building. These include topographical and environmental considerations as they relate to the immediate site and general location of the structure. Basic house design, good room planning, and a strong emphasis on energy conservation round out this section.

Section 5, Code Regulations and Legal Documents, serves as an introduction and comparison of the more widely used national model building codes. Most local code regulations are patterned after the model codes. The building codes are one of the major factors influencing building design. An understanding of contractual agreements and other related legal documents is important for managerial and supervisory persons in the construction industry.

Section 6, Related Mathematics, is important for students who need to review the mathematics of linear measurement. Also, at a time when both English and metric measurements are being used, metric measurement and the conversion process from metric to English measurement are important when studying and working with construction plans.

Section 7, Materials, discusses some more commonly used materials used in the construction of a building. This section is not meant to be a total comprehensive study of building materials, as that would require a separate text. The main emphasis in Section 7 is on materials referred to in the plans included in this book.

Section 8, Plan Assignments, contains lengthy assignments for a packet of three full sets of construction plans accompanying the book. These are plans for buildings that have already been built.

As the reader works his or her way through *Construction Print Reading,* the direct relationship among all the sections becomes more apparent. Study methods are very important to get the most value from the text. Be sure that each unit is well understood before going on to the next one. The Review Questions should be answered as best one can before checking against the Answer Keys provided in the appendix.

ACKNOWLEDGEMENTS

The author and publisher wish to thank the following companies, organizations, and individuals who have contributed photographs, tables, and other valuable information for this book.

Alexander Studio
American Institute of Architects
American Plywood Association
Anderson Windows, Inc.
Ayers Construction Company
Sarah Baker, Engineer
Josh Barclay, Architect
Building Officials and Code Administrators
   International (BOCA)
Michael Connell, architect
Construction Specification Institute
Garlinghouse Company, Inc.
Honeywell, Inc.
International Conference of Building Officials
Portland Cement Association
Timothy Pulver, architect
Southern Building Code Congress International
Western Wood Products Association

SECTION 1

Building Plans

Many groups and individuals are involved in the construction of a building: architect, general contractor (builder), subcontractors (electrical, plumbing, etc.), suppliers, construction workers. **Building plans,** or **construction plans,** make it possible for all these groups and individuals to coordinate their work by providing an overall view of the structure to be built and the necessary information to guide each stage of its construction.

Construction plans are often referred to as **blueprints.** This term derives from an older process of duplicating plans that produced white lines against a blue background. Modern reproduction methods show black or blue lines against a white background. Therefore, calling these drawings the *prints* or *plans* is more accurate.

To acquire the skill of plan reading, you must understand the *orthographic method* used to produce *working drawings.* (These terms will be explained in this chapter.) You must also be familiar with the basic **conventions** of construction plans—that is, the generally accepted practices for conveying information in the drawings. This is necessary because the drawings have very limited space for expressing information. Typical conventions include measurements, lines, scales, symbols, and abbreviations. ■

UNIT 1

Construction Drawings

A full set of construction plans includes a series of **working drawings,** which provide the contractors, trade persons, and all other individuals concerned with the building project information needed to complete every phase of the construction work. Working drawings go through a number of stages before they are completed. The process often begins with *pictorial* drawings that form the basis upon which the *orthographic* working drawings are developed.

Besides drawings, a full set of construction plans will contain specifications, giving additional written data not appearing in the drawings, and door and window schedules. Plans for larger and/or more complex projects may also feature schedules relating to electrical, plumbing, heating and cooling, and other mechanical drawings.

PICTORIAL AND ORTHOGRAPHIC DRAWINGS

Pictorial drawings are *three-dimensional* and can be compared with photographs of various views of the finished building. Architects use pictorial drawings of the exterior and interior of the building when developing plans for a client so that the client can visualize the information to be presented in the construction plans. The position of the building in relation to the surrounding terrain can

also be described in pictorial drawings. **(See Figure 1–1.)** However, a pictorial drawing cannot be used as a working drawing because it cannot include all the measurements and data necessary for construction.

Orthographic projection is the method used to create working drawings. *Orthographic* means "straight line," and projection refers to the "parts of an object drawn on a flat surface." Orthographic drawings are *two-dimensional,* showing only one side or surface of an object. Therefore, more than one orthographic drawing is needed to describe a single object. This makes it possible to show all dimensions and other information required for construction work. Orthographic projections for the pictorial drawing shown in Figure 1–1 would include separate orthographic drawings for the two exterior walls and the section of roof visible in the picture. **(See Figure 1–2.)**

WORKING DRAWINGS

In the series of working drawings that help make up a complete set of construction plans, each separate working drawing has a particular function. However, all the drawings are interrelated. A typical set of residential construction plans will contain the following:

■ Figure 1–1.
Pictorial drawings are three-dimensional and can provide a realistic picture of the building in relation to the surrounding terrain.

FRONT ELEVATION

SIDE ELEVATION

■ **Figure 1–2.**
An orthographic projection of the pictorial drawing shown in Figure 1–1 requires a separate drawing for the front and one side of the building.

✓ **Site Plan.** Information on preliminary groundwork, including grading, excavation, and building layout. This is also called a **plot plan.**

✓ **Foundation Plan.** Plan and section views that provide information on layout, components, and construction of the foundation.

✓ **Floor Framing Plan.** A guide to layout and construction of the floor unit above the foundation.

✓ **Floor Plan.** A major plan providing information on layout and construction of the walls above the floor unit. Window and door locations are also identified.

✓ **Door and Window Schedules.** Finish dimensions of doors and windows and the rough openings for them.

✓ **Ceiling Framing Plan.** A guide to layout and construction of the ceiling unit above the walls.

✓ **Roof Framing Plan.** Information on components, layout, and construction of the roof.

✓ **Section Views.** Vertical cuts made of parts of the building. These views provide essential structural information that cannot be clearly shown on other drawings.

✓ **Details.** Enlarged drawings of a small segment of a larger plan, providing structural information that is not clear on the larger and more general plan.

✓ **Elevation Plans.** A view from one side of an exterior or interior wall. The drawings provide information on the finish-wall materials and items attached to the walls. They also provide necessary vertical dimensions. Elevation plans are useful in clarifying information on the floor plans and other drawings.

✓ **Mechanical-Electrical-Plumbing Plan.** Information on electrical, plumbing, and heating installation.

Figure 1–3a-e shows a series of orthographic working drawings for a set of residential construction plans. These drawings have been considerably reduced from the original. Therefore, reading the measurements and other information is not possible. Each orthographic drawing is accompanied by a pictorial drawing.

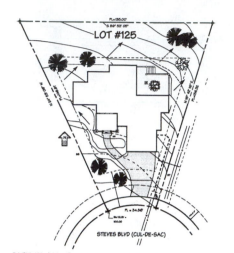

SITE PLAN: Ground work and building layout.

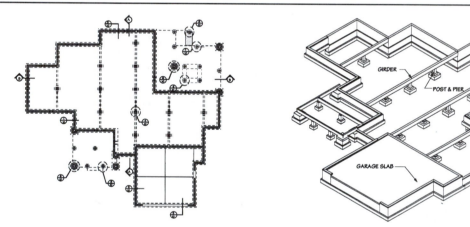

FOUNDATION PLAN: (PLAN VIEW): Layout and construction of foundation footings, walls, pier footings, posts & girders.

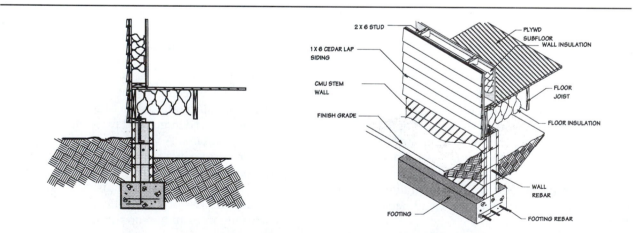

FOUNDATION SECTION VIEW: Provides measurements and information that does not appear in plan view.

■ **Figure 1–3a.**
Basic drawings guide the construction of a building.

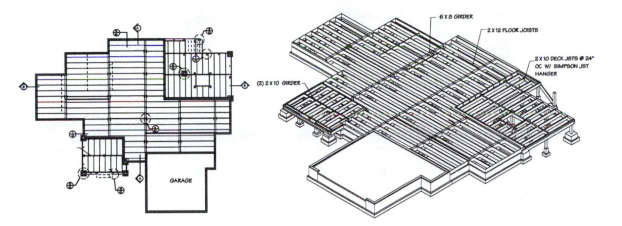

FLOOR FRAMING PLAN: Guides construction of floor unit above foundation.

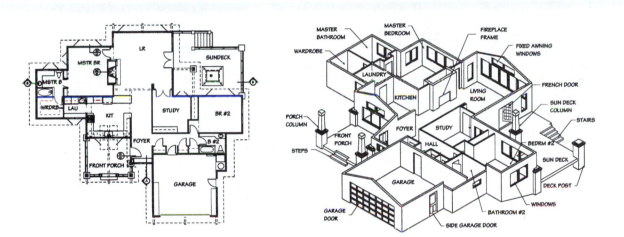

FLOOR PLAN: Major plan providing information to lay out and construct the walls over the floor unit.

■ **Figure 1–3b.**
Basic drawings guide the construction of a building (continued).

RELATED DRAWINGS

Studying several different drawings to obtain all the information needed to complete a section of a construction project is often necessary. For example, to lay out and construct the north window opening in the master bedroom shown on the floor plan in Figure 1–3, information must be obtained from the floor plan, window schedule, a section view, and an exterior elevation. **(See Figure 1–4.)**

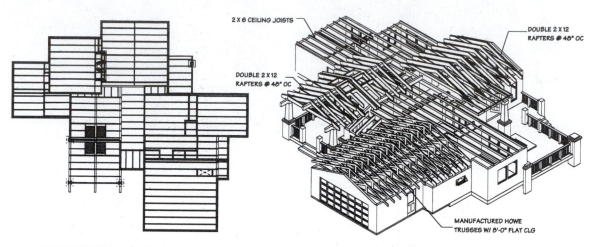

CEILING FRAMING PLAN: Shows components, layout, and construction of the ceiling over the walls.

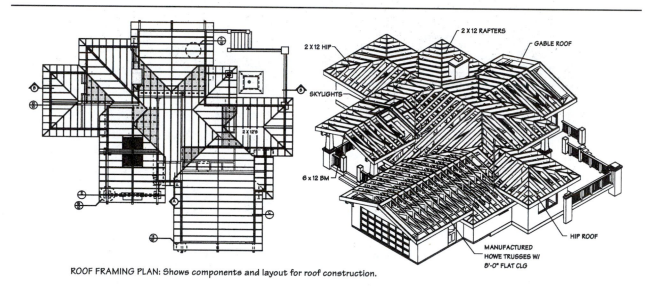

ROOF FRAMING PLAN: Shows components and layout for roof construction.

■ **Figure 1–3c.**
Basic drawings guide the construction of a building (continued).

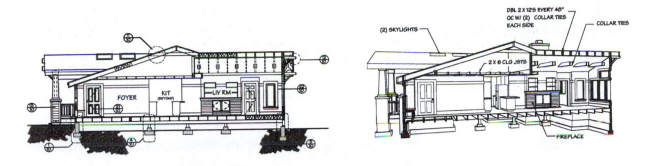

BUILDING SECTION: Vertical cut of the building providing essential structural information not seen on other plans.

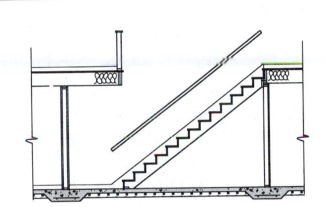
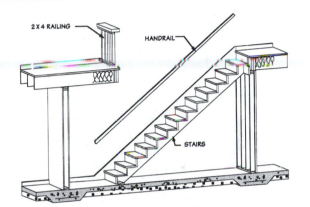

STAIR SECTION: Shows components and layout of stairway.

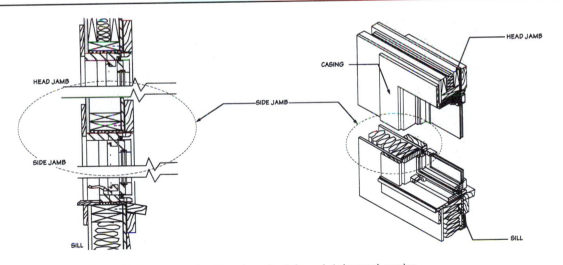

WINDOW SECTION: Provides detailed infromation about the window and window rough opening.

■ **Figure 1–3d.**
Basic drawings guide the construction of a building (continued).

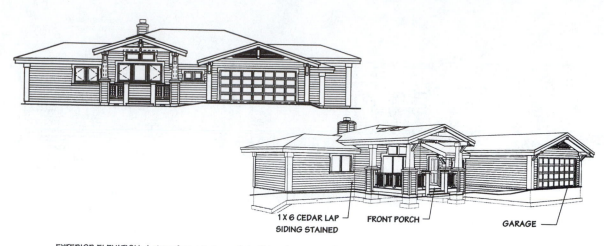

1 X 6 CEDAR LAP
SIDING STAINED

FRONT PORCH

GARAGE

EXTERIOR ELEVATION: A view of an exterior wall clarifying the exterior finish work.

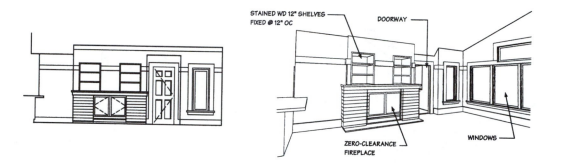

STAINED WD 12" SHELVES
FIXED @ 12" OC

DOORWAY

ZERO-CLEARANCE
FIREPLACE

WINDOWS

INTERIOR ELEVATION: A view of a finished interior wall and cabinets and other items attached to the wall.

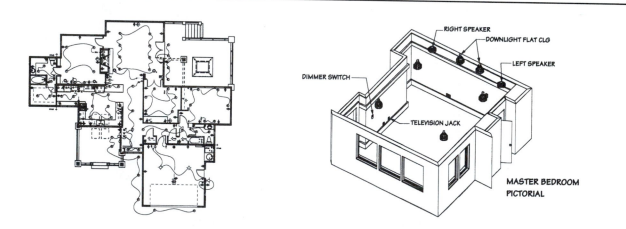

RIGHT SPEAKER

DOWNLIGHT FLAT CLG

LEFT SPEAKER

DIMMER SWITCH

TELEVISION JACK

MASTER BEDROOM
PICTORIAL

ELECTRICAL-PLUMBING-HEATING PLAN: Provides information regarding electrical, plumbing, and heating installation.

■ **Figure 1–3e.**
Basic drawings guide the construction of a building (continued).

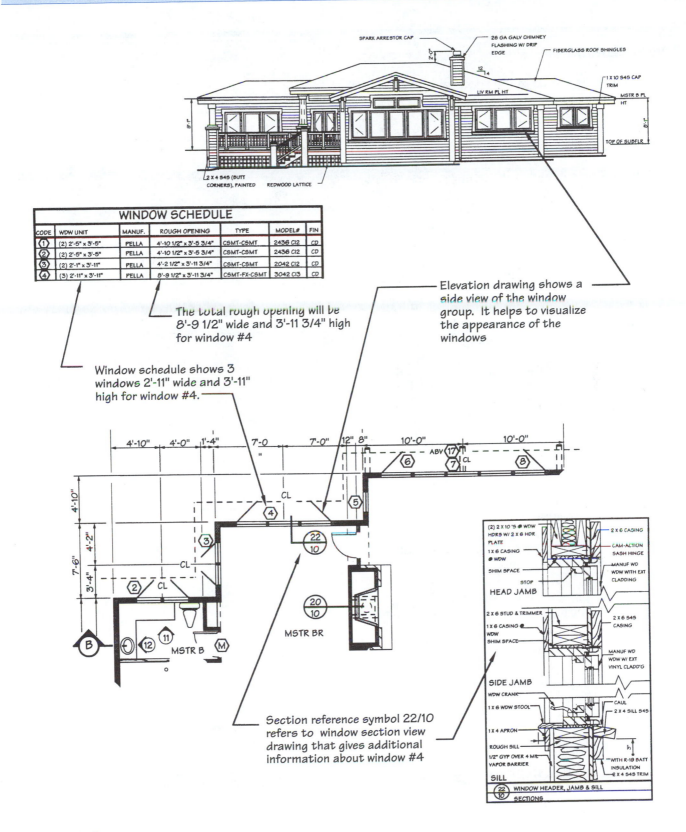

WINDOW SCHEDULE

CODE	WDW UNIT	MANUF.	ROUGH OPENING	TYPE	MODEL#	FIN
①	(2) 2'-5" x 3'-5"	PELLA	4'-10 1/2" x 3'-5 3/4"	CSMT-CSMT	2436 CI2	CD
②	(2) 2'-5" x 3'-5"	PELLA	4'-10 1/2" x 3'-5 3/4"	CSMT-CSMT	2436 CI2	CD
③	(2) 2'-1" x 3'-11"	PELLA	4'-2 1/2" x 3'-11 3/4"	CSMT-CSMT	2042 CI2	CD
④	(3) 2'-11" x 3'-11"	PELLA	8'-9 1/2" x 3'-11 3/4"	CSMT-FX-CSMT	3042 CI3	CD

The total rough opening will be 8'-9 1/2" wide and 3'-11 3/4" high for window #4

Window schedule shows 3 windows 2'-11" wide and 3'-11" high for window #4.

Elevation drawing shows a side view of the window group. It helps to visualize the appearance of the windows

Section reference symbol 22/10 refers to window section view drawing that gives additional information about window #4

■ **Figure 1–4.**
Several related drawings must be studied to understand the layout and construction of the #4 window opening in the north wall of the master bedroom.

REVIEW QUESTIONS

Enter the missing words in the spaces provided on the right.

1. A full set of construction plans includes a series of _____ drawings.

   _____

2. Three-dimensional drawings are called _____ drawings.

   _____

3. Pictorial drawings cannot be used as _____ drawings.

   _____

4. Pictorial drawings can be compared to _____ of different views of a building.

   _____

5. The method used to create working drawings is called _____ projection.

   _____

6. The term *orthographic* means _____ line.

   _____

7. Orthographic drawings are _____-dimensional.

   _____

8. Only _____ side(s) of an object is (are) shown in an orthographic drawing.

   _____

9. A single object on a plan may require more than one _____ drawing.

   _____

10. _____ give additional written data not appearing in the drawing.

    _____

11. Electrical, plumbing, and heating _____ are frequently included in plans for more complex projects.

    _____

UNIT 2

Linear Measurement

Linear measurement (or *linear dimension*) is a straight-line distance between two horizontal or vertical points. In the United States, the **English measurement system** (also referred to as **customary**) of yards, feet, inches, and inch fractions is used most often to express measurement. Dimensions on construction drawings are shown in feet, inches, and inch fractions. Elevations and grades are commonly expressed in feet and decimals of a foot and/or inch.

Direct measurement is taken with a measuring tool such as a tape or ruler. *Indirect or computed measurement* is a mathematical calculation made when it is not convenient or possible to take a direct measurement.

Metric measurement has not yet found wide acceptance in the U.S. construction industry, although it is the dominant form of measurement in the rest of the world. Canada uses metric measurement for all commercial and heavy construction work. However, most Canadian residential plans still feature English measurement.

ENGLISH LINEAR MEASUREMENT

As mentioned, English linear measurement is based on yards, feet, and inches. A *yard* contains three feet and a *foot* contains twelve inches. An *inch* for most architectural measurements is divided into sixteenths, eighths, quarters, and halves. **(See Figure 2–1.)**

The symbol for feet (') is placed to the right of the number. For example, 7 feet is written as 7'. The symbol for inches (") is also placed to the right of the number. For example, 3 inches is written 3". The abbreviations for feet (ft) and inches (in) may also be used. Therefore, 3" could also be written as 3 in. Fractions are shown with a numerator separated from the denominator by a slash mark. For example, three-fourths of an inch is written as 3/4".

Very often the dimensions on a plan combine all three units. The feet and inches are separated by a hyphen, and the inch fraction is written next to the inch quantity. It is recommended that a feet-only measurement include a hyphen and 0". This prevents the possibility of mistaking 35' (feet) for 35" (inches). Following are some examples:

21 feet, 9 and 5/8 inches = 21'-9 5/8"

35 feet = 35'-0"
11 feet and 1/2 inch = 11"-0 1/2"

When working with construction plans, it is sometimes necessary to convert inches to feet and feet to inches. It may also be necessary to add, subtract, and multiply and divide combinations of feet, inches and

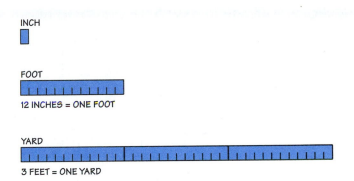

INCH

FOOT

12 INCHES = ONE FOOT

YARD

3 FEET = ONE YARD

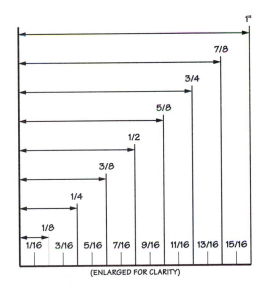

FRACTIONS OF AN INCH - ENGLISH MEASUREMENT

■ **Figure 2–1.**
English linear measurement is based on yards, feet, and inches. An inch is divided into sixteenths, eighths, fourths, and halves.

inch fractions. (See Unit 28 for math procedures and exercises.)

DECIMALS

When decimal measurements are shown on construction plans, it is usually to identify grades, heights, elevations, and sometimes property line dimensions. Most decimal dimensions are found in the plot or site plans of the working drawings.

In decimal measurement, a foot is divided into ten equal parts. Each part represents one-tenth of a foot. Each tenth is further divided into ten parts; each equal

to one-hundredth of a foot. Measurements in hundredths are used instead of tenths when greater accuracy is required. **(See Figure 2–2.)**

A tenth of a foot is written with a decimal point to the left of a single number. A hundredth of a foot is written with a decimal point to the left of two numbers. When feet and a decimal fraction of a foot are combined, they are separated by the decimal point, and a foot mark (') is placed to the right of the tenth or hundredth number. Following are some examples:

5/10 of a foot = .5'
43/100 of a foot = .43'
32 feet and 7/10 of a foot = 32.7'
15 feet and 43/100 of a foot = 15.43'

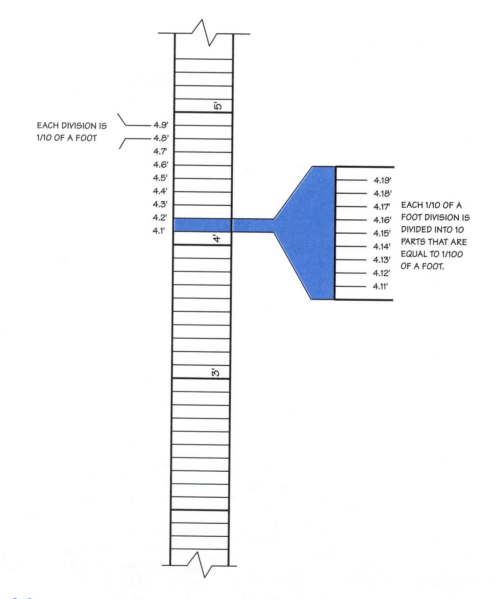

■ **Figure 2–2.**
In decimal measurement, a foot is divided into tenths and hundredths of a foot.

It is sometimes necessary to convert decimals to inches or inches to decimals. Tables are available to convert the more commonly used decimal-inch equivalents. **(See Table 2–1.)** These conversions can also be done mathematically. (See Unit 26 for conversion math procedures and exercises.)

METRICS

The metric system of measurement, called the *System International (SI)*, has already replaced English measurement in most of the world. A Metric Conversion Act was passed by Congress in 1975 and was later amend-

ed by the Omnibus Trade and Competitiveness Act in 1998. This legislation established the metric system as the preferred system of measurement in the United States. It mandated that wherever possible the metric system should be implemented in all federal procurement grants and business-related operations after September 30, 1992.

Thus far, the U.S. construction industry has been reluctant to convert to metric measurement. Much of this reluctance has been due to concern over the expected difficulty of making such a change. However, the experiences of other countries that have converted to metrics over the past 20 years (Britain, Australia, South Africa, and Canada) have been very positive. These countries found that the conversion process was much simpler than anticipated. There was no increase in design and construction costs. Architects and engineers found that metric dimensioning was less subject to error and easier to use than customary (English) dimensioning.

Once understood, metric units of measurement are more convenient to use than yards, feet, and inches. It is likely that some time in the near future the U.S. construction industry will begin the process of converting to metrics. Even now, understanding how to convert metric to English measurement and English to metric measurement can be important for persons in the construction industry.

LINEAR METRIC MEASUREMENT

The *meter (m)*, also spelled *metre*, is the basic linear unit in the metric system. The largest division of a meter is the *decimeter (dm)*, and there are ten decimeters in a meter. A decimeter is divided into ten *centimeters (cm)*. A centimeter is divided into ten *millimeters (mm)*, which is the smallest division of a meter. **(See Figure 2–3.)** Millimeters are the standard unit of measurement used on metric construction plans. **(See Figure 2–4.)** Following are some comparisons of English and metric measurements.

39.37" or 3'-3 3/8" = 1 m
1 foot = .3048 m
1 foot = 3.048 dm
1 foot = 30.48 cm
1 foot = 304.8 mm
1 inch = .0254 m
1 inch = .245 dm
1 inch = 2.54 cm
1 inch = 25.4 mm

English-to-metric conversions can be done mathematically. (See Unit 31 for metric-conversion procedures and exercises.)

DECIMAL CONVERSION TABLE

INCHES	DECIMAL OF AN INCH	INCHES	DECIMAL OF AN INCH
1/64	0.015625	33/64	0.515625
1/32	0.03125	17/32	0.53125
3/64	0.046875	35/64	0.546875
1/16	0.0625	9/16	0.5625
5/64	0.078125	37/64	0.578125
3/32	0.09375	19/32	0.59375
7/64	0.109375	39/64	0.609375
1/8	0.125	5/8	0.625
9/64	0.140625	41/64	0.640625
5/32	0.15625	21/32	0.65625
11/64	0.171875	43/64	0.671875
3/16	0.1875	11/16	0.6875
13/64	0.203125	45/64	0.703125
7/32	0.21875	23/32	0.71875
15/64	0.234375	47/64	0.734375
1/4	0.250	3/4	0.750
17/64	0.265625	49/64	0.765625
9/32	0.28125	25/32	0.78125
19/64	0.296875	51/64	0.796875
5/16	0.3125	13/16	0.8125
21/64	0.328125	53/64	0.828125
11/32	0.34375	27/32	0.84375
23/64	0.359375	55/64	0.859375
3/8	0.375	7/8	0.875
25/64	0.390625	57/64	0.890625
13/32	0.40625	29/32	0.90625
27/64	0.421875	59/64	0.921875
7/16	0.4375	15/16	0.9375
29/64	0.423125	61/64	0.953125
15/32	0.46875	31/32	0.96875
31/64	0.484375	63/64	0.984375
1/2	0.500	1"	1.000

■ **Table 2–1.**
A decimal conversion table is convenient for changing common fractions of an inch to decimal fractions of an inch.

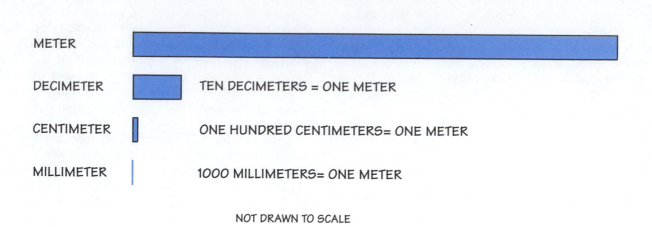

METER

DECIMETER   TEN DECIMETERS = ONE METER

CENTIMETER   ONE HUNDRED CENTIMETERS= ONE METER

MILLIMETER   1000 MILLIMETERS= ONE METER

NOT DRAWN TO SCALE

■ **Figure 2–3.**
The meter is the basic linear metric unit. It is divided into decimeters, centimeters, and millimeters.

REVIEW QUESTIONS

Enter the missing words in the spaces provided on the right.

1. The _____ measurement system is currently used in the United States.

   _____

2. The term _____ measurement means the same as English measurement.

   _____

3. Measuring tools are used to take a _____ measurement.

   _____

4. The units of English measurement are yards, feet, and _____.

   _____

5. Fourteen feet and 2 inches on a plan can be written as _____.

   _____

6. Grades, heights, and elevations are often written as _____ measurements.

   _____

7. In decimal measurement a foot is divided into _____ equal parts.

   _____

8. The basic linear unit in the metric system is the _____.

   _____

9. The smallest division of a meter is a _____.

   _____

10. One foot equals _____ mm.

   _____

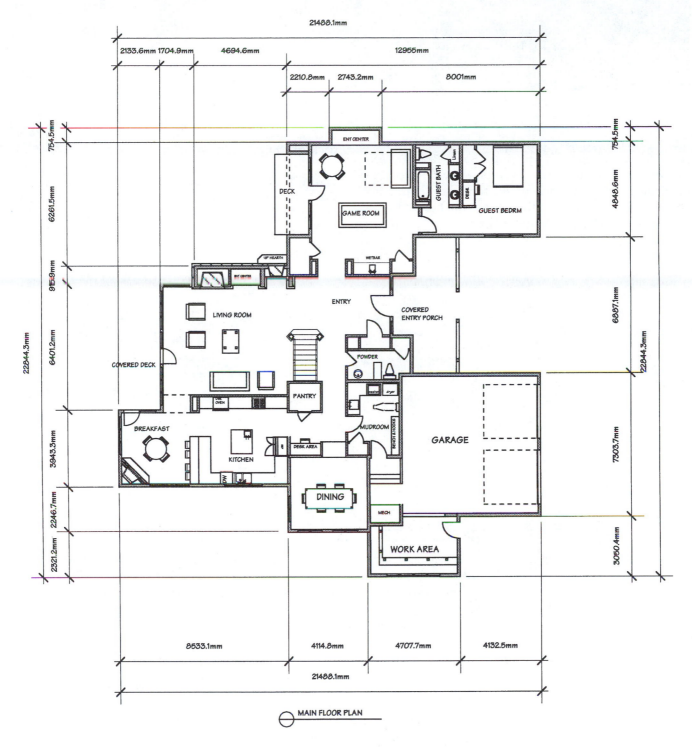

21488.1mm

2133.6mm 1704.9mm　4694.6mm　　　12955mm

2210.8mm　2743.2mm　　8001mm

754.5mm

6261.5mm

915.9mm

6401.2mm

22844.3mm

3943.3mm

2246.7mm

2321.2mm

ENT CENTER

DECK

GAME ROOM

18" HEARTH

ENT CENTER

LIVING ROOM

COVERED DECK

PANTRY

BREAKFAST

DBL OVEN

KITCHEN

DW

DINING

WETBAR

ENTRY

POWDER

DESK AREA

washer dryer

MUDROOM

BENCH & HOOKS

MECH

WORK AREA

GUEST BATH

Linen

DESK

GUEST BEDRM

COVERED
ENTRY PORCH

GARAGE

754.5mm

4848.6mm

6887.1mm

22844.3mm

7805.7mm

3050.4mm

8533.1mm　　4114.8mm　　4707.7mm　　4132.5mm

21488.1mm

 MAIN FLOOR PLAN

■ **Figure 2–4.**
Metric plans give their dimensions in millimeters.

UNIT 3

Lines

Different types of lines are used to identify different functions on drawings. Some lines are continuous and others are broken by dots, dashes, or zigzags. Some are heavy and some are light. Some show dimensions while others show the boundaries of property, buildings, and other objects.

IDENTIFYING LINES

Figure 3–1 includes a plan view of part of a foundation plan. The basic lines used in the drawings are shown separately in the illustration and related by numbers to the plan. More detailed explanations of the lines follow,

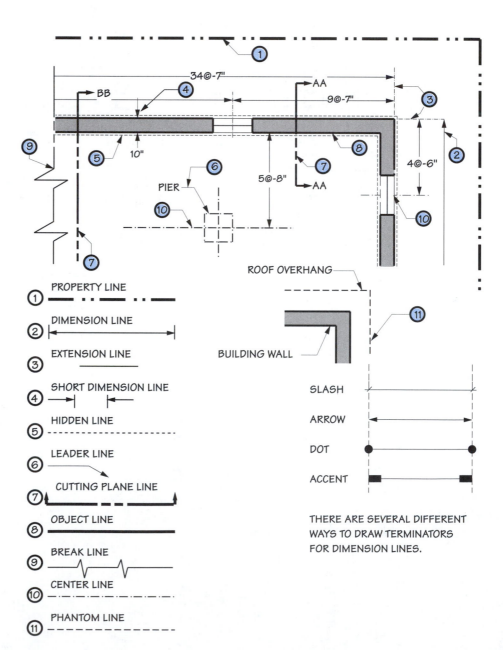

■ **Figure 3–1.**
Different types of lines identify measurements and objects on the plan.

identified by the same numbers that appear in the figure.

1. **Property line.** A heavy solid line with one long dash and then two short dashes or dots. It identifies the boundaries of the property.
2. **Dimension line.** A solid, thin line with terminators (arrows, dots, slash marks or accent marks) at each end. It gives the horizontal or vertical distance between two points. Dimension lines outside the drawing touch against extension lines.
3. **Extension line.** A thin line projecting at a right angle from walls or other objects. It is used with dimension lines.
4. **Short dimension line.** Used when there is not sufficient room to write the dimension between the arrows.
5. **Hidden line.** A series of short dashes. It shows the edges of the part of a structure that would not be seen from the view shown in a drawing. In Figure 3–1, hidden lines show the footing below the foundation walls and the piers below the steel columns. The footings and piers are covered with soil and are hidden from view in this drawing.
6. **Leader line.** A short line drawn at an angle to an object. It extends from a description, note, or measurement and points to the object being described.
7. **Cutting-plane line.** A solid line with short right angle lines and arrows at each end. An identifying letter appears in front of or next to the arrow. The cutting-plane line shows where a wall or object is cut, revealing a cross section. This makes it possible to show features and information within the wall or object. The arrows show the direction from which the wall is viewed. The identifying letters refer to the section view or detail that provides the desired information.
8. **Object line.** Used to identify the edges or outlines of object such as walls, buildings, and walks.
9. **Breaker line.** A long line broken by zigzags. It is used where space does not permit the continuation of a scaled drawing, or where the continuation of the drawing is not necessary to provide the needed information.
10. **Center line.** A series of alternating long and short dashes extending at a right angle from parts of the structure. It is used to find the centers of walls, door and window openings, pier footings, columns, equipment, and fixtures. In Figure 3–1, center lines extend through the window openings and pier footing. Dimension lines touch up against the center lines. Center lines are sometimes further identified by the abbreviation CL or the symbol ℄ .
11. **Phantom line.** These lines look the similar as hidden lines. They are used to identify an object that is *not shown* rather than hidden from view. A good example is the edge of a roof overhang on a floor plan.

REVIEW QUESTIONS

Enter the missing words in the spaces provided on the right.

1. The boundaries of a lot are identified by _____ lines.

   _____

2. _____ lines give horizontal and vertical distances between two points.

   _____

3. Dimension lines outside a drawing touch against _____ lines.

   _____

4. When there is not sufficient room to write a dimension between arrows, a _____ dimension line can be used.

   _____

5. _____ lines identify objects that could not be seen from the view shown in the drawing.

   _____

6. A _____ line extends from a written description to the object..

_____

7. A line showing where an object is cut to reveal information is called a _____ plane line.

_____

8. _____ appear at the end of cutting plane lines to show the direction of the view.

_____

9. A long line broken by zigzags is a _____ line.

_____

10. A series of alternating long and short dashes at a right angle to an object identify a _____ line.

_____

Identify these lines in the spaces provided on the right.

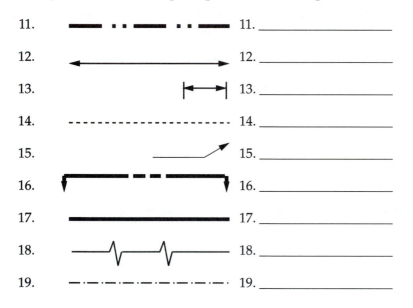

11.   11. _____

12.   12. _____

13.   13. _____

14.   14. _____

15.   15. _____

16.   16. _____

17.   17. _____

18.   18. _____

19.   19. _____

UNIT 4

Scale

Construction plans are drawn to *scale* so that the boundaries of the building site and all views of the building and its parts can appear on a set of plans. A **scale drawing** must be in exact proportion to the actual size of the building and its parts. Generally speaking, the larger the size of an object, the smaller the scale will be. For example, a smaller scale is used for site plans that show the full size of a lot. A larger scale can be used for the different plan views of a building. Individual detail drawings take up less space and can be drawn to the largest scale. The scale used will always be identified below or next to a drawing.

The three types of rules used to draw plans to scale are the architect's scale, engineer's scale, and metric scale. The rules are available in flat or triangular shapes. Persons reading the plans can use these rules to check or find dimensions omitted from the drawings. **(See Figure 4–1).**

ARCHITECT'S SCALE

The **triangular architect's scale** is used to scale plans drawn to English measurement. The ten choices of **scale** shown on it follow:

$$3/32'' = 1'\text{-}0''$$
$$1/8'' = 1'\text{-}0''$$
$$3/16'' = 1'\text{-}0''$$
$$1/4'' = 1'\text{-}0''$$
$$3/8'' = 1'\text{-}0''$$
$$1/2'' = 1'\text{-}0''$$
$$3/4'' = 1'\text{-}0''$$
$$1'' = 1'\text{-}0''$$
$$1\text{-}1/2'' = 1'\text{-}0''$$
$$3'' = 1'\text{-}0''$$

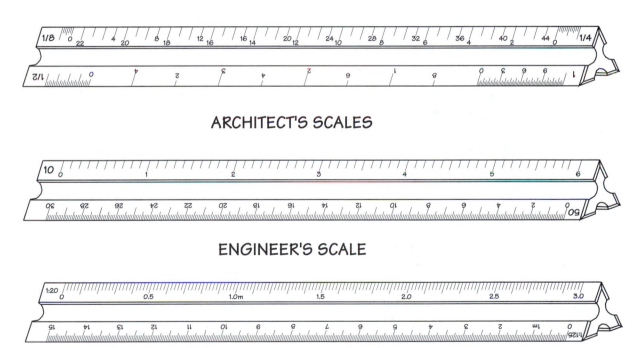

ARCHITECT'S SCALES

ENGINEER'S SCALE

METRIC SCALE

■ **Figure 4–1.**
The three types of scale rulers used for drawing linear measurements on construction plans are the architect's scale, engineer's scale, and metric scale. (Drawings here are enlarged slightly for easier reading.)

As an example, on the 1/4″ scale, each one-fourth of an inch in the drawing represents 1 foot (1′-0″) of the actual building size. This scale is written as 1/4″ = 1′-0″ and will usually appear directly below the drawing. The 1/8″ scale is often used on plot plans that show the size of a full lot. Floor plans and elevations are often drawn to 1/4″ scale. Detail drawings are usually proportionately larger and may be drawn to 1/2″, 3/4″, or 1″ scale.

The faces of the architect's scale rule contain scales at the upper and lower edges. Each edge has two scales. One scale begins from zero and reads from the left to right. A second scale that is twice as large begins from zero and reads from right to left. For example, the upper edge of one face shows a 1/8″ scale reading from left to right, and a 1/4″ scale reading from right to left.

The closer graduations to one side of the zero marks show inches and fractions of an inch. The divisions vary with the scaling size. For example, there is room for only six marks to the left of the zero mark on the 1/8″ scale. Therefore, each graduation represents 2 inches. There is room for 48 marks to the right of the zero mark of the 1″ scale. Therefore, each graduation represents one-fourth of an inch. **(See Figure 4–2a & b).** Scale can also be found mathematically. (See Unit 28 for math procedures and exercises.)

ENGINEER'S SCALE

The **engineer's scale** is used most often for dimensions covering a broader area such as site (plot) plans, survey plans, subdivision maps, and landscape plans. The different faces of the engineer's scale show inches divided into decimal parts. The six choices shown on it follow:

$$1″ = 10′$$
$$1″ = 20′$$
$$1″ = 30′$$
$$1″ = 40′$$
$$1″ = 50′$$
$$1″ = 60′$$

For example, on the 1″ = 10′ scale, each inch is equal to 10 feet, with each division of the inch representing 1 foot. The 10 feet can also be read as 100 feet, with each division representing 100 feet. As another example, the 1″ = 50′ scale means that each inch on the rule equals 50 feet, with each division representing 1 foot. The 50 feet can also be read as 500 feet, and each division would then represent 10 feet. **(See Figure 4–3.)**

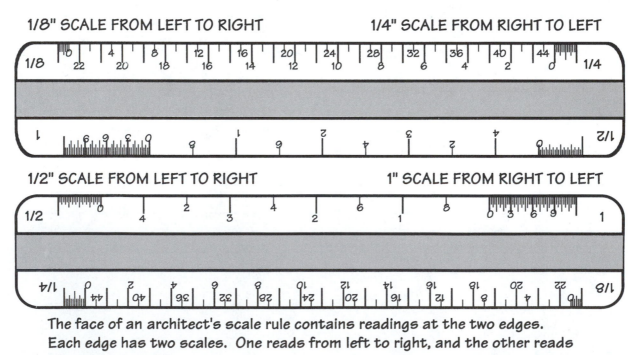

The face of an architect's scale rule contains readings at the two edges. Each edge has two scales. One reads from left to right, and the other reads from right to left.

■ **Figure 4–2a.**
Measuring with the architect's scale ruler.

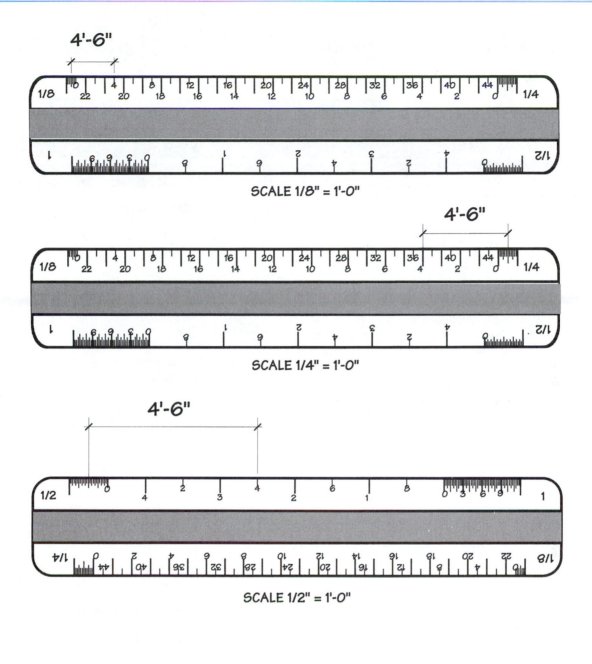

SCALE 1/8" = 1'-0"

SCALE 1/4" = 1'-0"

SCALE 1/2" = 1'-0"

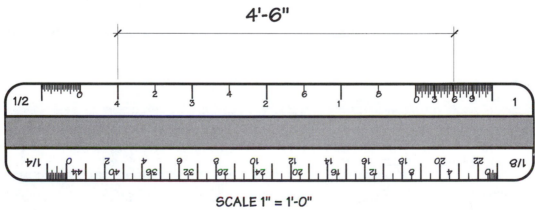

SCALE 1" = 1'-0"

■ **Figure 4–2b.**
The 1/8″, 1/4″, 1/2″ and 1″ scale on the architect's scale ruler are used here to measure 4′-6″.

18.00'

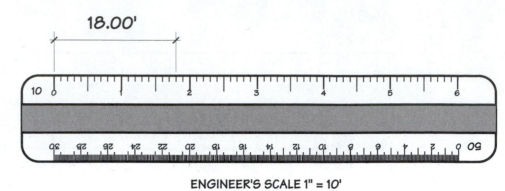

ENGINEER'S SCALE 1" = 10'

■ **Figure 4–3.**
The 1″ = 10′ scale on the engineer's scale rule can also be read as 1″ = 100′, and 1″ = 1000′. The measurement shown above is 18.00′ when the scale is 1″ = 10′.

METRIC SCALE

The **metric scale** is used when construction plans are drawn with metric linear measurements. Some metric drawings express longer measurements in meters carried out to three decimal places and millimeters for smaller section-view and detail drawings. However, the usual method is to show all linear measurements in millimeters. (A meter contains 1000 millimeters.) Following are some commonly used metric-scale ratios:

1:20
1:25

1:50
1:75
1:100
1:125

For example, the length of a wall drawn to a 1:20 scale means that 1 millimeter on the drawing is equal to 20 millimeters of the actual size of the wall. The length of a wall drawn to 1:125 scale means that 1 millimeter on the drawing is equal to 125 millimeters of the actual size of the wall; therefore, this scale would be used for drawings covering a much wider area than those using the 1:20 ratio. **(See Figure 4–4.)**

2600 MM

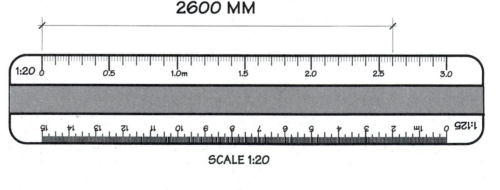

SCALE 1:20

■ **Figure 4–4.**
Both of these metric scales are measuring 2600 millimeters. The 1:20 ratio is used for drawing larger details and section views. The 1:125 ratio is used for plans that cover a large area such as site plans, foundation plans, and floor plans. (The 1:125 scale has been enlarged here for easier reading.)

2600 MM

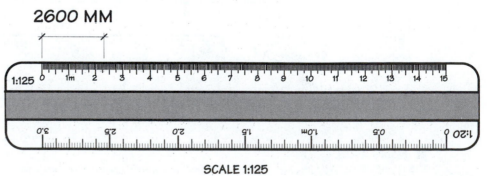

SCALE 1:125

REVIEW QUESTIONS

Enter the missing words in the spaces provided on the right.

1. A scale drawing must be in exact _____ to actual building size.

    _____

2. _____ plans are often drawn with a smaller scale.

    _____

3. The architect's scale, _____ scale, and engineer's scale are the three types used in drawing plans.

    _____

4. To scale plans to English measurement, a(n) _____ scale rule is used.

    _____

5. Fractions of an inch on the architect's scale are shown to one side of the _____ mark.

    _____

6. A scale in which 3/4 inch is equal to 1 foot would be written on a plan as _____.

    _____

7. The inches on an engineer's scale are divided into _____ parts.

    _____

8. A 1" = 60' scale on an engineer's rule means that 1 inch on the drawing equals _____ feet.

    _____

9. The _____ is the unit most often used for drawing measurements on metric plans.

    _____

10. A metric scale of 1:75 means that 1 millimeter on the drawing equals _____ millimeters of actual size.

    _____

UNIT 5

Symbols and Abbreviations

Construction plans are drawn to a very small scale, so it is not possible to show all the components of a building as they would be viewed by eye. For this reason, many parts of the drawings are identified by **standardized symbols** that are generally accepted by architects, contractors, drafters, and trade workers. Since there is also limited room for writing on the plans, many **abbreviations** are used instead of full words or phrases. A full set of construction plans will often include a list of the symbols and abbreviations appearing on those plans. The drawings of the one-story residence plan analyzed in Section 2 will include symbols and abbreviations pertaining to those drawings. More complete listings of commonly used abbreviations and symbols appear following Section 8 of the book.

SYMBOLS

A symbol may represent a plan view looking down at an object, a section view from an end, or an elevation view showing the side of the object. Symbols are used to identify a variety of materials and structural components. **Topographic symbols** indicate natural features on the job site. Electrical, plumbing, and HVAC (heating, ventilation, and air conditioning) plans contain many standard symbols. **(See Figure 5–1.)**

ABBREVIATIONS

Abbreviations are used extensively to identify objects and materials because of the limited space on the working drawings. Following are some general rules and examples of abbreviations:

1. Abbreviations are written with one or more uppercase (capital) letters.
   Bathroom (B)
   Brick (BRK)
   Damper (DMPR)
2. Two or more words may be abbreviated with the first letters of the words placed together. Traditionally, periods have not been placed between the letters of an abbreviation. However, it is not unusual today for architects or draftspersons to insert such periods. For example, Area Drain might be abbreviated as A.D.
   Area Drain (AD)
   Pounds Per Square Foot (PSF)
   Ground-Fault Circuit Interrupter (GFCI)
3. A period is not required at the end of an abbreviation unless the abbreviation spells out some word.
   Between (BET.)
   Combination (COMB.)
   Metal (MET.)
4. Abbreviations for two or more words are sometimes separated accordingly.
   Concrete Block (CONC BLK)
   Circuit Breaker (CIR BKR)
   AMERICAN NATIONAL STANDARD (AMER NATL STD)
5. Some words may be identified with more than one abbreviation.
   Closet (C), (CL), (CLOS)
   Bearing Plate (BPL), (BRG PL)
   Acoustical Tile (AT), (ACT)
6. The same abbreviation is sometimes used for different items; therefore, their meaning depends on where they are placed in the drawings.
   Riser (R)
   Ridge (R)
   Radius (R)
   Range (R)

■ **Figure 5–1.**
Symbols identify materials, components, fixtures, and many other items of a building plan. In the above examples, a section is a view from the end of an object, and an elevation is a side view. A plan view looks at an object from above.

REVIEW QUESTIONS

Enter the missing words in the spaces provided on the right.

1. Natural features are identified by _____ symbols.

2. _____ symbols show structural components of a building.

3. Many standard symbols are used on plumbing, HVAC, and _____ plans.

4. To save space, written information on a plan contain many _____.

_____

5. One or more _____ letters are used to write abbreviations.

_____

6. The abbreviation for the word *damper* is _____.

_____

7. CIR BKR is the abbreviation for _____.

_____

8. The word *closet* can be abbreviated as CL, CLOS, or _____.

_____

9. The abbreviation R may be used for the words *range, riser, radius,* and _____.

_____

10. The abbreviation for the word *combination* is _____.

_____

SECTION 1 TEST

Building Plans

Answer T (true) or F (false).

1. A set of construction plans consists of only pictorial drawings. _____

2. Pictorial drawings are two-dimensional. _____

3. A set of construction plans includes a series of working drawings. _____

4. Orthographic projection is the method used to create working drawings. _____

5. In the United States customary measurement is used. _____

6. In decimal measurement a foot is divided into 12 equal parts. _____

7. The decimal number .005 is the same as the fraction 5/10. _____

8. There are 10 decimeters in 1 meter. _____

9. An extension line projects at a 45-degree angle from an object. _____

10. A heavy solid line identifies a hidden object. _____

11. Dimension lines will always have arrows at each end. _____

12. A property line shows the boundaries of a building site. _____

13. Construction plans are drawn to scale. _____

14. A metric scale is used to draw plans to English measurement. _____

15. A smaller scale is generally used for smaller objects. _____

16. A 1″ = 10′ scale on an engineer's scale rule means 1 inch equals 10 feet. _____

17. Symbols represent objects shown in a construction plan. _____

18. Topographical symbols show structural components of a building. _____

19. AT is the abbreviation for acoustical tiles. _____

20. The common abbreviation for bathroom is BTH RM. _____

Choose one or more correct ways to finish each statement.

21. Pictorial drawings cannot be used in working drawings because
    a. they are too large.
    b. they are three-dimensional.
    c. it is not possible to show all the required dimensions and data on them.
    d. they are two-dimensional

    _____

22. Orthographic drawings
    a. show only one side of an object.
    b. are two-dimensional.
    c. make it convenient to include dimensions.
    d. none of the above.

    _____

23. A document that provides additional written information not appearing in the drawings is called the
    a. schedule.
    b. contract.
    c. construction notes.
    d. specifications.

    _____

24. A direct measurement is a dimension
    a. calculated mathematically.
    b. taken with a tape or rule.
    c. also known as a computed measurement.
    d. all of the above.

    _____

25. The dimension of ten feet, three and three-quarter inches is written as
    a. 10', 3", 3/4.
    b. 10'-3"-3/4'.
    c. 10'-3 3/4".
    d. 10' and 3 3/4".

    _____

26. Decimal measurements on a construction plan are
    a. never used.
    b. often used for the entire drawing.
    c. most often used to identify grades, heights, and elevations.
    d. used only for special calculations.

    _____

27. Fifty-three hundredths of a foot written as a decimal would be
    a. .53'.
    b. 5.3'.
    C. 53/100.
    d. .053'.

    _____

28. The smallest division of a meter is a
    a. decimeter.
    b. kilometer.
    c. centimeter.
    d. millimeter.

    _____

29. A line that gives the horizontal or vertical measurement of an object is called a(n)
    a. extension line.
    b. dimension line.
    c. leader line.
    d. object line.

    _____

30. A hidden line is shown by a
    a. heavy solid line.
    b. line with slash marks at each end.
    c. series of short dashes.
    d. series of long and short dashes.

31. Object lines are used to identify the outlines of
    a. buildings.
    b. walls.
    c. walks.
    d. all of the above.

32. A break line
    a. can be shown as a long line broken by zigzags.
    b. identifies the center of an object.
    c. is used where space does not permit the continuation of a drawing.
    d. all of the above.

33. The three types of scale rules used are called the
    a. architect's, customary, and metric scales.
    b. engineer's, architect's, and metric scales.
    c. metric, standard, and engineer's scales.
    d. architect's, engineer's, and customary scales.

34. The architect's scale rule is used to scale
    a. metric measurement.
    b. measurement shown in feet and decimal parts of a foot.
    c. dimensions shown in yards.
    d. feet, inches, and inch fraction measurements.

35. On an engineer's scale rule, the scale that represents 50 feet as being equal to 1 inch is identified as
    a. 1"-50'.
    b. 1-50.
    c. 1/50.
    d. 1" = 50'.

36. The 1:25 scale on a metric scale rule means that on the drawing
    a. 1 centimeter is equal to 25 millimeters.
    b. 1 inch is equal to 25 millimeters.
    c. 1 meter is equal to 25 millimeters.
    d. 1 foot is equal to 25 millimeters.

37. Because it is not possible to draw pictures of many building components on a set of plans, it is necessary to
    a. use abbreviations.
    b. make orthographic drawings.
    c. use symbols.
    d. make three-dimensional drawings.

38. Natural features on a job site are identified by
    a. architectural symbols.
    b. topographical symbols.
    c. pictorial sketches.
    d. HVAC symbols.

39. Abbreviations in construction plans
    a. are used because they make a neater appearance.
    b. are used because of limited space for written information on working drawings.
    c. are required by local building codes.
    d. help identify objects and materials.

    _____

40. The abbreviation R is used to identify the word
    a. riser.
    b. room.
    c. ridge.
    d. range.

    _____

Identify the lines in the drawing.

41. _____

42. _____

43. _____

44. _____

45. _____

46. _____

47. _____

48. _____

49. _____

50. _____

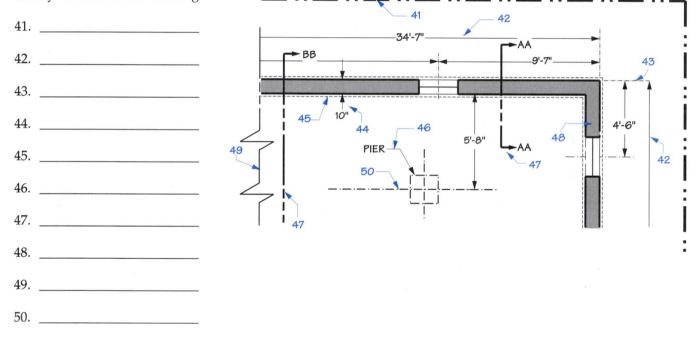

Write the abbreviation for each term.

51. Damper

    _____

52. Pounds per square foot

    _____

53. Concrete block

    _____

54. Closet

    _____

55. Radius

    _____

56. Combination

    _____

Identify the symbols in the drawings.

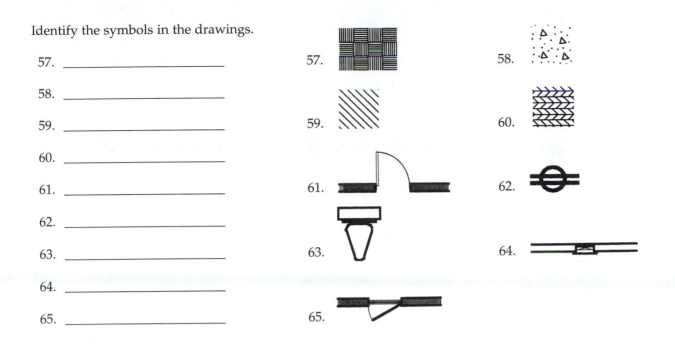

57. _____   57.

58. _____   58.

59. _____   59.

60. _____   60.

61. _____   61.

62. _____   62.

63. _____   63.

64. _____   64.

65. _____   65.

Find the length of each line using an architect's scale rule.

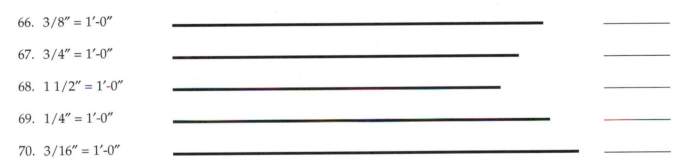

66. 3/8" = 1'-0"   _____

67. 3/4" = 1'-0"   _____

68. 1 1/2" = 1'-0"   _____

69. 1/4" = 1'-0"   _____

70. 3/16" = 1'-0"   _____

SECTION 2

Basics of Plan Reading

Section 2 presents the fundamentals of plan reading by thoroughly analyzing the working drawings of a full set of house plans. These plans will be identified throughout this section as the **"One- Story Residence."** The discussion of the different working drawings will, as much as possible, closely follow the actual steps in the construction of the building.

To understand a set of plans fully, it is necessary to have a general knowledge of the main structural components (parts) of a building and how it is constructed. Therefore, all of the units of this section include basic construction theory about the drawings being discussed. Each building component is illustrated and described.

The original drawings of the **One-Story Residence** have been reduced in size to fit the pages of this book. Therefore, the scale shown below each drawing is not accurate for the reduced drawing, although it was accurate for the original full-size drawing. Three-dimensional pictorial drawings are included with most of the orthographic drawings to help you visualize what is being presented in the plans.

Each separate working drawing has a particular function, and all the drawings are interrelated. If you were to observe a trade person working with the plans on a construction project, you would see this person constantly flipping from one sheet to another. Studying several different drawings at each stage of the construction work is usually necessary.

The plans for the **One-Story Residence** are **CADD (computer-aided design and drafting)** drawings produced with a computer program. The program also contains a library of symbols that are very similar to standardized hand-drawn symbols.

The mass of lines, dimensions, symbols, abbreviations, callouts, and other data on a set of plans can be overwhelming to a person inexperienced in reading construction drawings. In fact, plan reading is a special skill, but it is a skill that can be learned by the individual who is willing to make the effort. It should be said at this point that even a proficient plan reader does not immediately understand everything in a drawing at first glance, but will have to spend time studying and relating the various drawings to one another..

Units 6 through 14 will go through the steps of interpreting the drawings of the **One-Story Residence.** This house has a simple, basic floor plan. However, in the rest of the building the designer has attempted to blend elements of older traditional design with contemporary style. Thus, the **One-Story Residence** features exposed beams and collar ties inside the building. Wide and well- defined trim materials are used around door and window openings. Non-load-bearing outriggers are placed beneath some gable overhangs at the exterior of the building. Trim material 1 1/2″ thick is used around door and window openings and at the outside and inside corners of the building. On surface appearance, all this suggests a building style very common in the early years of the twentieth century. Emphasis on outdoor living space is also a notion developed during that period. However, the **One-Story Residence** is a very modern building in the sense that it incorporates the latest building and energy technology and the conveniences of contemporary construction.

In general, architects and designers use similar formats for drawing a set of plans. Yet, there may be individual differences in style and notation. The final section of this book contains three additional construction plans with accompanying assignments. These will expose the reader to some variations in drawing styles.

The most difficult thing for the beginner is to look at an orthographic drawing and visualize what it represents. As mentioned earlier, related three-dimensional pictorial drawings are provided for most of the orthographic drawings of the **One-Story Residence.** As you study each orthographic drawing, you should compare it with the related pictorial drawing. For example, the drawing of a crawl-space foundation plan will provide a plan view of the walls, piers, posts, and girders. Compare the orthographic views of these objects with the corresponding pictorial views of the same objects before proceeding to the next step.

Each unit of Section 2 includes a table of symbols and abbreviations. Memorizing them is very important. Starting from the top or bottom of the drawing, study all the printed data and begin listing all the abbreviations that can be found. Abbreviations can usually be recognized as an incomplete word. Write the full word next to each abbreviation listed. The text gives the figure number of the illustration containing the symbols and abbreviations. Study the symbols in the table so that they can be identified on the drawing.

Note that the blue arrows in the drawings point to different features and/or areas of the drawings. The number in an arrow corresponds to the explanation of that feature in the text. For example, the features pointed out in the plan view of a crawl-space foundation drawing are listed under "Crawl-Space Foundation Features—Plan View."

Be sure to answer the review questions at the end of each unit to make sure you understand the material of that unit before proceeding to the next. Correct answers to the review questions are found in Appendix A. ∎

Survey and Site Plans

Before construction can begin, the exact boundaries **(property lines)** of the building site **(lot)** must be established. The topographical features of the property, such as the differences between heights **(elevations)** and slopes, must also be recorded. This work is normally done by licensed surveyors or civil engineers. They obtain their initial information from a *plat plan* (discussed in this unit) covering the area where the construction will take place. Plat divisions are the result of a national system of rectangular township grids developed over many years by the federal government.

TOWNSHIP GRIDS AND SECTIONS

The **township grids** are based on north-to-south *meridian lines* and east-to-west *base lines* running at 24-mile intervals. Each 24-mile square is divided into 16 townships bordered by *township lines* running east to west, and *range lines* running north to south. Each township is divided into 36 sections that are each one mile square. A township section is further divided into smaller areas later identified by plat plans. **(See Figure 6–1).**

READING THE PLAT PLAN

Plat plans, also called **subdivision maps,** define the borders of land divisions in a given geographical area. (The term *plat plan* should not be confused with *plot plan.*) A plat plan is not part of a set of construction drawings. It is discussed here because the plat plan contains information directly affecting the design and construction of the building. Plat plans are usually available from the city or county recorder's office or other local government agencies. Plat plans are subject to change, so it is important to get information from the very latest plat plan available.

A plat plan shows the street locations within the boundaries of the plan. Each street is identified by its name, and the street width is often included. All the lots bordering the streets in the area are identified by number and are legally recorded. Their dimensions are shown, and reference points are identified from which the property lines can be measured.

The plat plan may also provide information about **utility easements** that allow for the passage of utility services through a property. Such services include sewer lines, gas lines, water pipes, and electrical lines. Another type of easement pertains to the right of passage over a particular property. From all this information a survey plan and/or site plan can be developed and included with a set of working drawings.

A section of a plat plan is shown in **Figure 6–2.** Concentrate on the following features while studying this plan.

- ✓ Lot identification.
- ✓ Property lines and their dimensions.
- ✓ Compass direction of property lines.
- ✓ Monuments.
- ✓ Easements.

Explanations of the plat plan features in Figure 6–2 follow:

1. **Compass direction.** A north-pointing arrow orients the position of the plat plan.
2. **Scale.** A scale of 1″ = 50′ is necessary to produce the entire drawing on one sheet.
3. **Curve data chart.** This chart gives the radius and the length of the curved section of a property line. For example, the code number for the front curved property line of Lot #125 is C6. Find C6 in the number column. Read across to the right to find out that the radius for the curve is 48.00′ and the length of the curved property line is 34.84′. (The Delta column gives the angle the original surveyor used to establish the curve.)
4. **Street center monument.** A permanent aluminum-capped street center monument has been established at the southeast end of Hemberg Drive.
5. **Street center monument.** A permanent monument identifies the center of Hemberg Drive and Steves Boulevard. By sighting through survey instruments from this monument to the street center monument in the cul-de-sac, you can establish center lines at any point of Steves Boulevard.
6. **Public utility easement (PUE).** This 8′-0″ PUE next to the street runs through all the properties shown in the plat plan.

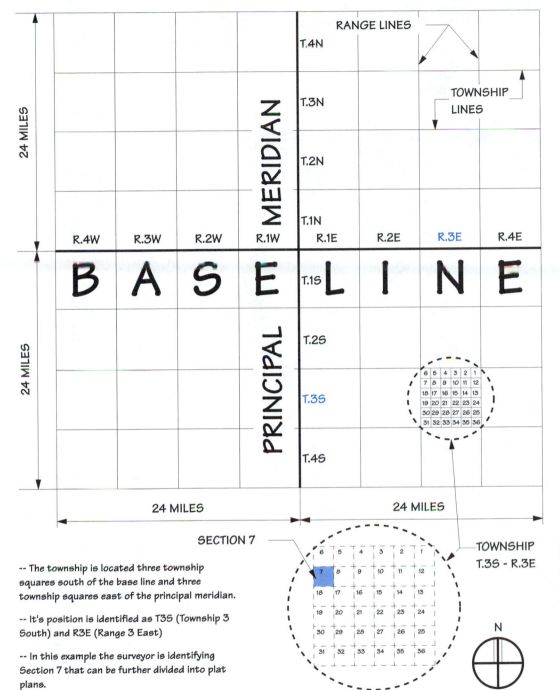

SECTION 7

-- The township is located three township squares south of the base line and three township squares east of the principal meridian.

-- It's position is identified as T3S (Township 3 South) and R3E (Range 3 East)

-- In this example the surveyor is identifying Section 7 that can be further divided into plat plans.

TOWNSHIP
T.3S - R.3E

N

■ **Figure 6–1.**
Township grids are the basis of a national system of land surveys. The above example shows how to identify the steps from township grid to township, and then from township to section.

7. **End-of-arc point.** This short line establishes the end of the arc at the southeast side of the cul-de-sac semicircle.

8. **Street centerline (CL).** Dimensions are given in both directions from the centerline. The front property line of Lot #120 is found by measuring 26'-0" from the street centerline. The straight property line sections of Lots #119, #121, and #127 are

also measured from this street centerline.

9. **Street name.** All streets adjoining the lots are named in the plat plan.

10. **Street center monument.** This is a permanent aluminum-capped monument embedded in the surface of the street. It identifies the center of Steves Boulevard in the area of the cul-de-sac.

11. **End-of-arc point.** This short line establishes the end

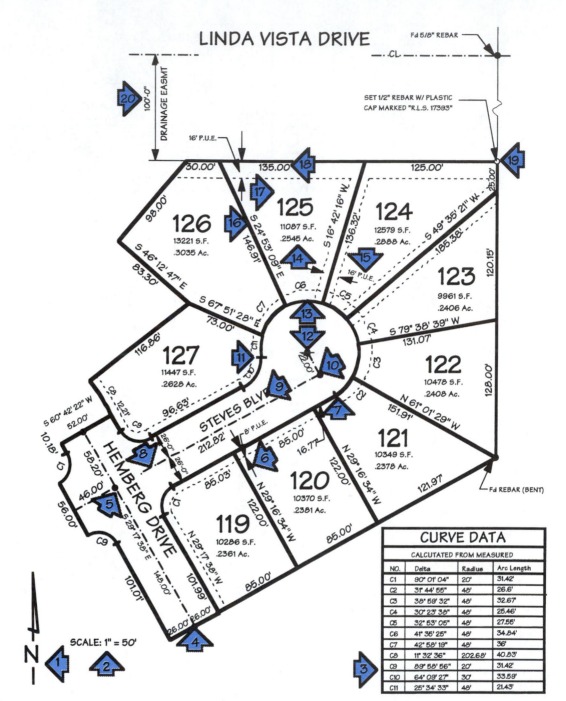

■ Figure 6–2.
This section of the plat plan provides information needed to survey the lot before construction of the building can begin.

CURVE DATA			
CALCUTATED FROM MEASURED			
NO.	Delta	Radius	Arc Length
C1	90° 01' 04"	20'	31.42'
C2	31° 44' 55"	48'	26.6'
C3	38° 59' 32"	48'	32.67'
C4	30° 23' 38"	48'	25.46'
C5	32° 53' 05"	48'	27.55'
C6	41° 35' 25"	48'	34.84'
C7	42° 58' 19"	48'	36'
C8	11° 32' 36"	202.68'	40.83'
C9	89° 58' 56"	20'	31.42'
C10	64° 09' 27"	30'	33.59'
C11	25° 34' 33"	48'	21.43'

of the arc on the northwest side of the cul-de-sac.

12. **Radius point.** A permanent aluminum-capped monument has been embedded in the street to identify the radius point from which the curve at the end of the cul-de-sac can be established. The radius length can also be found in the curve data chart.

13. **Front property line.** A code number (C6) refers to a curve data chart that gives the length of the front property line.

14. **Area of lot.** The total area of the lot equals 11,087

square feet. An acreage figure of .2545 is also given that is slightly more than 1/4 acre.

15. **Public utility easement (PUE).** A 16'-0" PUE easement extends 8' into each adjoining lot.

16. **Property line compass direction.** The bearing S 24° 53'09" E means the property line runs at 24 degrees, 53 minutes, 9 seconds east of south.

17. **Lot number.** The number of lot #125 is identified. The **One-Story Residence** will be constructed in this lot.

18. **Property line dimension.** The length of the rear

property line (PL) of Lot #125 is 135.00'. In a plat plan all property dimensions are given in feet and decimals of a foot.

19. **Corner monument.** A rebar with a plastic cap has been set to establish a corner monument. R.L.S. is the abbreviation for registered licensed surveyor, and the number 17393 is the registered number of the surveyor.

20. **Drainage easement.** The area of this easement is steeply sloped and subject to drainage from the adjoining properties.

SURVEY AND SITE PLANS

A **survey plan** provides information about the shape, size, and topographical features of the building site. This information also appears on a **site plan**, which includes information required to lay out the building and other structures. Construction plans for smaller projects do not often include a survey plan because a site plan is considered sufficient to contain all the necessary site information. However, a survey drawing is included with the plans for the **One-Story Residence** to help clarify site information for the reader.

Besides property lines, the survey and site plans show the following:

✓ Compass directions.
✓ Existing and finish elevations and contours.
✓ Locations of utility lines and easements.
✓ Roads bordering the property
✓ Tree and plan growth.

MEASUREMENTS AND ELEVATIONS

The surveyor or civil engineer will find the approximate lot location by its number on the plat plan and then stake the exact property corner. These measurements may be taken from street centers, and/or intersections, curbs, or established **points of beginning (POB)** identified by permanent markers called **monuments.** A monument can be a steel stake driven into the ground, an aluminum cap embedded in the street or sidewalk surface, or a concrete marker placed where it will not be disturbed.

The **elevations** of the property, found with a transit instrument **(see Figure 6–3),** are the different ground levels at the corners and within the property. The elevations are related to established city or county **datum points.** These datum points have been placed at convenient locations where they can be sighted by surveying instruments from building lots in the area. They are placed on building corners, concrete markers, or other

■ **Figure 6–3.**
A surveyor or civil engineer uses a transit instrument to establish property lines and elevations. *(Photo by Robert Zielkiewicz)*

types of stable objects that are not subject to movement. A datum point is often the distance above the **mean sea level (MSL)** at that point, which is based on the average sea level between high and low tides. So the elevations are usually the distance above MSL in relation to the datum points.

An elevation can be transferred to a **benchmark (BM),** which is a fixed point at the job site. It can be a stake driven into the ground, or it can be a mark on a street curb or a power pole or another immovable object. The benchmark can also be one corner of the actual property.

The elevation points on the survey plan are usually shown as feet and tenths or hundredths of a foot. Trade persons working on the job prefer to convert these measurements to feet and inches. Conversion charts are available for this purpose. **(See Table 6–1.)** Conversions can also be done mathematically. (See Unit 26 for mathematical conversion methods.)

When elevation points are later transferred to the site plan, architects will often simplify the elevation numbers by showing the benchmark as 100. Therefore, all the elevation points will be plus or minus 100. For example, an elevation reading of 101.5 feet on the plan is 1.5 feet above the benchmark (101.5 – 100 = 1.5' or 1'-6"). An elevation of 97.3 feet on the plan is 2.7 feet below the benchmark (100 – 97.3 = 2.7' or 2'-8 3/8").

ELEVATIONS AND CONTOURS

Elevation points on a site may be either existing or finish elevations. An **existing elevation** identifies the condition of the lot prior to beginning construction work. A

8TH	0"	1"	2"	3"	4"	5"	6"	7"	8"	9"	10"	11"
0	.00	.08	.17	.25	.33	.42	.50	.58	.67	.75	.83	.92
1	.01	.09	.18	.26	.34	.43	.51	.59	.68	.76	.84	.93
2	.02	.10	.19	.27	.35	.44	.52	.60	.69	.77	.85	.94
3	.03	.11	.20	.28	.36	.45	.53	.61	.70	.78	.86	.95
4	.04	.13	.21	.29	.38	.46	.54	.63	.71	.79	.88	.96
5	.05	.14	.22	.30	.39	.47	.55	.64	.72	.80	.89	.97
6	.06	.15	.23	.31	.40	.48	.56	.65	.73	.81	.90	.98
7	.07	.16	.24	.32	.41	.49	.57	.66	.74	.82	.91	.99

1. Find the figure .61 on the chart.
2. Follow the vertical column that contains the number .61 to the inches row at the top and write down the number of inches (7).
3. Follow the horizontal row that contains the number .61 to the 8ths column at the left and write down the number of 8ths (3).
4. Combine the inches (7) and the inch fraction (3/8).
5. .61′ = 7 3/8″

■ **Table 6–1.**
A conversion chart can be used to change tenths or hundredths of a foot to inches. In the above example .61′ is changed to inches.

finish elevation shows a change where soil has been removed or added to achieve a desired lot surface. Lots that are close to level will show little difference in their existing and finish condition. However, many lots are sloping or have uneven surfaces and require grading. In this case, besides identifying the corner elevations, the surveyor must record the existing elevations at different points on the property. Then the architect can establish contour lines on the survey plan as curved lines connecting the same elevations on the lot. The distances between contour lines may be, for example, 1, 5, 10, 15, or 20 feet, depending on the particular survey plan. The shorter the distance between contour lines, the steeper is the slope. **(See Figure 6–4.)**

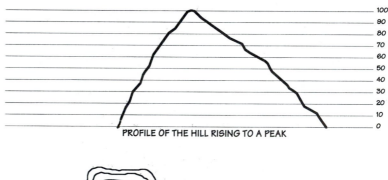

PROFILE OF THE HILL RISING TO A PEAK

■ **Figure 6–4.**
The contour lines in this drawing are ten feet apart. The closer lines indicate a steeper slope at the bottom section of the hill.

CONTOUR LINES AS THEY MIGHT APPEAR ON A HILL RISING TO A PEAK

After the existing elevations are known, the architect can determine the grading and the other groundwork that is necessary before building construction can begin. It is important that the ground be contoured (shaped) so that rainwater is directed away from the building. This may require removing (cutting) soil or adding fill soil in different areas of the lot. Dashed lines on the survey plan show the existing contours of the property before any grading has taken place. Solid lines identify the appearance of the contours after grading. Grading is usually done with earth-moving machinery such as graders, bulldozers, and backhoes.

COMPASS DIRECTIONS

A survey or site plan always has an arrow pointing north to orient the position of the building. Site boundaries are identified as north, south, east, and west property lines. The corners of the property are identified as southwest, northwest, southeast, and northeast corners. **(See Figure 6–5.)** It is very important to think in compass terms when reading construction plans to avoid mistakes in identification. For example, a particular wall should be called a north, south, west, or east wall rather than a front, back, right, or left wall.

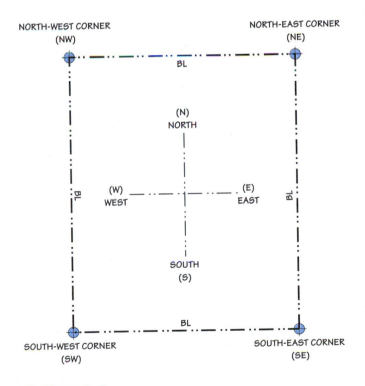

■ **Figure 6–5.**
Most construction plans use compass directions to show the corners of the property and building lines (BL), walls, and other parts of the structure.

ANGLES

On rectangular or square lots, the property lines will be at right (90-degree) angles to each other. However, on lots that are not rectangular or square, property lines often do not intersect at a 90-degree angle. In that case, the angles must be identified on the survey plan using the angle measurements of degrees (°), minutes (′), and seconds (″). For example, S 24°53′09″ E means that this angle is established by measuring 24 degrees, 53 minutes, and 09 seconds east of south. If an older type of transit instrument is being used, the surveyor adjusts the horizontal and vernier scale so that the zero reading is pointing due south. The surveyor then swings the instrument to the east so that it reads 24°53′09″ to lay out the angle of the property line. **(See Figure 6–6.)** Today more sophisticated devices are frequently used, such as a **total stations instrument** that can read distances and also angles.

READING THE SURVEY PLAN

The survey plan is a plan view looking down on the property. Its main function is to identify the exact property lines and show the main topographical features of the lot. Another of its important functions is to show the locations of the utility lines. Symbols and abbreviations appearing in the survey plan (and site plan) are shown in **Figure 6–7.**

A survey plan for the **One-Story Residence** is shown in **Figure 6–8.** Concentrate on the following when studying the plan:

✓ Symbols and abbreviations.
✓ Compass directions.
✓ Dimensions and angles of property lines.
✓ Existing and finish corner elevations.
✓ Existing and finish contour elevations.
✓ Locations of utility lines.
✓ Easements.
✓ Trees and shrubs.

Explanations of the survey plan features in Figure 6–8 follow:

1. **Scale.** The full-size survey plan is drawn to the engineer's scale of 1″ = 10′. The smaller scale is used because of the wide area covered by the lot.
2. **Curb radius.** The 48′ radius shown in the drawing is used to establish the street curve. The plus sign (+) shows the center point from which the radius is drawn. The location of this point is often identified by a monument marker in the street as shown on the plat plan.

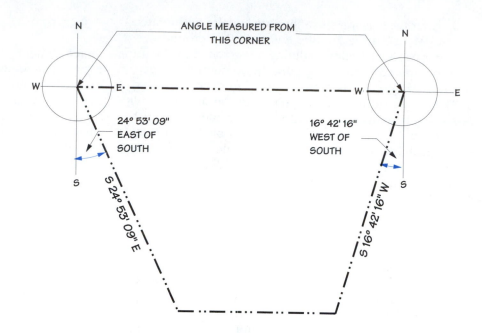

■ **Figure 6–6.**
The dashed lines represent Lot 125 on the plat plan in Figure 6–8. The angles are measured from the north side of the property.

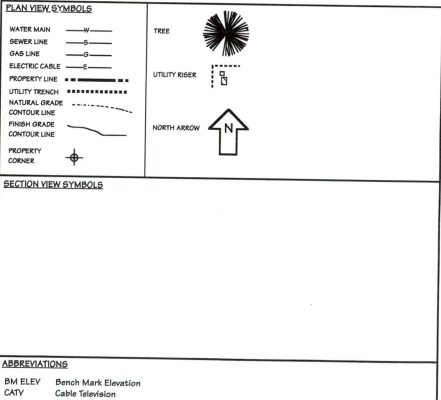

■ **Figure 6–7.**
Symbols and abbreviations found in the survey and site plans of the **One-Story Residence.**

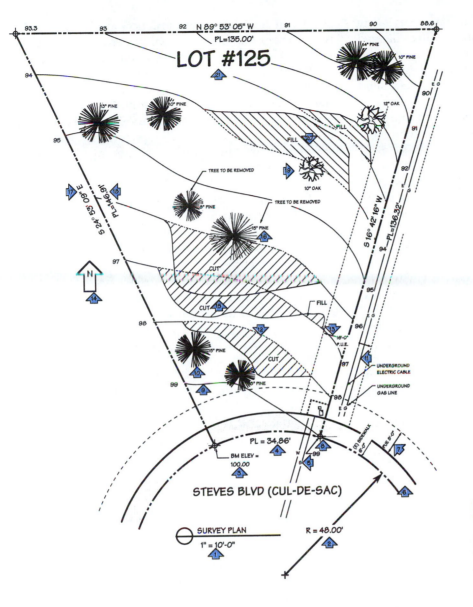

■ **Figure 6–8.**
The survey plan shows the exact property lines and provides other topographical information pertaining to the building site.

3. **Benchmark elevation.** The southwest corner of the property has been designated as the bench mark (BM) for the construction site. It is identified by a small circle at the intersecting point at that corner of the lot. The use of the 100.00′ figure rather than the number of feet above sea level simplifies comparisons between different points on the property. All recorded elevations are either below or above the 100.00′ figure.

4. **Property line (PL).** The length of the south curved property line is 34.86′. Long dashes separated by two short dashes identify this property line.

5. **Water and sewer lines.** The water (W) line and sewer (S) line leading to the building are identified. These lines hook up to the water and sewer mains already buried beneath the surface of the street.

6. **Sidewalk.** The existing sidewalk is 6′0″ wide.

7. **Public utility easement (PUE).** The width of the public utility easement is 8′. This easement runs through all the lots at the cul-de-sac. If necessary, additional utility lines serving the entire area may be placed within the bounds of the easement.

8. **Corner point.** The small circle identifies the southeast corner of the property lines. The height of the southeast elevation is 99′. This tells us that this corner of the property is 1′-0″ lower than the benchmark elevation at the southwest corner. (100.00′ − 99′ = 1′-0″.

9. **Contour line.** The solid line shows a finish-grade contour at this point of the lot. The elevation (99′) is identified at both ends of the line.

10. **Trees.** The tree species and trunk diameters are identified. These trees will not interfere with the construction and will not be removed.

11. **Electric cable and gas line.** An underground elec-

tric (E) cable and gas (G) line have previously been set in place. These locations are often flagged on the surface of the ground so they can be found more easily. When the construction begins, the building will tap into these lines for electric and gas power.

12. **Existing-grade contour line.** The dotted line identifies the existing contour in that area before any grading work has taken place.

13. **Public utility easement (PUE).** The width (16'-0") of another public utility easement is shown. On this construction project the electric cable and gas lines running through this easement also service adjoining buildings.

14. **North compass direction.** The arrow and letter (N) identify the north compass direction of the property.

15. **Cut.** The slanted lines show that earth will be removed in this area to even the slope over which the house is to be built.

16. **Trees.** These trees are in the building area and must be removed.

17. **Compass direction.** The compass direction (azimuth) of the west property line is given in degrees, minutes, and seconds. The reading is 24° 53' 09" east of north.

18. **Property line (PL).** The length of the west property line is 146.91'.

19. **Tree.** The sun deck at the rear of the building will later be built around this tree.

20. **Fill.** The slanted lines show where earth will be placed to reduce the slope of the lot in this area.

21. **Lot number.** This number corresponds to the lot number on the plat plan.

READING THE SITE PLAN

The site plan, also called a **plot plan,** contains all the information found in the survey plan, plus it shows the house and other structures on the property. The measurements around the building's perimeter may be given, and the distances from the front and side property lines establish the location of the house and other structures. All dimensions concerning concrete work are also shown on the site plan. Concrete work frequently includes driveways, walks, patios, terraces, and retaining walls. Site plans cover a much wider area than other working drawings on a set of plans. Therefore, the engineer's scale is commonly used to dimension the drawings of a site plan.

Based on the information provided by the site (or survey) plan, the initial groundwork can be concluded before the building construction begins. Trees and shrubs in the way of construction are removed. Any necessary grading, excavation, or trenching is completed. Wherever necessary, pipes and conduits are placed in the ground. Then building lines are set up to begin construction of the foundation. Symbols and abbreviations in the site are shown in Figure 6–7.

The site plan for the **One-Story Residence** is shown in **Figure 6–9.** A corresponding pictorial drawing of the site plan appears in **Figure 6–10.** Concentrate on the following when studying the plan:

✓ Symbols and abbreviations.
✓ Property line dimensions.
✓ House and other structural dimensions.
✓ Distances from property lines to the house and other structures.
✓ Dimensions of walks and driveways.
✓ Natural and finish-grade contours.
✓ Locations of utility lines.

Explanations of the site plan features in Figure 6–9 follow:

1. **Driveway width.** The width of the driveway at the curb is 11'-6", and it increases to 19'-0" in front of the garage.

2. **Front zoning setback.** The dotted line shows that 25'-0" is the minimum setback from the front property line to the building permitted by local zoning regulations. In this site plan the building is farther back than the minimum setback.

3. **Utility risers.** These are metal boxes that provide hookups for services such as water, telephone, electric, gas, and cable TV. Utility risers are usually found in rural and suburban neighborhoods.

4. **Walk.** The width of the walk leading to the front porch is 4'-8".

5. **Front building setback.** The actual distance from the front property line to the building is 30'-0".

6. **Utility trench.** The lines going from the utility risers to the building are placed in this trench.

7. **Utility service panels.** The main electrical panel containing circuit breakers and the electric meter are placed against the house wall at this location. The gas meter and sometimes the water meter is also placed against this wall.

8. **West side building setback.** The closest building corner to the property line measures 8'-0" from the property line. Note that here the zoning and building setbacks are the same distance from the west property line.

9. **Drainage swale.** A depression has been dug here to channel water toward the drainage flow at the rear of the building.

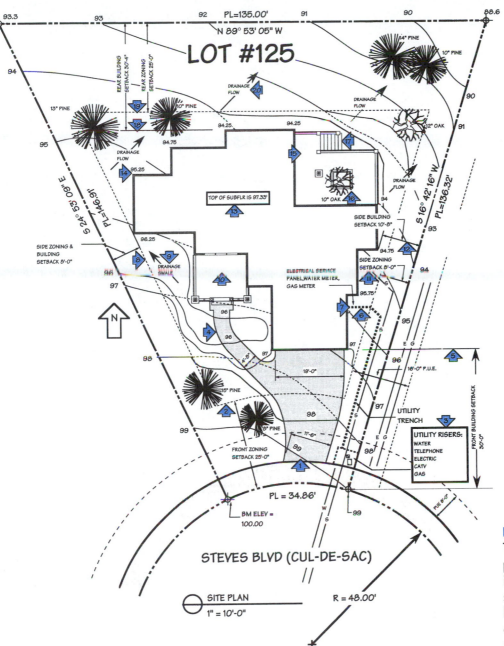

10. **Front porch.** The outline of the front porch area is shown.

11. **Side zoning setback.** The dotted line shows that 8'-0" is the minimum setback from the east property line to the building, as permitted by local zoning regulations. In this site plan the building is farther in than the minimum side setback.

12. **East side building setback.** The distance from the east side property line to the closest corner of the building is 10'-8".

13. **Top of subfloor.** The top of the main floor subfloor is 97.33' in relation to the benchmark (100.00') at the southwest corner of the property.

Therefore, the subfloor elevation is 2.67' lower than the benchmark (100.00' – 97.33' = 2.67').

14. **Corner ground elevation.** The ground elevation at this building corner is 95.25'. Ground elevations are shown at all the other building corners.

15. **Sun deck.** The rear sun deck area is identified.

16. **Tree.** The back porch will be constructed around this tree.

17. **Stairway.** The stairway to the sun deck is shown.

18. **Rear building setback.** The distance to the building line from the rear (north) property line is 30'-4".

19. **Rear zoning setback.** The dotted line shows that local zoning regulations permit a minimum set-

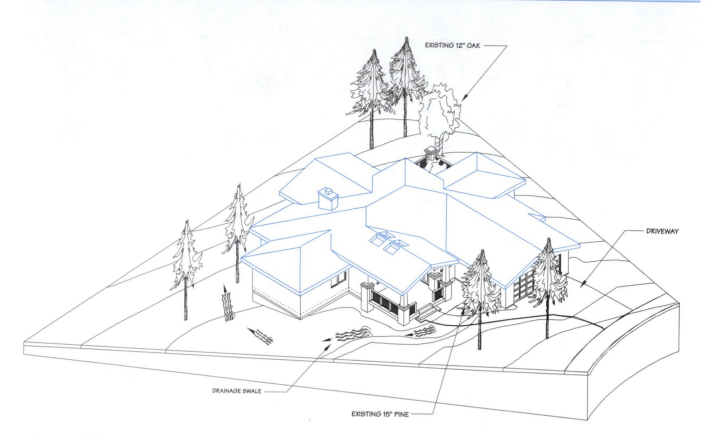

EXISTING 12" OAK

DRIVEWAY

DRAINAGE SWALE

EXISTING 15" PINE

■ **Figure 6–10.**
This pictorial drawing of the property and building is based on the site plan shown in Figure 6–9.

back of 25'-0" from the rear property line. In this site plan the rear of the building is farther back than the minimum zoning setback.

20. **Drainage flow.** The arrow points in the direction that water drainage will follow the slope of the ground.

REVIEW QUESTIONS

Enter the missing words in the spaces provided on the right.

1. Township grids enclose an area of _____ square miles.

2. There are 36 _____ in each township.

3. A _____ plan shows street locations and lots in a given area.

4. Survey and/or _____ plans get much information from the plat plan.

5. The property lines for a building site are shown on the _____ and/or plot plan.

6. The exact property corners are usually staked out by a _____.

7. Different ground levels identified on a lot are called _____.

8. A _____ mark is a fixed reference point on a job site.

9. _____ elevations identify points on the lot after grading has taken place.

10. Curved lines connecting the same elevations on plans are called _____ lines.

11. Non-right-angle property lines are measured in degrees, _____, and seconds.

12 In addition to the survey plan information, a site plan will show the _____.

13 A plot plan shows the _____ from the property lines to the house.

14 _____ and finish-grade contours are often shown on the plot plan.

15. The distance from a front property line to the building is called the front _____.

Match the letters next to the symbols with their descriptions.

16.	Utility trench	_____
17.	North arrow	_____
18.	Finish contour line	_____
19.	Electric cable	_____
20.	Water main	_____
21.	Tree	_____
22.	Natural contour line	_____
23.	Sewer line	_____
24.	Property line	_____
25.	Gas line	_____
26.	Property corner	_____
27.	Utility riser	_____

a.   ——W——

b.   ——S——

c.   ——G——

d.   ——E——

e.   ■ ■ ■■■■ ■ ■

f.   ■■■■■■■■■■

g.   - - - - - -

h.   〜〜

i.   ⊕

j.   ✹

k.   ⬚

l.   ⬆N

Choose the correct abbreviation for each term.

28. Benchmark elevation   _____   R

29. Cable television   _____   N

30. East   _____   PUE

31. Existing   _____   BM ELEV

32. North   _____   E

33. Property line   _____   CATV

34. Public utility easement _____   PL

35. Radius   _____   EXIST

Foundation Plans

The construction of a building begins with the foundation. Most of the information required to build the foundation is found in the **foundation plans.** A foundation must be designed to support its own weight plus the structural loads of the entire building. It must evenly distribute and transfer these loads to the ground below. The foundation also provides a base for leveling and anchoring the framework of the building.

FOUNDATION SYSTEMS

The foundation systems used most often in residential and other kinds of light construction are the crawl-space, full-basement, and slab. **(See Figure 7–1.)** The **One-Story Residence** discussed in this section can be constructed over any of these foundations. Working drawings for all three are included in this unit.

CRAWL-SPACE FOUNDATIONS

Crawl-space foundations do not have a basement area, but they provide a space between the framed first-floor unit and the ground. The foundation walls should be high enough to provide a minimum of 18" between the ground surface and the bottom of the floor joists resting on top of the foundation. Plans often call for 24" clearance to allow easier access for utility repairs. Concrete footings below wooden posts or masonry piers provide support for beams. The beams give central support to the floor joists.

FULL-BASEMENT FOUNDATIONS

A **full-basement foundation** extends below the ground and provides additional storage and work area beneath the first floor of the building. The walls must be high enough to allow sufficient clearance from the basement floor to the ceiling. Clearance requirements are designed by local codes and may range from 7 to 8 feet (7'-0" to 8'-0"). If required, concrete pier footings are constructed as a base for wood posts or for steel columns supporting wood or steel beams. The beams provide interior support for the floor unit above. Access to the basement is usually provided by a stairway from the first-floor level of the building. Some codes require a second entrance provided by an exterior stairway to the basement.

SLAB FOUNDATIONS

A **slab foundation,** also called **a slab-on-grade foundation,** consists of a concrete slab that receives its main support from the soil. Additional support is provided by footings around the perimeter and within the slab. Slab foundations are the least expensive foundation systems as they eliminate the necessity of constructing a wood framed floor unit.

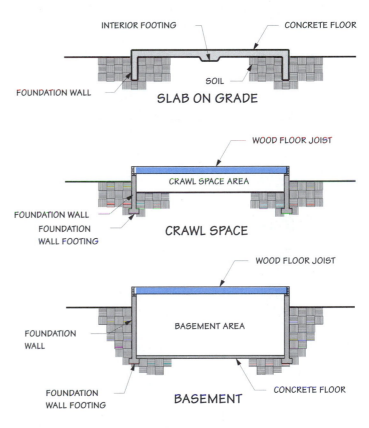

INTERIOR FOOTING   CONCRETE FLOOR

FOUNDATION WALL   SOIL   **SLAB ON GRADE**

WOOD FLOOR JOIST

CRAWL SPACE AREA

FOUNDATION WALL
FOUNDATION WALL FOOTING   **CRAWL SPACE**

WOOD FLOOR JOIST

BASEMENT AREA

FOUNDATION WALL

FOUNDATION WALL FOOTING   **BASEMENT**   CONCRETE FLOOR

■ **Figure 7–1.**
The three foundations systems used most often in residential and commercial construction are the full-basement, crawl-space, and slab (slab-on-grade) foundations.

TYPES OF FOUNDATIONS

The two foundation types used most are the spread foundations, shaped like an inverted T, and grade-beam foundations supported by piers or pilings. **Spread foundations,** frequently called **T foundations,** consist of a stem wall placed over a concrete footing. The **footing** is a wide concrete pad that spreads the building load over the supporting soil. **Stem walls** are vertical walls supported by the footings and constructed of poured-in-place concrete or concrete blocks. A formula sometimes used for footing design is that the height must be equal to the width of the foundation wall above, and the width must be twice the thickness of the wall. However, underlying soil conditions may require a larger footing in proportion to the wall than expressed in the formula.

Grade-beam foundations are used when there are unstable soil conditions close to the ground surface. Grade-beam foundations consist of a concrete foundation wall supported by piers that extend down to load-bearing soil. **(See Figure 7–2.)**

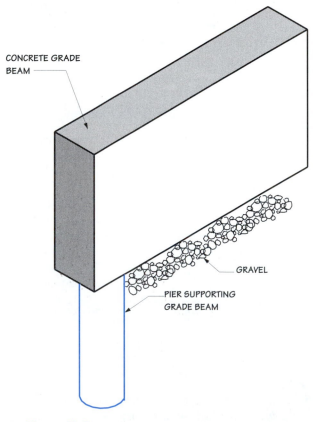

CONCRETE GRADE BEAM

GRAVEL

PIER SUPPORTING GRADE BEAM

■ **Figure 7–2.**
Grade beam foundations are used over unstable soil conditions. The main support of the walls is provided by piers extending down into load-bearing soil.

For buildings erected on steeply sloping lots, a **stepped foundation** may be used. This consists of a series of stepped-up walls and footings as shown in **Figure 7–3.** A low stud wall is framed over the steps to provide support for a level floor above. Stepped foundations are considered more economical as less concrete is required. Grade beams that follow the angle of the ground surface are another way to deal with steeply sloping lots. The grade beams are supported by piers.

FOUNDATION DESIGN FACTORS

Several key factors affect the design of a foundation. Vertical (down) pressures and loads, and lateral (side) pressures greatly determine its shape and size. Other important considerations are the type of soil beneath the foundation and the surface and below-surface groundwater conditions.

LOADS AND PRESSURES

Dead loads and live loads exert the greatest vertical pressure on a foundation. **Dead load** is the total weight of the building. This consists of the foundation, floors, walls, roof, plus any stationary mechanical equipment such as furnaces, hot water heaters, and air-conditioning units. **Live load** is the total anticipated weight of people in the building along with the equipment and furniture they use. The weight exerted by snow, ice, and water on the roof is also considered part of the live load.

The foundation walls must also be strong enough to resist the lateral pressures of earth and water pressing against the walls. **(See Figure 7–4).** In seismic areas a foundation must be designed to withstand the motions caused by earthquakes and other types of earth vibrations.

SOIL

Soil conditions strongly influence foundation design since they vary in their bearing capacity. Some settlement of the foundation can be expected over time as the soil around the foundation adapts to the loads of the building. This is not a problem if a small amount of equal settlement occurs under the entire foundation. Equal settlement occurs when the entire foundation rests on the same kind of soil.

However, uneven settlement can cause structural damage. It takes place when a foundation is constructed over ground that contains different types of soil all at the same depth. The result can be sloping floors, binding doors and windows, cracked glass, drywall cracks,

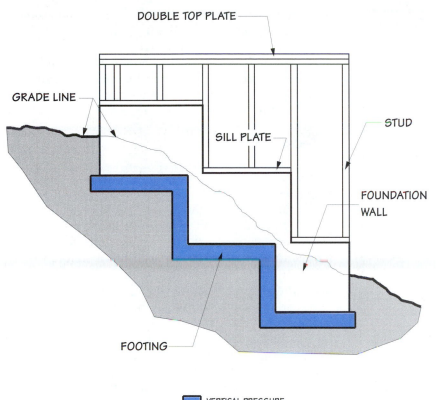

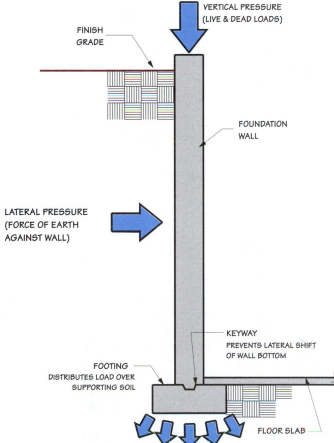

■ **Figure 7–3.**
Stepped spread foundations are used on steeply sloped lots.

■ **Figure 7–4.**
Foundation walls must be designed to resist lateral and vertical pressures.

and other problems. To prevent uneven or excessive settlement, it is often necessary to excavate through several layers (strata) of soil types to reach a layer of firm, load-bearing soil.

Soils are generally divided into coarse-grained and fine-grained categories. **Coarse-grained soils** such as sand and gravel provide the most stable base for a foundation. **Fine-grained soils** such as silt and clay provide a less stable base, and they sometimes provide inadequate load-bearing support. **(See Table 7–1.)**

Several different methods can be used to test soil conditions at the construction site. Shallow exploration can be done by digging open test pits. Mechanical drilling rigs are used for deeper exploration where test samples of soil can be brought to the surface for laboratory analysis. Soil testing is usually required for heavy construction projects that require deeper excavation. Whether or not analysis of subsurface soil conditions beneath residential and light construction projects is required depends on established practice in the area.

WATER DRAINAGE AND MOISTURE

Improper water drainage can cause water accumulation within and around the perimeter of the foundation. This can cause odors and wood decay above the stem walls of a crawl-space foundation. Seepage and flooding can occur in a full-basement foundation. Water can rise and penetrate the slab of a slab foundation.

SOIL BEARING CAPACITY TABLE	
TYPE OF SOIL	ASSUMED BEARING CAPACITY TONS PER SQ. FT.
Gravels or gravel-sand mixtures little or no fines	5
Gravel-sand-clay or gravel-sand-silt mixtures	2 to 2.5
Sands or gravelly sands, little or no fines	3 to 3.75
Sand-silt or sand-clay mixtures	2
Inorganic silts, very fine sands or inorganic clays	1

■ **Table 7–1.**
Different types of soils vary in their load-bearing ability.

Surface water problems are usually solved by grading the lot surface so that water is directed away from the building. Underground water can be diverted by surrounding the exterior of the foundation walls with a thick layer of gravel and by placing *drain tile* or *pipe* next to the footings of the foundation. **(See Figure 7–5.)** Underground water passes easily through gravel. Therefore, it will flow toward the bottom of a gravel layer and into the drain pipe, where it will be directed away from the building, sometimes into a sump pit.

Section views show this method in the full-basement foundation plans for the **One-Story Residence.**

It is generally recommended that a waterproofing material be placed on the exterior surface of full-basement foundation walls. Asphaltic mastics, synthetic rubber sheets, and plastic sheets are some materials used for this purpose.

FROST LINE

The bottoms of foundation footings should never rest above the frost line. The **frost line** is the depth to which the ground freezes below its surface during the winter season. Thus, the frost line is determined by the amount of water penetration that is accompanied by freezing temperature. The depths of frost lines therefore vary in different parts of the country because of climate variations. **(See Figure 7–6.)**

Soil heaving can occur when the base of a foundation is placed above the frost line. Soil heaving is a condition in which the soil freezes and expands, pushing the building upward. When the soil thaws, the building drops down. This up-and-down motion can cause structural damage.

REINFORCED CONCRETE AND CONCRETE-BLOCK WALLS

Most building foundations are constructed of poured-in-place concrete or concrete blocks resting on concrete footings. Building codes in most areas will require the addi-

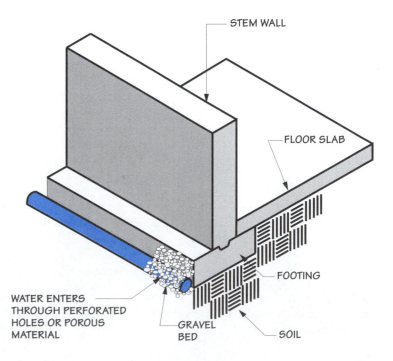

■ **Figure 7–5.**
Drain tile or pipe is placed next to the footing to direct water away from the foundation.

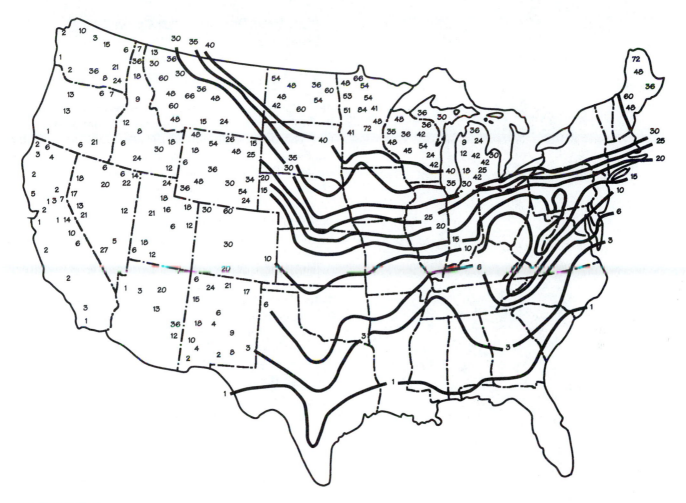

■ Figure 7–6.
The frost line depths vary greatly in different parts of the United States.

tion of steel reinforcement. Such reinforcement is particularly important in seismic (earthquake) risk areas.

Concrete is a mixture of cement, sand, gravel, and water that is placed in braced forms constructed to the shapes of the walls and footings. **Concrete blocks,** also called **concrete masonry units (CMUs),** are made of a stiff concrete mixture formed around a hollow interior. They are frequently placed over concrete footings. Concrete blocks are manufactured in a variety of shapes and sizes. See Unit 32 for further information about the composition of concrete and concrete block materials.)

STEEL REINFORCEMENT

The term **reinforced concrete** derives from the steel reinforcement placed into concrete walls and footings and concrete-block walls. Concrete has a great deal of compressive strength, enabling it to resist the vertical loads it must support. However, concrete by itself does not have sufficient tensile strength to adequately resist

lateral and stretching forces. For this reason, foundation walls and footings are usually reinforced with **the steel bars** commonly called **rebars** shown in **Figure 7–7(A).** They are round in shape, with surface ridges to give better bonding to the concrete. The section views of a foundation plan normally show the vertical and horizontal spacing of the bars. The cross-sectional size (diameter) of the bar is identified by numbers ranging from 3 to 18. The number multiplied by 1/8 gives **the** actual diameter of the bar in fractions and/or inches. For example, a #5 bar identifies the diameter of a 5/8" bar ($5 \times 1/8 = 5/8$").

Slab-foundation floors are reinforced with **either** rebars or **welded wire mesh (WWM),** also known as **welded wire fabric,** which is basically a grid of steel wires welded together 2 to 12 inches apart. WWM is shown in Figure 7–7(B). It comes in rolls. Section view drawings of a foundation plan specify which type of WWM to use. (See Unit 32 for additional information about rebars and wire mesh.)

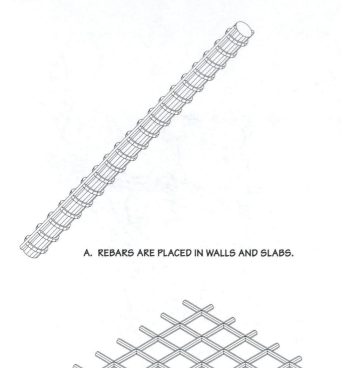

A. REBARS ARE PLACED IN WALLS AND SLABS.

B. WELDED WIRE MESH IS PLACED IN LIGHTER CONCRETE FLOOR SLABS, WALKS AND DRIVEWAYS.

■ **Figure 7–7.**
Steel bars (rebars) and welded wire mesh are used to reinforce concrete.

READING THE CRAWL-SPACE FOUNDATION PLAN

All types of foundation plans consists of a *plan view* and related *section views*. Plan view drawings are a view looking down at the foundation. Section view drawings can be compared with vertical cuts made at different sections of the foundation. They reveal much information that cannot be seen in the plan view. Some foundation plans include information about the spacing, size, and direction of first-floor framing components such as the joists and beams. However, this information can also be provided in a more detailed floor framing plan. Symbols and abbreviations appearing in the crawl space, basement, and slab foundation plans are shown in **Figure 7–8.**

PLAN VIEW DRAWINGS FOR CRAWL-SPACE FOUNDATION

The main purpose of the plan view drawings is to show the location and layout of the foundation walls and piers that support the building. A series of dimension line appears around the exterior of the plan view drawing. The outside dimension lines usually identify the total width and length of the building. Dimension lines closer to the building locate other corner points, interior walls, and door and window openings. The plan view also identifies the centers of pier footings and supporting posts and beams. Dimensions between rows of footings and other structural features may also appear inside the drawing. **(See Figure 7–9.)**

The plan view drawing for the **One-Story Residence** crawl-space foundation is shown in **Figure 7–10.** The footings below the stem walls and piers are constructed of poured-in-place concrete. The stem walls and piers are constructed of concrete blocks. A corresponding pictorial drawing of the plan view appears in **Figure 7–11.** Concentrate on the following when studying the drawing.

✓ Symbols and abbreviations.
✓ Dimensions establishing positions of the exterior and interior walls.
✓ Dimensions establishing the placement of beams and their supporting piers and footings.

Explanations of the crawl-space foundation plan view features in Figure 7–10 follow:

1. **Scale.** A frequently used scale for plan view drawings is 1/4″ = 1′-0″.
2. **Top of driveway.** The driveway elevation (96′-10″) is identified in front of the garage entrance.
3. **Outer dimension line.** The widest east-to-west point of the building is 68′-0.″.
4. **Inner dimension lines.** These measurements establish the following dimensions:

   ✓ Outside face of the west wall to the face of the post of the southwest porch column (16′-4″). The same dimension locates the outside face of the foundation wall in line with the posts of the west porch columns.
   ✓ Post center of the southwest porch column to the post center of the southeast porch column (14′-8″). The same dimension locates the outside face of the foundation wall in line with the east column.

SYMBOLS AND ABBREVIATIONS
FOUNDATION PLANS

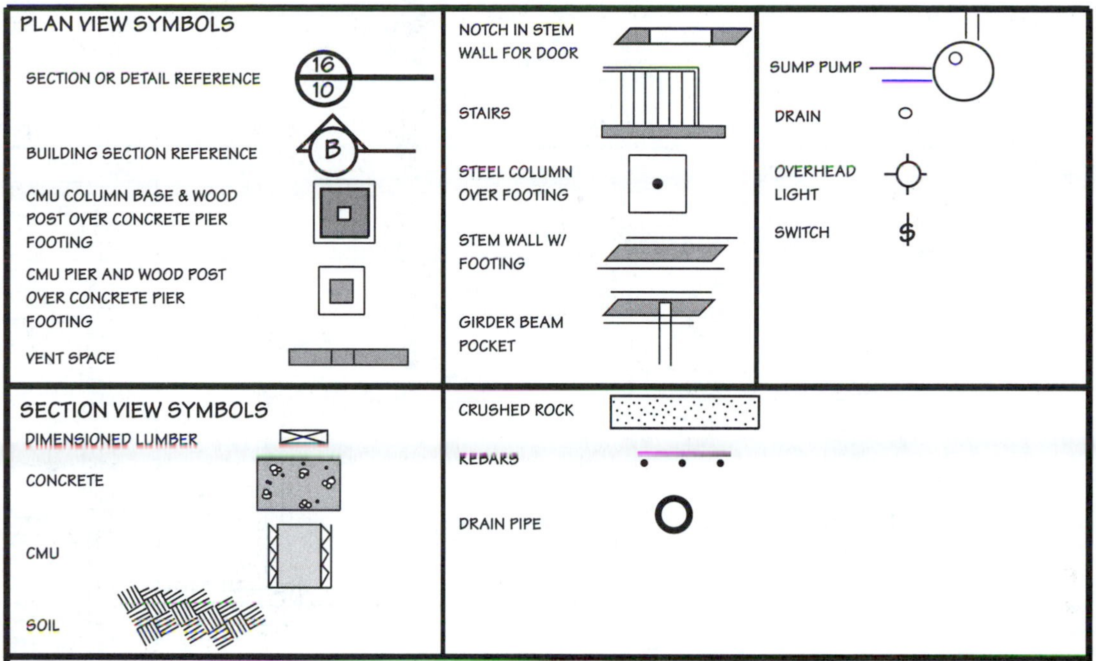

PLAN VIEW SYMBOLS

SECTION OR DETAIL REFERENCE

BUILDING SECTION REFERENCE

CMU COLUMN BASE & WOOD POST OVER CONCRETE PIER FOOTING

CMU PIER AND WOOD POST OVER CONCRETE PIER FOOTING

VENT SPACE

NOTCH IN STEM WALL FOR DOOR

STAIRS

STEEL COLUMN OVER FOOTING

STEM WALL W/ FOOTING

GIRDER BEAM POCKET

SUMP PUMP

DRAIN

OVERHEAD LIGHT

SWITCH

SECTION VIEW SYMBOLS

DIMENSIONED LUMBER

CONCRETE

CMU

SOIL

CRUSHED ROCK

REBARS

DRAIN PIPE

ABBREVIATIONS

@	At	MIN	Minimum
BET	Between	OC	On Center
BSMT	Basement	PERF	Perforated
BTM	Bottom	PERIM	Perimeter
BLW	Below	PVC	Polyvinyl Chloride
CL	Center Line	PT	Pressure Treated
CONC	Concrete	REBAR	Reinforcing Steel Bar
CONT	Continuous	SG	Solid Grout
CMU	Concrete Masonry Unit	SIM	Similar
Ø	Diameter	ST	Steel
FOW	Face of Wall	S4S	Surface Four Sides
FDN	Foundation	SQ FT	Square Foot
FL JSTS	Floor joists	TYP	Typical
FTG	Footing	W/	With
GALV	Galvanized	WD	Wood
INSUL	Insulation		
JTS	Joints		

■ **Figure 7–8.**
These are symbols and abbreviations found in the foundation plan of the **One-Story Residence.**

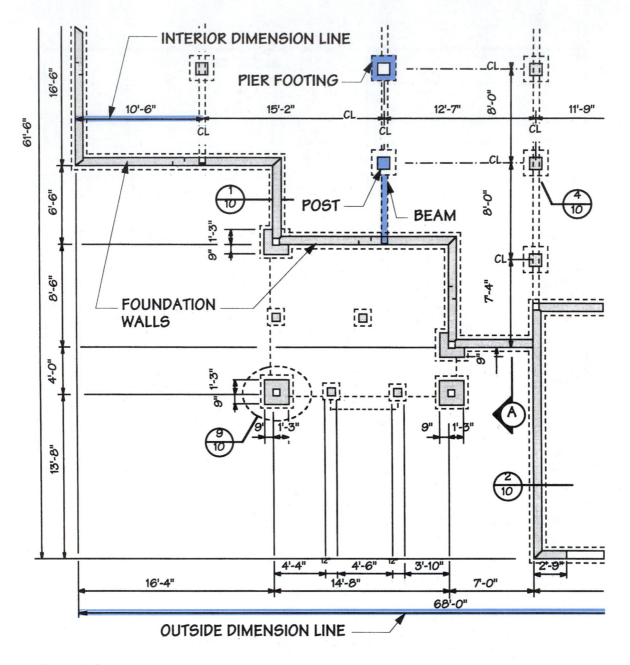

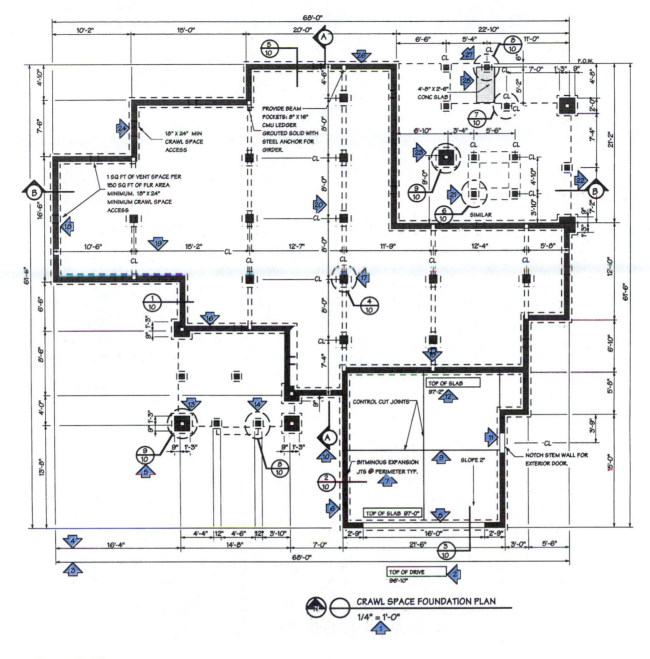

■ Figure 7–10.
The crawl-space foundation plan view identifies and provides information about the walls, piers, and footings. There is a pony (stud) wall at the north and west sides of the building.

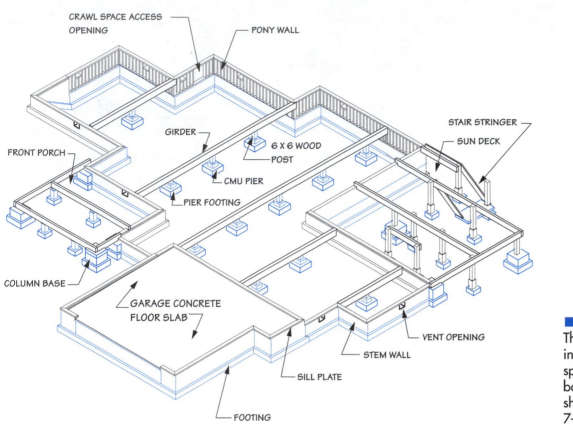

CRAWL SPACE ACCESS
OPENING

PONY WALL

GIRDER

6 X 6 WOOD
POST

CMU PIER

PIER FOOTING

FRONT PORCH

COLUMN BASE

GARAGE CONCRETE
FLOOR SLAB

FOOTING

SILL PLATE

STEM WALL

VENT OPENING

STAIR STRINGER

SUN DECK

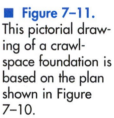

■ **Figure 7–11.**
This pictorial drawing of a crawl-space foundation is based on the plan shown in Figure 7–10.

✓ Post center of the southeast porch column to the outside face of the west garage wall (7′-0″).
✓ Outside face of the west garage wall to the outside face of the east garage wall (21′-6″).
✓ Outside face of the east garage wall to the outside face of the closest foundation wall (3′-0″).
✓ Outside face of the closest house foundation wall to the outside face of the farthest foundation wall (5′-6″).

5. **Section reference.** This symbol refers to a separate section drawing. A leader line points from the circle to the object being viewed. The number (9) at the top of the line within the circle is the number of the section drawing. The number (10) below the line is the number of the page containing the drawing. In the original plans the foundation sections appeared on page 10 of the plans. (Foundation section view drawings are discussed later in this unit.)

6. **Garage foundation wall.** The dashed (hidden) line identifies the concrete footing. The shaded area identifies the concrete-block stem wall that rests on top of the footing.

7. **Expansion joint.** This is a narrow space created wherever the slab butts against the foundation wall.

It is filled with a bituminous (asphalt) material. The expansion joint allows for expansion and contraction of the slab that will occur because of changing temperature conditions. This joint reduces the possibility of cracks developing in the slab.

8. **Garage entrance foundation.** The dotted lines identify the footings below the front of the garage. The width of the garage opening is 16′-0″. The front of the slab elevation is 97′-0″.

9. **Control joints.** These are grooves formed or cut into the slab. They help to control cracking resulting from expansion and contraction.

10. **Building section reference (A).** Identical symbols are shown at opposite sides of the building. If a cutting-plane line were drawn between the two symbols, it would identify the view shown on a separate building section drawing. (Building section drawings are discussed in Unit 11.)

11. **Side door to garage.** The top of the stem wall at the door opening must be notched to the slab level.

12. **Garage-slab rear elevation.** The slab elevation is 97′-2″ at this point. Therefore, there is a 2″ slope between the back and front of the garage floor. The slope helps prevent water penetration beneath the garage door from flowing toward the rear of the garage.

13. **Front porch column.** The dotted line identifies the concrete footing beneath a concrete-block pier.

14. **Deck newel-post pier.** The dotted line shows the footing below a concrete-block pier. Newel posts on each side of the stairway are fastened to the top of the pier. (A **newel post** is a vertical post at the head or foot of a stairway supporting an end of a handrail.)

15. **Beam pocket.** A rectangular pocket is provided at the top of the stem wall to receive the ends of girders supporting the floor joists. The bottom of the girder is secured to a steel anchor embedded in the concrete block below.

16. **House foundation wall.** The dashed lines show the footings extending from the shaded stem walls.

17. **Interior pier.** The dashed line shows the outline of the pier footing. The shaded square area is the pier over the footing.

18. **Vent openings.** These are placed throughout the foundation to allow air circulation in the crawl-space area. Such circulation is necessary to cut down on dampness leading to dry rot of wooden members.

19. **East-to-west pier layout.** The first dimension (10'-6") is from the outside face of the west foundation wall to the center of the first pier. The remaining dimensions are from the pier center lines (CL).

20. **South-to-north pier layout.** The first dimension (7'-4") is from the outside face of the south wall.

21. **Deck piers.** Four smaller piers help support the deck and bench surrounding the tree.

22. **Building section reference (B).** Identical symbols are shown at opposite ends of the building establishing an east-to-west cutting-plane line.

23. **Sun-deck column base.** The dashed line identifies the pier footing supporting the column. This column will help support a section of roof extending over the sun deck.

24. **Crawl-space access.** This opening provides entry into the crawl-space area from the outside.

25. **Concrete slab.** The bottom of the porch stairway rests on this small section of concrete slab.

26. **Stem wall and pony wall.** A wood stud pony wall is framed over the stem wall at the rear area of the house. Because of the slope of the lot, the north ground elevations are 6' to 10' lower than the south ground elevations. If the rear concrete-block walls were constructed with their tops level with the south walls, there would be much more exposed wall surface at the north end after the siding is placed. Constructing a stud wall over the block wall makes it possible to nail the siding much closer to the ground.

27. **Newel-post pier.** This pier is placed to support the newel post at the foot of the stairway leading up to the sun deck.

SECTION VIEW DRAWINGS FOR CRAWL-SPACE FOUNDATION

The crawl-space section view drawings give information not shown in the plan view drawings:

✓ The actual shapes of the stem walls and footings, with their widths and heights.
✓ Materials, such as concrete, concrete masonry units, sill plates, and framing components.
✓ The size and spacing of rebars and anchor bolts.

Before reviewing the section view drawings, place a bookmark at the page showing the crawl-space plan view drawing (Figure 7–10, page 55). While studying each section view drawing, relate the section number beneath the drawing to the section symbols on the plan view.

The section view drawings for the crawl-space foundation walls and piers, as well as the section drawings of the footings and piers supporting the front porch and sun deck, are shown in **Figure 7–12a-c.** Corresponding pictorial drawings are placed to the right of all the section drawings. Only the features of the foundation walls and piers are identified. (The framing members shown in the section views will be discussed in Unit 8). Concentrate on the following when studying the plans:

✓ Symbols and abbreviations.
✓ Vertical and horizontal dimensions.
✓ Material descriptions.
✓ Section views as related to the plan view.

Explanations of the crawl-space foundation section view features in Figure 7–12a-c follow:

1. **Foundation footing.** The foundation footings are 1'-4" wide and 10" high. Three #4 rebars run continuously along the length of the footings. Short #4 rebars extend across the width of the footing and are spaced 6'-0" on center.

2. **Frost line.** The bottom of the foundation footing must be at or below the frost line.

3. **Footing-base-to-grade distance.** The minimum distance from the bottom of the footing to grade is 2'-6".

4. **Stem wall.** The stem wall is constructed of concrete masonry units (CMUs). The hollow areas of these concrete blocks are filled with solid grout that is a special concrete mixture designed for this purpose.

Vertical rebars, spaced 48 inches on center (OC), extend from the top of the stem wall to 3″ from the bottom of the footing. Note that the bars are bent at a right angle where they end in the footing. These vertical rebars are the main tie between the wall and footing. A horizontal rebar runs continuously through the center block and this forms a bond beam along the length of the wall. The directions

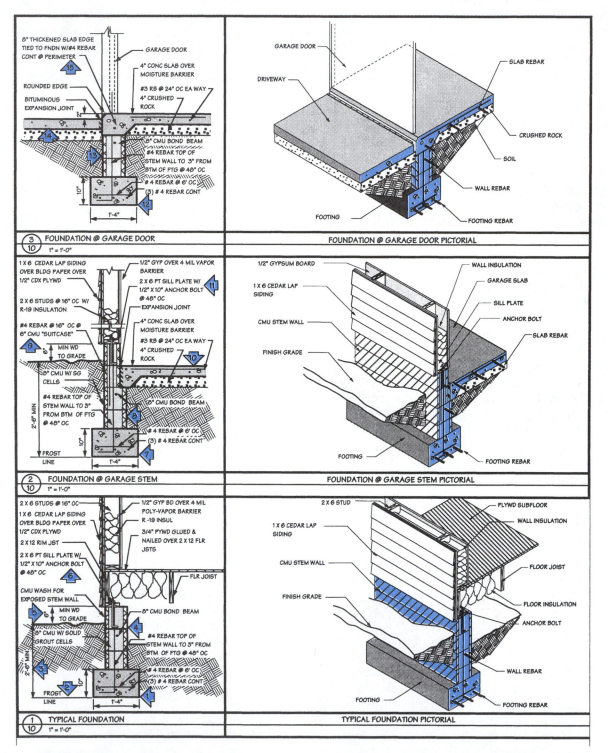

■ **Figure 7–12a.**
Crawl-space foundation section views show vertical views of the walls, piers, and footings. These views provide important information that cannot be seen in the plan view.

pointing to a CMU wash for the exposed area suggests a stucco-like finish on the face of the concrete blocks extending above grade.

5. **Minimum grade-to-wood distance.** A common

code requirement for this distance is a minimum of 6". This is to help prevent wood rot resulting from dampness and insect infestation.

6. **Sill plate and anchor bolt.** L-shaped bolts, spaced

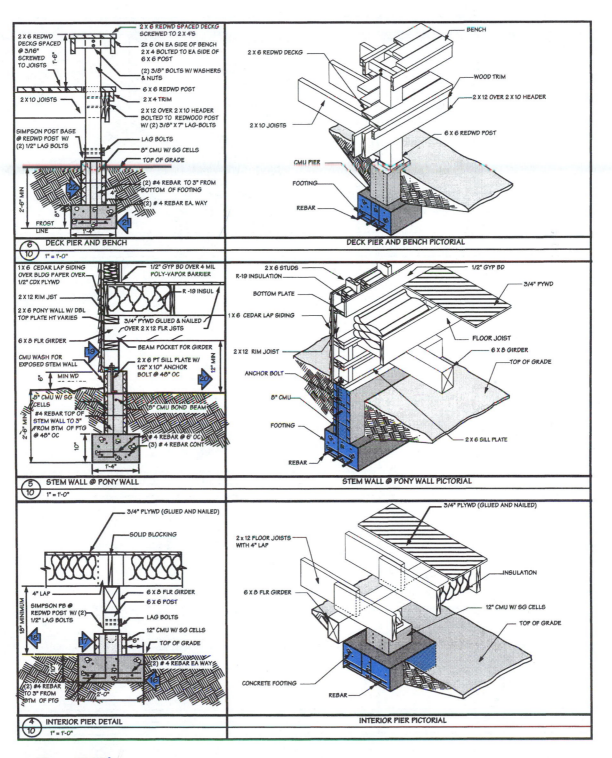

■ **Figure 7–12b.**

Crawl-space foundation section views show vertical views of the walls, piers, and footings. These views provide important information that cannot be seen in the plan view (continued).

48" OC, are embedded in the grout while it is still soft. The bend at the bottom of the bolt provides better holding action against uplifting pressures. After the grout has hardened, the sill plates are

fastened to the tops of the stem walls to provide nailing for the wooden joists. These **pressure-treated (PT)** sill plates have been chemically treated under pressure to resist damage from insect

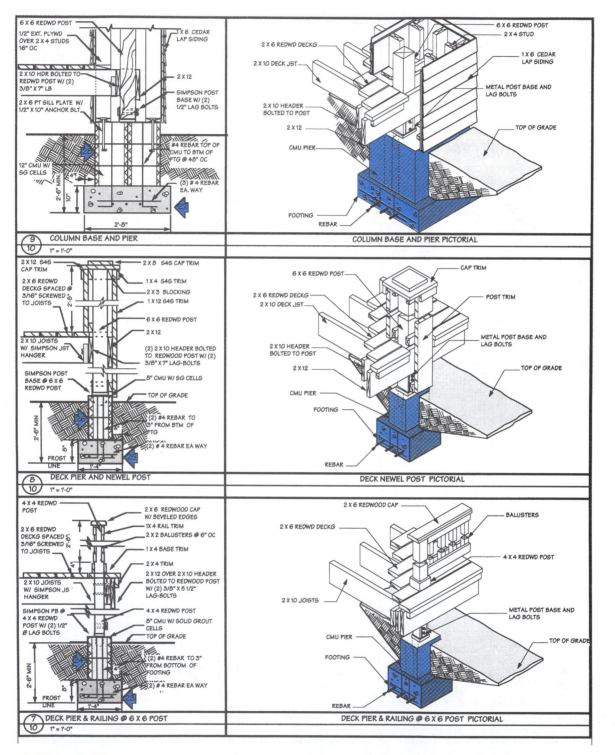

■ **Figure 7–12c.**
Crawl-space foundation section views show vertical views of the walls, piers, and footings. These views provide important information that cannot be seen in the plan view (continued).

infestation and dry rot.

7. **Foundation footing.** The dimensions and placement of rebars are the same as described in feature 1.

8. **Stem wall.** This section view describes the side and back walls around the garage area. A shoulder is formed in these walls to support the edge of the concrete garage slab. Vertical rebars, spaced 48" OC, extend from the top of the stem wall into the footing.

9. **Narrow "suitcase" wall section.** Smaller 6" concrete blocks are placed in this section of wall. The blocks create a 2" shoulder beneath the edge of the concrete slab. Vertical rebars placed 16" OC extend into the wider wall below.

10. **Garage concrete floor slab.** As previously indicated in the foundation plan view, an expansion joint is provided where the slab butts against the stem wall. The floor slab is 4" thick and is reinforced with #3 rebars running in two directions and spaced 24" OC. A moisture barrier is placed directly below the slab that rests on a 4" thick layer of crushed rock.

11. **Sill plate and anchor bolt.** Pressure-treated sill plates are secured to the tops of the walls with anchor bolts.

12. **Foundation footing.** The dimensions and placement of rebars are the same as described in feature 1.

13. **Stem wall.** This section view shows the stem wall beneath the slab at the front of the garage.

14. **Driveway.** The concrete driveway slab is 4" thick. It rests on top of a 4" layer of crushed rock. Note that the driveway surface is 2" below the garage slab surface. This helps control water seepage into the garage.

15. **Thickened slab edge.** The slab edge resting on the stem wall is 8" thick. A horizontal #4 rebar runs through the thickened section and extends into the stem walls on each side of the opening. This rebar ties the thickened edge to the foundation. The edge of the garage slab is rounded to prevent chipping.

16. **Pier footing.** The pier footing is 2'-0" square and 10" high. Two #4 rebars run in two directions.

17. **Pier.** A 12" concrete block serves as a pier resting on the footing. Two vertical #4 rebars extend down into the footing. The hollow core is filled with grout. A metal post base is embedded in the grout at the top of the pier. The upper section of the post base is bolted to the bottom of the post.

18. **Ground-to-joist clearance.** The minimum distance from the ground level to the bottom of the floor joist is 18".

19. **Pony wall.** This is a short stud wall built over the stem wall to extend the total height of the foundation wall. Here the pony wall is required at the rear of the building because the ground slopes down from south to north. Constructing the pony wall allows the exterior siding to be placed closer to the ground at the rear (south) wall.

20. **Ground-to-girder-bottom distance.** The minimum distance allowed between the ground and the bottom of the girder is 12".

21. **Deck pier and bench footing.** This footing is 1'-4" square and 8" high. It is reinforced with two #4 rebars running in two directions.

22. **Deck pier over bench footing.** The pier consists of two concrete blocks tied to the footing with two vertical rebars. A metal post base is imbedded in the top of the pier.

23. **Footing under pier supporting railing post.** This footing is also 1'-4" square and 8" high. It is reinforced with two #4 rebars running in two directions.

24. **Pier under railing post.** The pier consists of two concrete blocks tied to the footing with two vertical bars, and a metal post base is embedded in the top of the pier.

25. **Footing under pier supporting newel post.** The footing is 1'-4" square and 8" high. Two #4 rebars run in two directions.

26. **Pier under newel post.** The pier consists of two concrete blocks tied to the footing with two vertical rebars. A metal post base is embedded in the top of the pier.

27. **Footing under pier supporting column post.** The size of the pier above requires a footing 2'-8" square by 10" high. Three rebars run in opposite directions.

28. **Column-base pier.** Two vertical rows of 12" concrete blocks are placed side by side for this larger pier. The hollow cores are filled with grout, and vertical rebars extend into the footing. Short sill plates are attached to the top of the pier with anchor bolts.

CONSTRUCTING THE CRAWL-SPACE FOUNDATION

The **One-Story Residence** crawl-space foundation consists of a concrete footing and concrete-block stem walls. Trenches are dug for the concrete footings. The stem walls are constructed over the footing after the concrete has set.

FOOTING TRENCHES AND FORMS

Trenches for the footings are dug to the depth shown in the section views. A general rule is that footings extend down a minimum of 6" into undisturbed soil. If the frost line is lower than 6", the bottom of the footing must extend down to the frost line. If local ground conditions are firm enough to permit earth forms, the concrete for the footings can be placed directly into the trench. If conventional forms are required because of softer ground conditions, the trenches must be wide enough to allow for the construction of the footing forms.

LAYOUT OF BUILDING LINES

Building lines show the outside surface of the foundation walls. They are measured from the property lines, and the plot plan provides these measurements. Usually the layout of building lines is done with a transit instrument. Stakes are driven at all the building corners, and nails driven into the tops of the stakes show the exact corners. **(See Figure 7–13.)**

CONSTRUCTION OF FOOTINGS AND WALLS

If soil conditions are not suitable for earth-formed footings, a footing form is built. Horizontal rebars are placed, and vertical dowels are tied to the rebars. The concrete is placed. After the concrete has set up, vertical rebars are tied to the dowels. The concrete-block stem wall is then constructed over the footings. **(See Figure 7–14.)** If the stem wall is to be solid concrete, forms are constructed

over the footings. A section of a completed crawl-space concrete foundation is shown in **Figure 7–15.**

READING THE FULL-BASEMENT FOUNDATION PLAN

Full-basement foundation plans, like the crawl-space plans, consist of both plan views and section views. The drawings provide most of the information required to lay out the walls and the footings for piers and columns. In addition, full-basement foundation plans include windows and areaways. An **areaway** is a sunken space next to a foundation wall permitting light or air through a basement window. Also typically shown in full-basement foundation plans are (1) a stairway leading from the main floor to the basement, (2) framed walls if required, (3) floor drains, (4) overhead lights, and (5) electrical plugs and switches. Heating and air-conditioning equipment and water heaters are identified if they are to be placed in the basement.

PLAN VIEW DRAWINGS FOR FULL-BASEMENT FOUNDATION

The main purpose of the plan view drawings is to show the locations and layout of the walls and the footings

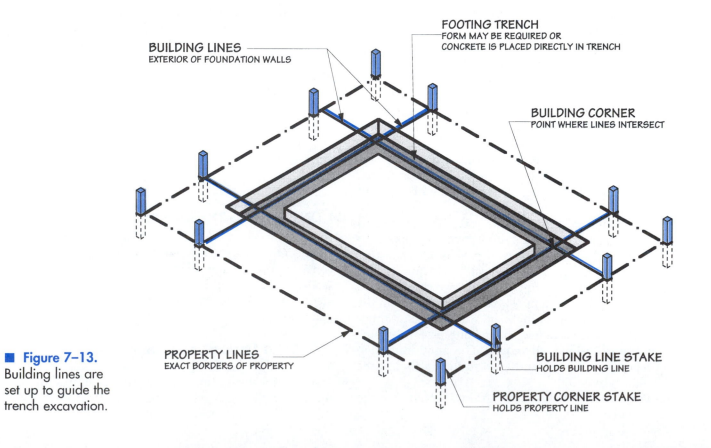

BUILDING LINES
EXTERIOR OF FOUNDATION WALLS

FOOTING TRENCH
FORM MAY BE REQUIRED OR
CONCRETE IS PLACED DIRECTLY IN TRENCH

BUILDING CORNER
POINT WHERE LINES INTERSECT

PROPERTY LINES
EXACT BORDERS OF PROPERTY

BUILDING LINE STAKE
HOLDS BUILDING LINE

PROPERTY CORNER STAKE
HOLDS PROPERTY LINE

■ **Figure 7–13.** Building lines are set up to guide the trench excavation.

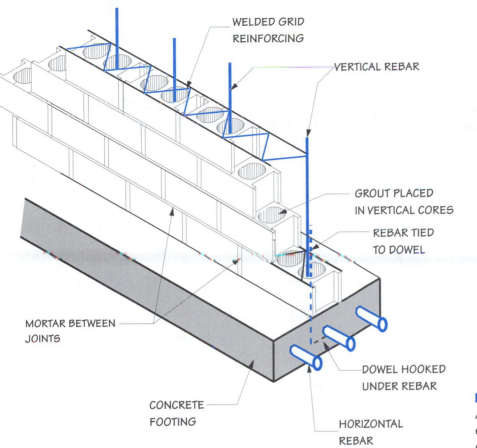

WELDED GRID REINFORCING

VERTICAL REBAR

GROUT PLACED IN VERTICAL CORES

REBAR TIED TO DOWEL

MORTAR BETWEEN JOINTS

DOWEL HOOKED UNDER REBAR

CONCRETE FOOTING

HORIZONTAL REBAR

■ **Figure 7–14.**
A concrete block wall is constructed over a poured-in-place concrete footing.

that support columns. Steel columns are used to support the floor beams because of the height between the floor slab and ceiling. The pier construction under the front porch and sun deck is the same as described for the crawl-space foundation. Dimension lines serve the same function as described for the crawl-space foundation. However, additional dimension lines are required to show the locations of the basement windows.

The plan view drawing for a full basement foundation for the **One-Story Residence** is shown in **Figure 7–16.** Note that the footings and walls are constructed of poured-in-place concrete. A corresponding pictorial drawing of this foundation appears in **Figure 7–17.** Concentrate on the following when studying the plans:

✓ Symbols and abbreviations.
✓ Dimensions establishing positions of the exterior and interior walls.
✓ Dimensions establishing the placement of beams and their supporting columns and footings.
✓ Section view reference numbers.

Explanations of the full basement plan view features shown in Figure 7–16 follow:

1. **Garage.** The garage construction is the same as that for the crawl-space foundation.
2. **Front porch.** The front porch construction is the same as that for the crawl-space foundation.

■ **Figure 7–15.**
This crawl-space foundation is constructed of concrete stem walls placed over concrete footings. Note the anchor bolts projecting from the top of the walls. Stem walls are also frequently constructed of concrete blocks.

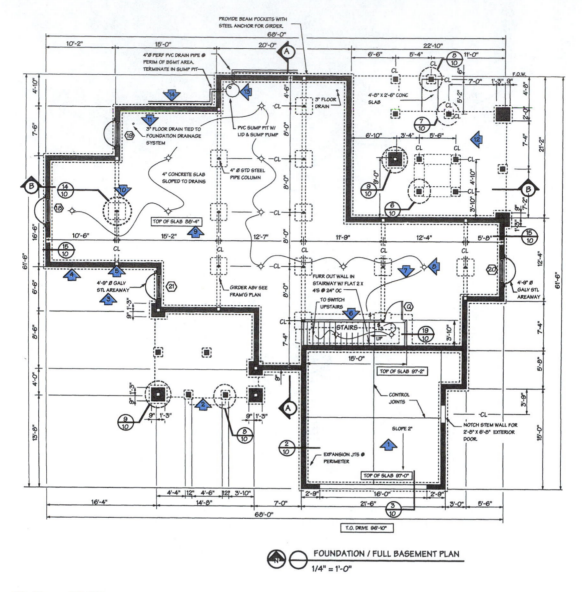

FOUNDATION / FULL BASEMENT PLAN
1/4" = 1'-0"

■ **Figure 7–16.**
The full-basement foundation plan view identifies and provides information about the walls, columns, footings, stairs, and electrical work.

3. **Areaway.** This sunken area around the basement window is protected by a 4'-9" diameter semicircular galvanized steel wall.

4. **Foundation.** Dotted lines show the footing below the foundation wall.

5. **Beam pocket.** Also called a girder pocket, this opening is prepared in the concrete wall to receive the end of the floor girder. The top of the girder will be at the same level as the top of the sill plate.

6. **Stairway.** The stairway descends from the main floor to the basement. Wood stud walls are shown beneath the west, north, and east side of the stairway. A door at the foot of the stairs opens to the basement. Two lights are overhead, controlled by a three-way switch next to the door. A second three-way switch is at the top of the stairway.

7. **Wiring symbol.** These lines identify concealed wiring that provides electricity to the overhead lights. The line begins at a light switch next to the door.

8. **Overhead light.** The symbol identifies an overhead light fixture.

9. **Slab elevation.** The elevation of the concrete basement floor is 88'-4" in relation to the outside benchmark (100'-0"). Therefore, the basement floor level is 11'-8" below the benchmark (100'-0" – 88'-4" = 11'-8").

10. **Column footing.** Dotted lines identify the foot-

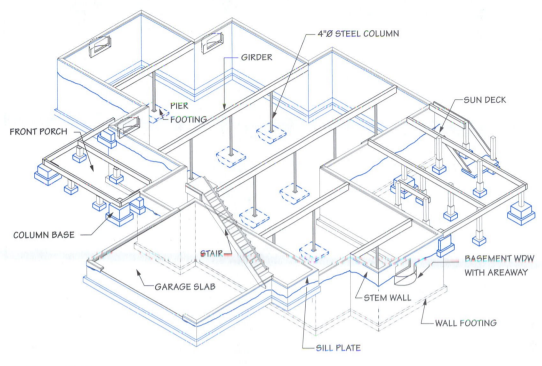

■ **Figure 7–17.**
This pictorial drawing of a full-basement foundation is based on the foundation plan shown in Figure 7–16.

ings. The small dark circle represents a steel pipe column.

11. **Floor drain.** A floor drain 3″ in diameter is placed close to the wall corner. Another drain is provided in the farthest northeast corner. The floors slope in the direction of the drains.

12. **Sun deck.** The construction of the sun deck is the same as that for the crawl-space foundation.

13. **Sump pit and pump.** These are provided in case of flooding in the basement. The water will flow into the pit and be pumped out of the basement.

14. **Drain pipe.** Perforated drain tiles are placed next to the foundation footings to prevent water collecting around the base of the foundation. Here the water is directed to the sump pit.

SECTION VIEW DRAWINGS FOR FULL-BASEMENT FOUNDATION

The full-basement section view drawings give information that does not appear in the foundation plan view drawing:

✓ The widths and heights of wall and column footings.
✓ The wall thicknesses and heights.
✓ The type of steel columns used to support the floor beams.

✓ The size and spacing of rebars and anchor bolts.
✓ Floor and lower wall framing components.

Before reviewing the section view drawings, place a bookmark at the page showing the full-basement plan view drawing (Figure 7–16, p. 64). While studying each section view drawing, relate the section number beneath the drawing to the section symbol on the plan view.

The section view drawings for the full-basement foundation, along with corresponding pictorial drawings, are shown in **Figure 7–18.** The pier construction beneath the front porch and sun deck is the same as already described for the crawl-space foundation. Concentrate on the following when studying the drawings.

✓ Symbols and abbreviations.
✓ Vertical and horizontal dimensions.
✓ Material descriptions.
✓ Section views as related to plan views.

Explanations of the full-basement foundation features in Figure 7–18 follow:

1. **Column footing.** This footing supports the weight transmitted from the steel column above. The bottom of the footing is 2′-6″ square and the top is 3′-4″ square. Three #4 rebars run in two directions.

2. **Steel-pipe column.** This type of column gives

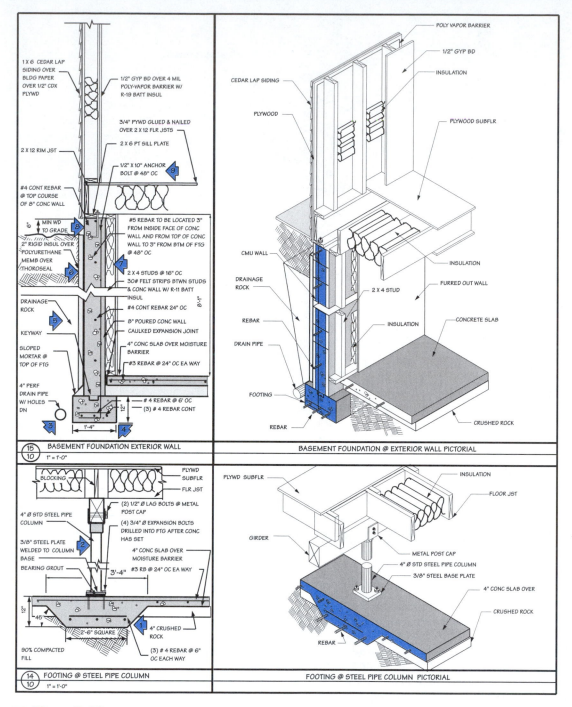

■ **Figure 7–18.**

Full-basement foundation section views show vertical views of the walls, piers, and footings.

intermediate support to the girders directly below the floor joists. A rectangular steel base plate with bolt holes is welded to the bottom of the column. Holes for expansion bolts are drilled into the hardened concrete, and the base plate is fastened with expansion bolts. A layer of grout placed beneath the base plate compensates for any unevenness below the plate.

3. **Drain pipe.** The pipe is placed next to the footing and extends the entire perimeter of the foundation to prevent water from collecting around the footing. Perforated holes at the bottom of the pipe allow water to seep into the pipe and flow into the sump pit inside the foundation.

4. **Foundation footing.** The footing is 1'-4" wide and 12" high. Short #3 rebars extend across the width of the footing and are spaced 6'-0" OC. Three #4 rebars run along the full length of the footing. A

keyway has been formed at the center of the top surface of the footing. A keyway is a groove at the top of the footing that helps hold in place the bottom of the stem wall resting on footing.

5. **Drainage rock.** Coarse gravel is placed next to the foundation wall and footing to help prevent water accumulation below the ground surface.

6. **Exterior insulation.** Two inches of thick rigid insulation are placed over a polyurethane membrane and a layer of waterproof covering (Thoroseal). This will help prevent water infiltration and will also insulate the wall.

7. **Furred wall.** A 2 × 4 stud wall is framed against the interior surface of the foundation wall. Batt insulation is placed between the studs. The wall is then finished off with gypsum board.

8. **Foundation wall.** The concrete foundation wall is 8" thick and must extend a minimum of 6" above grade. A vertical #5 rebar extends from the top of the wall to 3" from the bottom of the footing. Horizontal #4 rebars are placed 24" OC and run along the full length of the walls.

9. **Sill plates and anchor bolts.** Pressure-treated 2 × 6 sill plates are fastened to the top of the foundation walls with 1/2" × 10" anchor bolts placed 48" OC.

CONSTRUCTING THE FULL-BASEMENT FOUNDATION

The **One-Story Residence** full-basement foundation consists of concrete footings and high concrete walls placed on top of the footings. The footings are poured first. After the concrete for the footings has hardened, the wall forms are constructed over the footings. A full basement will require deep excavation because of the height of the basement walls.

EXCAVATION

Excavation for a full-basement foundation is determined by the height of the walls and footing and by how high the walls extend above the ground. Dimensions for the depth of the excavation are usually given in section view drawings of the foundation plan. The side walls of the excavation should be held back at least 2' from the building line to allow room for construction of the foundation forms. The floor of the excavation must be kept level.

LAYOUT OF BUILDING LINES

The building corners are established in the excavated area, usually with a transit instrument. Stakes are driven, and each exact building corner is identified by a nail driven on top of the stake. Another procedure is to fasten building lines to batter boards above the excavation and then plumb down from the intersecting corners to establish exact points on the building corner stakes. **(See Figure 7–19.)**

CONSTRUCTION OF FOOTINGS AND WALLS

Because of the high walls required for a full-basement foundation, footing forms are constructed separately

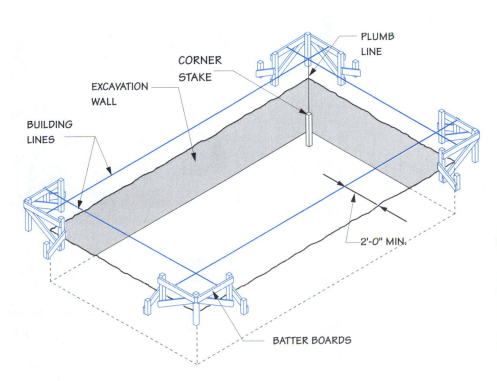

PLUMB LINE
CORNER STAKE
EXCAVATION WALL
BUILDING LINES
2'-0" MIN.
BATTER BOARDS

■ **Figure 7–19.**
Corner stakes for the building lines must be driven at the bottom of the excavation. They are positioned by plumbing down from the intersecting points of the building lines stretched from batter boards above the top of the excavation.

from the wall forms. Vertical rebars extending up from the footings are tied to horizontal rebars in the footings. A keyway may also be formed at the top of the footing. Together, the vertical rebars and keyways will help form a strong structural joint between the wall above and the footing. **(See Figure 7–20.)** A photograph of a full-basement concrete foundation is shown in **Figure 7–21.**

READING THE SLAB FOUNDATION PLAN

The slab foundation, also called a *slab-on-grade* foundation, may be constructed with either a monolithic or separately poured slab and footings. The **monolithic design** is the simplest to construct. Trenches for footings are dug around the outside perimeter and within the slab area where walls are to be placed above the slab. The concrete is then placed for the slab and footings at the same time. This type of construction is limited to areas where water tables are not close to the surface and frost lines do not extend deep into the ground. In a conventional *T foundation,* footings are constructed separately from the slab. The slab is then poured with its edges butting against or resting on a shoulder formed in the foundation wall. **(See Figure 7–22.)**

PLAN VIEW DRAWINGS FOR SLAB FOUNDATION

The plan view drawing for a slab foundation shows the outside walls and footings. Interior footings below the slab are identified with dashed lines. The front porch will have a concrete slab rather than a wooden deck. The sun deck at the rear of the building will be constructed the same way as for crawl-space and full-basement foundations.

A plan view drawing of a slab foundation for the **One-Story Residence** is shown in **Figure 7–23.** A corresponding pictorial drawing of this foundation appears in **Figure 7–24.** The outside walls are conventional concrete footings with concrete-block stem walls. The slab rests on a shoulder at the top of the wall. Concentrate on the following when studying the plan:

✓ Symbols and abbreviations.
✓ Dimensions of the outside footing and slab perimeters.
✓ Dimensions of the interior footings.

Explanations of the slab foundation plan views as shown in Figure 7–23 follow:

1. **Garage.** The garage construction is the same as shown in the crawl-space and full-basement foundation plans.
2. **Front porch.** The deck consists of a concrete slab.
3. **Porch column.** The same type of pier and footing is placed here as shown in section drawing 9 of the crawl-space foundation section views.
4. **Column footing.** The dotted lines identify a footing below a 6 × 6 wood post. The wood posts are positioned at points where heavier roof loads occur.
5. **Continuous footing.** All bearing walls above the

■ **Figure 7–21.**
This full-basement foundation has been constructed with high concrete walls over concrete footings. A dark, bituminous waterproofing material has been applied to the outside of the walls up to grade level. The vertical lines on the walls show that a panel forming system was used to construct the wall. *(Courtesy of Portland Cement Association).*

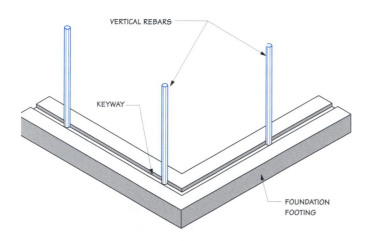

■ **Figure 7–20.**
A keyway and vertical rebars extending up from the footing help to form a strong structural joint between the wall and footing.

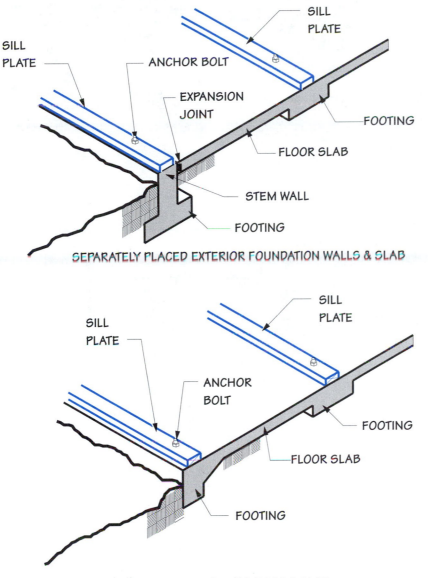

SILL
PLATE

SILL
PLATE

ANCHOR BOLT

EXPANSION
JOINT

FOOTING

FLOOR SLAB

STEM WALL

FOOTING

SEPARATELY PLACED EXTERIOR FOUNDATION WALLS & SLAB

SILL
PLATE

SILL
PLATE

ANCHOR
BOLT

FOOTING

FLOOR SLAB

FOOTING

MONOLITHIC EXTERIOR FOOTINGS & SLAB

■ **Figure 7–22.**
The footings and slab are placed together in monolithic slab foundation. When spread foundations are required for exterior walls, the slab and walls are placed separately.

slab must rest on a thickened footing.

6. **Sun deck.** The sun deck construction is the same as shown in the crawl-space and full-basement foundation plans.

7. **Foundation.** Dotted lines identify the footings below the stem walls.

SECTION VIEW DRAWINGS FOR SLAB FOUNDATION

Before reviewing the section view drawings, place a bookmark at the page showing the slab foundation plan view drawing (Figure 7–23, page 70). While studying each section view drawing, relate the section number beneath the drawing to the section symbol on the plan view.

READING THE SECTION VIEW DRAWINGS FOR SLAB FOUNDATION

The section view drawings of the slab foundation, along with corresponding pictorial drawings, are shown in **Figure 7–25.** Concentrate on the following when studying the section view drawings.

✓ Symbols and abbreviations.
✓ Exterior wall and footing dimensions.

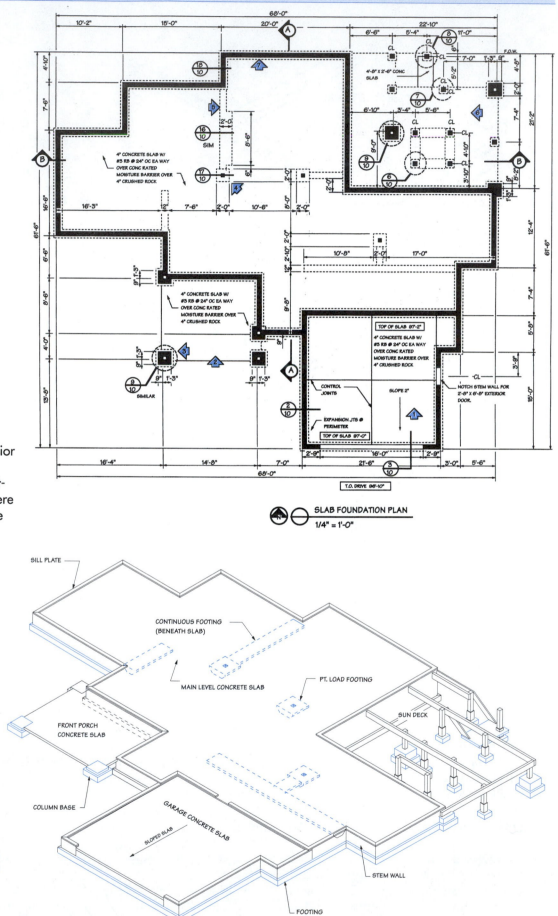

■ Figure 7–23.
The slab foundation plan view shows exterior walls and piers. Thickened interior footings are provided where framed walls are to be later constructed.

SLAB FOUNDATION PLAN
1/4" = 1'-0"

■ Figure 7–24.
This pictorial drawing of the slab foundation is based on the plan shown in Figure 7–23.

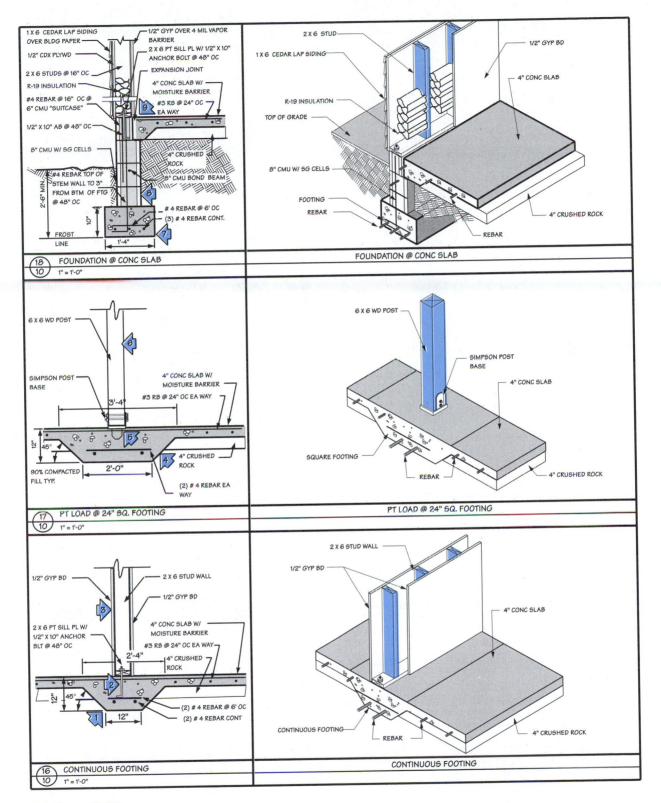

■ Figure 7–25.
Slab foundation section views show vertical views of the exterior walls, floor slab, and interior footings.

✓ Interior footing dimensions.
✓ Method of fastening sole (bottom) plates of exterior and interior walls.
✓ Method of fastening posts over piers.

Explanations of the slab foundation section view features in Figure 7–25 follow:

1. **Continuous footing.** A footing is required beneath the framed bearing walls above the floor slab. It is reinforced with two #4 rebars running the length of the footing and short #4 rebars, spaced 6'-0" OC, across the width of the footing.

2. **Sill plate and anchor bolts.** A 2 × 6 pressure-treated sill plate is bolted to the footing with 1/2" by 10" anchor bolts spaced 48" OC.

3. **Stud wall.** The stud wall is framed over the sill plate. Gypsum board, 1/2" thick, is nailed to the wall surfaces.

4. **Post footing.** The weight carried by the post above requires a supporting footing. It is 2'-0" square at the bottom and 3'-4" square at the top. The footing is reinforced by two #4 rebars running in both directions.

5. **Post base.** A metal post base is embedded in the concrete. The post above is bolted to the upper section of the post base.

6. **Wood post.** The 6 × 6 posts are required to help support heavier roof and wall loads.

7. **Footing.** A 1'-4" wide and 10" high footing supports the exterior stem walls.

8. **Stem wall.** A shelf at the top provides better support of the edge of the concrete slab.

9. **Slab edge.** The slab edge is thickened where it butts against the stem wall. There is an expansion joint between the slab edge and the stem wall.

CONSTRUCTING THE SLAB FOUNDATION

As with the full-basement and crawl-space foundations, the building corners are usually established with a transit instrument. Stakes are driven, and each exact building corner is identified with a nail driven on top of the stake. Another common procedure is to place batter boards behind the stakes to hold the building lines.

Trenches are dug for the perimeter and interior footings. The depths of the trenches are usually shown in the foundation section views. The base of the footings must extend to the frost line. The entire surface area below the slab is leveled and compacted if fill is placed to even the ground surface.

The exterior walls and footings are constructed in the same manner as described for the crawl-space foundations. A bed of gravel 4" to 6" deep is put down in the slab area, and a moisture-resistant vapor barrier is placed over the gravel. Before the slab is poured, rebars or welded wire mesh are put in place. Pipes and conduits are also positioned by the plumbers and electricians. See **Figure 7–26**. A completed slab-on-grade foundation system is shown in **Figure 7–27**.

REVIEW QUESTIONS

Enter the missing words in the spaces provided on the right.

1. There is no _____ area within a crawl-space foundation.

2. _____ inches is the minimum distance between the ground surface and floor unit resting on a crawl-space foundation.

3. Floor-to-ceiling clearance in full-basement foundations ranges from _____ to _____ feet.

4. In basement foundations, beams are supported by wood posts or _____ columns.

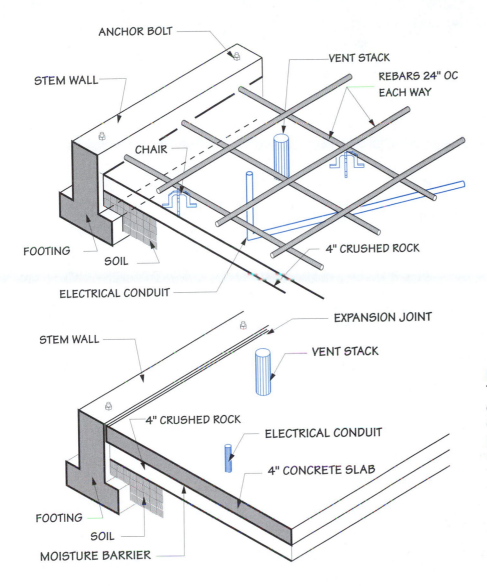

ANCHOR BOLT

STEM WALL

VENT STACK

REBARS 24" OC
EACH WAY

CHAIR

FOOTING

SOIL

4" CRUSHED ROCK

ELECTRICAL CONDUIT

EXPANSION JOINT

STEM WALL

VENT STACK

4" CRUSHED ROCK

ELECTRICAL CONDUIT

4" CONCRETE SLAB

FOOTING

SOIL

MOISTURE BARRIER

■ **Figure 7–26.**
The top drawing shows the stem wall and footing for the exterior walls of a slab foundation. In preparation for placing the slab, a 4" layer of crushed rock has been placed over the soil and a plastic moisture barrier spread over the crushed rock. Rebars are then set in place. The bottom drawing shows the poured slab. Note the expansion joint between the slab and the wall.

■ **Figure 7–27.**
Water pipes, waste stacks, and electrical conduit project from the floor slab of this slab-on-grade foundation. Note the anchor bolts for sill plates along the edges of the slab.

5. The soil below provides the main support for the floor in a _____ foundation.

_____

6. T-foundations are also called _____ foundations.

_____

7. A _____ supports the stem wall of a T-foundation.

_____

8. The width of a footing should be at least _____ times the stem wall thickness.

_____

9. Unstable soil conditions may require a _____-_____ foundation.

_____

10. _____ provide the main support for the walls of a grade-beam foundation.

_____

11. The total weight of a building is part of the _____ load of the building.

_____

12. Sand and gravel are examples of _____-grained soils.

_____

13. A lot surface should be graded to direct water _____ from the building.

_____

14. The depth to which the ground freezes during cold weather is called the _____ _____.

_____

15. _____, gravel, sand, and water make up a concrete mixture.

_____

16. Steel bars are placed in concrete to produce _____ concrete.

_____

17. A #7 rebar is _____ inch in diameter.

_____

18. Rebars or _____ _____ _____ are used to reinforce concrete slab floors.

_____

19. Plan and _____ views are included with the foundation drawings.

_____

20. On a drawing, the total width and length of a building is shown by the outside _____ lines.

_____

Match the letters next to the symbols with their descriptions.

21. Vent space _____

22. Girder beam pocket _____

23. Overhead light _____

24. CMU column base and wood post _____

25. Drain _____

26. Section or detail reference _____

27. Steel column over footing _____

28. Crushed rock _____

29. CMU pier and wood post _____

30. Soil _____

31. Sump pump _____

32. Building section reference _____

33. Stairs _____

34. Dimensioned lumber _____

35. Drain _____

36. Rebars _____

37. Concrete _____

38. Notch for door in stem wall _____

39. CMU _____

40. Light switch _____

41. Drain pipe _____

Plan View Symbols

a.

b.

c.

d.

e.

f.

g.

h.

i.

j.

k.

l.

m.

n.

Section View Symbols

o.

p.

q.

r.

s.

t.

u.

Choose the correct abbreviation for each term.

42. At _____

43. Between _____

44. Basement _____

45. Bottom _____

46. Center line _____

47. Concrete _____

48. Continuous _____

49. Concrete masonry unit _____

50. Diameter _____

51. Face of wall _____

52. Foundation _____

53. Floor joists _____

54. Footing _____

55. Galvanized _____

56. Insulation _____

57. Joints _____

58. Minimum _____

59. On center _____

60. Perforated _____

61. Perimeter _____

62. Polyvinyl chloride _____

63. Pressure-treated _____

64. Reinforced steel bar _____

65. Solid grout _____

66. Similar _____

67. Steel _____

68. Surfaced four sides _____

69. Square foot _____

70. Typical _____

71. With _____

72. Wood _____

W/	PERF
BTM	JTS
@	CONT
FDN	BET.
CONC	PERIM
SIM	PT
SG	GALV
MIN	CMU
TYP	FTG
WD	REBAR
SQ FT	ST
OC	PVC
FOW	BSMT
DIA	INSUL
CL	FL JSTS
S4S	

UNIT 8

Floor
Framing Plans

The framework of a house consists of the floors, walls, ceilings, and roof. The construction of the main-floor unit is the first framing operation over a crawl-space or full-basement foundation. A **floor framing plan** provides a plan view guiding the construction of the floor unit. However, related section views are necessary to completely understand the building procedures.

FLOOR FRAMING DESIGN FACTORS AND COMPONENTS

Platform framing is the method used to build the **One-Story Residence.** Here the wood frame floor unit provides a working platform for the wall construction. After the walls are built, the process is repeated for each additional story of the building. **(See Figure 8–1.)** This

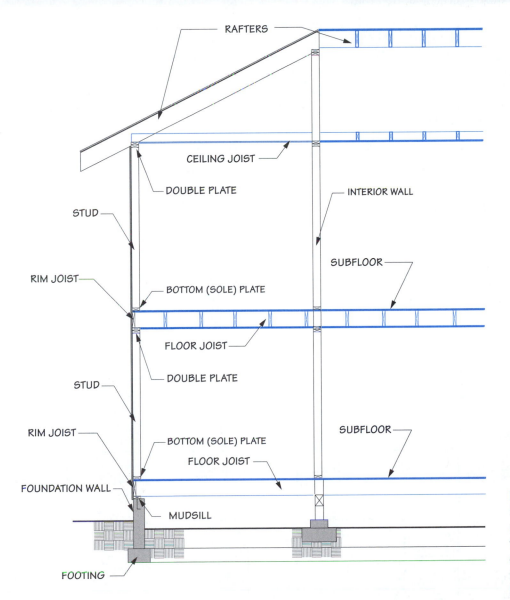

RAFTERS

CEILING JOIST

DOUBLE PLATE

INTERIOR WALL

STUD

SUBFLOOR

RIM JOIST

BOTTOM (SOLE) PLATE

FLOOR JOIST

DOUBLE PLATE

STUD

RIM JOIST

SUBFLOOR

BOTTOM (SOLE) PLATE

FLOOR JOIST

FOUNDATION WALL

MUDSILL

FOOTING

■ **Figure 8–1.**
In platform framing, the floor unit is constructed, and then the walls are framed on top of the floor. This process continues for each story of the building.

77

is the most widely used procedure for wood frame construction.

In platform framing the joists are the main structural components of the floor and ceiling units. A **joist** is a plank placed on its edge that spans the distance between walls and/or beams. Joists are spaced 16" or 24" OC (on center).

Typical floor joist sizes are 2 × 6, 2 × 8, 2 × 10, and 2 × 12. The joist material used is determined by its **allowable span,** which is the distance permitted between bearing points for the particular joist material. The distance is based on the strength and thickness of the material and the effect of the live and dead loads the joists must support. **(See Figure 8–2.)** Some lumber species used for joist material are stronger than others. Floor-joist span tables are available for the more frequently used species of framing lumber. (See Appendix G for allowable joist span tables.)

Other floor systems consisting of prefabricated *floor trusses* and *wood I-beams* (discussed in Unit 33) are gaining in popularity.

READING THE FLOOR FRAMING PLAN

The set of plans for the **One-Story Residence** includes a floor framing plan for a floor unit constructed over a full-basement or crawl-space foundation. The floor units over both foundations are very similar in construction. However, the framing plan over the full basement shows an opening for a stairway leading to the basement.

The floor framing drawings show a plan view of the floor unit, so it is necessary to refer to section views for additional structural information. Symbols and abbreviations appearing in the floor framing plan and rele-

vant section views are shown in **Figure 8–3.**

The first floor over a slab foundation is constructed of concrete, making a framed floor unit unnecessary.

PLAN VIEW DRAWING FOR FLOOR FRAMING

The floor framing plan view drawing provides information about joist size, direction, and spacing. Girder size and location, and the placement of blocking, are identified. The plan view drawing shows the outside ends of floor joists resting on exterior walls. The ends of the floor joists are also nailed to **rim joists** (also called **header joists**) placed along the outside edges of the walls. The interior ends of the joists rest upon beams or interior walls. Blocking is placed between the joists where they overlap. Joists may also butt up against beams and be supported by *metal joist hangers*. After the joist work has been completed, subfloor panels are glued and nailed over the joists.

The floor framing plan over a crawl-space foundation is shown in **Figure 8–4.** A corresponding pictorial drawing of this floor unit appears in **Figure 8–5.** Concentrate on the following when studying the plans:

✓ Symbols and abbreviations.
✓ Supporting beams.
✓ Joist placement and direction.
✓ Placement of blocking.
✓ Framing for fireplace support.

Explanations of the floor framing plan features in Figure 8–4 follow:

1. **Porch column.** This column supports the southwest corner of the porch roof.
2. **Stairs.** The dashed lines identify two steps lead-

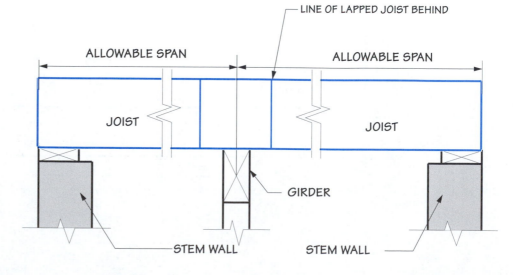

■ **Figure 8–2.**
The allowable span for joists is the distance permitted between bearing points. This is determined by joist size and loads they must support.

SYMBOLS AND ABBREVIATIONS
FLOOR FRAMING PLAN

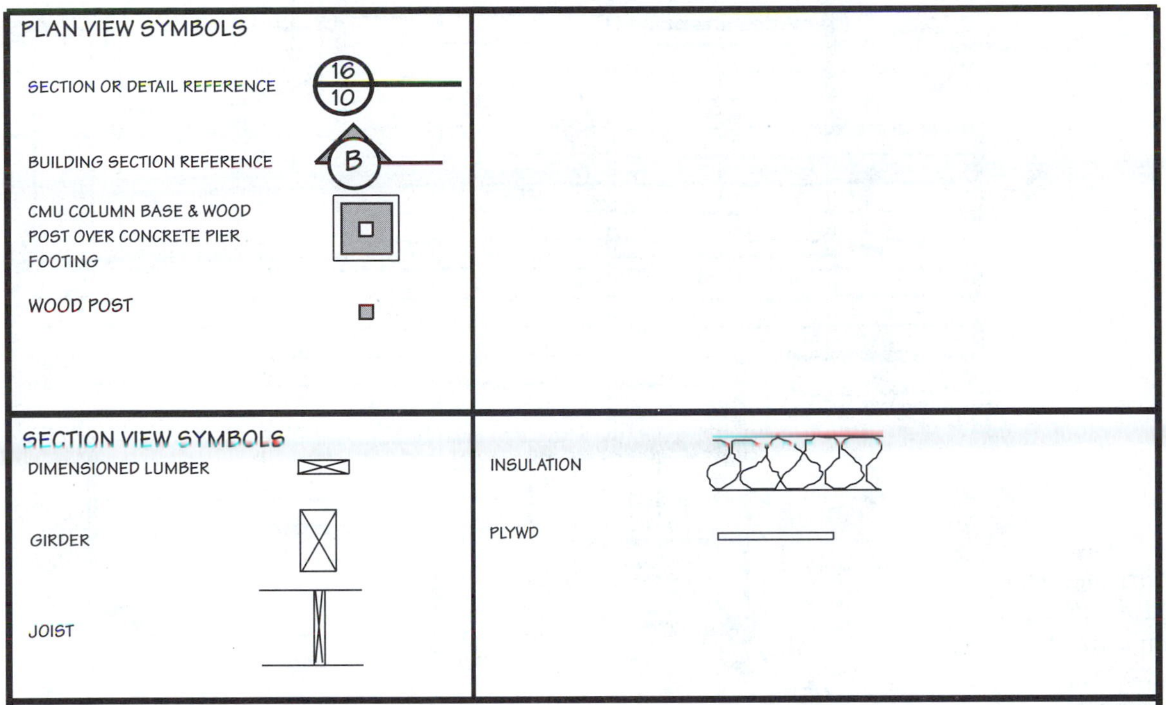

PLAN VIEW SYMBOLS

SECTION OR DETAIL REFERENCE

BUILDING SECTION REFERENCE

CMU COLUMN BASE & WOOD
POST OVER CONCRETE PIER
FOOTING

WOOD POST

SECTION VIEW SYMBOLS

DIMENSIONED LUMBER

GIRDER

JOIST

INSULATION

PLYWD

ABBREVIATIONS

BLDG	Building
DKG	Decking
DBL BLKG	Double Blocking
EA	Each
EXT	Exterior
FL	Floor
GYP BD	Gypsum Board
HT	Height
JSTS	Joists
MIL	Millimeter
OC	On Center
PLYWD	Plywood
POLY	Polyetheline
REDWD	Redwood

■ **Figure 8–3.**
These are symbols and abbreviations found in the floor framing plan of the **One-Story-Residence.**

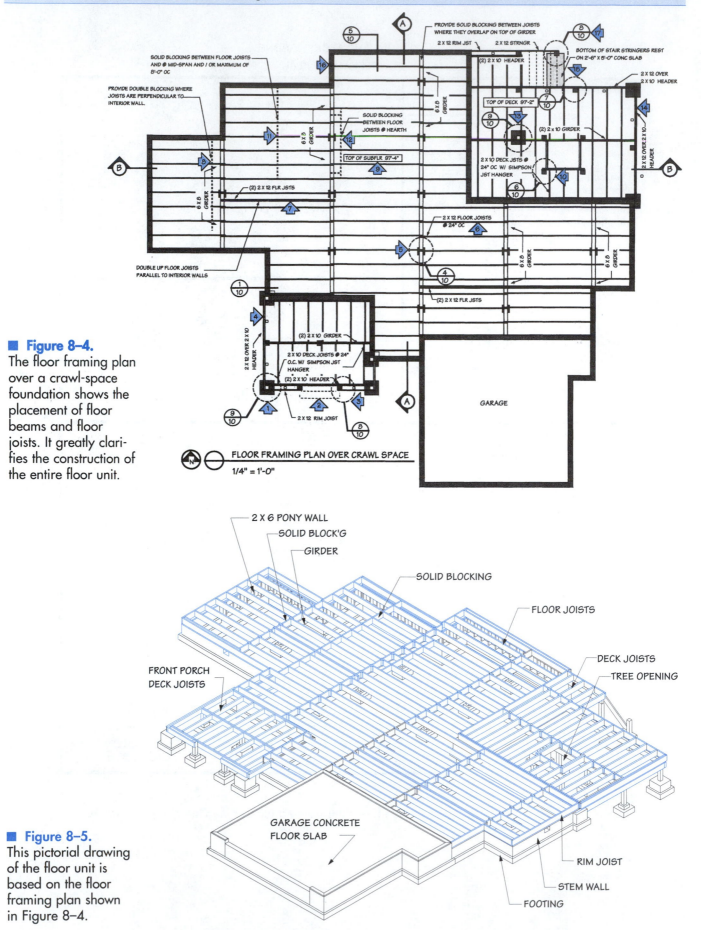

■ Figure 8–4.
The floor framing plan over a crawl-space foundation shows the placement of floor beams and floor joists. It greatly clarifies the construction of the entire floor unit.

FLOOR FRAMING PLAN OVER CRAWL SPACE
1/4" = 1'-0"

■ Figure 8–5.
This pictorial drawing of the floor unit is based on the floor framing plan shown in Figure 8–4.

ing up to the front porch.

3. **Deck pier and newel post.** The newel posts supported by a pier are placed at the top and each side of the stairway.

4. **Front porch.** A girder consisting of two 2 × 10s spiked together provides central support to the other joists. The 2 ×10 joists, spaced 24″ OC, butt up against the beam and are supported by metal joist hangers. Joist hangers also support the joists butting up against a double header at the front of the porch. The joists extending toward the foundation rest on top of the foundation wall.

5. **Post and floor girder.** A footing and pier are below the base of a short 6 × 6 post. The 6 x 8 floor girders rest on top of the posts. Five rows of floor girders supported by posts are placed beneath the entire floor frame.

6. **Floor joists.** The floor joists rest on the foundation walls and the floor girders. All floor joists within the building are 2 × 12s and spaced 24″ OC.

7. **Double joists.** The joists are doubled to provide added support to a parallel running wall above.

8. **Blocking under walls.** Solid blocking between joists is placed where partitions run at a right angle (perpendicular) to the joists.

9. **Top of subfloor.** After the plywood panels are placed over the floor joists, the subfloor elevation is 97′-4″ in relation to the 100′-00″ bench mark. The subfloor elevation is 2′-8″ lower than the benchmark (100′-00″-99′-4″ = 2′-8″). This occurs because of the sloping lot.

10. **Bench support.** Four posts resting on piers and footings support the bench around the tree within the sun deck.

11. **Blocking between spans.** Solid blocking is placed between the joists where spans are 8′-0″ or more. The blocking holds the joists in position and helps distribute the weight carried by the floor frame.

12 **Blocking beneath fireplace.** Solid blocking strengthens the floor area beneath the fireplace.

13 **Porch column.** This column helps support the deck and the roof section projecting over part of the sun deck.

14. **Sun deck.** Placement of joists and their supports is similar to that for the front porch.

15. **Sun-deck railing post.** This post rests on a supporting pier and footing and becomes part of the railing above the deck.

16. **Pony wall.** The height of the foundation is increased by the construction of a short pony wall. Pockets are framed into the pony wall to support the ends of floor girders.

17. **Newel post.** This post, supported by a pier and footing, is at the head of the stairway. It also supports the end of the handrail. The straight dotted

lines identify six steps leading up to the rear sun deck. The bottoms of the supporting stringers rest on a concrete slab.

SECTION VIEW DRAWINGS FOR FLOOR FRAMING PLAN

Section symbols on the floor framing plan over the crawl-space foundation refer to related section view drawings They show the floor joists and rim joists resting on the foundation walls. Components of the lower sections of the framed walls above the floor unit are also identified.

Before reviewing the section view drawings, place a bookmark at the page showing the floor frame plan view (Figure 8–4, p. 80). After studying each section view drawing, relate the section number beneath the drawing to the section symbol on the plan view.

The section view drawings containing floor framing information are shown in **Figure 8–6a–c.** (These are the same drawings as the foundation section views.) Corresponding pictorial drawings are placed to the right of each section drawing. Concentrate on the following when studying the plans:

✓ Symbols and abbreviations.
✓ Floor and wall framing members.
✓ Placement of insulation.
✓ Printed directions.

Explanations of the floor framing plan features in Figures 8–6a-c follow:

1. **Joist and insulation.** The end section view of a floor joist is shown. Blanket or batt insulation is placed between the joists.

2. **Rim joist.** Rim joists are placed at the outside perimeters of a building and nailed to the sill plate.

3. **Subfloor.** The subfloor material is 3/4″ thick plywood panels. The panels are glued and nailed to the 2 × 12 floor joists.

4. **Outside wall.** The section of 2 × 6 wall shown in the drawing is fastened to a bottom plate. The plate is nailed through the subfloor into the joist below. Studs are spaced 16″ OC, and R-19 insulation is placed between the studs.

5. **Exterior of wall.** The finish siding consists of 1 × 6 cedar boards. They are beveled and each board laps over the one below. Before the siding is applied, plywood sheathing is nailed to the 2 × 6 studs and building paper is placed over the plywood.

6. **Interior of wall.** R-19 batt insulation is fastened between the studs. A plastic vapor barrier is

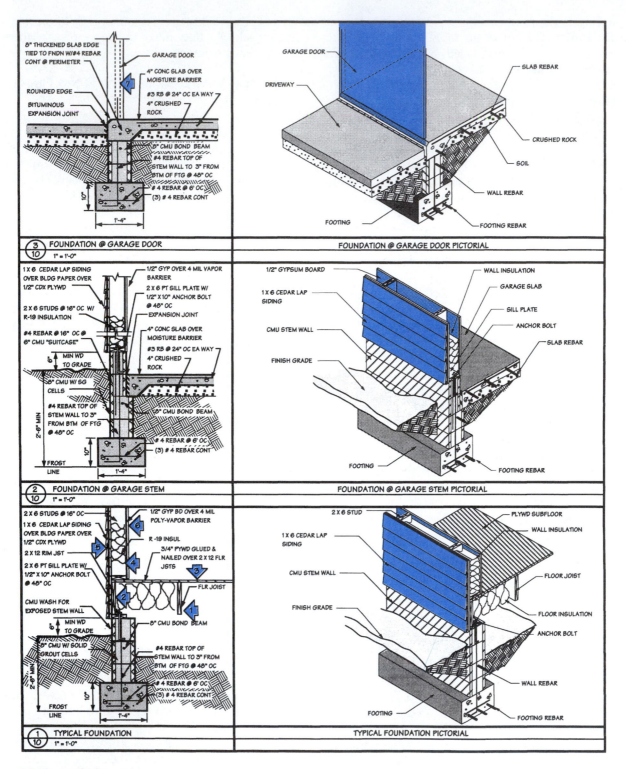

Figure 8–6a.

Section views related to the floor framing plan provide information that is not clearly visible on the floor framing plan.

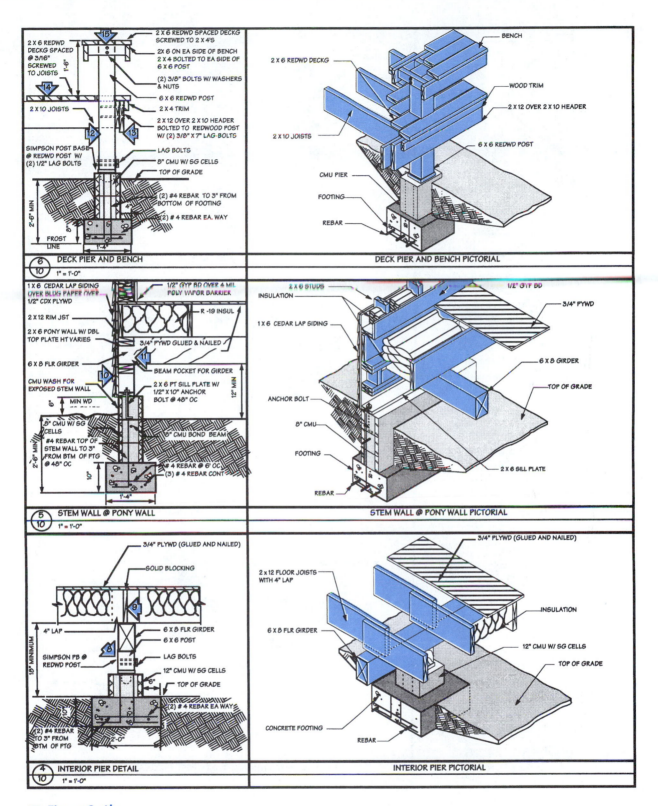

■ **Figure 8–6b.**

Section views related to the floor framing plan provide information that is not clearly visible on the floor framing plan (continued).

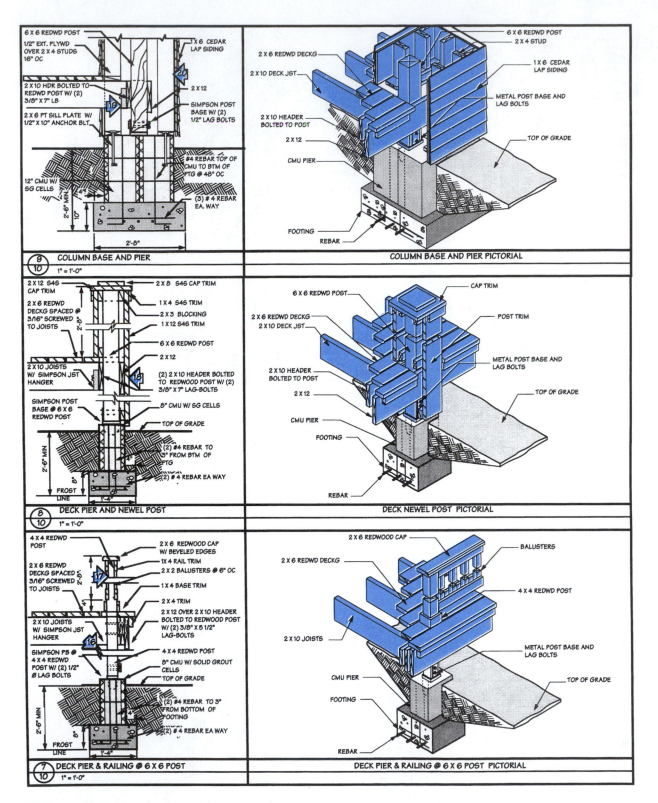

■ **Figure 8–6c.**

Section views related to the floor framing plan provide information that is not clearly visible on the floor framing plan (continued).

placed against the studs before 1/2' thick gypsum board is nailed into place.

7. **Garage door.** The dotted lines identify the garage door within the framed opening.

8. **Metal post base and post.** A Simpson (manufacturer) metal post base is embedded in the top of the pier. It is fastened with 1/2" diameter lag bolts to a short 6 × 6 post resting on the pier. The top of the post is toenailed to the 6 × 8 floor girder.

9. **Floor joist and blocking.** The interior ends of the floor joists rest on the girder. They lap past each other a minimum of 4". Solid blocking is placed between the lapped ends of the floor joists. They hold the joists in place and help stiffen the entire floor unit.

10. **Pony wall.** This short wood framed wall raises the height of the stem wall toward the rear of the building. The bottoms of the studs are nailed to the sill plate. The tops of the studs are nailed into double plates, and the 2 × 12 floor joists rest on top of the plates.

11. **Girder pocket.** The 6 × 8 girder is notched to fit under the double plates where the end of the girder comes against the pony wall. Short studs are nailed to each side of the girder, and two blocks are placed beneath the girder. Cripple studs may then be placed beneath the double blocks.

12. **Bench post.** Four 6 × 6 redwood posts are placed at the corners of the deck opening around the tree. The posts rest on footings and piers and extend upwards to support the bench. The posts also help support the sun deck.

13. **Header assembly.** A 2 × 10 header is placed against the posts supporting the bench. Deck joists butt up against the header and are fastened with metal joist hangers. The 2 × 12 placed over the header acts as a girder around the opening in the deck. Both pieces are fastened to the posts with two 3/8" × 7" lag bolts. A 2 × 4 trim piece is nailed against the 2 × 12 below the deck material extending past the posts.

14. **Decking.** The porch deck material consists of 2 × 6 redwood planks screwed into the floor joists. The 3/16" spaces between the planks allow for expansion and contraction by temperature changes. They also allow water to drip through the deck rather than collect on the deck surface.

15. **Bench.** Three 2 × 6s are used for the bench surrounding the tree opening. They are screwed to the 2 × 4s bolted to each side of the posts.

16. **Railing post—lower section.** The 4 × 4 redwood posts help support the deck. The 2 × 10 and 2 × 12 girder assembly is bolted to the outside of the posts. Deck joists are attached to the 2 × 10s with metal joist hangers.

17. **Railing Post—upper section.** This part of each 4 × 4 redwood post becomes part of the railing. These help to strengthen the stability of the railing.

18. **Newel post.** The 6 × 6 newel post is at the foot of the sun-deck stairs and helps support the stair stringer and handrail. A 2 × 8 and 2 × 12 piece cap is at the top of the post. Vertical 1 × 12s over spacer blocks are nailed over the posts.

19. **Corner porch column—lower section.** The deck joist seen in the drawing butts up against a 2 × 10 header bolted to the 6 × 6 post. It is fastened to the header with a metal joist hanger. The deck materials above the joists are spaced 2 × 6 planks for the front porch and also for the rear sun deck.

20. **Corner porch column—upper section.** The 6 × 6 post is at the center of this column assembly. These columns provide major support for the roof sections projecting over the front porch and the sun decks and the decks themselves. To give the appearance of a wider column, 2 × 4 studs are framed around the post and finished off with the same 1 × 6 siding placed on the walls.

FLOOR FRAMING PLAN OVER A FULL BASEMENT

The placement of the floor girders and joists over a full basement is the same as for the floor unit over a crawl-space foundation. However, the girders will be supported by metal columns rather than concrete-block piers. An opening for the stairway to the basement must also be framed in its proper location. **(See Figure 8–7.)**

CONSTRUCTING THE FLOOR UNIT

The layout and construction of the first-floor unit are done according to the information provided in the floor framing plan. The first-floor unit is framed over the masonry or concrete foundation walls. The beams and columns providing interior support for the floor joists must also be in place.

The first step in laying out the floor unit over a crawl-space or full-basement foundation is to place marks for the joists, 16" or 24" OC, on the sill plates, beams, and supporting double joists. Lines are squared at the marks, and an X is placed at the side of the mark where the joist is nailed. Double joists are marked with double Xs. They are required where a wall above the floor runs in the same direction as the joists. **Figure 8–8** illustrates a section of the floor joist layout below Bedroom #2 of the **One-Story Residence.**

The floor construction begins after the layout has been completed. The outside ends of the joists are toenailed into the mudsills and end-nailed into the rim

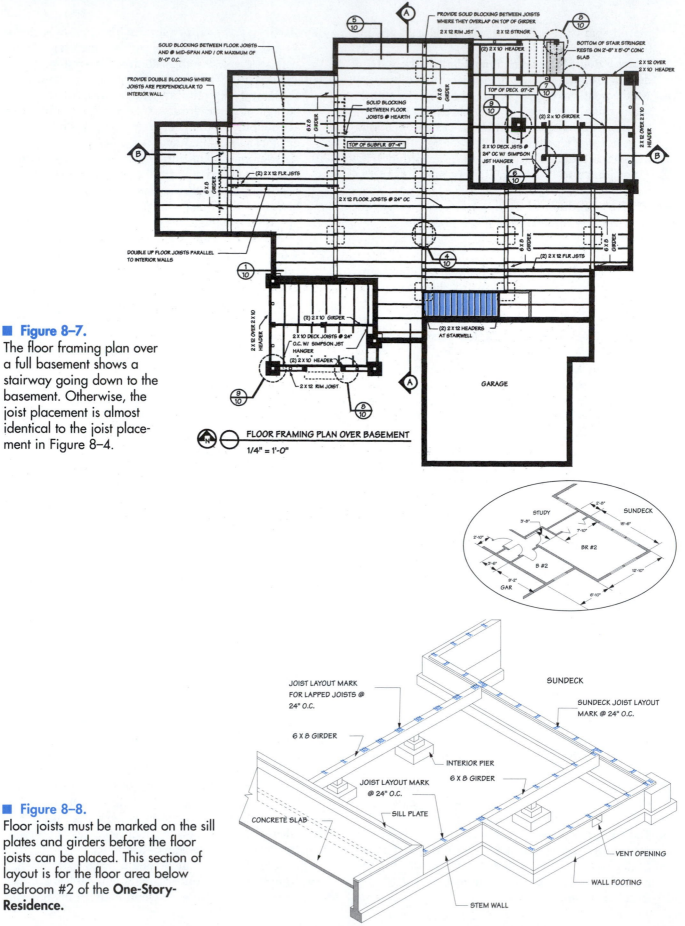

■ Figure 8–7.
The floor framing plan over a full basement shows a stairway going down to the basement. Otherwise, the joist placement is almost identical to the joist placement in Figure 8–4.

FLOOR FRAMING PLAN OVER BASEMENT
1/4" = 1'-0"

■ Figure 8–8.
Floor joists must be marked on the sill plates and girders before the floor joists can be placed. This section of layout is for the floor area below Bedroom #2 of the **One-Story-Residence.**

joists. The interior ends of the joist are toenailed to the supporting beams if the joists rest on top of the beam. Blocks are placed between the lapped ends. Joist ends butting against a double joist or beam are secured with steel joist hangers. **Figure 8–9** illustrates some floor construction according to the layout shown in Figure 8-8. Figure 8–7 shows a stairwell opening framed for the stairway leading to a basement.

The plans for the **One-Story Residence** call for tongue-and-groove 3/4" plywood panels placed over the floor joists. This information is given in several section view drawings. Other panel products, such as par-

ticleboard, waferboard, and oriented strand board, are also widely used. (Panel products are discussed in Unit 33). The required thickness of a floor panel is determined by the weight it must bear and the center-to-center spacing of the supporting joists. Industry guides are available giving specifications for Performance-Rated Floor Panels. (See Appendix F.)

The components, layout, and construction procedures for the first-floor unit over a full-basement foundation are the same as those for the floor unit over a crawl-space foundation. However, a stairwell opening will be framed for the stairway leading to a basement.

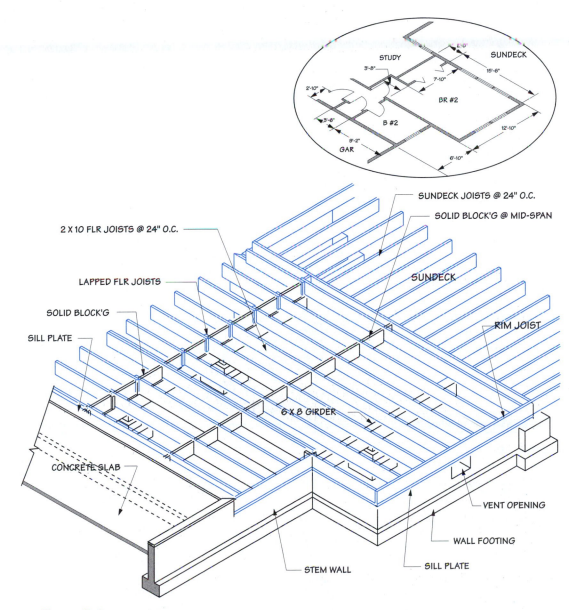

■ **Figure 8–9.**

The outside ends of the floor joists are toenailed to the sill plates and end-nailed to the rim joists over the foundation. The interior sections of the joists are toenailed to the girders and nailed to each other at their lapped ends.

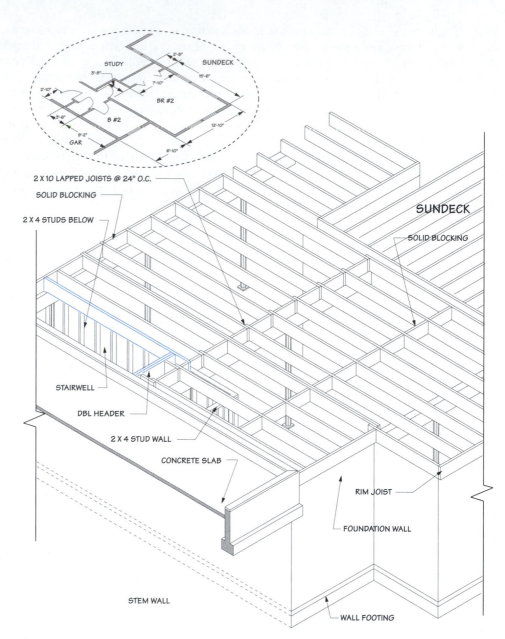

■ **Figure 8–10.**
The opening for the stairway to the basement must be constructed as shown on the floor framing plan for the **One-Story-Residence** over a full-basement foundation.

REVIEW QUESTIONS

Enter the missing words in the spaces provided on the right.

1. The most widely used procedure for wood frame construction is _____framing.

   _____

2. The main structural components of floor and ceiling units are the _____.

   _____

3. Floor and ceiling joists are usually spaced ___ or _____inches on center.

   _____

(See Figure 8–10.)

4. The distance permitted between the joist's bearing points is the _____ _____.

5. A _____ opening is shown on a floor framing plan over a full basement.

6. The ends of floor joists resting on exterior walls are nailed into _____ joists.

7. _____ and interior walls provide central support for the floor joists.

8. Where joists overlap interior beams or walls, _____ is placed between the joists.

9. _____ _____ support joist ends butting against beams.

10. Subfloor _____ are nailed over the joists after all the joists have been placed.

11. The floor joists are _____ beneath parallel running walls above.

12. Where joist spans are _____ feet or more, solid blocking is required between joists.

13. _____ or _____ batt insulation is commonly placed between floor joists.

14. The interior ends of floor joists should _____ past each other at least four inches.

15. _____ _____ butt up against the headers of stairwell openings.

Match the letters next to the symbols with their descriptions.

16. Insulation   _____

17. Girder   _____

18. Wood post   _____

19. Joist   _____

20. Section or detail reference   _____

21. CMU column base and wood post   _____

22. Plywood   _____

23. Dimensioned lumber   _____

24. Building section symbol   _____

a.

b.

c.

d.

e.

f.

g.

h.

i.

Choose the correct abbreviation for each term.

25. Building   _____   JSTS

26. Decking   _____   GYP BD

27. Double blocking   _____   BLDG

28. Each   _____   MIL

29. Exterior   _____   EA

30. Floor   _____   POLY

31. Gypsum board   _____   HT

32. Height   _____   REDWD

33. Joists   _____   PLYWD

34. Millimeter   _____   FL

35. On center   _____   DKG

36. Plywood   _____   DBL BLKG

37. Polyethylene   _____   OC

38. Redwood   _____   EXT

UNIT 9

Floor Plans

Wall framing begins after the floor unit has been completed. The **floor plan** is the main drawing guiding the layout and construction of the walls. It is a plan view looking down on the living or work area of a building. It is generally considered the most important drawing of a set of plans. However, other drawings and schedules must be studied to clarify some of the information on the floor plan. A single floor plan is required for a one-story building. A building with more than one story will have a separate floor plan for each level of the building.

FLOOR DESIGN FACTORS AND COMPONENTS

The main structural components of a wood frame wall are the plates, studs, corner posts, and bracing. The structural components of window and door openings are trimmers, cripples, and headers. **(See Figure 9–1).** Exterior panel sheathing is also a factor in bracing and strengthening the walls.

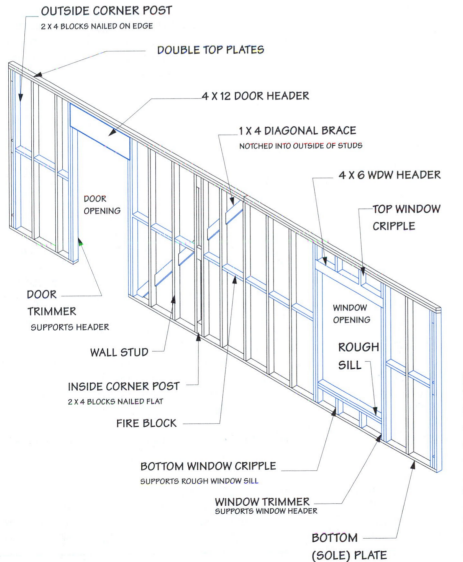

OUTSIDE CORNER POST
2 X 4 BLOCKS NAILED ON EDGE

DOUBLE TOP PLATES

4 X 12 DOOR HEADER

1 X 4 DIAGONAL BRACE
NOTCHED INTO OUTSIDE OF STUDS

4 X 6 WDW HEADER

TOP WINDOW CRIPPLE

DOOR OPENING

DOOR TRIMMER
SUPPORTS HEADER

WALL STUD

INSIDE CORNER POST
2 X 4 BLOCKS NAILED FLAT

FIRE BLOCK

WINDOW OPENING

ROUGH SILL

BOTTOM WINDOW CRIPPLE
SUPPORTS ROUGH WINDOW SILL

WINDOW TRIMMER
SUPPORTS WINDOW HEADER

BOTTOM (SOLE) PLATE

■ **Figure 9–1.**
The basic components of a typical framed wall consist of plates, studs, and corner posts. Window and door openings have headers, trimmers, and cripples. Two types of headers are shown for demonstration purposes. A solid 4 × 12 header is placed over the door opening. A 4 × 6 header is placed over the window opening, requiring short cripples between the header and top plates.

STUDS AND PLATES

Wall **studs** are the vertical pieces nailed between the top and bottom plates. They are most often placed 16" or 24" OC, depending on local code requirements. Stud material is usually 2×4 or 2×6 lumber. The size used is governed by local codes. A 2×6 stud wall allows the use of thicker and higher-rated insulation materials in the walls.

Plates are horizontal pieces nailed to the tops and bottoms of the studs. The bottom plates, often called **sole plates,** are spiked to the lower ends of the studs and are also nailed into the floor below. Top plates are nailed to the upper ends of the studs. A second plate is usually nailed to the top plate to strengthen the top of the wall and help tie the corners of the adjoining walls to each other. Together they are called the **double plates.**

CORNER POSTS

Framed walls are joined with outside and inside corner posts. An **outside corner post** is placed at the end of a wall that forms a corner with another wall. It often consists of two studs separated by short blocks placed with the block thickness (1 1/2") between the studs. This results in a strong corner tie between two walls and also provides nailing at the corner for inside finish materials. **Inside corner posts** may be constructed of two studs with the width side of blocks (3 1/2" or 5 1/2") nailed between the studs. The end stud of an inside wall can then be nailed to the blocks. The studs on each side of the blocks provide nailing for inside wall materials. Corner posts may also be constructed with a third stud in place of the blocks.

DIAGONAL BRACES

Diagonal braces may be required for a framed wall where structurally rated sheathing is not used. **Sheathing** is the panel material nailed to the outside surface of an exterior wall. **Diagonal braces** may consist of a 1×4 board notched into the studs, or a metal strap. They should be nailed as close to the ends of a wall as possible and at 25' intervals in longer walls. The recommended angle for diagonal braces is 45 to 60 degrees.

DOOR AND WINDOW OPENINGS

The main structural members of door and window openings are the header, studs, trimmers, and cripples. The **header** is nailed between two studs, and its thickness equals the width of the wall studs. For example, a header for a 2×4 wall is 3 1/2"thick. In this case the header can be two 2" thick pieces with a 1/2" plywood spacer in between. (The actual dimension of 2" thick

lumber is 1 1/2", which is why we need a 1/2" spacer.) The header can also be a solid 4×12 or 6×12 timber placed directly beneath the top plate of a standard 8'-1" high framed wall. This will allow the required clearance for the **door jamb** (finish frame) and a 6'-8" high door. A taller wall, or the use of a narrower header, requires cripple studs over the header. Another type of header frequently used in 2'6 framed walls is made up of two 2'10s on edge with a flat 2×6 nailed to the bottom edges of the 2'10s.

The ends of any type of header are supported by **trimmer studs.** A window opening is built in the same manner as a door opening, but it has a **rough sill** at the bottom of the opening. The sill is supported by bottom **cripple studs.**

READING THE FLOOR PLAN

The floor plan shows the shape, arrangement, and purpose for the rooms in the living or work area of the building. The floor plan is used more than any other drawing for the layout and construction of the exterior walls and the interior walls **(partitions).** It identifies the outside dimensions of the building and the necessary measurements to lay out and place the interior partitions.

The floor plan shows the locations of all the door and window openings. Some floor plans note the heights and widths of the openings. However, a common practice is to include a *door and window schedule* with the set of plans. The schedules provide the dimensions of the openings as well as other relevant information. Door and window schedules are included with the set of plans for the **One-Story-Residence** and are discussed later in this unit..

The locations of light switches, wall plugs, overhead lights, outlets, and electrically powered fixtures and equipment are sometimes shown on the floor plan. Often, though, this information is presented in a separate electrical-plumbing-heating plan. Such a plan is included with the **One-Story Residence** plans and is discussed in Unit 13. Larger residential and commercial projects usually feature *separate* mechanical, electrical, and plumbing plans. These plans are discussed in Section 3.

Floor plan symbols and abbreviations show the locations of sinks, bathtubs, furnaces, and other types of fixtures. The locations of heat and air outlets are included on some floor plans. Some also identify the floor finish materials in each room shown on the drawing. The floor plan for the **One-Story Residence** includes "Interior Finish Notes" below the drawing that give the wall and floor finish materials. Symbols and abbreviations appearing in the floor plan are shown in **Figure 9–2.**

The floor plan includes dimensions establishing the

locations of the walls, doors, and windows, as well as other units at that level of the building. A series of dimension lines appears around the exterior of the floor plan drawing. The outside dimension lines usually identify the total width and length of the building. Frequently the second row of dimension lines, closer to the building, locate the interior partitions and the offsets and recesses of the exterior walls. The third row of lines may show the placement of the doors, win-

dows, and other structural features. Measurements between partitions are often shown within the drawing. **(See Figure 9–3.)**

Total measurements are usually taken from the exterior faces of the outside corners. The first measurement to an interior partition may be from the exterior face of an outside corner to the face of (or center) of an interior partition. Dimensions between interior partitions may be from the surface of one partition to

SYMBOLS AND ABBREVIATIONS
FLOOR PLANS

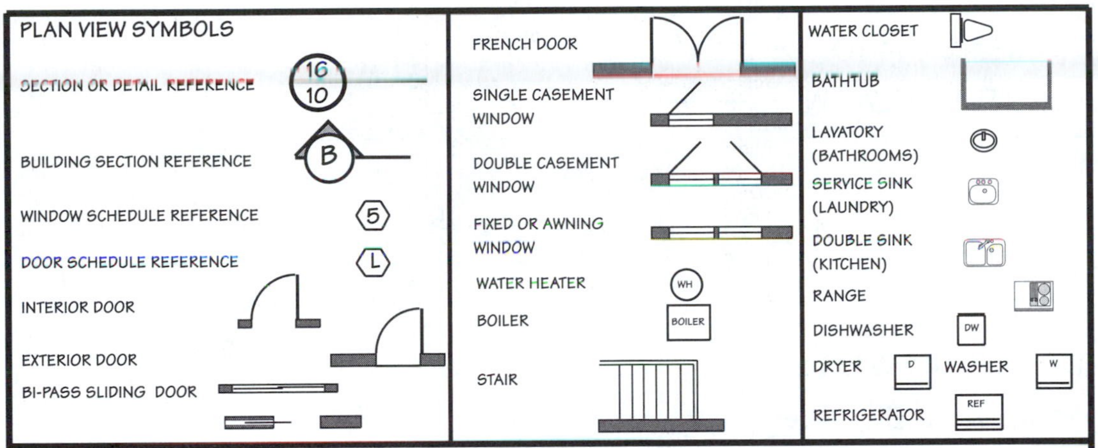

PLAN VIEW SYMBOLS

SECTION OR DETAIL REFERENCE

BUILDING SECTION REFERENCE

WINDOW SCHEDULE REFERENCE

DOOR SCHEDULE REFERENCE

INTERIOR DOOR

EXTERIOR DOOR

BI-PASS SLIDING DOOR

FRENCH DOOR

SINGLE CASEMENT WINDOW

DOUBLE CASEMENT WINDOW

FIXED OR AWNING WINDOW

WATER HEATER

BOILER

STAIR

WATER CLOSET

BATHTUB

LAVATORY (BATHROOMS)

SERVICE SINK (LAUNDRY)

DOUBLE SINK (KITCHEN)

RANGE

DISHWASHER

DRYER   WASHER

REFRIGERATOR

ABBREVIATIONS

@	At		KIT	Kitchen
ABV	Above		LAU	Laundry
B	Bathroom		LR	Living Room
BR	Bedroom		LS	Lazy Susan
BLW	Below		MA	Master
CL	Center Line		MECH	Mechanical
CNTR	Counter		MIN	Minimum
DBL	Double		RMS	Rooms
ELEC	Electric		PAN.	Pantry
EQ	Equal		SH	Shelf
GYP BD	Gypsum Board		SQ FT	Square Feet
HR	Hour		WDW	Window
INSUL	Insulation		WH	Water Heater

■ **Figure 9–2.**
These are symbols and abbreviations found on the floor plan of the **One-Story Residence.**

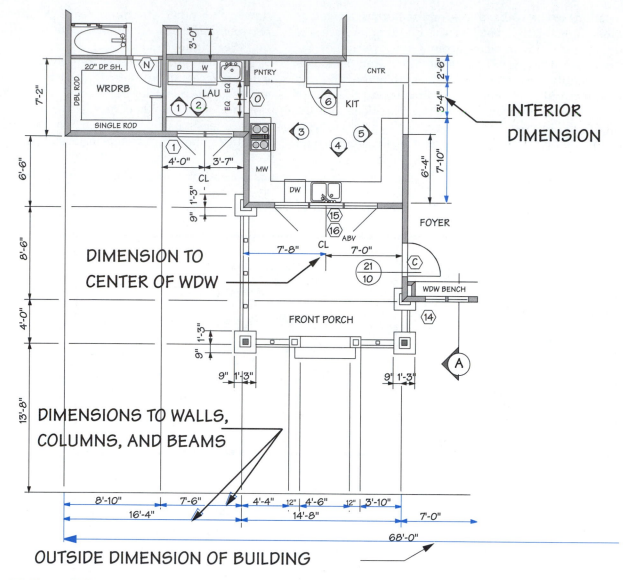

■ **Figure 9–3.**
Floor plan dimensions give the measurements to walls, beams, doors, and windows.

the surface of the opposite partition, or from the center of one partition to the center of the opposite partition. Another method calls for all measurements to be from wall to wall. **(See Figure 9–4.)**

The floor plan over a crawl-space foundation is shown in **Figure 9–5.** A corresponding pictorial drawing of this floor plan appears in **Figure 9–6.** Concentrate on the following when studying the plans.

✓ Symbols and abbreviations.
✓ Total dimensions of the exterior walls.
✓ Measurements to interior walls.
✓ Layout and placement of windows and doors.
✓ Locations of bathroom and kitchen fixtures

Explanations of the floor plan features shown in Figure 9–6 follow:

1. **Outer dimension line.** The greatest width of the building is 68′-0″.
2. **First inner dimension line.** The following measurements are obtained from the first inner dimension lines:

   ✓ The outside face of the west wardrobe wall to the west face of the porch and the kitchen wall (16′-4″).
   ✓ The west face of the porch and kitchen walls to the east face of the porch and kitchen wall (14′-8″).

✓ The face of the kitchen wall to the west face of the garage wall (7'-0").

✓ The outside face of the west garage wall to the outside face of the east garage wall (21'-6").

✓ The outside face of the east garage wall to the outside face of the mechanical area and Bathroom #2 walls (3'-0").

✓ The outside face of the Bathroom #2 wall to the outside face of the Bedroom #2 wall (5'-6").

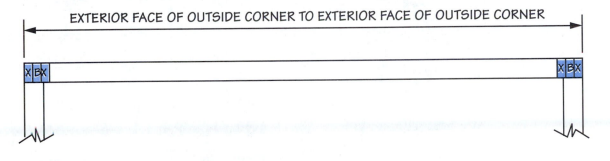

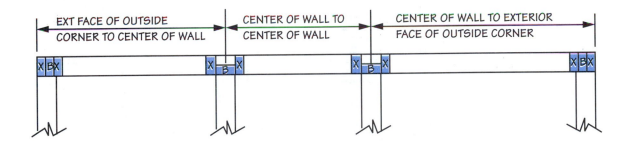

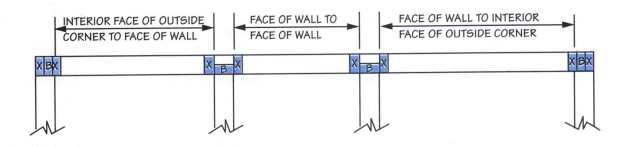

■ **Figure 9–4.**
Several methods can be used to lay out the dimensions to exterior and interior walls.

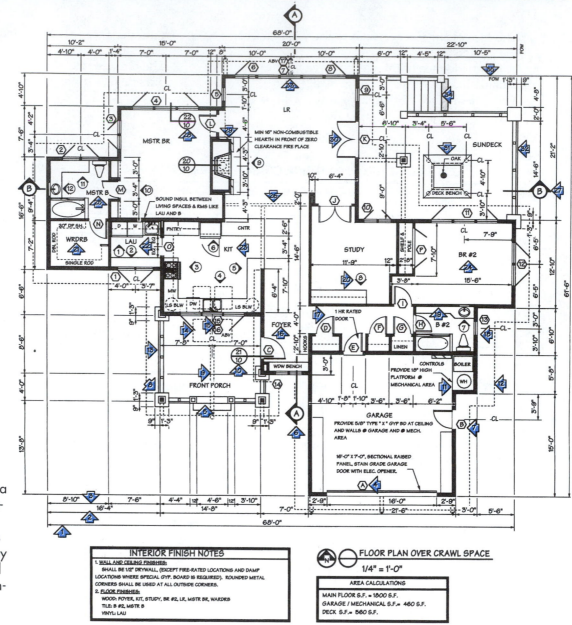

■ **Figure 9–5.**
The floor plan over a crawl-space foundation identifies the rooms and provides the dimensions to lay out the walls as well as the door and window openings.

3. **Second inner dimension line.** The following measurements are obtained from the second inner dimension line:

   ✓ The outside face of the west wardrobe and bathroom walls to the inside face of the east wardrobe and bathroom walls (8'-10").
   ✓ The remainder of the measurements are necessary to locate the newel posts on each side of the steps leading up to the front porch.

4. **Garage door.** The code letter (A) for the garage door is identified. Additional measurements and data are given in the door schedule.

5. **Front porch steps.** Two steps go from the walk up to the front porch.

6. **Building section reference (A).** A corresponding

symbol is also shown at the north end of the house. If a cutting-plane line were drawn between the two symbols, it would identify a longitudinal building section drawing. (Building section drawings are discussed in Unit 11.)

7. **Door.** The hexagon contains the code letter (B) for a side door to the garage. Measurements and additional data are given in the door schedule.

8. **Porch column.** This column supports the southwest corner of the front porch and that corner of the roof.

9. **Porch-roof beam.** The dashed lines identify a roof beam over the front porch.

10. **Door detail symbol.** The top number in the circle gives the number (21) of the detail drawing and the bottom number gives the page (10) where it is found. The straight leader line identifies the door

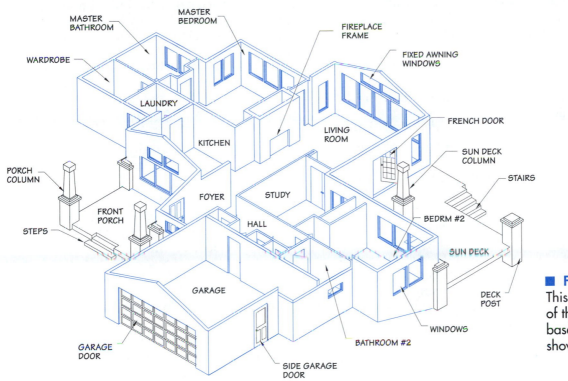

MASTER BATHROOM
MASTER BEDROOM
FIREPLACE FRAME
WARDROBE
FIXED AWNING WINDOWS
LAUNDRY
LIVING ROOM
KITCHEN
FRENCH DOOR
PORCH COLUMN
SUN DECK COLUMN
STAIRS
FRONT PORCH
FOYER
STUDY
BEDRM #2
STEPS
HALL
SUN DECK
GARAGE
DECK POST
WINDOWS
GARAGE DOOR
BATHROOM #2
SIDE GARAGE DOOR

■ **Figure 9–6.**
This pictorial drawing of the main floor is based on the floor plan shown in Figure 9–5.

shown in the detail. (Detail drawings are discussed in Unit 11.)

11. **Mechanical Room.** This area contains the water heater and the boiler that service the radiant floor heating system. The box containing the controls for the heating system is located to the left of the boiler.

12. **Roof overhang.** The dashed lines identify the overhang of the roof above the main floor.

13. **Porch Railing.** The west porch railing extends from the southwest porch column to the south wall of the kitchen. The three small squares identify posts in the railing.

14. **Skylights.** The dashed lines identify skylights in the porch roof and over the kitchen.

15. **Windows.** Two window code numbers appear next to each other. The code number closest to the windows (15) identifies the lower group of windows consisting of two swinging casements separated by a fixed window. The second code number (16) identifies two fixed transoms above the lower windows. The transom windows don't open, but they do allow additional light into the kitchen.

16. **Foyer.** This is the front-entrance hallway to the house. A window bench is located in the front of a pair of windows at the south end of the foyer. The entrance door from the porch is to the left of the window bench. The north end of the foyer leads into the living room and kitchen.

17. **Hallway.** This area opens up to a clothing closet (D), doorway (E) leading to the garage, and two linen closets (F & G). A bathroom door (H) is at the end of the hallway. Another door (I) opens to Bedroom #2.

18. **Bathroom window.** This single awning window (13) is hinged at the top and swings open from the bottom.

19. **Bathroom #2.** A built-in lavatory (sink) and counter are located along the north wall of this bathroom. The symbols identify an oval-shaped bathtub and water closet (toilet) along the south wall. Sliding glass doors in front of the tub are closed when the shower is in use.

20. **Interior elevation reference.** The circle points to the south wall of the study. The number (8) within the circle identifies an interior elevation drawing of that wall. (Interior elevations are discussed in Unit 12.)

21. **Wardrobe closet.** A large walk-in wardrobe closet is next to the master bathroom. The dotted lines identify shelves. Rods for hanging clothing are placed beneath the shelves.

22. **Laundry Room.** A pocket sliding door separates the laundry room from the kitchen. A dryer (D), washer (W), and utility sink are located at the north wall. A counter over base cabinets and drawers is at the south wall. Sound insulation is placed in the wall between the laundry and master bedroom.

23. **Kitchen.** Counters as well as base and wall cabinets are placed along all the walls of the kitchen. A dishwasher (DW) next to double sinks is located at the south wall. The cooking range and microwave (MV) are at the west wall. A refrigerator is between the pantry and a counter at the north wall. Elevation reference symbols and numbers point to all the walls.

24. **Bedroom #2.** A wardrobe closet with bypass sliding doors is located at the west wall. Windows are at the east and north walls.

25. **Master bathroom.** A recessed oval tub with sliding doors at the open side is located against the south wall. Counters above base cabinets and drawers are placed against the west and north walls. A built-in counter lavatory (sink) is installed in the west wall counter. The toilet is next to the north wall counter. A door opens from the master bedroom into the master bath. Note that sound insulation is also indicated in the wall between the bathroom and bedroom.

26. **Beams.** A 6 × 12 beam is placed here. Another 6 × 12 beam extends from the east kitchen wall. It butts up against the first beam and is supported by a metal beam hanger. The purpose of the second beam is to support the ends of a number of roof rafters over the kitchen area.

27. **Building section reference (B).** A corresponding symbol is shown at the west side of the house.

28. **Window section reference.** The reference number (22) indicates a section and detail drawing of this window. (Section and detail drawings are discussed in Unit 11.)

29. **Fireplace.** A zero-clearance fireplace is located at the west living room wall. The section reference symbol (20) refers to a structural section view drawing of the fireplace. (This section view is discussed in Unit 11.) The elevation reference symbol (9) in front of the fireplace refers to an elevation drawing of the wall. (Interior elevations are discussed in Unit 12.)

30 **French doors.** A pair of swinging french doors (K) open onto the rear sun deck. French doors usually have glass panes throughout their length. Measurements and other data are given in the door schedule.

31. **Deck bench.** The bench is constructed around the existing oak tree at this location.

32. **Sun-deck railing.** This type of railing is placed along the east and north sides of the sun deck.

33. **Living room windows.** Outswinging casement windows are located at each end of this group of windows. The slanted lines indicate the swing of the windows. The three center windows are fixed and do not open. Two fixed awning transom windows are located above the lower group.

34. **Stairway.** Five steps lead up to the sun deck. The bottom of the stairway rests on a small concrete slab.

35. **Face of Wall (FOW).** This extension line identifies the face of the north wall of the building.

FLOOR PLAN OVER FULL-BASEMENT FOUNDATION

The floor plan over a full-basement foundation is almost identical to the floor plan over a crawl-space foundation. However, a stairway goes from the main floor to the basement. This requires the elimination of the coat closet and two linen closets shown in the first plan. The door from the garage to the hallway also has to be repositioned. **(See Figure 9–7.)**

WINDOW AND DOOR SCHEDULES

Window and door schedules provide important information about the window and door openings shown on the floor plan. The code numbers and letters on the floor plan are listed in these schedules. Window and door units are manufactured in factories. They come with the window or door mounted in its frame and finish materials provided for nailing around the frames after they have been set into the rough wall opening. **(See Figure 9–8a, b.)**

Window and door schedules are included which are not exactly the same on all construction plans, but most present the same basic information. **Figure 9–9** shows a window schedule for the **One-Story Residence.** It is divided into columns that include the following information:

✓ CODE: The number on the floor plan that identifies the window.

✓ WDW (WINDOW) UNIT: The window unit consists of the manufactured finish frame and window. The first dimension is the width and the second dimension is the height. For example, Window #5 is 2'-ll" wide and 4'-ll" high.

✓ MANUF (MANUFACTURER): The manufacturer of the window units (Pella) is identified.

✓ ROUGH OPENING: These are the key dimensions for the wall framers. The rough opening must be larger than the window unit to allow for leveling and plumbing the window frame. For example, the width of the rough opening for Window #5 is 2'-ll" 1/2"; therefore, it is 1/2" wider than the finished window unit. The rough height is 4'-11 3/4"; therefore, it is 3/4" more than the height of the finished window unit.

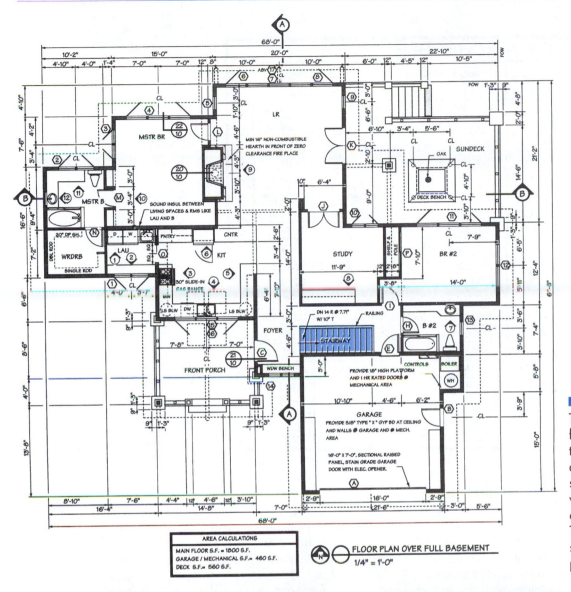

■ **Figure 9–7.**
The floor plan over a full-basement foundation identifies the rooms and gives the dimensions to lay out the walls and also the door and window openings. The dimensions for the stair opening are also provided.

■ **Figure 9–8a, b.**
Window and door units are ordered from the manufacturer. They are ready to be set in the rough openings. *(Courtesy of Andersen Windows, Inc.)*

WINDOW SCHEDULE

CODE	WDW UNIT	MANUF.	ROUGH OPENING	TYPE	MODEL#	FIN
1	(2) 2'-5" x 3'-5"	PELLA	4'-10 1/2" x 3'-5 3/4"	CSMT-CSMT	2436 CI2	CD
2	(2) 2'-5" x 3'-5"	PELLA	4'-10 1/2" x 3'-5 3/4"	CSMT-CSMT	2436 CI2	CD
3	(2) 2'-1" x 3'-11"	PELLA	4'-2 1/2" x 3'-11 3/4"	CSMT-CSMT	2042 CI2	CD
4	(3) 2'-11" x 3'-11"	PELLA	8'-9 1/2" x 3'-11 3/4"	CSMT-FX-CSMT	3042 CI3	CD
5	2'-11" x 4'-11"	PELLA	2'-11 1/2" x 4'-11 3/4"	FX	3054 CI	CD
6	2'-11" x 4'-11"	PELLA	2'-11 1/2" x 4'-11 3/4"	CSMT	3054 CI	CD
7	(3) 2'-11" x 4'-11"	PELLA	8'-9 1/2" x 4'-11 3/4"	FIX/FX/FX	3054 CI3	CD
8	2'-11" x 4'-11"	PELLA	2'-11 1/2" x 4'-11 3/4"	CSMT	3054 CI	CD
9	2'-11" x 4'-11"	PELLA	2'-11 1/2" x 4'-11 3/4"	FX	3054 CI	CD
10	(2) 2'-11" x 4'-11"	PELLA	5'-10 1/2" x 4'-11 3/4"	CSMT-CSMT	2054 CI2	CD
11	(3) 2'-11" x 3'-11"	PELLA	8'-9 1/2" x 3'-11 3/4"	CSMT-FX-CSMT	3042 CI3	CD
12	(2) 2'-11" x 3'-11"	PELLA	5'-10 1/2" x 3'-11 3/4"	CSMT-CSMT	3042 CI2	CD
13	2'-11" x 1'-5"	PELLA	2'-11 1/2" x 1'-5 3/4"	AWNING	3012 CI	CD
14	(2) 2'-1" x 2'-1"	PELLA	4'-2 1/2" x 2'-1 3/4"	FX/FX	2020 CI2	CD
15	(3) 2'-11" x 3'-11"	PELLA	8'-9 1/2" x 3'-11 3/4"	CSMT-FX-CSMT	3042 CI3	CD
16	(2) 2'-11" x 2'-1"	PELLA	5'-10 1/2" x 2'-1 3/4"	FX AWN TRANS	3020 CI21	CD
17	(2) 3'-11" x 2'-3"	PELLA	7'-10 1/2" x 2'-3 3/4"	FX AWN TRANS	4220CI2	CD
18	2'-11" x 2'-11"	PELLA	2'-11 1/2" x 2'-11 3/4"	CSMT	3030CI	CD
19	2'-11" x 2'-11"	PELLA	2'-11 1/2" x 2'-11 3/4"	CSMT	3030CI	CD
20	2'-11" x 2'-11"	PELLA	2'-11 1/2" x 2'-11 3/4"	CSMT	3030CI	CD
21	2'-11" x 2'-11"	PELLA	2'-11 1/2" x 2'-11 3/4"	CSMT	3030CI	CD

■ **Figure 9–9.**
Window schedules provide all the data needed to order the finished window units and frame the rough window openings.

✓ TYPE: The type of window is identified. For example, Window #5 is a fixed window that does not open.

✓ MODEL #: The model number for a window unit is included in the manufacturer's catalog. It is necessary to know this when ordering the unit.

✓ FIN (FINISH): The outsides of the frame and sash are vinyl clad (CD). This provides a protective covering and helps prevent deterioration due to weather exposure.

Figure 9-10 shows the door schedule for the **One-Story Residence.** It is divided into columns that include the following information:

✓ CODE: This is the letter on the floor plan that identifies the door.

✓ DR (DOOR) UNIT: These dimensions give the width and height of the door itself. For example, a (B) door is 2'-8" wide and 6'-8" high.

✓ MANUF (MANUFACTURER): Most of the doors are stock doors. However, the garage door (A) is manufactured by Raynor Corporation. The double French doors (K) leading from the living room to the sun deck are manufactured by Pella Corporation.

✓ ROUGH OPENING: These are the key dimen-

sions for the framers. The width of the rough opening must allow for the door width as well as the thicknesses of the jambs (door frame) on both sides of the door. In addition , 1/2" clearance is advisable behind each jamb to plumb the entire unit. For example, the width of the rough opening for door (B) is 2'-10 1/2"; therefore, it is 2 1/2" wider than the door width. This is calculated by adding the door width (2'-8"), two 3/4" thick side jambs (1 1/2"), and a 1/2" clearance on both sides (1"). The rough width is 2'-10 1/2" (2'-8" + 1 1/2"+ 1" = 2'-10 1/2").

The height of the rough opening must allow for the finish door height, the thickness of the head jamb, 1/2" clearance above the head jamb, and 3/4" below the door to allow for a threshold. For example, the height of the rough opening for door (B) is 6'-10"; therefore, it is 2" higher than the finish door height. This is calculated by adding the door height (6'-8"), the thickness of the head jamb (3/4"), the 1/2" clearance above the head jamb, and the clearance below the door (3/4"). The rough height is 6'-10" (6'-8" + 3/4" + 1/2" + 3/4" = 6'-10").

✓ TYPE: This refers to the surface appearance of the door. The surfaces of most of the doors are divided into panels.

DOOR SCHEDULE

CODE	DR UNIT	MANUF.	ROUGH OPENING	TYPE	FIN
Ⓐ	16'-0"x7-0"	RAYNOR	16'-0 1/2" x 7'-0 3/4"	GARAGE DR	WD
Ⓑ	2'-8"x6-8"	STOCK	2'-10 1/2" x 6'-10"	PNL/SASH	WD
Ⓒ	3'-0"x6-8"	STOCK	3'-2 1/2" x 6'-10"	PANEL DR	WD
Ⓓ	2'-4"x6-8"	STOCK	2'-6 1/2" x 6'-10"	PANEL DR	WD
Ⓔ *	2'-8"x6-8"	STOCK	2'-10 1/2" x 6'-10"	FLUSH DR	WD
Ⓕ	2'-4"x6-8"	STOCK	2'-6 1/2" x 6'-10"	PANEL DR	WD
Ⓖ	2'-4"x6-8"	STOCK	2'-6 1/2" x 6'-10"	PANEL DR	WD
Ⓗ	2'-4"x6-8"	STOCK	2'-6 1/2" x 6'-10"	PANEL DR	WD
Ⓘ	2'-6"x6-8"	STOCK	2'-8 1/2" x 6'-10"	PANEL DR	WD
Ⓙ	(2) 2'-4"x6-8"	STOCK	4'-10 1/2" x 6'-10"	PANEL DR	WD
Ⓚ	(2) 2'-6"x6-8"	PELLA	5'-2 1/2" x 6'-10"	FR.DR/ FR.DR	WD/GL
Ⓛ	2'-6"x6-8"	STOCK	2'-8 1/2" x 6'-10"	PANEL DR	WD
Ⓜ	2'-4"x6-8"	STOCK	2'-6 1/2" x 6'-10"	PANEL DR	WD
Ⓝ	2'-4"x6-8"	STOCK	2'-6 1/2" x 6'-10"	PANEL DR	WD
Ⓞ	2'-6"x6-8"	STOCK	2'-8 1/2" x 6'-10"	POCKET DR	WD
Ⓟ	(2) 3'-0"x6-8"	STOCK	6'-2 1/2" x 6'-10"	BI-PASS DR	WD
Ⓠ **	2'-8"x6-8"	STOCK	2'-10 1/2" x 6'-10"	PANEL DR	WD

*SELF-CLOSING 1 HOUR RATED DR (1 3/8" SOLID CORE)
**SEE FOUNDATION / FULL BASEMENT PLAN

■ **Figure 9–10.**
Door schedules provide all the information needed to order the finished door units and frame the rough door openings.

✓ FIN (FINISH): All of the doors have a wood finish with the exception of the French doors (K) that are wood and glass.

CONSTRUCTING THE WALLS

The main layout and construction of the wall are done according to information provided in the floor plan. The window and door schedules provide the dimensions for the rough openings. The walls are laid out and built over the subfloor of the floor units. Walls over a slab-on-grade foundation are built on top of the slab. They are fastened to the slab with bolts set in the concrete or are nailed down with powder-actuated fasteners.

The procedure for laying out walls constructed over a wood subfloor or concrete slab begins with snapping lines for the locations of the walls. The outside walls are marked by measuring in the thickness of the wall plates and then snapping the lines. . Lines are snapped next for the interior walls. If a measurement is to the center of a wall, the line is snapped a distance of one-half the width of the wall from the center mark. An X is marked to show which side of the line to nail the wall. **(See Figure 9–11).**

Wall plates are cut and tacked next to the lines. The outside and inside corners are marked. The locations of the door and window rough openings are established. Measurements to openings away from corners of wood framed walls are taken to the centers of the openings. The widths of the openings are obtained from the door and window schedules. The opening is laid out by measuring a distance of one-half the opening in opposite directions from the center line. If an opening is located next to a wall corner, the measurement is taken from the corner. **(See Figure 9–12.)**

The studs and cripples are marked on the plates. Studs are usually spaced 16" OC, although codes in some areas permit 24" OC. Door and window cripples follow the same layout as the studs. The marking of the studs and cripples completes the wall layout. **Figure 9–13** illustrates a portion of the layout of the walls of Bedroom #2 of the **One-Story Residence.**

Wall construction begins after the layout has been completed. The walls are framed in a horizontal position on the subfloor and then raised into place. The bottom (sole) plates are nailed through the subfloor into the joists below. Corners of adjoining walls are nailed together, and the walls are plumbed and held in position with temporary braces. **Figure 9–14** shows the north and east walls of Bedroom #2 framed according to the layout described in Figure 9–11 and Figure 9–13.

STUDY

2'-8"

SUNDECK

3'-8"

7'-10"

15'-6"

2'-10"

BR #2

3'-6"

B #2

9'-2"

12'-10"

GAR

6'-10"

4'-2"

2'-8"

B

15'-6"

2'-10"

7'-10"

3'-8"

3'-6"

A

12'-10"

9'-2"

6'-10"

■ **Figure 9–11.**
The first step in wall layout is to snap lines for the bottom plates according to dimensions shown on the floor plan. The X mark shows the side of the line to place the wall. This section of layout is for the walls around Bedroom #2 of the **One-Story-Residence.**

A. Measure back the thickness of the wall plate and snap line for the exterior walls.
B. Measure to the interior walls. Snap lines and place an X mark at the side of the line where the wall will be placed.

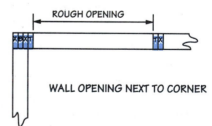

ROUGH OPENING

WALL OPENING NEXT TO CORNER

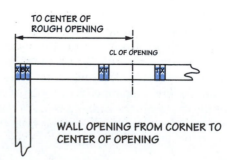

TO CENTER OF ROUGH OPENING

CL OF OPENING

WALL OPENING FROM CORNER TO CENTER OF OPENING

■ **Figure 9–12.**
Except for openings at wall corners, measurements are taken to the centers of window and door openings.

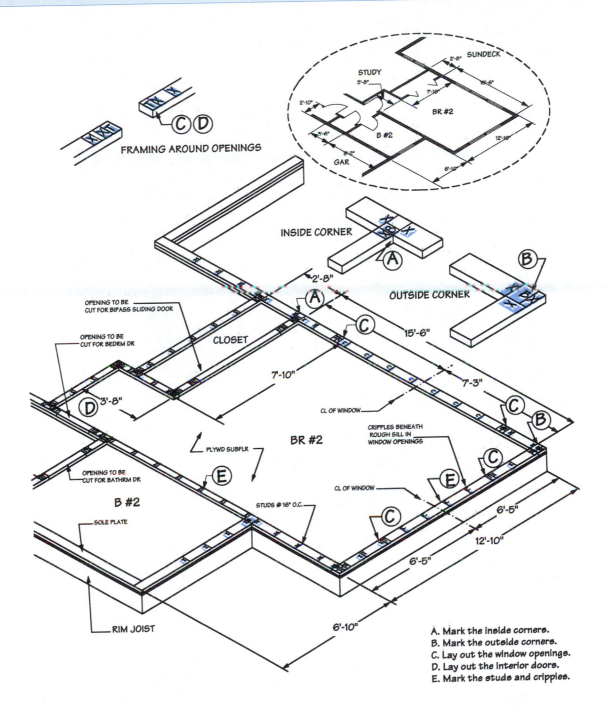

FRAMING AROUND OPENINGS

INSIDE CORNER

OUTSIDE CORNER

SUNDECK

STUDY

BR #2

B #2

GAR

2'-8"

OPENING TO BE CUT FOR BIPASS SLIDING DOOR

OPENING TO BE CUT FOR BEDRM DR

CLOSET

15'-6"

7'-10"

7'-3"

CL OF WINDOW

D 3'-8"

CRIPPLES BENEATH ROUGH SILL IN WINDOW OPENINGS

BR #2

PLYWD SUBFLR

OPENING TO BE CUT FOR BATHRM DR

CL OF WINDOW

B #2

STUDS @ 16" O.C.

6'-5"

SOLE PLATE

12'-10"

6'-5"

RIM JOIST

6'-10"

A. Mark the inside corners.
B. Mark the outside corners.
C. Lay out the window openings.
D. Lay out the interior doors.
E. Mark the studs and cripples.

■ **Figure 9–13.**
The bottom plates are tacked temporarily on the side of line identified by the X mark. The plates are then laid out for vertical framing members of the wall. This section of layout is for the walls around Bedroom #2 of the **One-Story-Residence.**

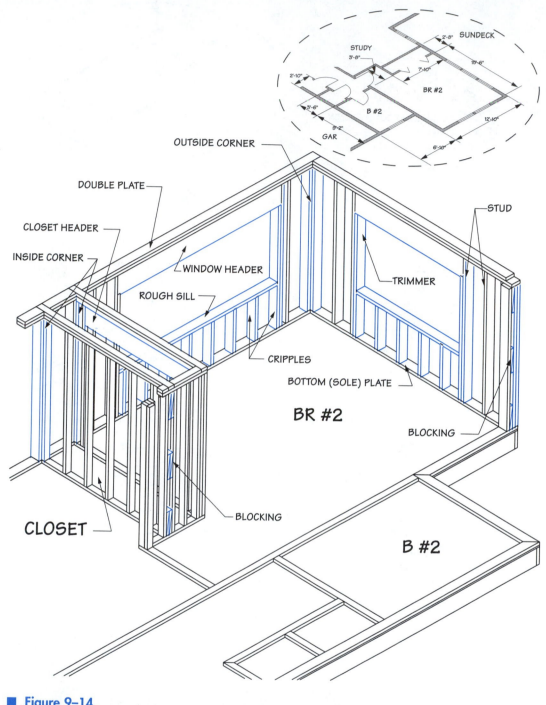

■ **Figure 9–14.**
This section of walls is framed according to the layout in Figure 9–13.

REVIEW QUESTIONS

Enter the missing words in the spaces provided on the right.

1. Vertical pieces nailed between the top and bottom plates are called ____ ____.

   _____

2. _____ and _____ lumber is used most often for wall studs.

   _____

3. A sole plate is also called a _____ plate.

   _____

4. A _____ _____ strengthens the top of a wall and helps tie adjoining walls together.

   _____

5. An _____ corner post is placed at the ends of a wall.

   _____

6. The recommended angle for a diagonal brace is ___ to ____ degrees.

   _____

7. Headers in door and window openings are supported by _____ studs.

   _____

8. A window opening has a ____ ____ at the bottom of the opening.

   _____

9. Bottom _____ studs are placed below the rough sill.

   _____

10 A view looking down at the living or work area is provided by the _____ plan.

   _____

11. ____ numbers or letters identify doors and windows on door and window schedules.

   _____

12 A window unit consists of a window mounted in its ____.

   _____

13. The first step in wall layout is to snap _____ for the wall locations.

   _____

14. Openings away from wall corners are usually measured to the _____ of the opening.

   _____

15 Studs are most often spaced _____ and on center.

   _____

Match the letters next to the symbols with their descriptions

16. Range   _____

17. Fixed or awning window   _____

18. Pocket sliding door   _____

19. Bathtub   _____

20. French door   _____

21. Section or detail reference   _____

22. Lavatory (in bathroom)   _____

23. Stair   _____

24. Window schedule reference   _____

25. Water heater   _____

26. Building section reference   _____

27. Interior door   _____

28. Double sink (kitchen)   _____

29. Single casement window   _____

30. Double casement window   _____

31. Service sink (laundry)   _____

32. Bypass sliding door   _____

33. Water closet   _____

34. Dryer   _____

35. Washer   _____

36. Dishwasher   _____

37. Door schedule reference   _____

38. Exterior door   _____

39. Refrigerator   _____

40. Boiler   _____

a.

b.

c.

d.

e.

f.

g.

h.

i.

j.

k.

l.

m.

n.

o.

p.

q.

r.

s.

t.

u.

v.

w.

x.

y.

Choose the correct abbreviation for each term.

41. At	_____	RMS
42. Above	_____	INSUL
43. Bathroom	_____	LS
44. Bedroom	_____	HR
45. Below	_____	@
46. Center line	_____	LR
47. Counter	_____	MA
48. Double	_____	EQ
49. Electric	_____	ABV
50. Equal	_____	CL
51. Gypsum board	_____	MIN
52. Hour	_____	SH
53. Insulation	_____	SQ FT
54. Kitchen	_____	KIT
55. Laundry	_____	WH
56. Living room	_____	CNTR
57. Lazy susan	_____	LAU
58. Master	_____	BR
59. Mechanical	_____	DBL
60. Minimum	_____	BLW
61. Rooms	_____	MECH
62. Pantry	_____	PAN.
63. Shelf	_____	WDW
64. Square feet	_____	B
65. Window	_____	GYP BD
66. Water heater	_____	ELEC

UNIT 10

Ceiling and Roof Framing Plans

Ceiling construction for a one-story building begins after the walls have been raised, plumbed, aligned, and braced. Information related to ceiling construction is sometimes shown on the floor plan. However, a separate **ceiling framing plan** is often included to provide this information in greater detail. The drawings for the **One-Story-Residence** include a ceiling framing plan.

Roof framing plans guide the construction of the roof. They provide a plan view looking down on the roof of a building. A very simple roof plan would show only the roof outline along with ridges and key rafters. For a more complex design, such as the roof for the **One-Story Residence,** a more detailed roof framing plan would be included with the drawings.

CEILING DESIGN FACTORS AND COMPONENTS

The thicknesses and widths of ceiling joists are determined by their span and the load they must support. Therefore, different sizes of joists may be used over smaller or larger spans in the ceiling. As with floor joists, tables are available giving the allowable spans for the more commonly used sizes of framing lumber. (See Appendix G.) Unless there is a high enough attic for storage or living space, the main load supported by the ceiling joists is the finish materials nailed to the bottom of the joists.

Ceiling joists are supported by walls or beams whose locations often determine the direction of the joists. A series of ceiling joists may run north to south in one section of the ceiling and east to west in another section. There is a structural advantage to joists running in the same direction as the roof rafters. The ends of the joists can then be nailed to the heels of the rafters as well as the wall plates. This results in a stronger tie between the outside walls of the building.

READING THE CEILING FRAMING PLAN

The ceiling framing plan gives the size and direction of the ceiling joists. It will show where the joists butt against a supporting interior beam or rest on top of the beam. Sloped ceilings are identified. Framed openings for the attic access and the fireplace flue are also identified. Symbols and abbreviations appearing in the ceiling framing plan (and the roof framing plan) are in shown in **(Figure 10–1.)**

The ceiling framing plan for the **One-Story Residence** is shown in **Figure 10–2.** A corresponding pictorial drawing of this ceiling appears in **Figure 10–3.** Concentrate on the following when studying the plan:

- ✓ Symbols and abbreviations.
- ✓ Direction and size of ceiling joists.
- ✓ Locations of supporting walls and beams.
- ✓ Sloped areas of the ceiling.
- ✓ Location of attic access.
- ✓ Location of flue opening.

Explanations of ceiling framing plan features shown in Figure 10–2 follow:

1. **Notes.** The notes provide some general information about the ceiling unit. This reduces the amount of writing on the drawings.
2. **Trusses.** Prefabricated roof trusses are placed over the garage. (More complete information about roof trusses is contained in the roof framing plan.)
3. **Ridge beam.** This beam supports the peak of the sloping roof over the front porch and kitchen.
4. **Double 2 × 12 rafter.** Every other set of rafters in the porch and kitchen area is doubled. Collar ties are placed on each sides of the doubled rafters.
5. **Sloped ceiling at end of foyer.** The short joists extend from the beam down to a 7' wall at the end of the foyer. The joists follow the same slope (4/12) as the roof rafters.
6. **Attic access.** A removable panel is placed over this opening to the attic area.
7. **Skylights.** These are placed in the roof over the front porch and kitchen areas.
8. **Ceiling height over foyer and hallway.** The flat ceilings over the foyer and the hallway leading to Bedroom #2 are 8'-0" from the subfloor.
9. **Ceiling height over Bedroom #2.** The flat ceilings over Bedroom #2 and over the adjoining study are 9'-0" from the floor.
10. **Sun-deck roof section.** The ceiling under the roof

SYMBOLS AND ABBREVIATIONS
CEILING AND ROOF FRAMING PLANS

PLAN VIEW SYMBOLS

SECTION OR DETAIL REFERENCE

BUILDING SECTION REFERENCE

ATTIC ACCESS

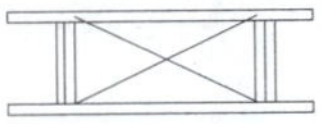

SKYLIGHTS

ABBREVIATIONS

BM	Beam
CHM	Chimney
CLG JSTS	Ceiling Joists
DN	Down
EA	Each
GALV SH MTL	Galvinized Sheet Metal
GYP BD	Gypsum Board
GLU-LAM	Glue-Laminated
HDR	Header
HT	Height
OH	Overhang
PL	Plate
LB	Pound
RS	Rough Surfaced
SF	Square Feet
W/	With

■ **Figure 10–1.**
These are symbols and abbreviations found in the ceiling and roof framing plans of the **One-Story-Residence.**

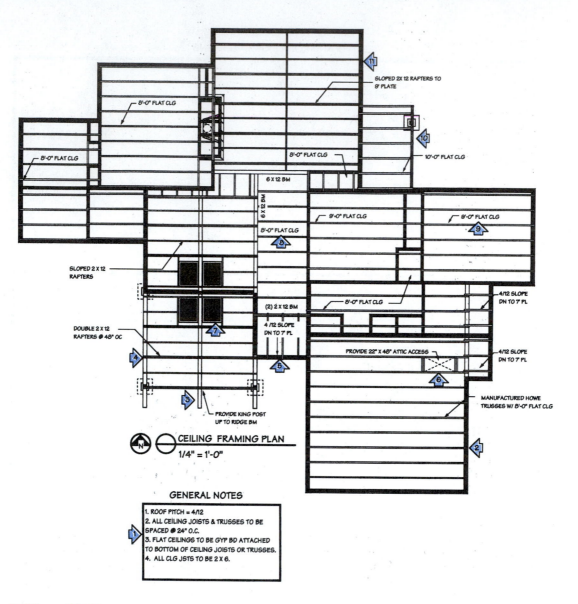

■ **Figure 10–2.**
The ceiling framing plan shows the layout and placement of the ceiling joists and the exposed sloping roof rafters that can be seen from below.

projection over one corner of the sun deck is 10'-0" from the deck.

11. **Living room ceiling.** Exposed sloped rafters are placed over the living room area. Every other rafter is doubled, with collar ties placed on each side.

CONSTRUCTING THE CEILING UNIT

The layout and construction of the ceiling unit are similar to the procedures described for the floor unit. However, ceiling joists do not nail into rim joists, because they are later fastened to the roof rafters. Blocking between the lapped interior ends is usually not required.

If the spacing of the ceiling joists is the same as for the studs below, the ceiling joists are marked directly over the wall studs. If it is different, the ceiling joists are marked according to the spacing shown on the plans. For example, the wall studs of the **One-Story Residence** are 16" OC, and the ceiling joists above are 24" OC. In this case every third ceiling joist will fall over a stud. The ceiling joists are also laid out where they rest on interior walls or beams, and they follow the same layout as the exterior walls. **Figure 10–4** illustrates the layout of some of the ceiling joists over Bedroom #2 of the **One-Story Residence.**

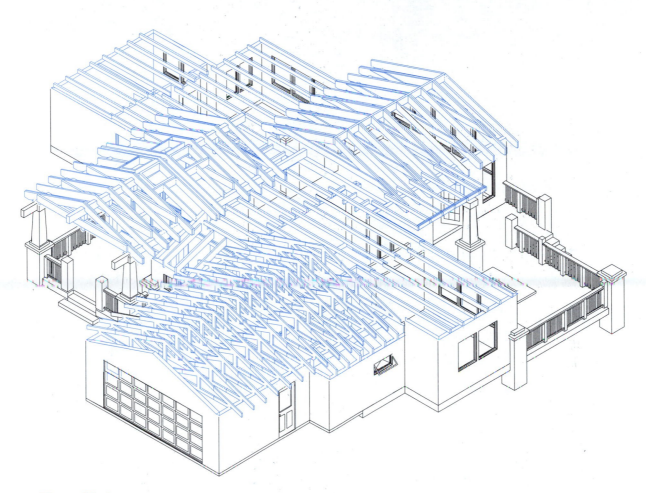

■ **Figure 10–3.**
This pictorial drawing of the ceiling frame is based on the ceiling framing plan shown in Figure 10–2.

The ceiling construction begins after the layout has been completed. The joists are toenailed into the top wall plates and beams, and are nailed to each other where they lap over a wall or beam. **Figure 10–5** illustrates some of ceiling construction according to the layout shown in Figure 10-4.

ROOF DESIGN FACTORS AND COMPONENTS

A roof framing drawing shows the layout and placement of the roof rafters and ridges and also includes special instructions where necessary. A roof of very simple design might require only a plan view showing the basic roof members such as the ridges and major rafters. For more complicated roof structures, such as the roof for the **One-Story Residence**, a detailed framing plan is necessary. Related section views and details must also be studied for complete roof information. The symbols and abbreviations found in the roof framing plan were shown in Figure 10–1.

There are a variety of individual roof styles and combinations. Whatever the style, a roof must be designed to sustain the pressures, loads, and strains to which it may be subjected. The slope of the roof is another important factor.

ROOF STYLES

The basic roof styles used most often are the flat, shed, gable, hip, gambrel, and mansard. These basic styles are frequently combined to produce more complex roof designs. The **shed roof** slopes in one direction. The **gable roof** slopes in two directions with a ridge at the center. The **hip roof** has a ridge at its center and slopes in four directions. The **gambrel roof** resembles a gable roof with its slope broken at the center. The **mansard roof** has a similar relationship to a hip roof. **(See Figure 10–6.)**

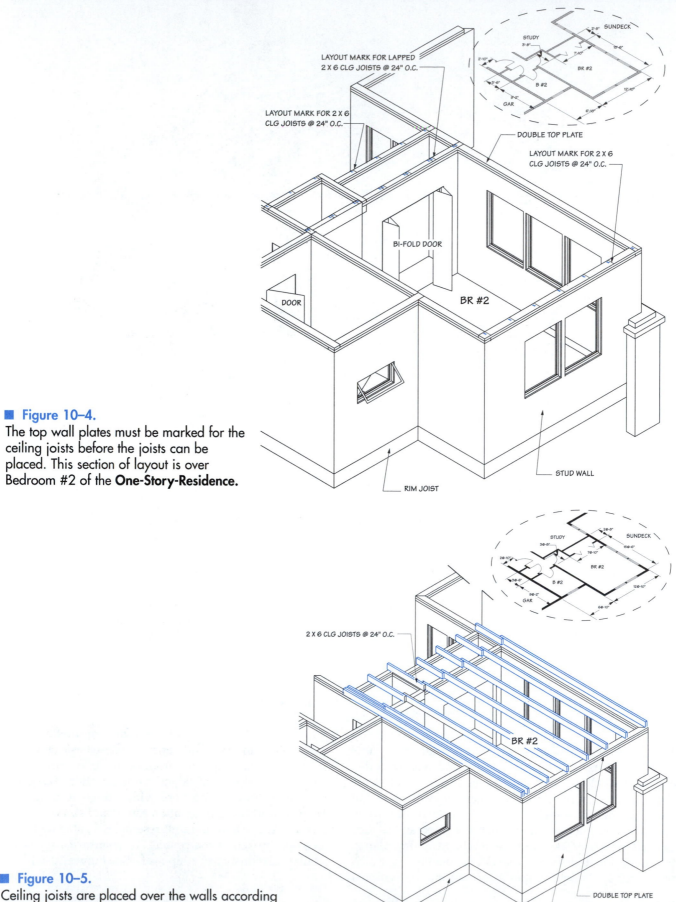

LAYOUT MARK FOR LAPPED
2 X 6 CLG JOISTS @ 24" O.C.

LAYOUT MARK FOR 2 X 6
CLG JOISTS @ 24" O.C.

STUDY
SUNDECK
BR #2
B #2
GAR

DOUBLE TOP PLATE

LAYOUT MARK FOR 2 X 6
CLG JOISTS @ 24" O.C.

BI-FOLD DOOR

DOOR

BR #2

STUD WALL

RIM JOIST

■ **Figure 10–4.**
The top wall plates must be marked for the ceiling joists before the joists can be placed. This section of layout is over Bedroom #2 of the **One-Story-Residence.**

STUDY
SUNDECK
BR #2
B #2
GAR

2 X 6 CLG JOISTS @ 24" O.C.

BR #2

DOUBLE TOP PLATE

STUD WALL

RIM JOIST

■ **Figure 10–5.**
Ceiling joists are placed over the walls according to the layout shown in Figure 10–4.

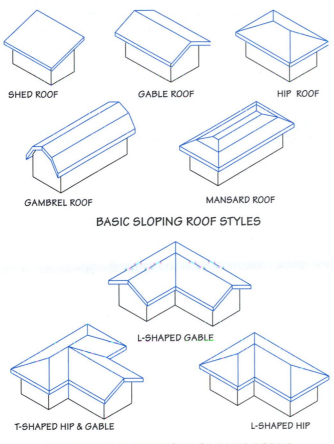

SHED ROOF   GABLE ROOF   HIP ROOF

GAMBREL ROOF   MANSARD ROOF

BASIC SLOPING ROOF STYLES

L-SHAPED GABLE

T-SHAPED HIP & GABLE   L-SHAPED HIP

FREQUENT COMBINATIONS OF BASIC STYLES

■ **Figure 10–6.**
Basic roof styles are the shed, gable, hip, gambrel, and mansard roofs. They are frequently combined as intersecting roofs.

ROOF DESIGN

Live and dead loads are major factors in the structural design of a roof. **Dead load** is the total weight of the materials used to construct the roof. This includes all the framing components and the finish roof covering. **Live load** for the roof is the weight of snow on the roof and wind pressures. Snow load and wind pressures are not the same for all regions of the country. Local building codes usually provide this information.

ROOF PITCH

Roof pitch is the angle of the roof from the ridge board to the top wall plates. Therefore, roof pitch determines the actual slope of the roof. The roof slope should conform with the rest of the house design and have enough incline to shed water and snow. Roof pitch is determined by the **unit rise** of the roof, which is the number of inches the roof rises for every foot of **unit run**. Unit rises range from 2" to 12". Usually unit rise and run are identified in the elevation drawings and/or section drawings by two short lines at right angles to each other. The longer horizontal line identifies the unit run (always 12"). The shorter vertical line is the unit rise. The unit rise of the **One-Story Residence** is 4". To express this as roof pitch, place the unit rise number (4") over 24 and reduce the answer. This roof has a 1/6 pitch (4/24 = 1/ 6). The slope of a roof becomes steeper as the unit rise increases.

The total height of a roof is called the **total rise.** This can be determined by multiplying the **total run** (total length of the roof) times the unit rise. The total run is one half of the **total span**, which is the horizontal width of a building beneath the run of the roof rafters. **(See Figure 10–7.)** Unit 29 contains math exercises for find-

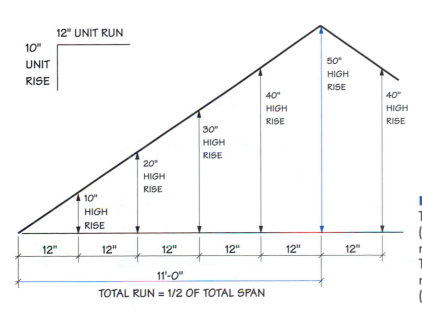

12" UNIT RUN

10" UNIT RISE

50" HIGH RISE

40" HIGH RISE   40" HIGH RISE

30" HIGH RISE

20" HIGH RISE

10" HIGH RISE

12"   12"   12"   12"   12"   12"

11'-0"

TOTAL RUN = 1/2 OF TOTAL SPAN

■ **Figure 10–7.**
The unit rise determines the angle of the slope (pitch) of the roof. The total rise (height) of the roof is equal to the total run times the unit rise. The unit rise shown above is 10" and the total run is 11'-0". Therefore, the total rise is 110" (11 × 10" = 110" or 9'-2").

ing total roof rise. (The terms *unit rise, unit run, total rise,* and *total run* are also used in relation to stairs, as shown in Unit 30.)

ROOF COMPONENTS

The main structural components of a framed roof are the ridge boards, roof rafters, and jack rafters. Additional support may be provided by collar ties, purlins, and braces. **Ridge boards** are nominally 2" thick pieces (1 1/2" actual thickness) placed at the center and high points of the roof. They provide nailing for the upper ends of the roof rafters.

A **roof rafter** extends from the outside wall plate to a ridge. A major factor in choosing the size of rafter material is the projected live and dead loads the rafters must support. Other factors are the allowable span and grade of the lumber being used. Rafter span tables are available for this purpose. (See Appendix G for roof framing rafter tables.)

The three types of rafters are the common, hip, and valley rafters. The **common rafter** runs at a right angle from the plate line to the ridge board. **Hip rafters,** used with hip roofs, and **valley rafters,** used with intersecting roofs, run at a 45-degree angle from the plate lines to the ridge board. **Jack rafters** fill the spaces between ridge boards and hip and valley rafters. **(See Figure 10–8.)**

Collar ties are nailed to opposite rafters and help to further stiffen the roof rafters. Braces may also be included to provide central support to the roof rafters. The lower ends of the braces rest on top of a wall, and the upper ends nail to the roof rafters. **Purlins** run horizontally below the roof rafters and are nailed to the braces. **(See Figure 10–9.)**

Another framing method used extensively with intersecting roofs is **blind valley construction.** This procedure eliminates the need for valley rafters. One roof section is framed and sheathed, and then the intersecting roof section is framed over the sheathed area.

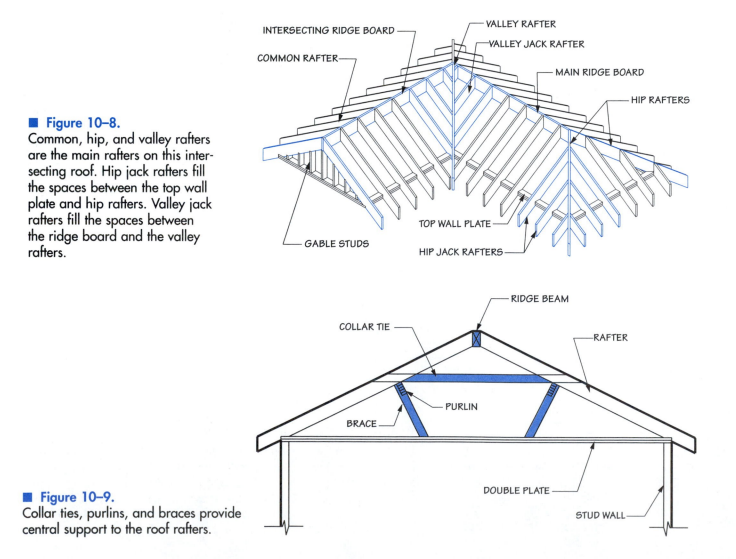

■ **Figure 10–8.**
Common, hip, and valley rafters are the main rafters on this intersecting roof. Hip jack rafters fill the spaces between the top wall plate and hip rafters. Valley jack rafters fill the spaces between the ridge board and the valley rafters.

■ **Figure 10–9.**
Collar ties, purlins, and braces provide central support to the roof rafters.

(See Figure 10–10.) This method is used in the roof construction of the **One-Story Residence.**

TRUSS ROOFS

Roof trusses are widely used for roof construction. A roof can consist entirely of trusses, or trusses can be combined with conventional roof framing, as in the roof framing plan for the **One-Story-Residence.** Here trusses are placed over the garage area, and rafters are framed over the rest of the main roof section.

In most cases, roof trusses are prefabricated in a plant and delivered to the job site. They are then set in place over the top wall plates. There are a variety of **truss** designs, but all have the basic components of bottom chords, top chords, and web members. The **bottom chords** act as ceiling joists. The **top chords** serve the same purpose as roof rafters. The **web members** may be wood or metal and are designed to stiffen and support the top and bottom chords. The chords and webs are usually tied together with metal connector plates. **(See Figure 10–11.)** (See Appendix H for examples of truss designs.)

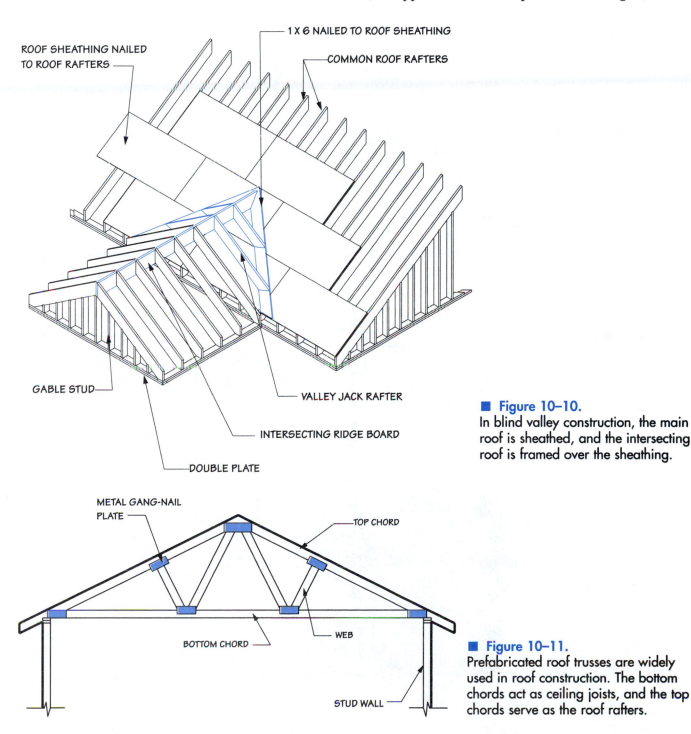

■ **Figure 10–10.**
In blind valley construction, the main roof is sheathed, and the intersecting roof is framed over the sheathing.

■ **Figure 10–11.**
Prefabricated roof trusses are widely used in roof construction. The bottom chords act as ceiling joists, and the top chords serve as the roof rafters.

THE ROOF FRAMING PLAN

The **roof framing plan** is a plan view looking down on the drawing. It indicates all the ridges and beams, as well as major hip and valley rafters. The placement and spacing of common and jack rafters is also shown. Special detail drawings are often included, and reference may be made to section drawings on other pages of the plans.

READING THE ROOF FRAMING PLAN

A simple roof plan for the **One-Story-Residence** identifies only ridges and major rafters. **(See Figure 10–12.)** A complete roof framing plan is shown in **Figure 10–13.** A corresponding pictorial drawing of the roof frame appears in **Figure 10–14.** Concentrate on the following when studying the plan:

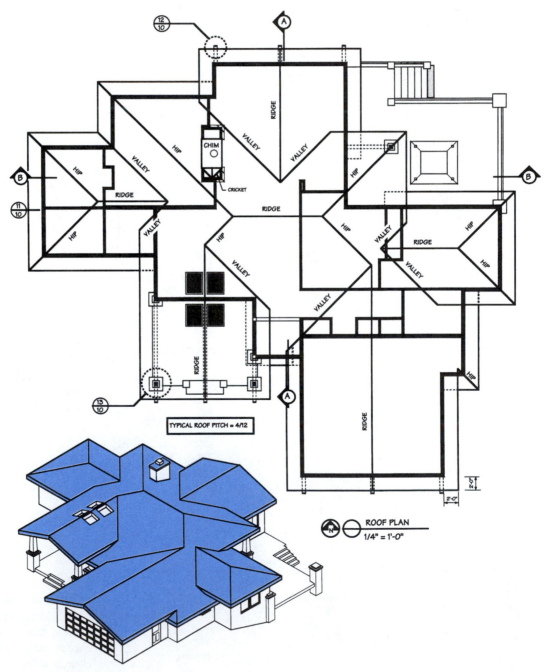

■ **Figure 10–12.**
The top drawing is a simple orthographic roof plan. The bottom pictorial drawing illustrates the roof plan. Compare these numbers with the numbers in the pictorial drawing.

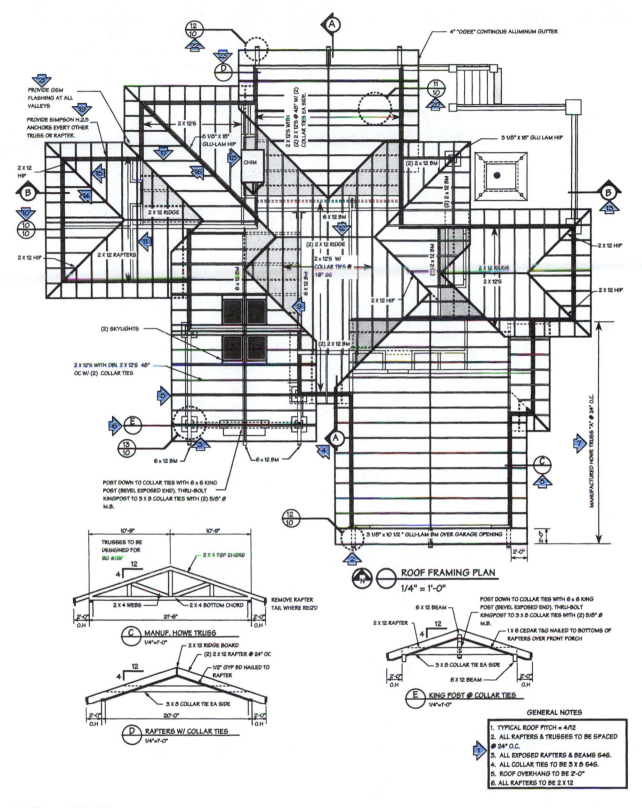

■ Figure 10–13.

The roof framing plan guides the construction of the roof. It identifies the placement of ridges, rafters, and other roof frame components.

✓ Symbols and abbreviations.
✓ Outline of building.
✓ Construction over framed areas.
✓ Placement of ridge boards
✓ Placement and direction of common rafters.
✓ Placement of valley rafters and valley jack rafters.
✓ Detail drawings as related to plan view drawings.

Explanations of the roof framing plan features shown in Figure 10–13 follows:

1. **Notes.** The notes provide some general information about the roof. This reduces the amount of writing on the drawings.

2. **Outrigger.** Nonstructural, decorative outriggers are placed beneath the garage roof overhang. One is below the ridge beam, and the other two are below the lower outside beams. Similar outriggers are placed at the south front porch roof overhang and the overhang over the outside of the north living room wall.

3. **Porch column.** A porch column supports the end of the 6 × 12 beam holding up one side of the roof. An identical column is beneath the end of the opposite beam.

4. **Building section reference (A).** A corresponding symbol is also shown at the north end of the roof. If a cutting-plane line were drawn between the

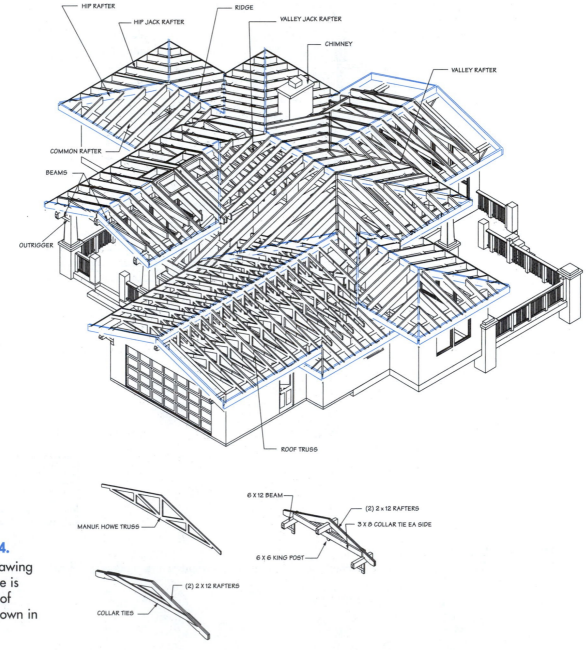

■ **Figure 10–14.**

This pictorial drawing of the roof frame is based on the roof framing plan shown in Figure 10–13.

two symbols, it would identify a longitudinal building section. (Building section drawings are discussed in Unit 11.)

5. **Truss detail reference (C).** Manufactured Howe trusses are placed from the front end of the garage to the south walls of Bedroom #2 and the study. The term "Howe" refers to the truss design. Refer to detail C at the bottom of the page.

6. **King post and collar tie detail reference (E).** The 6 × 6 king post supports the front end of the 6 × 12 ridge beam. Refer to detail E at the bottom of the page. The bottom of the king post is bolted to the collar ties below. The ends of the collar ties rest on the outside beams and are bolted to the lower ends of the rafters. The ridge beam runs the entire length of the front porch and kitchen. The bottoms of the rafters over the porch are finished off with 1 × 6 cedar tongue-and-groove boards.

7. **Truss placement.** The dimension line shows that the trusses will be positioned from the front of the garage to the walls of Bedroom #2 and the study.

8. **Front porch rafters and collar ties.** This assembly is very similar to detail E for the front end of the porch roof. However, there is no king post below the ridge beam.

9. **Under framing.** The shaded areas on the plan identify where one section of the roof frames over another. In this method of roof construction, the roof section over the front porch and kitchen is framed, then plywood panel sheeting is nailed on top of the rafters. Next, part of the adjoining section of roof is framed over the sheathed section of the first roof section.

10. **Roof overhang detail reference (10).** This symbol refers to a detail (10) that provides an enlarged drawing of the overhang section. This drawing will be shown with additional roof-related drawings later in this unit.

11. **Common rafter.** The common rafter runs at a 90-degree angle from the wall plate to the ridge.

12. **Ridge.** All ridges on the roof are 2 × 12s. A ridge provides nailing for the top ends of the rafters.

13. **Building section reference (B):** A corresponding symbol is also shown at the west side of the house.

14. **Hip rafter.** These rafters run at a 45-degree angle. They extend from a corner of the wall to the ridge.

15. **Hip jack rafter.** These rafters extends from the plate to the hip rafter. They increase in length as the hip rafter rises toward the ridge.

16. **Glu-Lam hip rafter.** A heavier material is used for this hip because of its length and the load it must support.

17. **Hip-valley jack rafter.** These rafters extend from the hip rafter to the valley formed in that section of the roof

18. **Chimney.** The location of the chimney is identified.

19. **Rafter anchors.** These metal anchors tie the rafters to the wall plates. Simpson is the manufacturer of the anchor, and H.2.5 is the identifying item number of the product.

20. **Detail reference symbol (11).** This symbol refers to a detail drawing of the rafter and collar-tie construction.

21. **Valley flashing.** Galvanized sheet metal (GSM) flashing is placed at all the valley areas of the roof. Valleys are formed where two sloping areas intersect and are particularly vulnerable to water leakage. The metal flashing is placed to prevent this. The fiberglass finish roof material will lap over the flashing.

22. **Detail reference symbol (12).** See explanation regarding outriggers in feature #2.

23. **Detail reference symbol (D).** Every other rafter assembly over the living room is doubled, and collar ties are bolted to each side of the doubled rafters. The detail drawing (D) toward the bottom of the page shows that the collar ties are held up from the wall, unlike the collar ties over the front porch and kitchen.

ROOF-RELATED DETAIL DRAWINGS

In addition to detail drawings included on the same page as the roof framing plan, there are several more roof-related detail drawings for the **One-Story Residence.** They are essential to illustrate materials being used and also construction methods that cannot be clearly drawn on the larger roof framing plan. They show the following:

✓ Fastening methods between the walls and the roof.
✓ The components and design of the roof overhangs.
✓ More information about collar-tie and rafter connections.
✓ More information about major roof-supporting columns.

Before reviewing the detail drawings, place a bookmark at the page showing the roof framing plan in Figure 10–13, page 117. While studying each detail drawing, relate the detail number beneath the drawing to the detail number and symbol on the roof framing plan.

READING THE ROOF-RELATED DETAIL DRAWINGS

The roof-related detail drawings, along with corresponding pictorial drawings, are shown in **Figure 10–15a, b.** Concentrate on the following when studying the drawings:

✓ Symbols and abbreviations.
✓ Materials used.
✓ Structural design.
✓ Dimensions of materials

Explanations of the roof-related detail drawings in Figure 10–15 follow:

1. **Window header.** The detail drawing includes the top section of a window beneath the overhang. The header is a 2 × 6 nailed to the bottoms of two 2 × 10s. Insulation is placed between the 2 × 10s.

2. **Window.** The double-pane glass window panes are identified.

3. **Frieze Block.** This 1 × 4 trim piece butts up against the 3/8″ plywood. Cedar lap siding shown beneath the frieze block rests on top of the window trim.

4. **Soffit.** The 3/4″ thick rough-sawed (RS) plywood soffit is fastened to the bottom of the overhang.

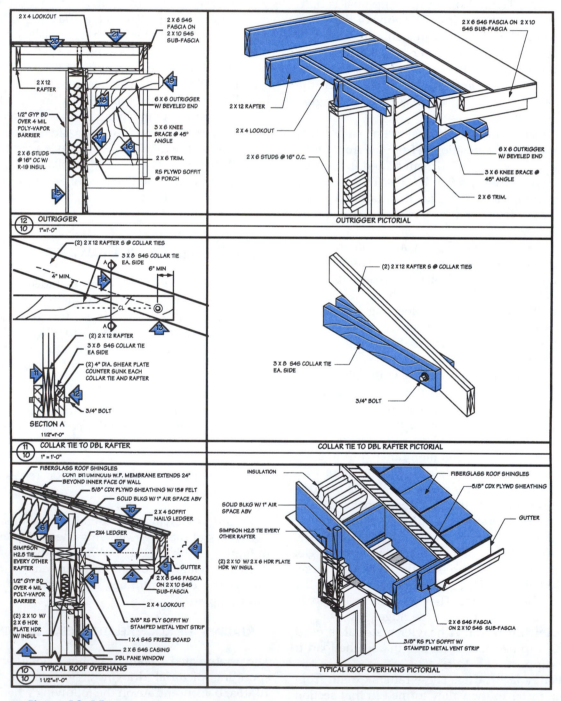

■ **Figure 10–15a.**
Roof-related detail drawings provide information that cannot be seen on the roof framing plan.

The term **rough-sawed** means that the surface of the lumber has not been planed to a smooth finish. A vent opening runs the full length of the soffit and allows air to circulate between the rafters. A metal vent strip screens out insects. This type of *closed-cornice* design is used around the entire roof with the exception of the gable overhangs that include outriggers. The term **cornice** generally applies to the area under the overhang where the roof and sidewalls meet.

5. **Fascia.** These boards finish off the ends of the rafters. They consist of a 2 × 6 fascia board nailed to a 2 × 10 sub-fascia board.

6. **Metal rafter tie.** This device greatly strengthens the tie between the rafters and wall plates.

7. **Solid blocking.** Wooden blocking is placed between the rafters. The 1″ air space above the blocking allows air coming through the soffit vent to pass over the blocking and circulate between the rafters.

8. **Lookout.** This 2 × 4 piece is nailed along the sides of the joists and provides additional nailing for the soffit.

9. **Gutter.** Vinyl gutters are nailed to the fascia where they are required. Gutters receive water runoff from the roof and direct the water to the downspouts.

10. **Roof surface finish.** Fiberglass roof shingles are nailed to the 3/4″ roof sheathing. Waterproof roofing paper is placed over the plywood before the shingles are nailed down. A bituminous material extends from the ends of the rafters to 24″ past the inner face of the wall. This added material placed over the roofing paper acts as an **eaves flashing** and helps protect the roof sheathing from ice dams that form in cold weather.

11. **Rafter section view.** This is an end view of the doubled rafters.

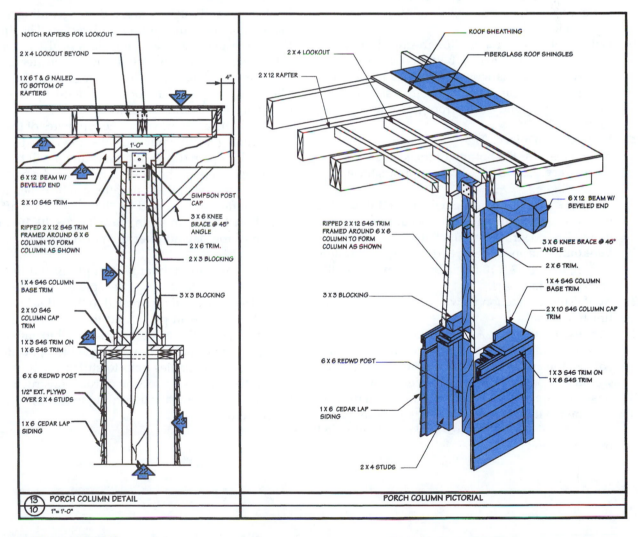

■ **Figure 10–15b.**
Roof-related detail drawings provide information that cannot be seen on the roof framing plan (continued).

12. **Rafter–collar tie connection (section view).** Detail A shows a 3/4″ diameter machine bolt that fastens the collar ties to each side of the double rafters. Round metal shear plates are grooved into the inside face of each collar tie. The shear plate strengthens the tie between the collar tie and rafters by helping to prevent movement where they are bolted together.

13. **Rafter–collar tie connection (side view).** The bolt running through the collar tie and rafter must be in the center of the collar tie and 6″ from the end. It must also be a minimum of 4″ from the bottom edge of the rafter.

14. **Detail reference symbol (A).** The symbols show the direction of viewing the connection described in the section view below.

15. **Stud wall.** A typical stud wall is shown. The interior of the wall is finished off with 1/2″ gypsum board placed over a vapor barrier. R-19 insulation is placed between the studs. The exterior finish is cedar lap siding over building paper and plywood.

16. **Gable soffit overhang.** Rough-sawed plywood material is nailed to the bottoms of the rafters. This type of overhang is shown outside the north living room. A similar overhang design is used over the south end of the front porch and garage roof.

17. **Vertical trim piece.** The 2 × 6 board provides a nailing base for the bottom of the knee brace.

18. **Knee brace.** A 2 × 6 knee brace supports the end of the outrigger.

19. **Outrigger.** The 6 × 6 outrigger is placed under the roof beam. The outrigger is for decorative purposes and has no structural function.

20. **Lookout.** The lookout is tied to the two rafters shown, and the 2 × 10 fascia board is nailed to the ends of the lookouts.

21. **Roof.** This is a section view of the rafters and roof finish materials.

22. **Porch column post.** A 6 × 6 redwood post supports the end of the outside roof beam. If it extended past the cutoff line at the bottom of the post, it would rest on a concrete-block pier below.

23. **Base frame.** The lower column section is widened with a 2 × 4 frame. The outside of the frame is finished off with 1 × 6 lap siding nailed over plywood.

24. **Cap trim.** The shoulder of the frame is capped off with a 2 × 10 board. A 1 × 3 piece is placed over 1 × 6 trim beneath the edge of the cap trim.

25. **Vertical trim.** The upper section of the column is finished off with 2 × 12 material. The boards slant in from bottom to top; therefore, a 3 × 3 spacer block is placed between the lower end of the 2 × 12 and the post. A thinner 3 × 3 spacer block is placed at the upper end of the post.

26. **Roof beam.** A 6 × 12 roof beam rests on top of the post.

27. **Porch rafter finish.** Tongue-and-groove (T&G) 1 × 6 cedar boards are nailed to the bottom of the rafters in the roof area.

28. **Roof.** This is a section view of the roof rafters and roof finish.

CONSTRUCTING THE ROOF

Roof layout and construction is done according to information provided in the roof framing plan. The drawing shows the placement and the sizes of the rafters and ridge boards. Refer to section views and detail drawings for additional information regarding collar ties. The roof framing plan for the **One-Story Residence** does not show purlins and braces.

The rafters are marked on the top plates of the walls. If the ceiling joist layout indicates the joists to the left side of the squared line, the roof rafters will be placed to the right side of the line. In this way the end of the joist and the heel of the roof rafter can be nailed together. The rafters are marked on the ridge board after the ridge board has been set in place. **Figure 10–16** shows a section of roof layout over the north and east walls of Bedroom #2.

The common rafters are set in place, and the lower ends are nailed to the top wall plates and the ceiling joists. The upper ends are nailed to the ridge board. The hip rafters are set in place, and the hip jack rafters are nailed between the ridge boards and hip rafters. The collar ties are installed and are nailed to opposite common rafters. Collar ties help to stiffen the roof rafters. **Figure 10–17** shows a section of roof construction over the north and east walls according to the layout shown in Figure 10–16.

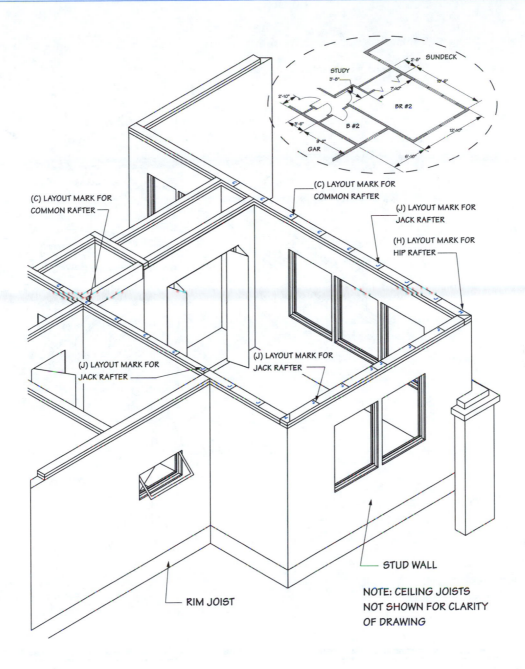

■ **Figure 10–16.**
The roof rafters and jacks must be laid out before placing the rafters and jacks. This section of layout is for the intersecting hip roof over Bedroom #2 of the **One-Story-Residence.**

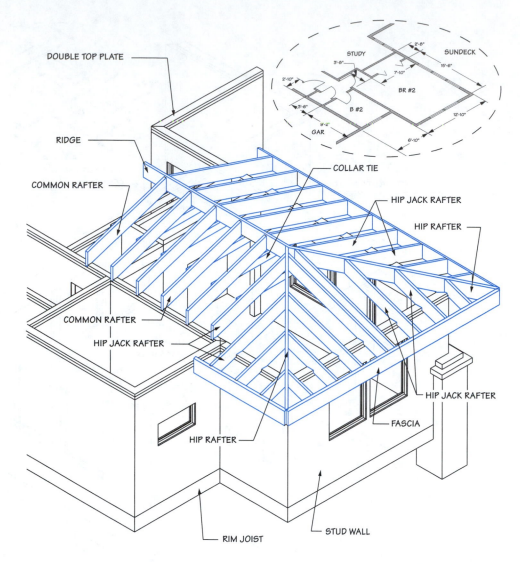

■ Figure 10–17.
This section of roof is constructed according to the layout shown in Figure 10–16.

NOTE: CEILING JOISTS NOT SHOWN FOR CLARITY OF DRAWING

REVIEW QUESTIONS

Enter the missing words in the spaces provided on the right.

1. _____ and load support determine the size of joists used for the ceiling.

   _____

2. If possible, the ends of ceiling joists should be nailed to the heels of roof _____.

   _____

3. The ceiling framing plan will show the framed opening for the _____ access.

   _____

4. Ceiling joists are _____ into the top of the walls and beams

   _____

5. If ceiling joists are laid out 24" OC and the studs below are laid out 16" OC, every ___joist will fall over a stud.

_____

6. A gable roof has a ridge at the center and slopes in _____ directions.

_____

7. A roof sloping in four directions is called a _____roof.

_____

8. The weight of snow and wind pressure are factors affecting ___load of the roof.

_____

9. The slope of the roof is determined by the _____ _____.

_____

10. The number of inches a roof rises for every foot of run is the _____ _____.

_____

11. _____inches is the length of the unit run.

_____

12. Total _____ is the total of height of a roof.

_____

13. The upper ends of roof rafters are nailed into the _____ boards.

_____

14. Hip and valley rafters run at a ___angle from the plate line to the ridge board.

_____

15. Bottom chords, top chords, and web members are the basic components of roof _____.

_____

Match the letters next to symbols with their descriptions.

16. Attic access _____

17. Skylights _____

18. Building section reference _____

19. Section or detail reference _____

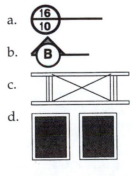

Choose the correct abbreviations for each term.   PL

20. Beam   _____   RS

21. Chimney   _____   LB

22. Ceiling joists   _____   GYP BD

23. Down   _____   CHM

24. Each   _____   GALV SH MTL

25. Galvanized sheet metal _____   GLULAM

26. Gypsum board   _____   EA

27. Glue-laminated   _____   W/

28. Header   _____   HDR

29. Height   _____   HT

30. Overhang   _____   BM

31. Plate   _____   DN

32. Pound   _____   SQ FT

33. Rough surfaced   _____   CLG JSTS

34. Square feet   _____   OH

35. With

UNIT 11

Building Frame Section and Detail Drawings

Section and detail drawings are used with other plans during the entire construction process to provide additional information that cannot be clearly shown on plan view drawings. Such drawings include the floor plan and the framing plans for the floors, ceilings, and roof.

Building section drawings provide a view of the entire length or width of a building, as if a vertical cut had been made through it. Individual wall, roof, and ceiling section drawings may also be provided. A set of plans will frequently contain stair, fireplace, door, window, and other section and detail drawings.

Detail drawings provide an enlarged picture of smaller features that are also a part of a larger plan. They are necessary because the smaller scale of larger drawings does not allow for marking dimensions on these smaller features or providing a clear view of the components within the area of the detail.

READING THE BUILDING SECTION AND DETAIL DRAWINGS

The section symbols that refer to particular section drawings are commonly shown in the foundation plan, floor plan, the various framing plans, and sometimes in the elevation drawings. A standard section symbol consists of a circle joined to an arrowhead. The letters within a circle identify the building section being viewed. The arrowheads point in the direction of the view. Section symbols are placed at opposite sides of the plan view drawings. On some plans opposite section symbols are joined with a visible cutting-plane line running through the drawing. **(See Figure 11–1.)**

The two basic cutting planes are the **longitudinal section,** cut along the length of a building, and the **transverse section,** cut along the width of a building. **(See Figure 11–2.)**

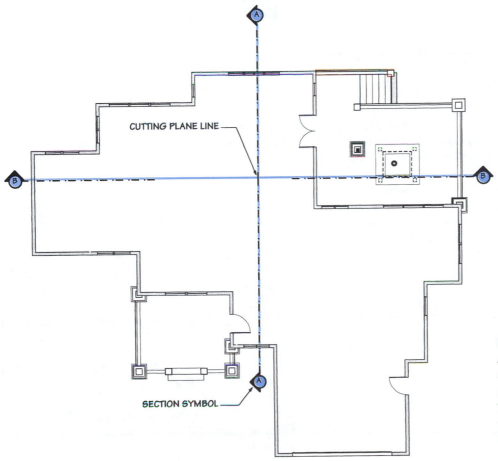

CUTTING PLANE LINE

SECTION SYMBOL

■ **Figure 11–1.**
A standard section symbol consists of a circle and arrowhead pointing in the direction being viewed. Some plans show a cutting-plane line drawn between the symbols.

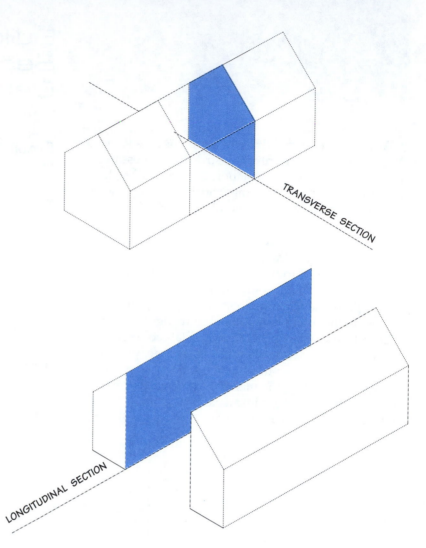

■ **Figure 11–2.**
A longitudinal section line runs along the length of a building. A transverse section line runs across the width of the building.

Building section A-A of the of the **One-Story Residence** plans is a transverse cut along the width of the building. It provides a view looking west and cuts through the front porch, foyer, kitchen, and living room. **(See Figure 11–3.)** Building section B-B is a longitudinal cut along the length of the building. It provides a view looking north and cuts through the master bathroom, master bedroom, living room, and the rear sun deck. Symbols and abbreviations appearing in the building, stair, and roof section and detail drawings are shown in **Figure 11–4.**

Building section A-A, along with a corresponding pictorial drawing, is shown in **Figure 11–5.** Building section B-B, along with a corresponding pictorial drawing, is shown in **Figure 11–6.** Concentrate on the following when studying the drawings:

✓ Symbols and abbreviations.
✓ Foundation and pier sections.
✓ Floor levels.
✓ Floor-to-ceiling heights.
✓ Wall locations.

✓ Window locations.
✓ Ceiling joists.
✓ Roof components.
✓ Placement of floor, wall, and ceiling insulation.

Explanations of the building section features in Figure 11–5 follow:

1. **South foundation wall.** The drawing shows a section view of a foundation at the south side of the building. It runs from east to west beneath the front wall of the foyer.
2. **Piers.** Three piers supporting the floor girder can be viewed looking west from the cutting-plane line A-A.
3. **North foundation wall.** The drawing shows a section view of the foundation running east to west beneath the north living room wall.
4. **Floor girder.** A 6 × 8 floor girder is identified. It helps to support the joists above.
5. **Joists.** End views of the floor joists resting on the girder are shown.

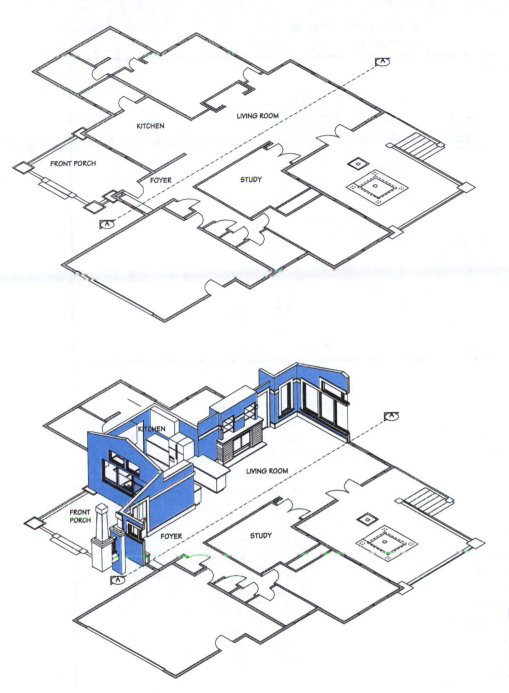

■ **Figure 11–3.**
A building section drawing must be related to its proper position on the floor plan.

6. **Porch column**. The column shown here supports the southeast corner of the roof over the front porch.

7. **Foyer.** The entrance panel door opens from the front porch into the foyer. Note the slant of the ceiling toward the front of the foyer. To the left of the door is a side view of a window bench. A stained 1 × 6 trim piece is nailed to the wall. This trim piece can also be seen in sections of the living room wall area.

8. **Kitchen.** This space identifies the open area leading into the kitchen (beyond). A side view of

kitchen base cabinets can be seen when looking in that direction.

9. **Living room.** From left to right is the fireplace with open shelves above, the door leading into the master bedroom, and a window at the right of the door.

10. **Foyer ceiling.** A flat ceiling is shown over the foyer area. The 2 × 6 ceiling joists are spaced 24" OC. Gypsum board is to be nailed to the bottom of the joists.

11. **Living room ceiling.** The sloped ceiling over the living room is framed with 2 × 12 rafters. Every

SYMBOLS AND ABBREVIATIONS
BUILDING SECTIONS AND DETAILS

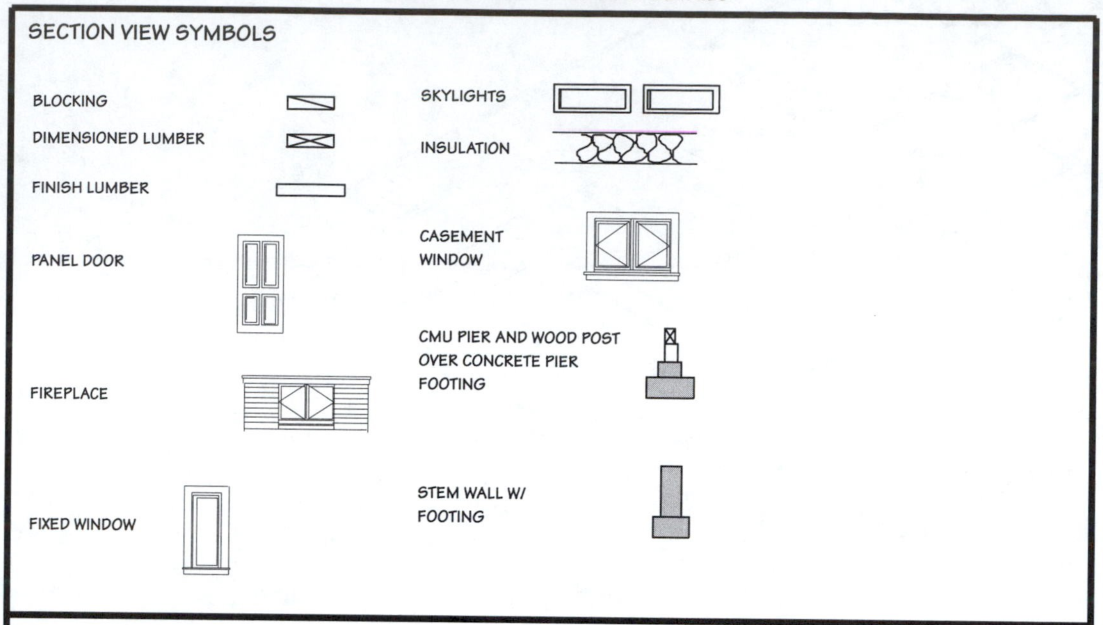

SECTION VIEW SYMBOLS

BLOCKING		SKYLIGHTS	
DIMENSIONED LUMBER		INSULATION	
FINISH LUMBER			
PANEL DOOR		CASEMENT WINDOW	
FIREPLACE		CMU PIER AND WOOD POST OVER CONCRETE PIER FOOTING	
FIXED WINDOW		STEM WALL W/ FOOTING	

ABBREVIATIONS

@	At		JSTS	Joists
BM	Beam		LR	Living Room
BLKG	Blocking		LS	Lazy Susan
BTM	Bottom		MA B	Master Bathroom
CLG	Ceiling		MA BR	Master Bedroom
DBL	Double		MANU	Manufactured
DIA	Diameter		MIL	Millimeter
DR	Door		PL	Plate
EA	Each		PLYWD	Plywood
EXT	Exterior		POLY	Polyethelene
FLR JSTS	Floor joists		S4S	Surface Four Sides
GA	Gauge		SIM	Similar
GALV	Galvanized		SQ FT	Square Feet
GYP BD	Gypsum Board		TYP	Typical
HDR	Header		WDW	Window
HT	Height		W/	With
INSUL	Insulation			

■ **Figure 11–4.**

These are symbols and abbreviations found in the building section drawings for the **One-Story-Residence.**

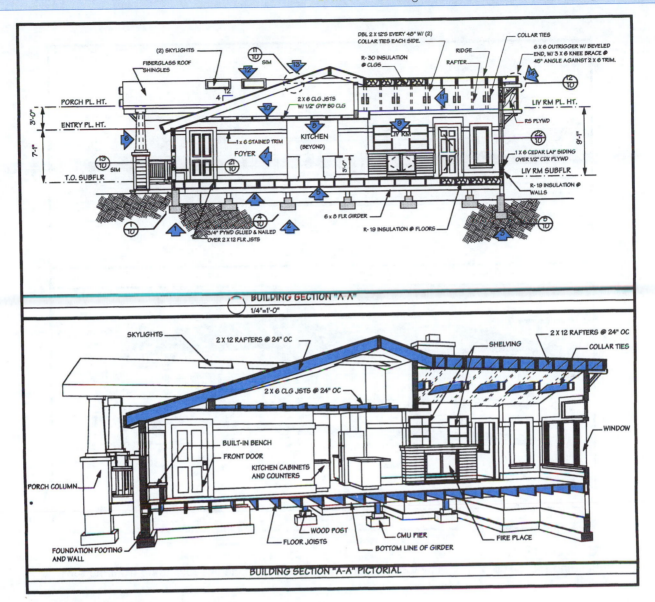

BUILDING SECTION "A-A"
1/4"=1'-0"

BUILDING SECTION "A-A" PICTORIAL

■ **Figure 11–5.**
Building section "A-A" is a view looking west from a cutting-plane line extending from the foyer to the living room.

other rafter is doubled, and collar ties are nailed on each side. The collar ties will remain exposed after gypsum board is nailed to the underside of the rafters.

12. **Skylights.** These two skylights are found in the east slope of the roof section over the front porch and kitchen. An opposite pair is on the west slope.

13. **Collar tie.** This type of single collar tie is placed toward the upper end of the rafter and is visible only in the attic area.

14. **Outrigger.** The outriggers can be seen below the roof ridge and the lower end of one side of the roof.

Explanations of the building section features shown in Figure 11–6 follow:

1. **West foundation wall.** The drawing shows a section of the foundation wall below the west wall of the master bathroom.

2. **Piers.** Two rows of piers supporting the floor girders can be seen north of cutting-plane line B-B. Another pier to the left is identified with dashed (invisible) lines because it is behind the cutting-plane line. (For more clarification, refer to the floor framing plan shown in Figure 9–5.

3. **East foundation wall.** A section view of the foundation wall below the exterior east living room wall is shown. The design of this wall is the same as under the west foundation wall.

4. **Bench piers.** These piers are below the four posts that support the bench on the sun deck.

5. **Plate height.** The rough height of the master bath-

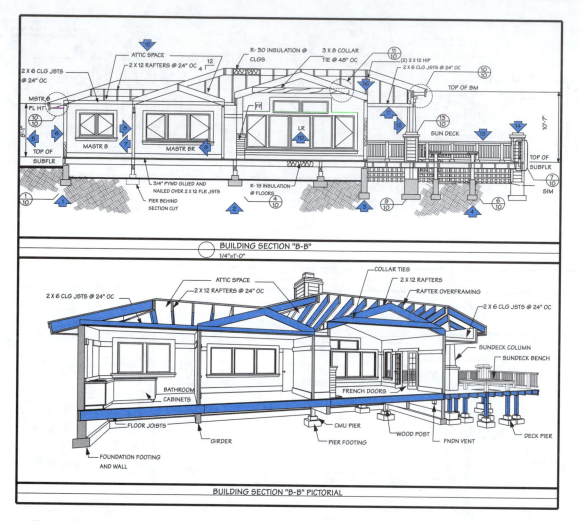

■ **Figure 11–6.**
Building section "B-B" is a view looking north from a cutting-plane line extending from the sun deck to the master bathroom.

room wall is 8′-1″. Gypsum board will be nailed to the ceiling joists, and hardwood flooring will be placed over the subfloor. This will produce the 8′-0″ finish height referred to in the ceiling framing plan.

6. **Exterior wall.** The west exterior wall of the master bathroom will contain batt insulation.

7. **Master bathroom.** Double casement windows can be viewed in the north wall. Ceiling joists are 2 × 6 s, spaced 24″ OC.

8. **Interior partitions.** This wall separates the master bathroom and master bedroom. Another interior wall separates the master bedroom and living room.

9. **Master bedroom.** Two casement windows with a center fixed window are in the north wall. Ceiling joists are 2 × 6s, spaced 24″ OC.

10. **Living room.** A series of five windows (three fixed and two casements) is in the north wall of the liv-

ing room. Two fixed transom windows are above the lower windows. The sloped ceiling shows exposed collar ties. An end view of the fireplace (FP) can be seen to the left of the window.

11. **Sun-deck ceiling joists.** These 2 × 6 joists are below the section of roof extending over part of the sun deck.

12. **Porch column.** The view looking north shows the column that helps support the section of roof projecting over part of the sun deck.

13. **Porch railing.** The posts, balusters, and cap can be seen in this view of the railing

14. **Corner sun-deck post.** The southeast corner of the sun deck is supported by this post. The end of the railing is fastened to the side of the post.

15. **Attic space.** Attic spaces will occur over the flat ceilings.

16. **Porch roof rafters.** These rafters extend from the

hip rafter to the beam supporting the section of roof that projects over the porch.

READING THE STAIR SECTION DRAWINGS

Stair section drawings provide the data needed to construct the stairway. They may be separate drawings or part of a building section drawing. They give the size and number of stringers used to support the stairway. **Stringers** are planks placed on their edges and cut out, or dadoed, to receive the finish treads and risers. The **tread** is the horizontal part of a step. The **riser** is the piece nailed against the vertical surface of a step.

STAIR DESIGN AND LAYOUT

The design of a stairway must conform to local code requirements and be constructed at a safe and comfortable walking angle. The recommended angles of stairways are between 30 and 35 degrees. Preferred riser heights range from 7″ to 7 1/2″. However, many codes allow a range from 6″ to 8 1/2″. Recommended tread widths range from 9″ to 12″.

Riser and tread measurements are called the **unit rise** and **unit run.** They are computed from the **total rise,** the vertical measurement of the stairway from floor to floor, and the **total run,** the horizontal measurement of the stairway. **(See Figure 11–7.)** (You will recall these terms

are also used in relation to roofs.) Additional information for computing riser and tread size is provided in Unit 30. Symbols and abbreviations appearing in the stair section drawing are shown in Figure 11–4, page 130.

A basement stair section drawing, along with a corresponding pictorial drawing, is shown in **Figure 11–8.** Concentrate on the following when studying the drawings.

✓ Symbols and abbreviations.
✓ Total rise and run.
✓ Tread and riser dimensions.
✓ Framing around stairwell opening.
✓ Railing height.

Explanations of the stair section features shown in Figure 11–8 follow:

1. **Footing.** The wall supporting the landing at the head of the stairway requires a footing at its base.
2. **Total rise.** The total distance from the basement floor to the main floor is 9′-0″.
3. **Total run.** The total horizontal distance of the stairway is 12′-10″.
4. **Head room.** This is the vertical distance (6′-8″) between the nosing of a stair tread and any part of the ceiling surface above the stairway. Nosing is the edge of a stair tread that projects beyond the face of the riser below.
5. **Stair stringer.** Also called *carriages,* the stringers

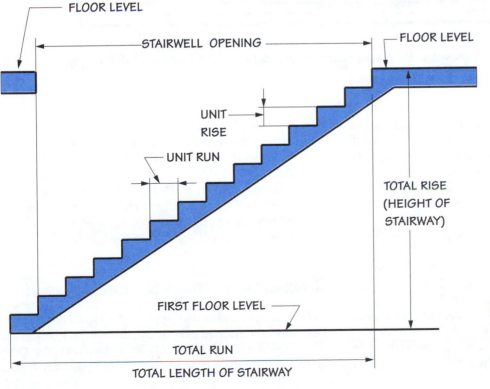

■ Figure 11–7.
Unit rise and run determine the size of the steps. These measurements are derived from the total rise and run.

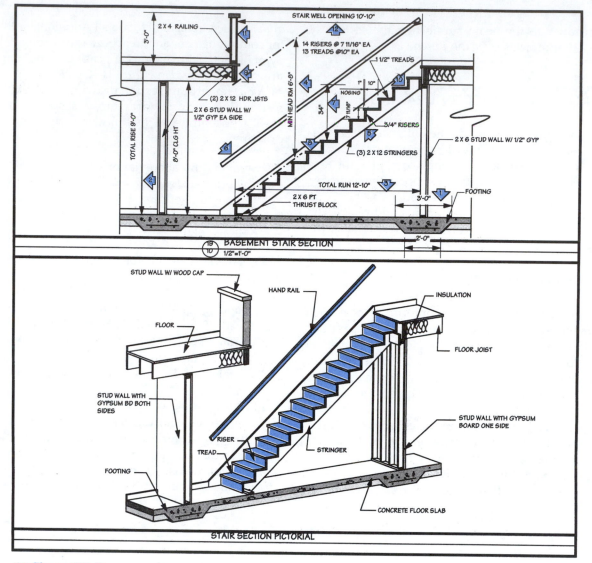

■ **Figure 11–8.**
The stair section drawing provides dimensions and structural details of the stairway and stairway opening.

are the main support of the stairway. They are either cut out or dadoed to receive the finish treads and risers. A **dado** is a rectangular groove cut into a board so that a like dimensioned piece can fit into it. The drawing indicates three 2 × 12 stringers, one on each side of the stairway and one in the middle.

6. **Railing.** The stair railing runs parallel with the stairway.

7. **Railing height.** The vertical distance between a tread and the top of the railing is 34".

8. **Tread.** The finish tread material is 1-1/2" thick and 10" wide. A nosing projects 1" past the riser material below. This creates a 11" walking surface.

9. **Double header.** The joists butting against the double header on this side of the stairwell opening are

fastened with joist hangers.

10. **Riser.** The finish riser material is 3/4" thick. The riser height from a tread surface to tread surface is 7.71" or 7 11/16".

11. **Stairwell Railing.** The 2 × 4 railing around the north and east side of the stairway is 3'-0" high.

12. **Stairwell opening.** The length of the stairwell opening framed in the floor unit is 10'-10".

READING THE FIREPLACE SECTION DRAWINGS

A fireplace section drawing provides details of the fireplace and the framework for the fireplace and chimney. The fireplace section drawing included with the **One-Story Residence** plans shows a **zero clearance fire-**

place, a prefabricated insulated metal unit and flue that can be placed close to the framework around the fireplace opening. The hearth rests directly on the floor. Symbols and abbreviations appearing in the fireplace section drawing are shown in Figure 11–4, page 130.

A fireplace section drawing, along with a corresponding pictorial drawing, is shown in **Figure 11–9.** Concentrate on the following when studying the drawing:

✓ Symbols and abbreviations.
✓ Fireplace dimensions.
✓ Flue placement.
✓ Frame components.
✓ Top of chimney details.

Explanations of the fireplace section features shown in Figure 11–9 follow:

1. **Pier, post, and beam.** A short post rests on a con-

crete-block pier over a concrete footing. The bottom of the post is bolted to a metal-post base imbedded in the concrete block. The post supports a 6 × 8 floor girder.

2. **Floor joist.** A side view of a floor joist is shown. The rear wall behind the fireplace unit is a bearing wall. Therefore, double blocking is placed between the joists under the fireplace. The front fireplace wall is nonbearing, so a single row of blocking is adequate.

3. **Hearth.** Ledgestone masonry veneer is used for the hearth material at the base of the fireplace opening.

4. **Zero-clearance fireplace.** Glass doors are placed at the opening. A gas outlet is provided for fire lighting convenience. An inlet for outside air combustion is shown.

5. **Fireplace header.** A double header supports the framing above the fireplace opening.

6. **Double wall pipe.** The flue pipe is doubled to cut down on heat radiation and danger of fire.

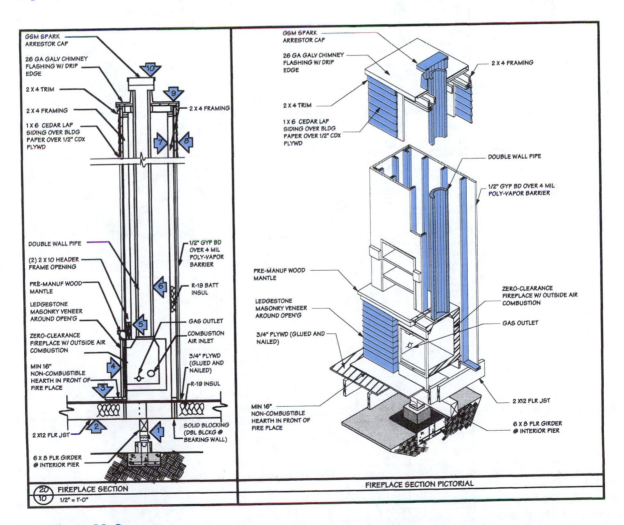

■ **Figure 11–9.**
The fireplace section drawing provides dimensions and structural details of the framing around the fireplace unit. Information about the fireplace itself is also included.

7. **Stud.** The frame around the fireplace is constructed of 2 × 4 studs.

8. **Chimney exterior.** The outside surface of the chimney is finished off with lap siding over building paper and plywood.

9. **Flashing.** Metal flashing is required to prevent water leakage at the shoulder of the chimney.

10. **Metal spark arrester cap.** This helps reduce flying sparks. It also prevents rainwater and snow from entering the chimney.

READING THE DOOR AND WINDOW SECTION DRAWINGS

Door and window section drawings provide additional information about the doors and windows and also their framed openings. Window section drawings combine a horizontal and vertical cut taken through the window and the window frame. They show the type of window and the window sash and glass. The materials used for the top jamb, side jamb, finish sill, and trim around the opening are also identified. Door details show the door, the door frame, and trim around the opening. Door sills are shown beneath exterior doors. The top and bottom drawings of a window and door section show a side view. The center drawing is a view looking down at the window or door. Symbols and abbreviations appearing in the door and window section drawings are shown in Figure 11–4.

Door and window detail drawings are shown in **Figure 11–10.** Corresponding pictorial drawings are placed to the right of each detail drawing. Concentrate on the following when studying the drawings.

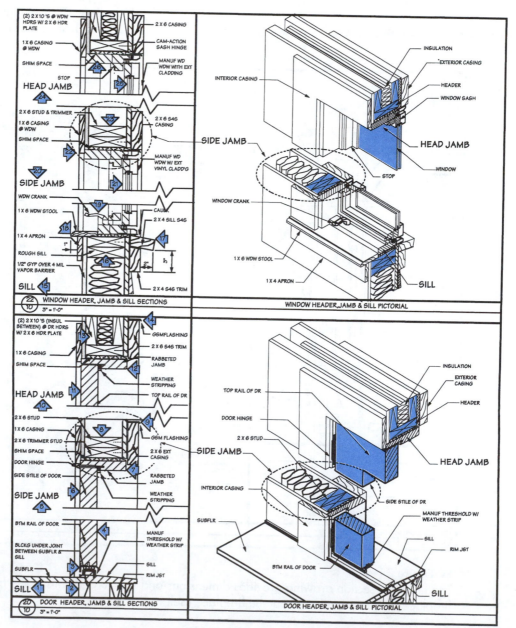

■ **Figure 11–10.**
The door and window details describe the components of the doors, windows, and also the frames to which they are attached.

✓ Symbols and abbreviations.
✓ Structural members at top and bottom of openings.
✓ Finish materials.
✓ Dimensions.

Explanations of the door and window detail features shown in Figure 11–10 follow:

1. **Sills.** The term *sill* is used here to identify the lower drawing. It provides a *side* view of the bottom section of the door.

2. **Sill and floor support.** Blocking is placed beneath the subfloor and sill joint. The outside edge of the sill is supported by the rim joist.

3. **Thresholds and weatherstripping.** Air leakage beneath the door is prevented by an interlocking type threshold.

4. **Bottom rail.** The bottom rail of a panel door is the bottom section of the frame.

5. **Side jamb.** The term *side jamb* is used here to identify the middle drawing. It provides a view looking *down* on the door and the side door jamb.

6. **Stiles.** The stile of a panel door is a side section of the door frame.

7. **Rabbeted jamb.** The jamb is the finish frame around the door. The **rabbet** is a recessed groove, and the door is hinged in this part of the jamb. Weatherstripping is placed between the door and the edge of the rabbet.

8. **Stud and door trimmer.** The side members of a door rough opening are the studs and trimmers. The door trimmer is nailed to a stud. The main purpose of the trimmer is to support the ends of the door header over the rough opening.

9. **Side flashing.** Metal flashing is placed along the edge of the side casing to prevent water leakage. Flashing is installed before the siding is nailed in place.

10. **Head jamb.** The term *head jamb* is used here to identify the top drawing. It provides a *side* view detail of the top section of the door.

11. **Top rail.** The top rail of a panel door is the top section of the door frame.

12. **Rabbeted jamb.** A side view of the rabbeted jamb is shown. Weatherstripping is placed between the door and the edge of the rabbet.

13. **Header.** The header assembly consists of a 2 × 6

nailed to the bottoms of two 2 × 10s. Insulation has been placed between the 2 × 10s.

14. **Top flashing.** Metal flashing is placed at the top of the exterior casing material before the siding is nailed in place.

15. **Sill.** The term *sill* is used here to identify the lower drawing. It provides a *side* view of the bottom section of the window.

16. **Rough window sill.** Double 2 × 6s are shown for the rough sill that frames the bottom of the rough opening. Bottom window cripples will be nailed below the rough sill.

17. **Finished exterior window sill.** The finished window sill at the outside surface of the window rests on a 2 × 4 trim piece. Caulking is required between the finished sill and the bottom of the window frame.

18. **Window stool.** This is a flat trim piece inside the window. A 1 × 4 apron butts up against the bottom of the stool.

19. **Window crank.** The casement window can be opened or closed by turning the crank.

20. **Side jamb.** The term *side jamb* is used here to identify the middle drawing. It is a view looking *down* on one side of the window and finish window frame.

21. **Window panes.** Double glass panes are used for increased insulation. The *sash* (frame) that holds the glass is 1 1/2″ thick and covered (*clad*) with a vinyl material at the weather side of the window.

22. **Side window frame.** This view shows exterior and interior casing nailed on both sides of the frame.

23. **Stud and window trimmer.** The side of a rough window opening is framed with a stud and trimmer. The trimmer is nailed to the stud, and its main purpose is to support the ends of the header over the rough opening.

24. **Head jamb.** The term *head jamb* is used here to identify the top drawing. It is a *side* view of the top of the window and window frame.

25. **Window panes.** Doubled window panes are identified.

26. **Top of window frame.** The top of the casement window is hinged to the top of the frame with cam-action sash hinges. These hinges are rabbeted into the top and bottom of the window sash.

REVIEW QUESTIONS

Enter the missing words in the spaces provided on the right.

1. A cut through the entire length or width of a house is shown by _____section drawings.

   _____

2. Enlarged pictures of smaller features of a larger plan are shown by _____drawings.

   _____

3. A _____joined to an arrowhead is a standard section symbol.

   _____

4. A cut along the length of a building is called a _____section.

   _____

5. A transverse section is a cut along the _____of the building.

   _____

6. Routed or cut planks placed on edge to support a stairway are called _____

   _____

7. ____ to ____ degrees is the recommended angle for a stairway.

   _____

8. Preferred riser heights are from ____ to _____inches.

   _____

9. The total measurement from floor to floor of a stairway is the ____ ____.

   _____

10. The total rise is the _____ measurement of a stairway.

    _____

11. The total run is the _____measurement of astairway.

    _____

12. The framework for the chimney is shown in the _____section drawings.

    _____

13. The _____of a zero-clearance fireplace rests directly on the floor.

    _____

14. The top and bottom of the window and door section drawings provide a _____view.

    _____

15. A view looking down is shown in the _____door and window section drawings.

    _____

Match the letters next to symbols with their descriptions.

16. CMU and wood post _____

17. Fixed window _____

18. Insulation _____

19. Fireplace _____

20. Stem wall w/ footing _____

21. Blocking _____

22. Panel door _____

23. Casement window _____

24. Finish lumber _____

25. Skylights _____

26. Dimensioned lumber _____

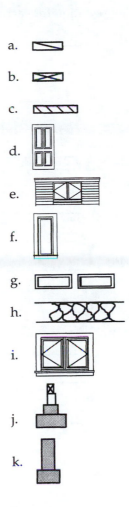

a.

b.

c.

d.

e.

f.

g.

h.

i.

j.

k.

Choose the correct abbreviation for each term.

27. At _____ PL

28. Beam _____ HT

29. Blocking _____ DIA

30. Bottom _____ EXT

31. Ceiling _____ INSUL

32. Double _____ DBL

33. Diameter _____ BLKG

34. Door _____ MANU

35. Each _____ BTM

36. Exterior _____ GA

37. Floor joists _____ TYP

38. Gauge _____ CLG

39. Galvanized _____ FLR JSTS

40. Gypsum board _____ SIM

41. Header _____ GYP BD

42. Height _____ POLY

43. Insulation _____ MIL

44. Joists _____ WDW

45. Living room _____ MA BR

46. Master bathroom _____ @

47. Master bedroom _____ BM

48. Manufactured _____ LR

49. Millimeter _____ W/

50. Plate _____ JSTS

51. Plywood _____ MA B

52. Polyethylene _____ DR

53. Surfaced 4 sides _____ PLYWD

54. Similar _____ S4S

55. Typical _____ GALV

56. With _____ EA

57 Window _____ HDR

UNIT 12

Elevation Plans

An **elevation plan** provides a view of one side of an object. A set of plans will include exterior elevation drawings of all the outside walls, and interior elevation drawings of many inside walls. Since elevation plans are flat orthographic drawings, any projecting surfaces on a wall are not clear on the drawing. However, projecting surfaces can easily be clarified by relating the elevation drawing to the floor plan.

READING THE EXTERIOR ELEVATION PLANS

Exterior elevations are drawings of the outside surfaces of the exterior walls. Separate elevation drawings for each side of the building are included in a set of plans. The drawings show vertical building dimensions and identify floor levels. They also name the finish materials used on the outside surfaces of the walls. Some common exterior finishes include a variety of board and panel products, masonry, stucco, and aluminum and vinyl siding. (Additional information about exterior wall finish materials is provided in Units 34, 35, and 36.)

Doors and windows are shown on exterior elevations, and the trim (finish) materials around the openings are identified. Finished corner boards are shown as well.

An exterior elevation will also include the roof section above the wall. The roof slope is given by its unit rise. The finished roof covering is identified. Commonly used coverings are asphalt composition shingles, wood shingles and shakes, and clay tiles. In addition, the roof area of an elevation drawing will describe the type and placements of vents for air circulation in the attic. The gutters fastened at the roof overhangs are shown, along with the downspouts that drain water from the gutters.

Elevation drawings are usually identified by the compass direction of the building surface they represent: *north elevation, south elevation, east elevation,* and *west elevation.* Each elevation is a drawing of the side being viewed, and visualizing this view in its actual compass position is necessary. For example, if the front wall of a building is identified as the south elevation, the view is the same as shown on the drawing. However, the north wall elevation drawing must be reversed to conform to the floor plan. **(See Figure 12–1.)** Symbols and abbreviations appearing in the exterior

and interior elevation plans are shown in **Figure 12–2.**

The exterior south elevation and west elevation drawings for the **One-Story Residence,** along with a corresponding pictorial drawing, are shown in **Figure 12–3.** The exterior north and east elevation drawings, along with a corresponding pictorial drawing, are shown in **Figure 12–4.** Concentrate on the following when studying the plans:

- ✓ Symbols and abbreviations.
- ✓ Relationship of elevation to floor plan.
- ✓ Vertical dimensions.
- ✓ Exterior wall finishes and trim.
- ✓ Roof pitch and design.
- ✓ Roof finish covering and flashing.
- ✓ Gutters and downspouts.

Explanations of the south and west elevation features in Figure 12–3 follow:

1. **Lap siding.** The siding material is beveled cedar 1" thick and 6" wide. It will have a stained finish.
2. **Column corners.** Surfaced-four-sides (S4S) 2 × 4s butt against each other to finish off the corners of the lower section of the front porch columns. The top section is tapered from bottom to top, and an outrigger is placed under the roof.
3. **Stairs.** Two steps lead up to the front porch. Newel posts are on each side of the lower step.
4. **Porch railing.** The main construction of the railing consists of a 2 × 6 redwood cap (top piece) supported by vertical 2 × 4 posts and 2 × 2 balusters.
5. **Garage door.** The size, design, and finish of the overhead garage door is given.
6. **Subfloor.** The dashed line identifies the top of the subfloor. All the plate heights are measured from this level.
7. **Plate heights.** The **plate height** is the distance from the subfloor to the top of the framed wall. It is also called the *rough height* of the wall. The vertical dimension lines identified by the arrow give the plate heights of Bedroom #2 and Bathroom #2. The low bathroom plate height of 7'-1" is due to a slope at the east end of the bathroom ceiling that is necessary to carry through the slope of the roof. The plate height of Bedroom #2 can be calculated by adding 2'-0" to the 7'-1" bathroom plate height

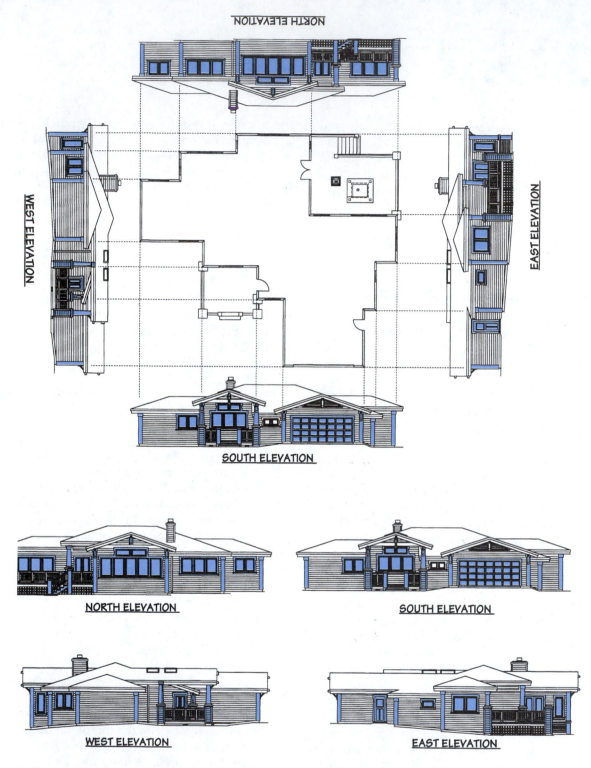

■ **Figure 12–1.**
A set of plans includes exterior elevations for all the outside walls. Note that the north elevation must be reversed when relating the drawing to the floor plan.

SYMBOLS AND ABBREVIATIONS
ELEVATION DRAWINGS

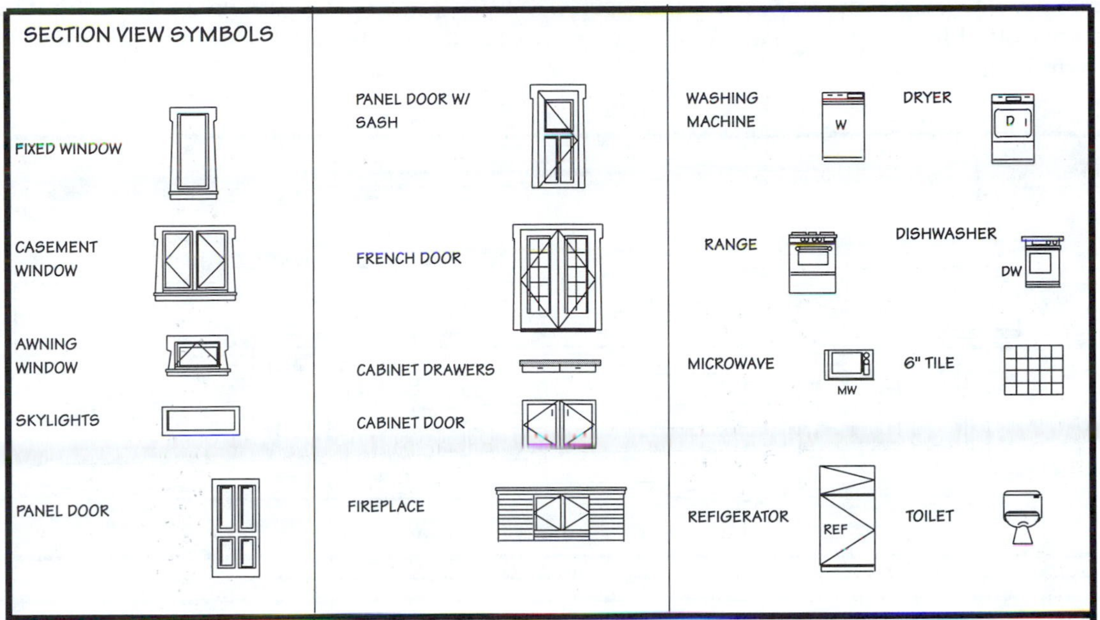

SECTION VIEW SYMBOLS

FIXED WINDOW	PANEL DOOR W/ SASH	WASHING MACHINE DRYER
CASEMENT WINDOW	FRENCH DOOR	RANGE DISHWASHER
AWNING WINDOW	CABINET DRAWERS	MICROWAVE 6" TILE
SKYLIGHTS	CABINET DOOR	
PANEL DOOR	FIREPLACE	REFRIGERATOR TOILET

ABBREVIATIONS

@	At		MAS VEN	Masonry Veneer
ABV	Above		MC	Medicine Cabinet
ADJ	Adjustable		MSTR B	Master Bedroom
BRD BD	Bread Board		MW	Microwave
CNTR	Counter		PL HT	Plate Height
CMU	Concrete Masonry Unit		PREFAB	Prefabricated
D	Dryer		RWD	Redwood
DP	Depth		SFS	Surface Four Sides
DW	Dish Washer		W/	With
CAB	Cabinet		WD	Wood
DRWR	Drawer		WM	Washing Machine
ELEC	Electric			
FLR	Floor			
GA	Gauge			
GALV	Galvanized			
LAM	Laminated			
LR	Living Room			

■ **Figure 12–2.**
These are symbols and abbreviations found in the exterior and interior elevation drawings for the **One-Story Residence.**

$(2'-0'' + 7'-1'' = 9'-1'')$.

8. **Double casement windows.** These windows are in the south wall of the laundry room. The point where the dotted lines meet shows the hinge side of the windows.

9. **Skylights.** The fronts of the skylights over the front porch are identified.

10. **Chimney.** A south view of the chimney is shown.

11. **Roof.** This is an orthographic view of the south roof section.

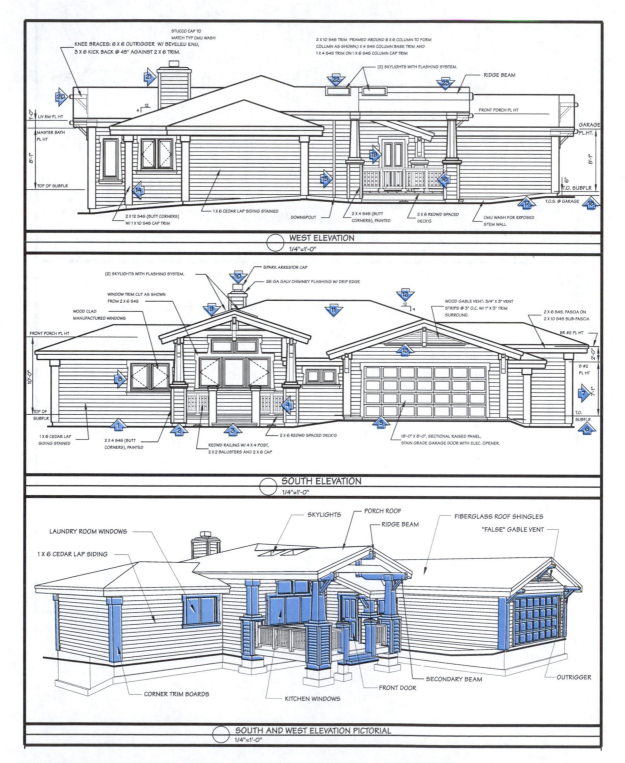

■ **Figure 12–3.**
The south and west exterior elevation drawings provide information regarding finish materials used on those walls. Doors, windows, and railings are shown. Vertical wall dimensions are also identified.

12. **Roof pitch.** This determines the slope of the roof; it rises 4″ for every foot of run.
13. **Gable vent.** This vent helps provide air circulation in the closed attic areas. The 3/4″ thick by 3″ wide vent strips (*louvers*) are slanted and overlap each other. The borders of the vent are finished off with 1 × 3 trim.

14. **Building corner.** All exterior building corners are finished off with butted 2 × 12s. The tops are capped with 1 × 10 trim.
15. **Downspout.** The downspouts drain water from the gutters to the ground.
16. **Railing.** A broader view of the railing is shown here. The 2 × 4 posts are spaced evenly, and the 2

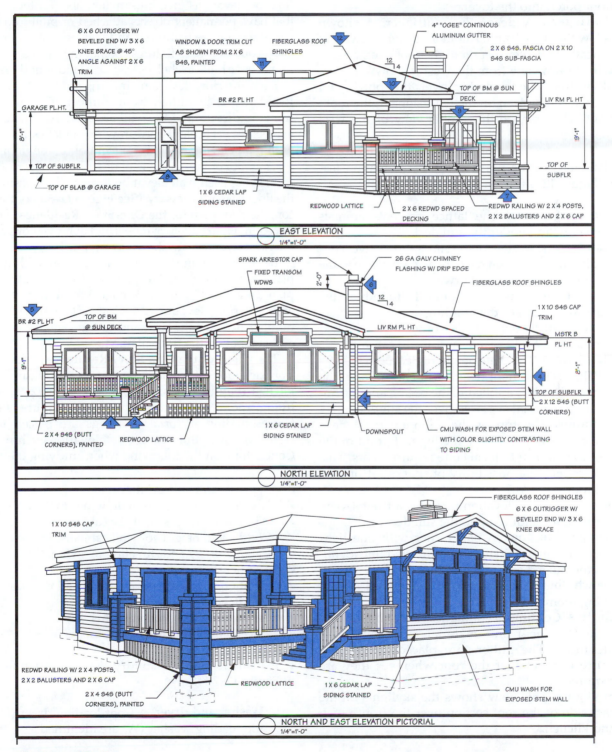

■ **Figure 12–4.**
The north and east exterior elevation drawings show the exterior finish materials used on those walls. Doors, windows, and railings are shown. Vertical wall dimensions are also provided.

× 2 balusters are placed between the posts.

17. **CMU wash.** A stucco-like finish is placed on the surface of the stem wall where it shows above the ground level.

18. **Garage slab level.** The surface of the garage floor is 6" below the subfloor level.

19. **Front door.** A panel entrance door opens from the front porch into the foyer.

20. **Outriggers.** A decorative outrigger is shown below the ridge beam. Another outrigger is visible beneath the lower section of the roof overhang.

21. **Chimney.** A view from the west is shown.

22. **Skylights.** The skylights over the porch and kitchen can be seen on the west slope of the roof.

23. **Roof.** This is an orthographic view of the west roof section.

Explanations of the north and east evaluation features in Figure 12–4 follow:

1. **Redwood lattice.** The lattice prevents animals from crawling under the sun deck.

2. **Stairs.** The side view of the stairs shows five steps leading up to the sun deck. Newel posts are at the top and bottom of the stairway.

3. **Downspout.** This downspout is found at the north side (rear) of the building.

4. **Corner finish:** All outside corners are trimmed with butted 2 × 12 planks and 1 × 10 cap trim.

5. **Plate height.** The plate height of Bedroom #2 is 9'-1". The finished floor height will be 9'-0" after the gypsum board is applied to the ceiling and the finish floor material is placed over the subfloor.

6. **Chimney.** The top of the chimney must be a minimum of 2'-0" above the highest point of the roof to conform to local code requirements. This is to reduce the possibility of sparks settling on the roof.

7. **Stairs.** The east elevation provides a frontal view of the stairs leading up to the sun deck.

8. **Garage side door.** A panel door with one **light** (section of glass) is shown. The dotted line identifies the hinged side.

9. **French doors.** Double glass doors open from the living room to the sun deck.

10. **Gutters.** Continuous rain gutters are shown. The style (4" ogee) and material (aluminum) are identified. The gutters are placed around the entire perimeter of the roof wherever they are required.

11. **Skylights.** This view shows the skylights on the east side of the roof section over the front porch and kitchen.

12. **Finish roof covering.** Fiberglass shingles are to be used on the entire roof.

READING THE INTERIOR ELEVATION PLANS

Interior elevations give information about interior walls that have permanent objects attached or built into them. Usually these are bathroom and kitchen walls with cabinets, counters, and fixtures. Sometimes walls in other rooms have built-in shelves, cabinets, and sound systems.

Each interior elevation drawing for the **One-Story Residence** is identified by a number. The room and compass direction of the wall are also given. The number below each elevation drawing matches a number within an elevation reference symbol on the floor plan. The symbol points to the wall described in the drawing.

Viewing each wall elevation in its proper position on the floor plan is necessary. **(See Figure 12–5.)** For this reason, the floor plan for the **One-Story Residence** is reproduced in **Figure 12–6**. Symbols and abbreviations for the interior elevations are shown in Figure 12–2, page 143.

Before reviewing the interior elevation drawings, place a bookmark at the page that contains the floor plan drawing (Figure 12–6, page 148). While studying each interior elevation drawing, relate the elevation number beneath the drawing to the elevation number and symbol on the floor plan.

The plans for the **One-Story Residence** contain interior elevations for the laundry room **(Figure 12–7)**, kitchen **(Figure 12–8)** bathrooms **(Figure 12–9)**, study, living room and master bedroom **(Figure 12–10)**. Corresponding pictorial drawings are placed to the right of all the exterior elevations drawings. Concentrate on the following when studying the plans:

✓ Symbols and abbreviations.
✓ Relationship of elevation drawings to the floor plan.
✓ Widths and heights of base cabinets.
✓ Widths and heights of wall cabinets.
✓ Cabinet doors and direction of swing.
✓ Cabinet shelves.
✓ Locations of refrigerator, range, and dishwasher in the kitchen.
✓ Cabinet drawers in bathrooms
✓ Placement of mirrors in bathrooms.

Explanation for the laundry room elevation features in Figure 12–7 follow:

1. **Washer and dryer.** A front-loading dryer (D) and top-loading washer (W) are identified.

2. **Base area north wall.** The 2′-4″ wide by 3′-0″ high base cabinet has a plastic laminate counter with a bull-nosed (rounded) edge at the front of the counter. A 4″ splash is against the wall and rests on the counter. Its purpose is to prevent water leakage at the back of the counter. The points of the dotted lines show the hinged sides of the cabinet doors. Base cabinets usually have a recessed **toe space** at the bottoms of the cabinets.

3. **Vertical measurements.** The distance from the floor to the bottom of the open shelves is 4′-10″, with an additional measurement of 2′-2″ to the top of the shelves. The shelves are 12″ deep.

4. **Utility sink.** A 22″ deep by 25″ wide **self-rimming** utility sink is set in the countertop. In this type of sink the rim resting on top of the counter is part of the sink itself.

5. **Drawers.** A series of drawers is placed at one side and at the top of the base cabinets. The short dash at the center of each drawer identifies a drawer pull.

6. **Ceiling height.** The distance from floor to ceiling

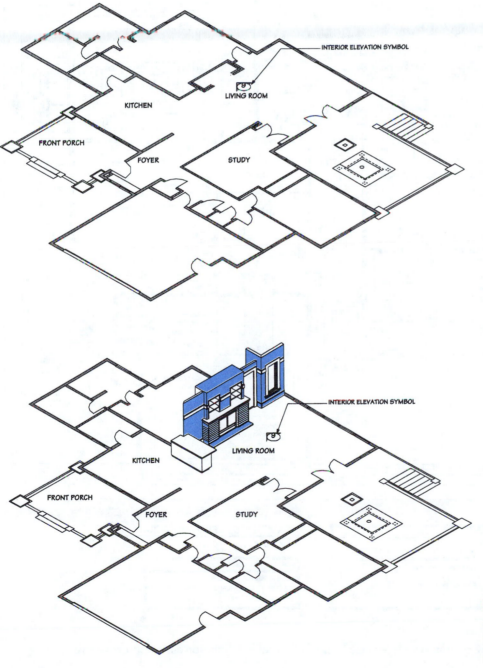

■ **Figure 12–5.**
An interior elevation drawing must be related to its proper position on the floor plan.

in the laundry room is 8'-0".

7. **Windows**. A pair of casement windows open out from the south laundry room wall.

8. **Window trim.** The casing material around the top and sides of the window is 1×6. The stool material is also 1×6, with an apron underneath.

Explanations for the kitchen elevation features in Figure 12–7 follow:

1. **Base area west wall.** The gas range is identified. A base cabinet is to the left of the range, and a row of drawers is at the right.

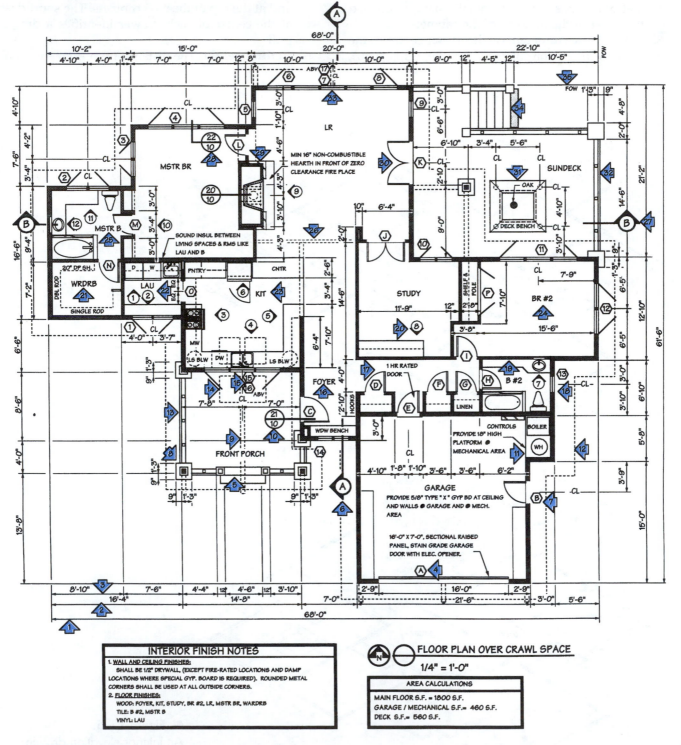

■ **Figure 12–6.**
The floor plan includes many section and reference symbols that refer to interior elevation drawings.

2. **Pocket sliding door.** Roller hangers are attached to the top of the door. The wheels on the hangers fit into an overhead track enabling the door to slide in and out of the pocket.

3. **Ceramic tile.** A section of tiles is visible between the base and wall cabinets of the adjoining wall.

4. **Wall cabinets.** Open shelves are shown next to the wall cabinets. An appliance closet with a rolling door and a microwave (MW) oven is below the cabinets. The wall space above the stove is finished with ceramic tiles.

5. **Soffit.** The purpose of the soffit on the west wall is to close off the space between the top of the wall cabinets and the ceiling. The crossed lines at the right identify a soffit on the north wall.

6. **Base area south wall.** A countertop double sink is at the top of the center base cabinet. The dishwasher (DW) is set below the counter, to the right of the center base cabinet. A row of drawers is to the left. Lazy susans are at the bottoms of the two end cabinets.

7. **Tile.** The countertop and open-wall areas are finished off with 6" × 6" ceramic tiles.

8. **Windows.** Fixed transom windows are located above two swinging casements and a fixed window. The trim around the windows is 1 × 6 material.

9. **Sloped ceiling.** The ceiling of the south kitchen wall is sloping. The notch at the top identifies the placement of a roof beam.

10. **Base area east wall.** Cabinets and drawers are shown. To the left is the open area leading into the foyer. An end view of the north wall counter is shown at the far left.

11. **Valance and counter light.** The valance extends down 3" from the face of the wall cabinet. A fluorescent light is placed behind the valance.

12. **Wall cabinets.** The wall cabinets have adjustable shelves. A row of open shelves is to the right.

13. **Pantry.** The lower pantry has pullout shelves. The upper pantry has shelves on the insides of the doors, and also standard shelves.

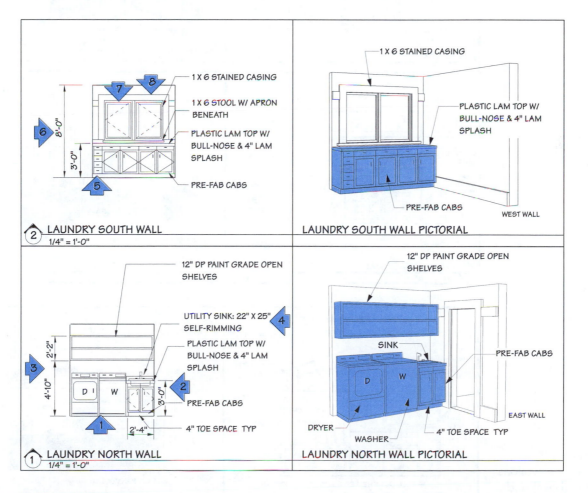

■ Figure 12–7.
Laundryroom interior elevation drawings must be related to its proper position on the floor plan.

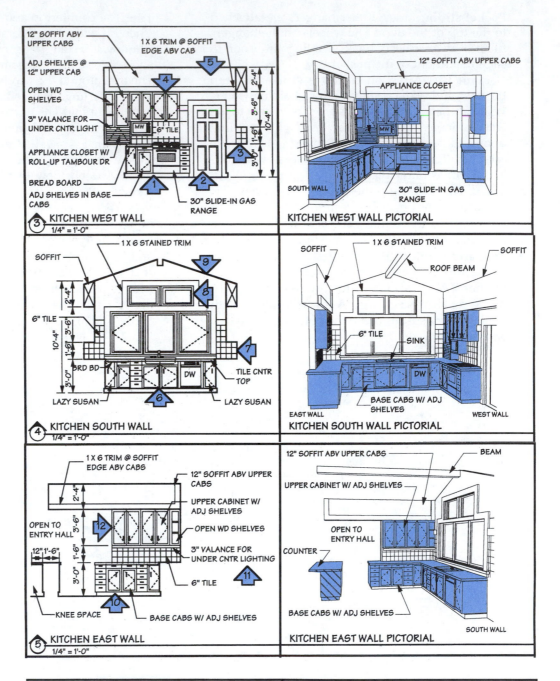

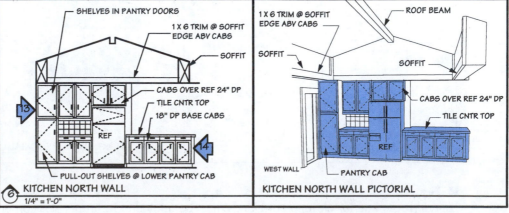

■ **Figure 12–8.**
Kitchen interior elevations show bottom (base) and top cabinets, sinks, refrigerator, and stove.

14. **Base area north wall.** The base cabinets and drawers at the right of the refrigerator are 18″ deep.

Explanations for the bathroom elevation features in Figure 12–9 follow:

1. **Base area north wall of Bathroom #2.** Base cabinets and drawers are shown. A built-in counter sink is over the right cabinet.

2. **Tile counter and splash.** Ceramic tiles are used on the counter and splash.

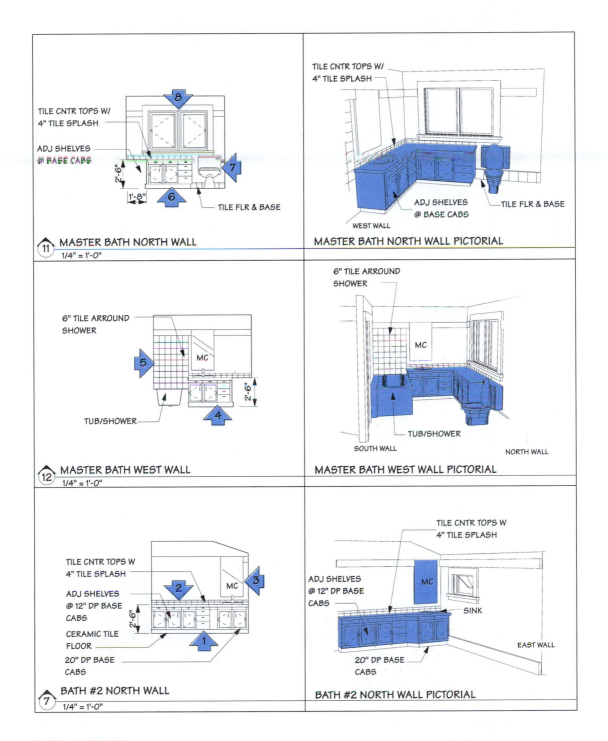

■ **Figure 12–9.**
Bathroom interior elevations identify cabinets and fixtures in the bathrooms. Wall areas receiving ceramic tile are also shown.

3. **Medicine cabinet.** The medicine cabinet (MC) is positioned over the sink and recessed in the wall.
4. **Base area west wall of master bathroom.** A base cabinet with drawers and a built-in counter sink is shown. The surfaces of the counter and splash are finished with ceramic tiles.

5. **Tub and shower.** Wall tiles are installed around three sides of the tub and shower.
6. **Base area north wall of master bathroom.** A base cabinet with drawers is shown. The countertop and splash are finished off with tiles.
7. **Toilet.** The toilet is to the right of the base cabinet.

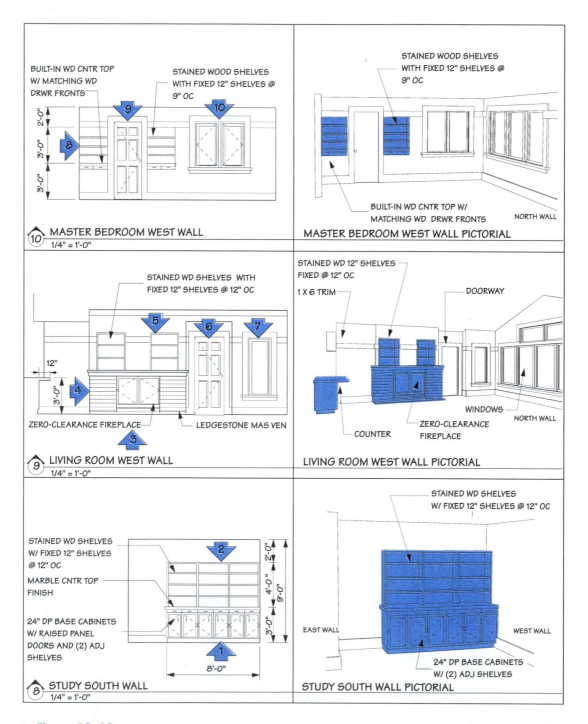

■ **Figure 12–10.**

Study, living room, and master bedroom interior elevations describe special wall features in those rooms.

8. **Windows.** Double casement windows are above the counter.

Explanation for the study, living room, and master bedroom features in Figure 12–9 follow:

1. **Base area south wall of study.** Base cabinets with a marble countertop extend 8'- 0" from the west wall. A row of drawers is at the top of the cabinets.
2. **Bookcase.** An open bookcase is above the cabinets.
3. **Zero-clearance fireplace.** A pair of glass doors is shown at the fireplace opening.
4. **Fireplace facing.** Ledgestone masonry veneer finishes off the surface area at each side of the opening. **Ledgestone** is a design that gives the appearance of rows of flat stone.
5. **Stained shelves.** A pair of open-shelf units is at each side of the mantle.
6. **Door.** A panel door opens to the master bedroom.
7. **Window.** This fixed window is located to the right of the door.
8. **Shelves and counter.** These units, one on each side of the door, are attached to the walls. They consist of open shelves with a pair of drawers at the lower end.
9. **Door.** The panel door opens to the master bathroom.
10. **Windows.** A pair of casement windows is located in the west wall.

REVIEW QUESTIONS

Enter the missing words in the spaces provided on the right.

1. One side of an object is shown in an _____ drawing.

   _____

2. Elevation drawings are orthographic and do not show _____ surfaces.

   _____

3. Elevation drawings should be studied as they relate to the _____ plan.

   _____

4. Each _____ of a building is shown in a set of exterior elevation plans.

   _____

5. The _____ section above the wall is shown in exterior elevations.

   _____

6. The type and placement of _____ for attic air circulation appears on the elevations.

   _____

7. Elevation plans show gutters and _____ that drain water from the gutters.

   _____

8. Exterior elevations are usually identified by their _____ direction.

   _____

9. Walls within a building may be shown as _____ elevations.

   _____

10. Interior elevations are identified on the floor plan by a number in a reference ___.

   _____

Match the symbols with their descriptions.

11. Microwave _____

12. French door _____

13. Panel door _____

14. Range _____

15. 6″ tile _____

16. Fireplace _____

17. Fixed window _____

18. Cabinet drawers _____

19. Toilet _____

20. Refrigerator _____

21. Washing machine _____

22. Casement window _____

23. Dryer _____

24. Skylights _____

25. Cabinet doors _____

26. Awning window _____

27. Dishwasher _____

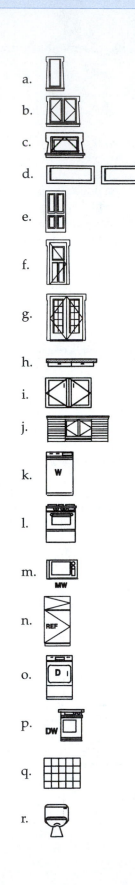

a.

b.

c.

d.

e.

f.

g.

h.

i.

j.

k.

l.

m.

n.

o.

p.

q.

r.

Choose the correct abbreviation for each term.

28.	At	_____	RDWD
29.	Above	_____	ELEC
30.	Adjustable	_____	WM
31.	Bread board	_____	MAS VEN
32.	Counter	_____	WD
33.	Concrete masonry unit	_____	DRWR
34.	Dryer	_____	MW
35.	Depth	_____	CAB
36.	Dishwasher	_____	LR
37.	Cabinet	_____	DW
38.	Drawer	_____	BRD BD
39.	Electric	_____	MC
40.	Floor	_____	CNTR
41.	Gauge	_____	FLR
42.	Galvanized	_____	GALV
43.	Laminated	_____	MSTR BR
44.	Living room	_____	GA
45.	Masonry veneer	_____	@
46.	Medicine cabinet	_____	CMU
47.	Master bedroom	_____	LAM
48.	Microwave	_____	ABV
49.	Plate height	_____	W/
50.	Prefabricated	_____	ADJ
51.	Redwood	_____	PREFAB
52.	Surfaced four sides	_____	PL HT
53.	With	_____	DP
54.	Wood	_____	D
55.	Washing machine	_____	S4S

Combined Electrical-Plumbing-Heating Plan

A combined electrical-plumbing-heating plan is usually adequate for a project the size of the **One- Story Residence**. Larger construction projects require separate mechanical, electrical, plumbing, and heating plans, with additional schematic drawings. (These types of plans and drawings are discussed in Section 3.) A combined-electrical-plumbing-heating plan is basically a floor plan whose main function is to provide information about the electrical, plumbing, and heating systems above the basement level of the building. The electrical and plumbing information for the basement area of the **One-Story Residence** has already been shown in the full-basement foundation plan (Unit 7). Symbols and abbreviations appearing in the mechanical-electrical-plumbing-heating plan are shown in **Figure 13–1.**

ELECTRICAL SYSTEMS

Electric energy for a building comes from aboveground wires extending from power poles or from underground wires encased in cables. Whatever the origin, the wires are connected to a main service panel. The different electrical circuits run from the service panel to different parts of the building, where they supply power to lights, outlets, fixtures, and a variety of appliances. Each circuit has a fuse or circuit breaker in the service panel to protect the circuit from a power overload.

PLUMBING SYSTEMS

Plumbing work consists mainly of the hot- and cold-water pipes and the waste pipes that connect to fixtures and appliances. If gas fuel is required for water and heat, pipes must also be installed for that purpose. (The electrical-plumbing-heating plan for the **One-Story Residence,** however, does not include pipe diagrams.) The plumber learns the locations of all fixtures and appliances by studying the floor plan of the building. It is the plumber's responsibility to install the piping according to local plumbing code requirements and good construction practices..

HEATING SYSTEMS

A **radiant heating system** is used to heat the **One-Story Residence.** Radiant heat is a system for heating space by means of heated surfaces. It is more expensive to install than some heating systems. However, it is very efficient and cost effective to operate, and its use continues to increase in new construction.

The radiant heating system in the **One-Story Residence** consists of thermoplastic tubing laid over the floor area. The tubing is then covered and encased in a 1 1/4" layer of a flowable gypsum-type material. **(See Figure 13–2.)** Finish floor covering such as wood, ceramic tiles, or carpeting can later be fastened to the surface of the gypsum. The heat is created by warm water flowing through the tubing A control system makes it possible to regulate heat to individual rooms.

READING THE ELECTRICAL-PLUMBING-HEATING PLAN

The electrical-plumbing-heating plan for the **One-Story Residence** is shown in **Figure 13–3.** A pictorial drawing of a section of the plan is shown in **Figure 13–4.** Concentrate on the following when studying the plan.

- ✓ Symbols and abbreviations.
- ✓ Locations of wall plugs.
- ✓ Locations of light switches.
- ✓ Locations and types of lighting.
- ✓ Relation between switches and light systems.
- ✓ Elements of heating system and heating zones.

Explanations of the electrical-plumbing-heating plan features in Figure 13–3 follow:

1. **Exterior down light.** The symbols R in the circle and F in the triangle identify recessed exterior lights pointing down toward the ground. These lights are recessed in the roof overhang around the garage.
2. **Overhead garage door opener.** This symbol identifies the motor that operates the garage door opener and the light. A power plug is found to the

rear of the opener. A wiring line leads to the push-button that operates the opener.

3. **Service panel.** Electricity from the outside cable power source is hooked up to this panel. The electrical circuits for the entire house will then originate from the panel.

4. **Exterior duplex outlet.** This receptacle can hold two electric plugs at once. GFI is the abbreviation for **ground fault interrupt.** It is a safety device to protect against the possibility of electric shock, particularly in damp areas. WP is the abbreviation for waterproof, a requirement for any exterior outlet.

5. **Interior duplex outlet.** Also called a **convenience outlet,** this is the most frequently used type of outlet in the building.

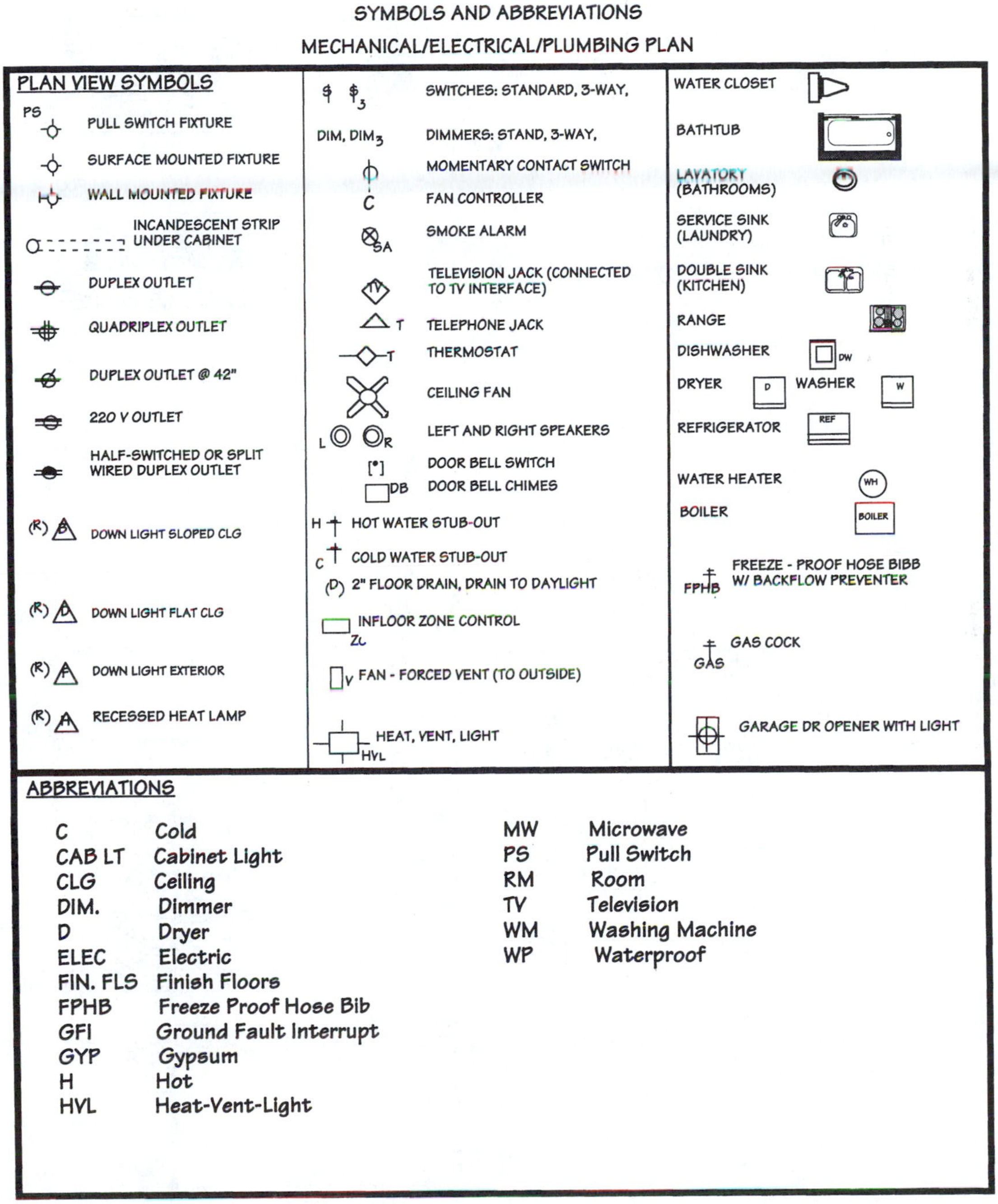

SYMBOLS AND ABBREVIATIONS
MECHANICAL/ELECTRICAL/PLUMBING PLAN

PLAN VIEW SYMBOLS

PS — PULL SWITCH FIXTURE	SWITCHES: STANDARD, 3-WAY,	WATER CLOSET
SURFACE MOUNTED FIXTURE	DIM, DIM₃ DIMMERS: STAND, 3-WAY,	BATHTUB
WALL MOUNTED FIXTURE	MOMENTARY CONTACT SWITCH	LAVATORY (BATHROOMS)
INCANDESCENT STRIP UNDER CABINET	FAN CONTROLLER	SERVICE SINK (LAUNDRY)
DUPLEX OUTLET	SMOKE ALARM	DOUBLE SINK (KITCHEN)
QUADRIPLEX OUTLET	TELEVISION JACK (CONNECTED TO TV INTERFACE)	RANGE
DUPLEX OUTLET @ 42"	TELEPHONE JACK	DISHWASHER
220 V OUTLET	THERMOSTAT	DRYER WASHER
HALF-SWITCHED OR SPLIT WIRED DUPLEX OUTLET	CEILING FAN	REFRIGERATOR
	LEFT AND RIGHT SPEAKERS	WATER HEATER
DOWN LIGHT SLOPED CLG	DOOR BELL SWITCH / DOOR BELL CHIMES	BOILER
DOWN LIGHT FLAT CLG	HOT WATER STUB-OUT	FREEZE - PROOF HOSE BIBB W/ BACKFLOW PREVENTER
DOWN LIGHT EXTERIOR	COLD WATER STUB-OUT	GAS COCK
RECESSED HEAT LAMP	2" FLOOR DRAIN, DRAIN TO DAYLIGHT	GARAGE DR OPENER WITH LIGHT
	INFLOOR ZONE CONTROL	
	FAN - FORCED VENT (TO OUTSIDE)	
	HEAT, VENT, LIGHT	

ABBREVIATIONS

C	Cold	MW	Microwave
CAB LT	Cabinet Light	PS	Pull Switch
CLG	Ceiling	RM	Room
DIM.	Dimmer	TV	Television
D	Dryer	WM	Washing Machine
ELEC	Electric	WP	Waterproof
FIN. FLS	Finish Floors		
FPHB	Freeze Proof Hose Bib		
GFI	Ground Fault Interrupt		
GYP	Gypsum		
H	Hot		
HVL	Heat-Vent-Light		

■ **Figure 13–1.**

These are symbols and abbreviations found in the electrical-plumbing-heating plan of the **One-Story Residence.**

6. **Wiring.** The line identifies the wire running from the switch or switches to the light outlets.
7. **Surface-mounted fixture.** An overhead light is placed here.
8. **Wall-mounted fixture.** This exterior light is mounted on the wall.
9. **Hot water heater.** Cold water © and gas (GAS)

connections (*stub-outs*) are identified. The boiler for the radiant heating system is next to the hot water heater. Water and gas connections to the boiler are also shown.
10. **Three-way switch.** The overhead lights can be turned on and off with this switch and with a second three-way switch next to the side garage door.

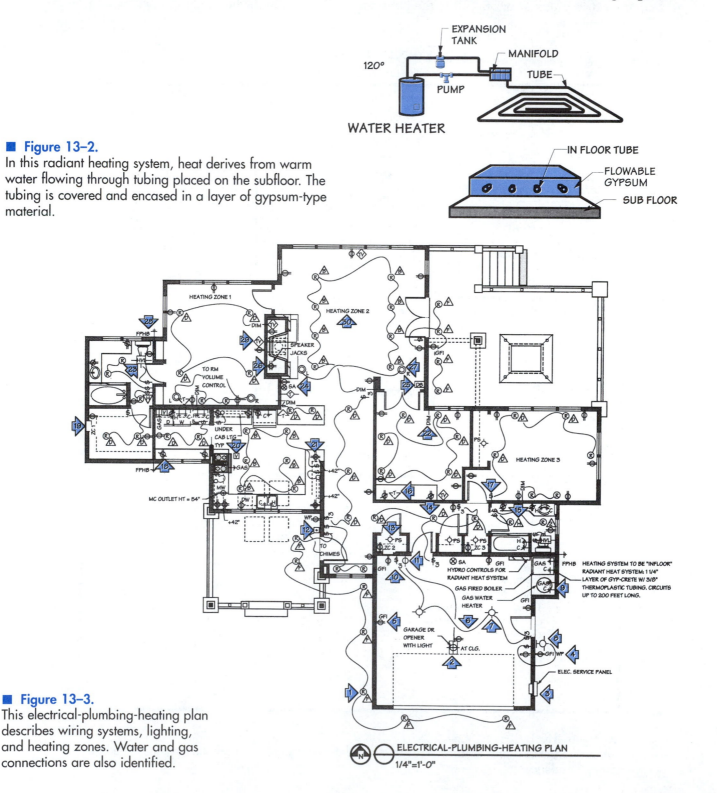

■ **Figure 13–2.**
In this radiant heating system, heat derives from warm water flowing through tubing placed on the subfloor. The tubing is covered and encased in a layer of gypsum-type material.

■ **Figure 13–3.**
This electrical-plumbing-heating plan describes wiring systems, lighting, and heating zones. Water and gas connections are also identified.

11. **Overhead garage door pushbutton.** This button is pressed to open and close the garage door.

12. **Door button.** This door button outside the front-entrance door operates the interior doorbell chimes.

13. **Pull-switch fixture.** A pull chain extends from this closet light to turn the light on and off.

14. **Down light in flat ceiling.** This recessed fixture projects a light from a flat ceiling in the hallway.

15. **Heater.** An electric heating unit is recessed in the ceiling of the bathroom.

16. **Telephone jack.** This outlet provides for a telephone connection.

17. **Light switch.** The overhead lights in Bedroom #2 are operated by this switch. Light switches should always be placed opposite the hinge side of a door.

18. **Duplex outlet.** This outlet is placed above the laundry room counter. The slash mark on the sym-

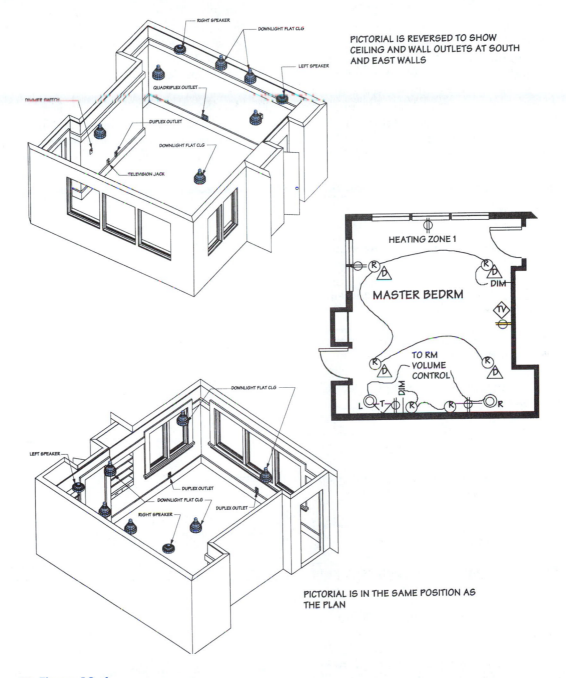

■ **Figure 13–4.**

These pictorial drawings show the wall outlets, ceiling lights, and wall switch for the master bedroom in the electrical-plumbing-heating plan shown in Figure 13–4. Space does not allow a pictorial drawing of the whole plan.

bol shows that the outlet is raised 42″ off the floor.

19. **In-floor zone control.** The radiant heating system in the building is divided into three zones. The zone control (ZC) station allows the heat to be adjusted to different levels of warmth in different areas of the building.

20. **Vent.** Kitchen odors and smoke can be sucked up through this vent. Although the symbol appears to be to one side of the range, the vent is actually directly above the range.

21. **Quadruplex outlet.** This outlet above a kitchen counter can hold four plugs at once.

22. **Dimmer switch.** This type of light switch can lessen or increase the brightness of the overhead lights.

23. **Heat-vent light.** The electrically powered overhead unit in the master bathroom can be controlled to provide heat or pull air out through the vent. The unit also includes an overhead light.

24. **Smoke alarm.** A loud beeping sound is activated by smoke in the air in case of fire. Another smoke alarm is found at the back wall of the garage.

25. **Doorbell chime.** The chimes are activated by the pushbutton at the left of the front-entrance door.

26. **Speaker jack.** Speaker wires will plug into a speaker outlet.

27. **Speaker.** A speaker is pointed out. Another speaker is toward the front of the fireplace. They are identified as right (R) and left (L). A pair of speakers are also located in the Master Bedroom.

28. **Hose bibb.** The symbol identifies a *freeze-proof hose bibb* (FPHB). This is a hose connection for outside watering. Additional hose connections are outside the south laundry room wall and the east wall of Bathroom #2.

29. **Television jack.** A cable connection is provided here.

30. **Heating zone 2.** This is the third of three heating zones that make up the radiant heating system. Each zone can be operated independently by separate zone controls.

REVIEW QUESTIONS

Enter the missing words in the spaces provided on the right.

1. The main _____ panel receives the wires from the electrical source.

   _____

2. Electrical _____ run from the service panel to different parts of the building.

   _____

3. Above-ground power sources supply electricity through wires extending from _____ _____.

   _____

4. _____ or _____ _____ in the service panel protect each circuit from a power overload.

   _____

5. The plumber locates fixtures and appliances by studying the ____ plans.

   _____

6. The local plumbing _____ must be strictly followed when installing plumbing work.

   _____

7. Hot and cold water pipes and ____ pipes make up most of the rough plumbing work.

   _____

8. Radiant heating systems are very efficient and _____ effective.

   _____

9. Thermoplastic tubing is covered with a flowable _____ material in a radiant system.

   _____

10. _____ _____ flowing through thermoplastic tubing provides heat in radiant heating.

    _____

Match the letters next to the symbols with their descriptions.

11.	Freeze-proof hose bib	_____
12.	Ceiling fan	_____
13.	Split-wired duplex outlet	_____
14.	Heat-vent light	_____
15.	Gas cock	_____
16.	Floor drain	_____
17.	Smoke alarm	_____
18.	Duplex outlet	_____
19.	Pull-switch fixture	_____
20.	Garage door opener with light	_____
21.	Left and right speakers	_____
22.	220-volt outlet	_____
23.	Doorbell switch	_____
24.	Television jack	_____
25.	Quadruplex outlet	_____
26.	Doorbell chimes	_____
27.	Incandescent light	_____
28.	Hot water stub-out	_____
29.	Wall-mounted fixture	_____
30.	Cold water stub-out	_____
31.	Telephone jack	_____
32.	Duplex outlet at 42″	_____
33.	Thermostat	_____
34.	Surface-mounted fixture	_____
35.	Dimmer	_____
36.	Standard switch	_____
37.	Three-way switch	_____

a. PS

b.

c.

d.

e.

f.

g.

h.

i.

j. $

k. $₃

l. DIM,

m. ⊗SA

n.

o. △T

p. ◇T

q.

r. L◎ ◎R

s. [•]

t. ☐DB

u. HVL

v.

w. FPHB

x. GAS

y. H✝

z. c✝

aa. (P)

Choose the correct abbreviation for each term.

38.	Cold	_____	RM
39.	Cabinet light	_____	WP
40.	Ceiling	_____	GFI
41.	Dimmer	_____	C
42.	Dryer	_____	CAB LT
43.	Electric	_____	PS
44.	Finish floors	_____	D
45.	Freeze-proof hose bib	_____	DIM
46.	Ground fault interrupt	_____	FIN. FL
47.	Gypsum	_____	FPHB
48.	Hot	_____	H
49.	Heat-vent light	_____	TV
50.	Microwave	_____	ELEC
51.	Pull switch	_____	GYP
52.	Room	_____	CLG
53.	Television	_____	HVL
54.	Washing machine	_____	MW
55.	Waterproof	_____	WM

UNIT 14

Specifications

Specifications are a legal document included with a set of construction plans. They are commonly called *specs* by persons working in the construction industry. Specifications contain important written information that is not noted or fully explained in the drawings. Specifications may also repeat information appearing on the plans. Therefore, the agreement and accuracy of this data are essential.

The specifications serve as a guide for general contractors and subcontractors to bid on the job and explain the contractors' liability during construction. The specifications are also useful in settling disputes that may arise between the owner, architect, contractor, and subcontractors.

FORMATS FOR SPECIFICATIONS

The specifications often list the major building operations as *divisions,* and then subdivide the divisions into *sections.* A division may contain information regarding size, grade, and quality of materials used on the construction site. Brand names may be provided for windows, doors, fixtures, appliances, and many other items. Specifications also set the standards for the work to be done. All materials and operations must conform to the local building codes.

The American Institute of Architects (AIA) with the Construction Specifications Institute (CSI) have developed general guidelines called the **CSI Master Format** for writing specifications. **(See Appendix L for the CSI Master Format.)** However, it should be stressed that the formats for specification and/or notes vary with different architects. Some architects write up the specifications in a separate document as discussed above. Other architects place the same information on one sheet of a set of plans as "General Notes." Some plans only include a "Description of Materials" form such as the one published by the U.S. Department of Housing and Urban Development. Another method employed by architects consists of detailed notes related to major drawings (foundation, floor plan, etc.), appearing on the same sheet with the drawings.

The following specifications are written as a separate document for the **One-Story Residence.** The major areas are divided into *divisions,* and the divisions are broken down into *sections.* Code requirements are governed by the Uniform Building Code (UBC). The text in bold type at the end of each division is not part of the written specifications but is a general commentary about the information in the division as it relates to the **One-Story Residence.**

SPECIFICATIONS FOR THE ONE-STORY RESIDENCE

DIVISION 1: GENERAL CONDITIONS

Scope: This division discusses some of the obligations of the building contractor at the job site.

Section 1. The work encompasses the general construction, including electrical and mechanical work, of a one-story residence as shown in the plans. All work will comply with applicable building codes and local zoning ordinances.

Section 2. The general contractor must verify that he or she has visited and examined the site and is familiar with any and all conditions that might affect the implementation of the contract or any changes that might affect the job bid.

Section 3. If any inconsistencies are found among parts of the drawings, between the drawings and the specifications, or between contract documents and conditions uncovered on the job site, such items should be brought to the immediate attention of the architect.

Section 4. It is the responsibility of the contractor to protect the building and related structures from damage during construction. Any damage that does occur must be repaired to the satisfaction of both the architect and owner.

Section 5. The contractor will furnish and assume the cost of all labor and materials and required tools and equipment used during construction. In addition, the contractor will pay for water, light, heat, power, and transportation related to the building project.

Section 6. The construction will be completed within 90 calender days after written Notice to Proceed has been granted to the contractor.

Section 7. The general contractor guarantees to correct any defects related to the construction work for one year after the completion of construction. This guarantee includes defects resulting from labor, material, shrinkage, or settlement.

Division 1 clarifies some legal responsibilities of the owner, contractor, and architect. These issues are discussed in greater detail in the actual contract between owner and contractor.

DIVISION 2: SITE WORK

Scope: This division includes information on excavation, grading, filling, and all the other factors related to ground preparation for construction.

Section 1. All elevations (heights) are relative to the building design. See site plan for relation of elevations to existing grade.

Section 2. The contractor will verify all grades and top-of-slab elevations and notify the architect of any discrepancies.

Section 3. The contractor will verify locations of all utilities before construction.

Section 4. All workers and subcontractors are to be careful during construction to avoid damage to trees and plants designated to be saved.

Section 5. To the extent possible, all utility trenches will be routed so that they do not encroach into any required setback, except where crossing the front setback to utility hookup locations. The lines should be parallel to the driveway alignment, if practical, to minimize site scars.

Section 6. Earth-formed footings are permissible if the soil is sufficiently firm to prevent caving while the concrete is placed.

Section 7. The backfill is to consist of soil removed during excavation.

Section 8. At the completion of construction, the ground surface is to be returned (as much as possible) to its previous natural condition.

Division 2 relates to the survey and site plans and presents information not adequately described in these plans. Several issues related to groundwork are discussed. Other major concerns are the locations of public utilities such as electrical hookups, water mains, gas lines, and sewer lines. Care must be taken during excavation not to damage any of these lines, and the plumbers and electricians must know where to hook up to the different utility lines.

DIVISION 3: CONCRETE AND MASONRY

Scope: This division includes information on the design and placement of all concrete, concrete masonry units, and steel reinforcements.

Section 1. The concrete mix will conform to the design requirements of the American Concrete Institute (ACI) and the Uniform Building Code (UBC). Compressive-strength requirement is 3000 PSI.

Section 2. Concrete masonry units (CMUs):
  a. *CMUs* will conform to the design set forward in amended Chapter 24 of the UBC.
  b. **CMUs** for foundation walls will be a nominal 8" wide, 8" high, and 16" in length with solid-grout cells.

Section 2. Reinforcing:
  a. *Wall footings* beneath cast-in-place and/or CMU walls are to be reinforced with three #4 rebars running the continuous length of the footing, and #4 rebars 6'-0" on center (OC) across the width of the footing. Bottom hooked vertical rebars extending into the walls above are to be placed 48" OC and to within 3" of the bottom of the footing.
  b. *Pier footings* beneath CMU piers are to be reinforced with two #4 rebars running in two directions. Two #4 vertical rebars placed within 3" of the bottoms of the footings will extend up into the pier blocks. Footings beneath steel pipe columns will contain three #4 rebars, 6" OC, running in two directions.
  c. *Concrete slabs* are to be reinforced with #3 rebars, 24" OC, running in two directions.
  b. *Hooked anchor bolts,* 1/2" × 10" and spaced 48" OC, are to be used to fasten sill plates to the tops of the foundation walls. Anchor bolts placed in CMU walls must be grouted solidly.

Section 2. The *concrete driveway* will be 4" thick and have a brushed aggregate concrete finish.

Section 3. *Bituminous expansion joints* will be provided at the perimeters of all slabs that butt up against foundation walls.

Section 4. *Control joints* will be cut in two directions at the center of the garage slab.

Section 5. *Girder pockets* will be placed in CMU walls with steel anchors for securing girders.

Section 6. *Vent-opening design* is based on 1 square foot per 150 square feet of crawl-space area.

Section 7. *Crawl-space access* is to be 18" × 24" minimum.

Division 3 should be reviewed when studying the foundation plan and section views. It refers to the ACI and UBC sources for design requirements for the concrete footings and the concrete-block foundation walls. This division also elaborates on the size and placement of rebars in the foundation footings, walls, piers, and slabs. Additional information is given about the expansion and control joints, girder pock-

ets, vent openings, and crawl-space access.

DIVISION 4: CARPENTRY

Scope: This division includes information on types and grades of framing materials used and their installation. Interior and exterior finish materials will be discussed in Division 6: Millwork.

Section 1. Materials:
a. *Floor joists* of the main floor are to be 2 × 12 , Douglas fir #2 or better.
b. *Subfloor* is to be 3/4" thick APA-Rated Sturd-I-Floor.
c. *Front porch* and *rear sun-deck joists* are to be 2 × 10s, Douglas fir #2 or better.
d. *Porch decks* are to be 2 × 6s, redwood, spaced 3/16" apart and screwed to the joists.
e. *Exterior walls* are to be 2 × 6s, Douglas fir #2 or better.
f. *Interior walls* to be 2 × 4s' Douglas fir #2 or better. The wall between the master bedroom and laundry room is to be constructed of 2 × 6s to allow for adequate sound insulation.
g. *Ceiling joists* are to be 2 × 6s, Douglas fir #2 or better.
h. *Roof rafters* to be 2 × 12s, Douglas fir #2 or better.

Section 2. Rough framing:
a. For spans over 8'-0", place *solid blocking* between the floor joists at the center of the spans. For longer spans, install solid blocking 8'-0" OC.
b. Double up floor joists below and parallel to interior walls. Place blocking below walls running perpendicular to the joists.
c. *Simpson* joist hangers will be used where shown.
d. The general contractor will verify all rough door and window openings.
e. *Door and window headers* are two 2 × 10s on edges with a flat 2 × 6 nailed to the bottoms of the 2 × 10s.
f. *Glu-lam beams* are to be 24F-V3 DF/DF.
g. *Simpson post caps* are required at all post-and-beam connections.
h. *Shaded areas* on the roof plan show over framed areas.
i. *Manufactured trusses* are to be designed for 60 pounds per square foot of roof load and should have heel heights that match the 2 × 12 rafters.
j. *Simpson H.2.5 anchors* are to be placed at every other truss or rafter and installed according to manufacturer's instructions.

Division 4 contains information concerning the floor plan, building section plans, and also the floor, ceiling, and roof framing plans. Lumber species and grades are identified, as well as the brand name of the metal anchors (Simpson). Some information regarding framing procedures is also included.

DIVISION 5: MILLWORK

Scope: This division includes information on all millwork for doors, windows, interior trim, exterior trim, and porch railings.

Section 1. All interior trim materials are to be appearance-grade Douglas fir unless otherwise noted. Exterior trim is paint-grade pine.

Section 2. Interior trim:
a. *Door and window casings* are to be 1 × 6.
b. *Wall base* is to be 1 × 6.
c. *Interior window sills* are to be 1 × 6 stools with routed-edge treatment over a 1 × 4 apron.

Section 3. Exterior millwork:
a. *Siding* is to be 1 × 6 cedar lap siding.
b. *Window and door casings* are to be 2 × 6, S4S.
c. *Building corners* are to be butted 2 × 12s, S4S, with a 1 × 10 cap trim.
d. *Porch and deck column corners* are to be butted 2 × 4s with a 1 × 6 cap trim.
e. *Railing components* for the front porch and rear sun deck (posts, caps, and balusters) are to be S4S redwood.
f. *Front porch roof* underside is to be finished with 1 × 6 T&G cedar boards nailed to the bottoms of the roof rafters.

Section 4. Windows
a. *Pella* manufactured windows are to be used throughout the building. Refer to the window schedule for dimensions and model numbers.
b. *Tempered glass* is to be used in all patio doors and windows placed less than 18" from the floor or within 12" of any door. This also applies to any window within a shower enclosure or over a tub.
c. The *glazing system* must comply with UBC requirements for Glass and Glazing (Chapter 24).
d. *Vinyl-clad* wood frame windows will be used throughout the structure.
e. *Snap-in grids* used on windows will be placed 12" OC vertically and horizontally.

Section 5. Doors:
a. *Interior doors* (except the mechanical room door) will be six-panel Douglas fir. The panel designs are to be confirmed with the owner.
b. *Front-entrance door* will be four panels, 1 3/4" thick.
c. *Side garage door* will be two panels, 1 3/4" thick.
d. *One-hour fire-rated door* will be used at the entrance to the mechanical area.
e. *Overhead garage door* will be 16'-0" × 7'-0" with raised sectional panels.

f. *French doors* leading from the living room to the sun deck will have wooden rails and glass lights.

Section 6. Cabinets:

a. *Raised panel design* will be a feature of all the cabinets throughout the house.

b. *Countertops and splashes* in the kitchen and bathrooms will be finished off with 6" ceramic tiles.

c. *Countertops and splashes* in the laundry room will be finished off with plastic laminate. Edges should be bull-nosed.

d. *Marble countertop* will be placed over the cabinet in the study.

e. *Built-in wood countertop* will be placed beneath the open shelves in the master bedroom.

f. The owner is to decide the styles and colors concerning all of the above.

Division 5 furnishes information about the millwork shown in the floor plan, building section drawings, and interior and exterior elevation drawings. The term *millwork* refers to ready-made products manufactured at wood-planing mills or door, window, and cabinet assembly plants. The grades and species of inside and outside trim materials are identified. Additional information about door and window design is added to the data presented in the door and window schedules. Cabinet information also includes descriptions of counter material and splashes.

DIVISION 6: INTERIOR SURFACE MATERIALS

Scope: This division includes information on the finish surface covering of the interior walls, ceilings, and floors.

Section 1. Walls and ceilings:

a. *Wall and ceiling surfaces* in the living areas will be 1/2" gypsum board.

b. *Type X 5/8" gypsum* will be applied in fire-rated and damp locations such as the garage and mechanical room.

c. *Rounded metal corners* are to be used at all outside corners.

d. *All joints* will be taped and coated according to the manufacturer's specifications.

Section 2. Floors:

a. *Straight-line oak, select grade,* will be placed in the foyer, kitchen, study, living room, master bedroom, Bedroom #2, and wardrobe room.

b. *Floor tiles* are to be placed in the master bathroom and Bathroom #2. The floor tile will be set on minimum 1/2" wonder board or 1/4" hardy backer and applied with a thin-set mortar. Subflooring must be increased to a 1" minimum thickness.

c. *Vinyl tiles* over a plywood underlayment will be placed in the laundry room.

Division 6 elaborates on the type of surface finish materials shown in the interior and exterior elevation drawings and the section and detail drawings. Some additional information regarding preparation for and placement of these materials is provided.

DIVISION 7: FLASHING, UNDERLAYMENT, AND SHINGLES

Scope: This division includes information on flashing, underlayment, shingles, and other items related to the finish roof work.

Section 1. Valley and ridge flashing will be 26-gauge galvanized iron and extend 10" in each direction from the valley or ridge.

Section 2. Underlayment

a. Before the roof shingles are laid, the roof surface must be covered with #15 bituminous-saturated roofing felt.

b. An additional layer of bituminous material must be placed to cover 24" from the outside edge of the overhang past the inner face of the wall below.

Section 3. Shingles

a. *Black fiberglass shingles* of "Elk Prestique" are to be placed over the roof and installed according to the manufacturer's directions.

b. *Zinc-coated nails* are to be used for fastening the shingles.

Section 4. *Ogee vinyl gutters* are to be used where required.

Section 5. *Roof skylights* are to be Velux FX-304 with 4" flashing system.

Section 6. *Vinyl downspouts* are to be provided where required.

Section 7. All exposed metal, including skylight frames, flashing, vent pipes, a chimney cap, etc., will be painted to blend with adjacent roofing material.

Division 7 provides information regarding the roof areas shown in the exterior elevation plans, building section drawings, and the roof plans. It specifies the type, and application, of flashing, underlayment, and roof shingles. Vinyl gutter materials are identified. The type of roof skylights and the manufacturer's name are also identified.

DIVISION 8: PAINTING AND STAINING

Scope: This division includes information on all interior and exterior finishes. All colors are to be approved by the owner.

Section 1. Interior finishes:

a. A delicate, spray knock-down texture is to be

applied to the gypsum board on the walls and ceilings. The surfaces are to be finished with two coats of paint.

b. Stain is to be applied to all kitchen and bathroom cabinets. Laundry cabinets are to be painted.

Section 2. Exterior finishes:

a. The lap siding is to be stained.

b. All butt corners are to be painted.

c. Trim around doors and windows is to be stained.

d. Exposed stem walls will be mortar-washed and painted with a color slightly contrasting with the siding stain.

Division 8 identifies which surfaces are to receive a stained or painted finish. The colors will be decided after consultation with the owner. The cabinet finishes are also identified.

DIVISION 9: FINISH HARDWARE

Scope: This division includes information on all finish hardware for doors, windows, cabinets, and other related items.

Section 1. Doors:

a. *Exterior swing doors* will have three pairs of 4 1/2″ × 4 1/2″ bronze butt hinges.

b. *Interior swing doors* will have two pairs of 3 1/2″ × 3 1/2″ bronze butt hinges.

c. *Cylinder door locks* for all exterior and interior doors will be Schlage A series with a bronze finish.

Section 2. Windows:

a. *Pella* operators and catches are to be used for casement windows.

b. *Pella* operators and catches are to be used for transom windows.

Section 3. Cabinet hinges, catches, and drawer pulls will be *Amerock* or of equal quality.

Division 9 identifies the type of hardware to be used with the doors, windows, and cabinets. The finish color, series numbers, and manufacturer's name are identified.

DIVISION 10: MOISTURE PROTECTION AND THERMAL INSULATION.

Scope: This division includes information on moisture barriers and thermal and sound insulation.

Section 1. A 4-mil polypropylene vapor barrier is placed against the inside surface of the studs at all exterior walls.

Section 2. Insulation:

a. Provide *R-19 batt or blanket insulation* between the studs in all the exterior walls.

b. Provide *R-19 batt or blanket insulation* between the floor joists of the living area.

c. Provide *R-30 batt or blanket insulation* between the ceiling joists of the flat ceilings and between the rafters at the vaulted ceilings.

d. To control sound, provide *R-ll batt or blanket insulation* between the 2 × 6 studs of the wall separating the master bedroom from the laundry room and kitchen.

Division 10 identifies the type and thickness of the wall vapor barriers. Also identified is the type of insulation placed in the walls and floors and the "R" values of the insulation.

DIVISION 11: HEATING AND HOT WATER

Scope: This division includes information about the heating system and hot water supply.

Section 1. Heating:

a. Heating is to be supplied by an in-floor radiant heating system placed over the subfloor. The manufacturer's installation procedures and specifications are to be followed.

b. The tubing is 3/8″ thermoplastic with circuits up to 200′ long.

c. The tubing is embedded in a 1 1/4″ layer of gyp crete placed over the floor.

d. Water is supplied to the tubing from a gas-fired boiler.

Section 2. Hot water for general use is supplied from a 75-gallon gas-fired LPG conventional water heater.

Division 11 provides some information about the radiant heating system and the type of water heater tank but refers to the manufacturer's instructions for installing the system.

DIVISION 12: ELECTRICAL SYSTEM

Scope: This division includes information on interior wiring, electrical fixtures and appliances, and other items related to the entire electrical system.

Section 1. All electrical installation is to conform with the National Electric Code.

Section 2. The electric service supplied to the building by an underground cable will be 150 amps, 60 cycles, three wires.

Section 3. Wall outlets:

a. *Interior duplex outlets* are to be set 12″ above the floor in a horizontal position.

b. *GFI outlets* are to be installed in the kitchen,

garage, bathrooms, laundry room, and all outdoor locations.

Section 4. Ceiling outlets:

a. *Recessed cans* for ceiling outlets must be IC type throughout for Halo or Juno brand light fixtures.

b. *Recessed lights* at sloped ceilings are to have special housings with tilt trim.

Section 5. *Smoke alarms* will be interconnected.

Section 6. *Security system* power will be provided.

Section 7. Electrical appliances:

a. The owner is to confirm current choices.

b. *UL approval* is required for all appliances.

c. *Dishwasher* will be *Kitchen Aid, Superba Series,* model #KUDS230Y, color white.

d. *Refrigerator/freezer* will be *Amana,* Model #SBD20N, color white.

e. *Washer* will be *Amana,* model #LW 4302 L, color white.

f. *Dryers* will be *Amana,* model #LG 4409 L, color white.

g. *Garbage disposal* will be *Waste King,* model #2600.

Division 12 contains some information about the wiring system and ceiling and plug outlets. The brand names and model numbers of the electrical appliances to be ordered are specified. All appliances must be approved by the Underwriters' Laboratories (UL) Inc. UL Inc. is a nonprofit and nongovernment organization sponsored by the National Board of Fire Underwriters. Its purpose is to inspect and test electrical devices to assure that they conform with the National Electric Code.

DIVISION 13: PLUMBING

Scope: This division includes information on plumbing installation and fixtures.

Section 1. All plumbing must conform to the local plumbing code.

Section 2. Installation:

a. *Water lines* are to be copper, and all sewer lines are to be cast iron.

b. *Vent piping* installed through the roof is to extend not less than 1' above the roof, and #4 lead flashing is to extend 24" in all directions.

c. *Pipe cleanouts* are to be cast iron with brass plugs.

d. *Waste and vent piping* is to be galvanized wrought iron.

e. *Soil pipes* are to be cast iron with coated fittings.

Section 3. Plumbing fixtures:

a. The owner is to confirm the current choices.

b. *Utility sinks* will be *Floorstone* 22" × 25" self-rimming fiberglass with an acrylic weave.

c. *Kitchen sink* will be *Kohler "Executive Chef."*

d. *Master bathroom* and *Bathroom* #2 fixtures:
Sink: *Kohler "Farmington"* K-2905.
Toilet: *Kohler "Wellworth Lite"* K3420.
Bathtub: *Kohler "Village"* K-715.

Division 13 identifies the main materials used with the plumbing system and specifies the manufacturer's name, model numbers, and brand names to be ordered.

REVIEW QUESTIONS

Enter the missing words in the spaces provided on the right.

1. A guideline developed for writing specifications is called the _____ _____.

   _____

2. Specifications are often broken down into _____ and sections.

   _____

3. The _____ _____ division of the specifications discusses legal responsibilities.

   _____

4. Protecting against building damage during construction is the responsibility of the _____.

   _____

5. Excavation work is usually discussed in the _____ _____ division of the specifications.

   _____

6. If the soil is sufficiently firm, _____ ____ footings are generally acceptable.

7. Steel reinforcement for concrete is discussed in ____ ___ division of the specifications.

8. ACI and UBC requirements offer design requirements for the concrete ____.

9. The _____ division of the specifications discusses some of the framing procedures.

10. All rough door and window openings should be verified by the _____ ____.

11. The Carpentry division of the specifications often gives the brand names of metal _____.

12. Doors and windows are discussed in the _____ division of the specifications.

13. _____-grade lumber should be used for interior trim.

14. Siding comes under the category of _____ millwork.

15. Specification data about window units often include the ____ of the manufacturer.

16. Countertop information is found in the _____ section of the Millwork division.

17. The Interior Surface Materials division includes finish coverings for ____, ____, and ____.

18. A surface material placed in fire-rated and damp locations is type _____ ____.

19. The Flashing, Underlayment, and Shingles Division discusses ____ and ____ roof flashing.

20. The _____ instructions should be used when installing roof shingles.

21. The Painting and Staining division specifies that all colors are to be approved by the ____.

22. The _____ ____ division provides information about door hinges and locks.

23. ____ and ____ insulation is in the Moisture Protection and Thermal Insulation division.

24. The size of the hot water heater is given in the _____ and _____ division.

    _____

25. The _____ division discusses interior wiring.

    _____

26. All electrical appliances require _____ approval.

    _____

27. Water and waste pipe materials are described in the _____ division.

    _____

28. The choice of plumbing _____ must often be confirmed by the owner.

    _____

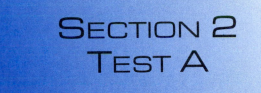

Plat Plan–Survey Plan– Site Plan–Foundation Plan

Supply the following information from the plat plan (Figure 6–2, page 36).

1. Scale. _____

2. Length of front property line of Lot 122. _____

3. Distance from street center line to the front property line of Lot 119. _____

4. Width of Public Utility Easement (PUE) running along all lots shown in the plat plan. _____

5. Total square foot area of Lot 120. _____

6. Acreage of Lot 127. _____

7. Name of streets adjoining Lot 119. _____

8. Distance from street center line to the radius point of the arc of the cul-de-sac. _____

9. Total width of PUE extending from the property line between Lots 125 and 124. _____

10. Angle bearing of northeast property line of Lot 121. _____

Supply the following information from the survey plan (Figure 6–8, page 41).

11. Scale. _____

12. Radius of curved street curb. _____

13. Elevation at southeast corner of lot. _____

14. Width of existing sidewalk. _____

15. Distance of front PUE from front property line. _____

16. Compass direction of benchmark lot corner. _____

17. Length of front property line. _____

18. Angle of west property line. _____

19. Length of north property line. _____

20. Width of east PUE. _____

21. Compass direction of rear property line. _____

22. Lot number. _____

Supply the following information from the site plan (Figure 6–9, page 43).

23. Distance of front zoning setback. _____

24. Lot corner location of utility risers. _____

25. Narrowest width of front driveway. _____

26. Distance of front building setback. _____

27. Elevation of southwest corner of garage. _____

28. Widest width of driveway. _____

29. Building setback from east property line. _____

30. Width of concrete walk to front porch. _____

31. Elevation at rear of building. _____

32. Top of subfloor elevation. _____

33. Difference between benchmark and subfloor elevations. _____

34. Corner elevation at farthest northeast corner of building. _____

35. Building setback from west property line. _____

36. Utility lines placed in utility trench. _____

37. Elevation of first contour line from the front property line. _____

38. Direction of drainage flow at west side of lot. _____

39. Diameter of tree trunk closest to front property line. _____

40. Difference in elevation between southeast and northeast lot corners. _____

Supply the following information from the plan view crawl-space foundation plan (Figure 7–10, page 55).

41. Scale. _____

42. Type of expansion joint around perimeter of garage slab. _____

43. Distance between centers of front-porch columns. _____

44. Distance between front-porch stair newel posts. _____

45. Distance from outside face of west foundation wall to center of first pier. _____

46. Center-to-center distance from above pier to the next pier. _____

47. Dimensions of crawl-space access. _____

48. Difference between front and back elevations of garage slab. _____

49. Distance from front of garage wall to the outside face of the rear building wall. _____

50. Distance from outside face of rear building wall to center of first pier. _____

51. Number of square feet of vent space required for every 150 square feet of crawl-space area. _____

52. Measurement from north building wall next to sun deck to center of bench post. _____

53. East-to-west center-to-center measurement between bench posts. _____

54. Name of foundation wall cavities at beam ends. _____

55. Total distance from east foundation wall to center of the second row of piers. _____

56. Width of south front-porch columns. _____

57. Total east-to-west width of building. _____

58. North-to-south dimension of front porch. _____

59. East to west dimension of sun deck. _____

60. Dimensions of concrete slab to provide support at foot of sun-deck stairway. _____

Supply the following information from the crawl-space foundation section view drawings (Figure 7–12 a-c, pages 58-60).

61. Height and width of foundation wall footing. _____

62. Size and spacing of rebars running across the width of the wall footings. _____

63. Size and spacing of vertical rebars in CMU foundation wall. _____

64. Minimum distance of any wood material from the ground. _____

65. Thickness and width of sill plates fastened to top of foundation wall. _____

66. Diameter, length, and spacing of bolts used to fasten sill plates to the top of the foundation wall. _____

67. Number of and size of rebars running along the length of the foundation footings. _____

68. Minimum height from bottom of footing to top of foundation wall. _____

69. Thickness of concrete slab in garage. _____

70. Thickness of crushed rock layer beneath concrete slab. _____

71. What is the depth of the frost line? _____

72. How far do vertical rebars extend into the footings? _____

73. Size and spacing of rebars in garage floor slab. _____

74. Thickness of slab edge under garage door opening. _____

75. Height and width of interior pier footing. _____

76. Width of CMU pier over pier footing. _____

77. Minimum distance from grade to bottom of floor joists. _____

78. Quantity and size of vertical rebars in deck pier. _____

80. Width of pier under porch column base. _____

Supply the following information from the full basement foundation plan (Figure 7–16, page 64) and the full basement section views (Figure 7–18, page 66).

81. Top of driveway elevation. _____

82. Difference between top of driveway and front garage slab elevation. _____

83. Diameters of galvanized steel walls around areaways. _____

84. Number of overhead lights in basement. _____

85. Diameter and number of floor drains. _____

86. Elevation of basement floor slab. _____

87. Corner compass direction of sump pump. _____

88. Top and bottom widths and height of footing under pipe column. _____

89. Thickness of basement floor slab. _____

90. Thickness of steel plate welded to base of column. _____

91. Diameter of steel column. _____

92. Height and width of foundation wall footing. _____

93. Diameter of perforated drain pipe. _____

94. Size and spacing of rebars in floor slab. _____

95. Thickness of concrete foundation wall. _____

96. Size and spacing of horizontal rebars in foundation wall. _____

97. Thickness of rigid insulation against outside of foundation wall. _____

98. Size and spacing of vertical rebars in foundation wall. _____

99. Thickness, width, and spacing of studs against foundation wall. _____

100. Minimum distance from grade to bottom of wood siding. _____

Supply the following information from the slab foundation plan (Figure 7–23, page 70) and the slab foundation section views (Figure 7–25, page 71).

101. Thickness of concrete slab. _____

102. Top and bottom widths of continuous footings. _____

103. Metal device securing bottom of wood post. _____

104. Height and width of flooring under foundation wall. _____

105. Thickness of CMU foundation wall.

SECTION 2 TEST B

Floor Framing Plan–Floor Plan–Ceiling and Roof Framing Plans

Supply the following information from the floor framing plan (Figure 8–4, page 80) and the related section views (Figure 8–6a-c, pages 82-84).

1. Size of floor joists in living area.   _____

2. Spacing of floor joists.   _____

3. Girder size supporting floor joists in living area.   _____

4. Girder size at center of front-porch area.   _____

5. Joist size at front porch and rear sun deck.   _____

6. Joist-fastening device against center beam in porch area.   _____

7. Number of steps up to the front porch.   _____

8. Post size over piers supporting beams in living area.   _____

9. Difference in elevation between the house subfloor floor and sun deck.   _____

10. Rating of insulation between floor joists.   _____

11. Type and size of subfloor material in house.   _____

12. Name of manufacturer of metal-post base.   _____

13. Type and diameter of bolts securing post to post base.   _____

14. Name of lumber used for posts.   _____

15. Size of decking material.   _____

16. Spacing between decking material.   _____

17. Type of fasteners used to secure decking to joists.   _____

18. Size of rim joists over house foundation.   _____

19. Device for fastening joists to center beam under front porch.   _____

20. Compass direction and view indicated by section symbol A.   _____

21. Type of framed wall over north foundation stem wall.   _____

22. Height from decking to top of newel-post cap trim on front porch.   _____

23. Material placed between joists under walls perpendicular to floor joists. _____

24. Elevation at decking level of rear sun deck. _____

25. Height from decking to top of bench at sun deck. _____

26. Size and spacing of railing balusters. _____

27. Length and width of concrete slab at foot of sun-deck stairs. _____

28. Material placed between lapped ends of floor joists. _____

29. Size of headers around stairwell over full basement. (Refer to Figure 8–7, page 86.) _____

30. Diameter and type of column supporting beams of full-basement foundation. (Refer to Figure 7–16, page 84). _____

Supply the following information from the floor plan (Figure 9–5, page 96), the window schedule (Figure 9–9, page 100), and the door schedule (Figure 9–10, page 101).

31. Longest east-to-west dimension of building. _____

32. Width and height of garage door. _____

33. Outside width of south end of garage. _____

34. Outside width of north garage wall. _____

35. West-to-east measurement of wardrobe closet off master bedroom. _____

36. Dimension from west railing of front porch to center of fixed kitchen window. _____

37. Measurement from south garage wall to south side of front porch. _____

38. Height of mechanical room platform in garage area. _____

39. Code letter of front-entrance door. _____

40. Width of linen closet next to Bathroom #2. _____

41. Dimension from south wall of porch to south wall of wardrobe closet off master bedroom. _____

42. Width of south front-porch columns. _____

43. Dotted rectangles on each side of south kitchen wall. _____

44. Code number of window at south end of foyer. _____

45. Measurement from south foyer wall to north garage wall. _____

46. Width and height of door leading into Bathroom #2. _____

47. Measurement from east wall of Bedroom #2 to face of closet. _____

48. Width and height of window in Bathroom #2. _____

49. Width of hallway leading to Bathroom #2. _____

50. South-to-north dimension of the study. _____

51. East-to-west dimension of master bathroom. _____

52. Wall location (compass direction) of double sink in kitchen. _____

53. Total width and length measurements of window group in north wall of master bedroom. _____

54. Width and height of rough opening for north study window. _____

55. Rough opening of double doors leading into study. _____

56. Walls (compass directions) where cabinets and counters are located in master bathroom. _____

57. Width and length of fireplace hearth in living room. _____

58. Room containing washer and dryer. _____

59. Difference between the rough and finish width and height openings of the lower window group in north wall of living room. _____

60. Compass-view direction of B section symbol. _____

61. Total width and height of transom windows at south kitchen wall. _____

62. Dimension from north railing of sun deck to center of stairway. _____

63. Dashed line running parallel to outside walls. _____

64. North-to-south center-to-center measurement of deck bench. _____

65. Plumbing fixtures along south wall of Bathroom #2. _____

66. Measurement from south face of north garage wall to north face of stairway railing (Figure 9–7, page 99). _____

67. Length of stairway leading down to basement (Figure 9–7, page 99). _____

Supply the following information from the ceiling framing plan (Figure 10–2, page110).

68. Spacing of all joists and roof trusses. _____

69. Size of all ceiling joists. _____

70. Ceiling height over master bedroom. _____

71. Width and length of attic access. _____

72. Ceiling height over study. _____

73. Size of exposed rafters over living room. _____

74. Size of ceiling beam extending from short east wall of kitchen. _____

75. Shaded rectangles on both sides of south kitchen wall. _____

Supply the following information from the roof framing plan (Figure 10–13, page 117) and the roof-related detail drawings (Figure 10–15a and b, page 120).

76. Typical roof pitch. _____

77. Type and size of beam over garage opening. _____

78. Type of trusses over garage area and size of chord and web material. _____

79. Horizontal measurement from center of garage trusses to outside walls. _____

80. Distance of roof and truss overhangs. _____

81. Size of all roof rafters. _____

82. Size of roof beam over front porch and kitchen. _____

83. Spacing of doubled rafters over front porch. _____

84. Type of post supporting end of beam at the front end of porch. _____

85. Size of exposed collar ties fastened to each side of doubled rafters over living room. _____

86. Type and thickness of material used for roof sheathing. _____

87. Name of metal tie used to fasten rafters to top plates. _____

88. Material placed over sheathing before applying shingles. _____

89. Type and thickness of overhang soffit material. _____

90. Diameter of sheer plate between collar ties and rafters. _____

91. Minimum bolt distance from end of collar tie. _____

92. Thickness and width of fascia and sub-fascia boards. _____

93. Thickness, width, and angle of knee brace under outrigger. _____

94. Cross-section size of outrigger and type of end finish. _____

95. Thickness and width of rafter lookout block. _____

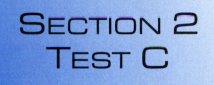

Section and Detail Drawings–Elevation Plans

Supply the following information from the Building Section A drawing (Figure 11–5, page 131).

1. Thickness and width of floor girder.

2. Thickness and width of floor joists.

3. Rating of insulation in floors and exterior walls.

4. Living room plate height.

5. Porch-plate height.

6. Rating of insulation in living room ceiling.

7. Difference between plate height of living room and porch plate height.

8. Room above which rafter collar ties are exposed.

Supply the following information from the Building Section B drawing (Figure 11–6, page 132).

9. Height from subfloor to top of beam extending over sun deck.

10. Spacing of collar ties over living room.

11. Quantity and types of windows in master bedroom.

12. Type of windows over lower windows in living room.

13. Thickness, width, and spacing of ceiling joists at roof extension over sun deck.

Supply the following information from the basement-stair section drawing (Figure 11–8, page 134).

14. Total run of stairway.

15. Tread to top of railing height.

16. Number, thickness, and width of stair stringers.

17. Thickness of finish riser material.

18. Total width of tread including nosing.

19. Total rise of stairway.

20. Thickness of finish tread material.

21. Amount of nosing projection.

22. Height of headroom.

23. Riser height. _____

24. Total number of treads. _____

25. Total number of risers. _____

26. Length of stairwell opening. _____

Supply the following information from the fireplace section drawing (Figure 11–9, page 135).

27. Added support placed between joists below the bearing wall behind fireplace. _____

28. Type of flue. _____

29. Material used for spark arrestor cap. _____

30. Type of masonry veneer around opening. _____

31. Size of header material over fireplace opening. _____

32. Exterior finish material around chimney above roof line. _____

33. Material used for chimney flashing. _____

Supply the following information from the door and window detail drawings (Figure 11–10, page 136).

34. Name of bottom section of door frame. _____

35. Thickness and width of interior casing around door jamb. _____

36. Name of side part of door. _____

37. Recessed groove cut along edge of door frame that acts as a doorstop. _____

38. Assembly of door opening with two 2 × 10s on edge and a flat 2 × 6 below. _____

39. Doubled 2 × 6s at bottom of window opening. _____

40. Flat trim piece below bottom of window. _____

41. Clad material over wooden window sash. _____

42. Type of window hinge. _____

43. Flat trim piece beneath window stool. _____

Supply the following information from the west and south elevation plans (Figure 12–3, page 144).

44. Size and type of exterior siding. _____

45. Thickness and width of railing cap. _____

46. Width and height of garage door. _____

47. Height from subfloor to Bedroom #2 plate height. _____

48. Front-porch plate height. _____

49. Window types at left of front porch. _____

50. Roof pitch. _____

51. Material used for butt corners at porch columns. _____

52. Type of finish on lap siding. _____

53. Pipe used to drain water from gutters to ground. _____

54. Number of skylights seen on drawing. _____

55. Height from garage slab to garage plate. _____

56. Material used for butt corners at building corners. _____

57. Shaded rectangles below ridge of south roof surface. _____

Supply the following information from the east and north elevation plans (Figure 12–4, page 145).

58. Finish on 2 × 4 butt corners of porch column. _____

59. Material used for lattice. _____

60. Type of windows over lower windows at north living room wall. _____

61. Measurement from highest point of roof to top of chimney. _____

62. Roof finish material. _____

63. Gutter design. _____

64. Window and door trim finish. _____

65. Thickness and widths of fascia and sub-fascia. _____

66. Railing material. _____

Supply the following information from the Laundry Room interior elevations (Figure 12–7, page 149).

67. Height of cabinet counters. _____

68. Countertop material. _____

69. Width of splash board. _____

70. Size of utility sink. _____

71. Depth of toe space. _____

72. Floor to ceiling height. _____

73. Height and depth of open shelves. _____

74. Type of windows _____

Supply the following information from the Kitchen interior elevations (Figure 12–8, page 150).

75. Distance between top of floor cabinets and bottom of wall cabinet. _____

76. Wall finish material between floor cabinets and wall cabinets.   _____

77. Height of wall cabinets.   _____

78. Total height from floor to the bottom of the soffit.   _____

79. Countertop finish material.   _____

80. Thickness and width of trim material at bottom of soffit and around windows.   _____

81. Width of valance in front of counter light in the west wall.   _____

82. Type of appliance closet door.   _____

83. Width of slide-in gas range.   _____

84. Depth of cabinets above the refrigerator.   _____

85. Type of shelves at lower pantry cabinet in north wall.   _____

86. Number of drawers at top of cabinets to the right of the refrigerator.   _____

Supply the following from the Bathroom interior elevations (Figure 12–9, page 151).

87. Floor and base finish material in Master Bathroom.   _____

88. Height of base cabinets.   _____

89. Type of shelves in base cabinets.   _____

90. Wall finish around tub and shower.   _____

91. Height of tile splash.   _____

92. Depth of cabinet under sink in Bathroom 2.   _____

93. Wall showing medicine cabinet in Bathroom 2.   _____

Supply the following information from Master Bedroom, Living Room, and Study interior elevations (Figure 12–10, page 152).

94. Material used for built-in countertops in Master Bedroom.   _____

95. Number of rows of shelves at each side of door in Master Bedroom.   _____

96. Hinge side of door in west wall (right or left).   _____

97. Finish material at front of fireplace.   _____

98. Finish of wood shelves over fireplace.   _____

99. Material used for countertop finish along the south wall of the Study.   _____

100. Height from floor to the top of shelves in the Study.   _____

101. Total width of floor cabinets in the Study.   _____

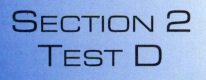

SECTION 2 TEST D

Mechanical–Electrical–Plumbing Plan Specifications

Supply the following information from the mechanical-electrical-plumbing plan (Figure 13–3, page 158). When identifying or referring to wall locations, include the room name and compass direction.

1. Location of electric service panel. _____

2. Type of plug at outside of east garage wall. _____

3. Type of hose bib outside east bathroom wall. _____

4. Objects housed in the mechanical room. _____

5. Number of recessed exterior lights placed around garage area. _____

6. Wall closest to location of smoke alarm in garage. _____

7. Number of wall plugs in Bedroom 2. _____

8. Type of light switch next to side door of garage. _____

9. Number of overhead lights in interior of garage. _____

10. Number of flat-ceiling down lights in study. _____

11. Study room wall where television jack is located (compass direction). _____

12. Rooms containing ceiling heater-ventilating-light units. _____

13. Type of lights in hallway closets. _____

14. Wall where overhead garage-door switch is placed. _____

15. Type of heating system. _____

16. Type of switch next to door leading to master bedroom. _____

17. Rooms containing left and right speakers. _____

18. Number of identified heating zones. _____

19. Height from floor of wall outlets along the south kitchen wall. _____

20. Height from floor of microwave outlet. _____

21. Number of plugs received by each outlet over south-wall and east-wall kitchen counters. _____

22. Room wall where doorbell chimes are placed. _____

23. Room containing washer and dryer. _____

24. Wall location of doorbell. _____

25. Room and wall location of telephone jack. _____

26. Room and wall location of heating system's thermostat. _____

Answer T (true) or F (false) for the following questions about the specifications for the **One-Story Residence** (Unit 14, pages 163-168).

27. The specifications for a set of plans are not legally binding. _____

28. Specifications are often divided into sections further broken down into divisions. _____

29. The contractor will pay for all electrical power used during construction. _____

30. After Notice to Proceed has been received, the construction must be completed within 60 days. _____

31. All grades and slab elevations will be verified by the contractor. _____

32. Concrete masonry units for foundation walls are 8" wide, 8" high, and 18" long. _____

33. Sill plates are fastened to the foundation with hooked anchor bolts. _____

34. The garage slab will have control cut joints 12'-0" OC. _____

35. Rough framing materials are discussed in the Carpentry division. _____

36. Roof trusses must be designed for a load of 50 pounds per square foot. _____

37. The Millwork division includes information about door and window units. _____

38. Gypsum board, 1/2" thick, will be applied in fire-rated and damp locations. _____

39. The laundry room will receive rubber floor tiles over a plywood underlayment. _____

40. Vinyl roof flashing will be used at the valleys and ridges. _____

41. All exposed metal on the roof will be painted to blend with rest of the roof covering. _____

42. Two pair of 4 1/2" by 4 1/2" butt bronze hinges are used for the exterior doors. _____

43. Schlage A series locks with bronze finish are used on all doors. _____

44. Vaulted ceilings will receive R-19 insulation. _____

45. The National Electric Code will be followed for all electrical installation. _____

46. Galvanized wrought-iron pipe will be used for all waste and vent piping. _____

Choose one or more correct ways to finish each statement about the specifications for the **One-Story Residence** (Unit 14, pages 163-168).

47. Inconsistencies found between any drawings in the construction plans should be brought to the immediate attention of the _____
    a. code officials.
    b. architect.
    c. zoning officials.
    d. subcontractors.

48. The correction of structural defects related to the construction is guaranteed by the contractor for _____
    a. 90 days.
    b. 180 days.
    c. one year.
    d. two years.

49. The Site Work division includes _____
    a. filling.
    b. excavation.
    c. grading.
    d. all of the above.

50. Information in the Site Work division mostly relate to the _____
    a. foundation plan.
    b. plat plan.
    c. site plan.
    d. floor framing plan.

51. Backfill will consist of _____
    a. gravel.
    b. cinders.
    c. earth brought from a different location.
    d. soil removed during excavation.

52. Design of the concrete mix is based on requirements established by the _____
    a. Uniform Building Code.
    b. Portland Cement Association.
    c. American Concrete Institute.
    d. Standard Building Code.

53. Wall footings are to be reinforced with _____
    a. #4 rebars across the width of the footing 12″ OC.
    b. three #4 rebars running the length of the footing.
    c. welded wire mesh.
    d. #4 rebars across the width of the flooring 6′ OC.

54. An entry to the crawl space beneath the house is _____
    a. accessed through an 18″ × 24″ minimum opening in the foundation wall.
    b. accessed through a scuttle in the floor above.
    c. not available.
    d. accessed through a 24″ by 48″ opening in the foundation wall.

55. The floor joists are to be
    a. 2 × 10 Douglas fir.
    b. 2 × 12 redwood.
    c. 2 × 10 pine.
    d. 2 × 12 Douglas fir.

56. Solid blocking is installed
    a. under walls running parallel with the joists.
    b. at the center of joist spans over 8'-0".
    c. between doubled joists.
    d. at the center of joist spans over 12'-0".

57. The interior walls constructed with 2 × 6 studs are
    a. the west living room wall.
    b. the wall between the master bedroom and master bathroom.
    c. the south master bedroom wall.
    d. none of the above.

58. Door and window headers at the exterior walls are constructed of
    a. two 2 × 10s spiked together.
    b. two 2 × 12s on edge with a 2 × 6 nailed to the bottoms.
    c. two 2 × 10s on edge with a spacer block in between.
    d. two 2 × 10s on edge with a 2 × 6 nailed to the bottoms.

59. Redwood lumber is used to construct the
    a. exterior walls of the building.
    b. front porch.
    c. rear sun deck.
    d. roof.

60. The Millwork division includes
    a. doors and windows.
    b. interior and exterior trim.
    c. porch railings.
    d. all of the above.

61. Vinyl-clad wood framed windows are
    a. used only in the living room.
    b. used throughout the building.
    c. not used at all.
    d. used where there are two rows of windows, such as in the south
       kitchen wall.

62. Six-inch ceramic tile is used for the counters and splashes
    a. in the kitchen and both bathrooms.
    b. only in the bathrooms.
    c. in all of the building's cabinet counters.
    d. only in the kitchen.

63. Oak flooring will be placed
    a. in all the rooms of the living area.
    b. in the living room and study only.
    c. in all the rooms except the laundry room and bathrooms.
    d. in all the rooms except the laundry room and kitchen.

64. When nailing down the roof shingles
    a. copper valley and ridge flashing must first be placed.
    b. the roof sheathing must be covered with #10 bituminous-saturated roof felt.
    c. aluminum-coated nails must be used.
    d. none of the above.

65. Rainwater and snow melt is drained from the roof with
    a. roof flashing.
    b. a system of vinyl gutters and downspouts.
    c. slightly sloped gutters with a drain hole at the lower end.
    d. downspouts alone.

66. The final approval of the interior paint color is given by the
    a. painting contractor.
    b. general contractor.
    c. owner.
    d. architect.

67. Regarding the exterior finish of the building,
    a. the lap siding is to be painted.
    b. all butt corners are to be stained.
    c. trim around doors and windows is to be stained.
    d. all of the above.

68. Two pair of 3 1/2″ × 3 1/2″ bronze butt hinges are used with the
    a. interior swing doors.
    b. casement windows.
    c. closet sliding doors.
    d. exterior swing doors.

69. Amerock or other hardware of equal quality is required for the
    a. butt hinges.
    b. cylinder door locks.
    c. cabinet catches.
    d. drawer pulls.

70. The vapor barrier material placed against the inside of the exterior wall studs is
    a. blanket insulation.
    b. 4-mil polypropylene.
    c. bituminous-saturated felt.
    d. kraft building paper.

71. R-11 batt or blanket insulation is
    a. placed between the studs of the exterior walls.
    b. placed between the floor joists.
    c. used for sound insulation in the south wall of the master bedroom.
    d. placed between the ceiling joists.

72. The in-floor radiant heating system
    a. includes 3/8″ thermoplastic tubing.
    b. has tubing circuits up to 250′.
    c. has copper tubing.
    d. receives heated water from the hot water tank.

73. Hot water for general use is supplied by a
    a. gas-fired boiler.
    b. 65-gallon water heater.
    c. solar-heated panel.
    d. 75-gallon LPG water heater.

    _____

74. Unless otherwise noted, electrical duplex outlets are placed
    a. in a vertical position 4" from the floor.
    b. along the walls and in the floors.
    c. 12" above the floor and in a horizontal position.
    d. in a horizontal position 6" above the floor.

    _____

75. Ground-fault circuit interrupter outlets are
    a. installed in all the rooms of the building.
    b. installed for all outdoor outlets.
    c. not required in the interior of the building.
    d. installed in the bedrooms only.

    _____

76. Smoke alarms in the building are
    a. not required.
    b. interconnected.
    c. operated independently of each other.
    d. none of the above.

    _____

77. The underwriter's laboratories
    a. are a government-sponsored agency.
    b. conduct original research.
    c. underwrite funding for research.
    d. inspect and test electrical devices.

    _____

78. The material used for water lines is
    a. lead.
    b. cast iron.
    c. copper.
    d. galvanized wrought iron.

    _____

79. Vent piping
    a. is to be galvanized wrought iron.
    b. is to be the same material used for waste pipes.
    c. is to extend a minimum of 1' above the roof.
    d. all of the above.

    _____

80. Regarding the plumbing fixtures,
    a. the Kohler brand is to be used for all types of sinks.
    b. the owner must confirm the choices.
    c. the Floorstone brand is to be used for the bathtubs.
    d. only Kohler brand fixtures are to be used in the bathrooms.

    _____

SECTION 3

Electrical and Mechanical Plans

The electrical and mechanical information for most small to medium houses is frequently included on the floor plan. However, larger residential and commercial buildings require a separate series of plans to show the electrical and mechanical installations. Commercial construction includes projects such as retail stores, multifamily buildings, office complexes, and other larger structures related to trade or commerce. A set of commercial plans will include the same type of basic drawings required for residential plans plus the electrical and mechanical plans with their schedules and related data. Included in the mechanical system category are plumbing and HVAC (heating, ventilation, and air conditioning) plans. Sometimes drawings are also provided for such mechanical systems as elevators and escalators. The tradespeople involved in basic construction must not only understand the plans concerning their specific work, but should also have some understanding of the drawings related to other crafts. For example, carpenters must know where to place backing and/or frame special openings for electrical and plumbing fixtures. Electricians, plumbers and HVAC workers must closely coordinate their installations. ■

UNIT 15

Electrical Plans

A set of construction drawings for residential and commercial buildings with more complicated electrical systems will include a separate electrical plan. Also, additional electrical diagrams are often prepared by an electrical engineer.

All electrical work in a building must conform to local code rulings. Most cities and counties use the *National Electric Code (NEC)* as their reference. Some localities may include modifications of the NEC when conditions call for a stricter requirement.

Most electrical plans are floor plans focusing on the wiring systems to the lights and outlets. If the building has more than one story, an electrical plan is provided for each floor level. Electrical plans are used by the electrical contractor to estimate the cost of electrical work and later supervise the work. The electricians on a job must read the plans and be familiar with local code requirements in order to install the circuits and wiring systems for the building. The electricians must also have a working knowledge of how the entire building is constructed in order to run the wiring for the circuits and place the outlet boxes correctly.

THE POWER SOURCE

The electrical power source for a building originates from an overhead power pole or underground cable provided by the local electric utility company. Electrical wires extend from the overhead or underground power source to the building where they will pass through a meter and connect to the service panel. All the circuits of the building originate and are controlled from the service panel. This combination of equipment and wiring bringing electricity into the building is called the **service entrance.** The electrical plan identifies all the wiring to switches and wall and ceiling outlets.

UNITS OF ELECTRICAL MEASUREMENT

Basic units of measurement for electrical energy are volts, amperes, and watts. The interconnection of these units must be understood when designing the electrical system for a building. A **volt** is the unit used for measuring the force causing the electric current to flow through a conductor such as an electric wire. The higher the voltage is, the greater the flow. The ampere is the unit for measuring the amount of electrical flow (current) through the electric wire. Amperes have a direct bearing on the wire size, circuit breakers, and fuses to be used in an electrical system. A watt is the unit measuring electrical power. Total wattage is found by multiplying volts times amperes. For example, 120 volts times 35 amps equals 4200 watts. The unit used by electric power companies to measure the consumption of electricity is the **kilowatt-hour** (1000 watts).

OVERHEAD POWER SOURCE

Overhead electrical power is supplied to a community through high wires attached to the tops of wooden poles or metal towers. When a new building is constructed, the electric utility company in the area will connect three wires between the building and a **transformer** at the top of the pole. The transformer is necessary to change the voltage coming through the high wires to the lower voltage required for the electrical system of the building. The usual voltage for residential electrical systems is 120 and 240 volts.

Feed wires extending from the transformer to the building are fastened to an insulator bracket attached to the building. This is known as the **service drop.** Another set of service-entrance wires is spliced to the feed wires and advances through a conduit attached to the side of the building. After passing through a meter, the lines are attached to the *service disconnect* and a *service panel* (discussed in detail in this unit). **(See Figure 15–1.)** The plot plan of a set of drawings often identifies the location of the power pole and shows the wires extending to the building.

UNDERGROUND POWER SOURCE

Underground electrical power is supplied through utility cables buried beneath the ground surface and connected to grade-level transformers. Wires, placed in rigid conduits, are connected to the underground cable at one end and to the service panel at the other end. **(See Figure 15–2.)**

The plot plan of a set of drawings will often show the locations of the underground utility cable and the

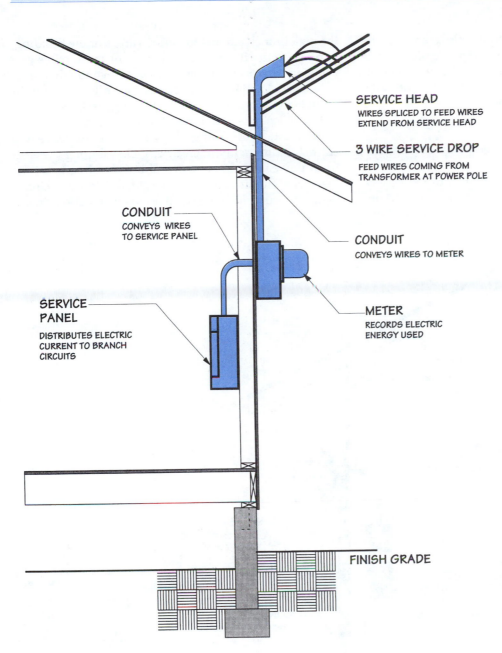

SERVICE HEAD
WIRES SPLICED TO FEED WIRES
EXTEND FROM SERVICE HEAD

3 WIRE SERVICE DROP
FEED WIRES COMING FROM
TRANSFORMER AT POWER POLE

CONDUIT
CONVEYS WIRES TO METER

METER
RECORDS ELECTRIC
ENERGY USED

CONDUIT
CONVEYS WIRES
TO SERVICE PANEL

**SERVICE
PANEL**
DISTRIBUTES ELECTRIC
CURRENT TO BRANCH
CIRCUITS

FINISH GRADE

Figure 15–1.
Feed wires from the transformer
of an overhead power source
connect to the service head.

grade-level transformer, as well as the service panel. **(See Figure 15–3.)**

THE SERVICE PANEL

A **service panel,** also called a **distribution panel,** receives the electrical current from the power source. It then distributes the electrical current through a series of *branch circuits* (discussed shortly). Before reaching the service panel, the electric wires must pass through a watt-meter and service disconnect. The **watt-meter** records the amount of electrical energy used by the consumer. It contains a small motor whose speed is determined by the amount of electric current flowing at any

given time. A gear mechanism connects the rotating shaft of the motor to dials that register the total number of kilowatt-hours used. The monthly electrical cost to the consumer is then based upon the meter reading. **(See Figure 15–4.)**

After passing through the meter, the wires are connected to the **service switch,** also known as the **disconnect switch.** The service switch may be placed in a separate box or in the service panel. The service panel contains fuses or circuit breakers. A metal strip in the **fuse** will melt and open the electric current if an overload of electrical current occurs. A **circuit breaker** accomplishes the same thing by opening and thus shutting down the electrical current. The electric code requires that the service panel be fully enclosed, with an outside lever within reach and in plain view. The lever makes it possible to shut off the electrical current to the entire building. Service panels should be placed where they can be reached conveniently. They are frequently put outside the building or in the garage, basement, or laundry room.

BRANCH CIRCUITS

The service panel receives the wiring from the service disconnect switch and distributes the electricity through branch circuits to various areas of the building. A **branch circuit** is made up of two or more wires carrying an electrical current from its source to where it is to be used. Each branch circuit is controlled by an individual circuit breaker or fuse. **(See Figure 15–5.)**

If the electrical system of a building consisted of only one circuit, overloading the circuit would leave the entire building without electricity. Distributing the electrical power through a series of branch circuits means that each circuit can be protected by its own circuit breaker or fuse. If an overload occurs in a branch circuit,

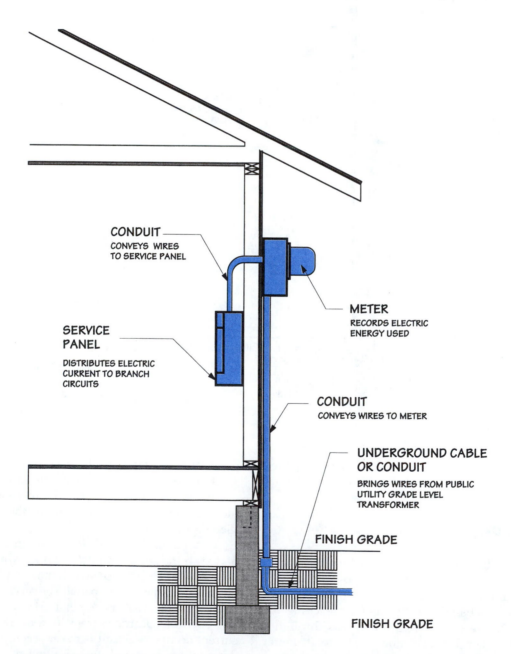

CONDUIT
CONVEYS WIRES
TO SERVICE PANEL

METER
RECORDS ELECTRIC
ENERGY USED

SERVICE PANEL

DISTRIBUTES ELECTRIC
CURRENT TO BRANCH
CIRCUITS

CONDUIT
CONVEYS WIRES TO METER

UNDERGROUND CABLE OR CONDUIT

BRINGS WIRES FROM PUBLIC
UTILITY GRADE LEVEL
TRANSFORMER

FINISH GRADE

FINISH GRADE

Figure 15–2.
Underground electric power is supplied through utility cables buried beneath the ground surface.

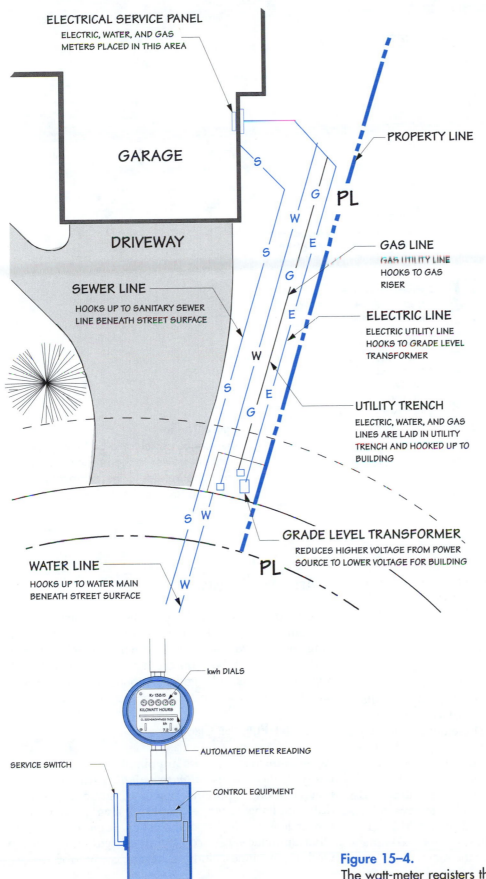

ELECTRICAL SERVICE PANEL
ELECTRIC, WATER, AND GAS
METERS PLACED IN THIS AREA

GARAGE

DRIVEWAY

PROPERTY LINE

PL

GAS LINE
GAS UTILITY LINE
HOOKS TO GAS
RISER

ELECTRIC LINE
ELECTRIC UTILITY LINE
HOOKS TO GRADE LEVEL
TRANSFORMER

SEWER LINE
HOOKS UP TO SANITARY SEWER
LINE BENEATH STREET SURFACE

UTILITY TRENCH
ELECTRIC, WATER, AND GAS
LINES ARE LAID IN UTILITY
TRENCH AND HOOKED UP TO
BUILDING

GRADE LEVEL TRANSFORMER
REDUCES HIGHER VOLTAGE FROM POWER
SOURCE TO LOWER VOLTAGE FOR BUILDING

PL

WATER LINE
HOOKS UP TO WATER MAIN
BENEATH STREET SURFACE

Figure 15–3.
The plot plan will show the locations of underground utility cables, conduits, water and gas lines, and the grade level transformer.

kwh DIALS

KILOWATT HOURS

AUTOMATED METER READING

SERVICE SWITCH

CONTROL EQUIPMENT

WATT METER AND SERVICE CONTROL BOX

Figure 15–4.
The watt-meter registers the number of kilowatt-hours used during a monthly period.

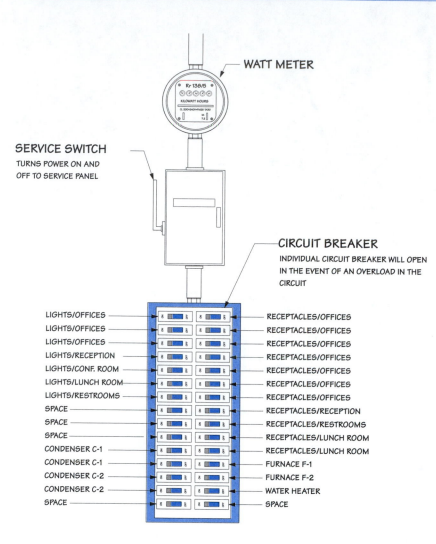

WATT METER

SERVICE SWITCH
TURNS POWER ON AND
OFF TO SERVICE PANEL

CIRCUIT BREAKER
INDIVIDUAL CIRCUIT BREAKER WILL OPEN
IN THE EVENT OF AN OVERLOAD IN THE
CIRCUIT

LIGHTS/OFFICES	RECEPTACLES/OFFICES
LIGHTS/OFFICES	RECEPTACLES/OFFICES
LIGHTS/OFFICES	RECEPTACLES/OFFICES
LIGHTS/RECEPTION	RECEPTACLES/OFFICES
LIGHTS/CONF. ROOM	RECEPTACLES/OFFICES
LIGHTS/LUNCH ROOM	RECEPTACLES/OFFICES
LIGHTS/RESTROOMS	RECEPTACLES/OFFICES
SPACE	RECEPTACLES/RECEPTION
SPACE	RECEPTACLES/RESTROOMS
SPACE	RECEPTACLES/LUNCH ROOM
CONDENSER C-1	RECEPTACLES/LUNCH ROOM
CONDENSER C-1	FURNACE F-1
CONDENSER C-2	FURNACE F-2
CONDENSER C-2	WATER HEATER
SPACE	SPACE

Figure 15–5.
The service panel for a building provides electrical power to the branch circuits. This panel relates to the electrical plans discussed toward the end of this unit.

the circuit breaker opens and cuts off electrical power for only that individual circuit. The three main categories of branch circuits are *general-purpose, appliance,* and *special-purpose.*

GENERAL-PURPOSE CIRCUITS

General-purpose circuits, also called **lighting circuits,** control the electrical current going to *convenience outlets* (discussed later under "Outlets"). These outlets are usually placed 12" to 18" from the floor. They provide electricity for lamps, clocks, television sets, radios, and portable appliance objects such as vacuum cleaners. General-purpose circuits are also used for ceiling and wall lighting outlets. It is common practice to have lights in the same room connected to different circuits. This way, if one circuit overloads and trips the circuit breaker, the other circuit can still supply light for the room.

APPLIANCE CIRCUITS

Appliance circuits usually serve the convenience outlets in rooms such as the kitchen, laundry room, pantry, dinning area, and workshop. In a kitchen many convenience outlets are placed above the cabinet counters so that small appliances can be plugged into them, such as toasters, toaster ovens, mixers, blenders, or electric frying pans.

SPECIAL-PURPOSE CIRCUITS

Special-purpose circuits are mainly used for individual outlets called *special-purpose outlets* (discussed further later on) supplying power to stationary, heavy-duty appliances. For example, a single circuit may power just an electric dishwasher. Some examples of other appliances serviced by an individual circuit are washing machines, clothing dryers, countertop ranges, water heaters, water-pump garbage disposals, air conditioners, and furnaces.

Outlets

Electrical outlets are generally found in the walls and ceilings of the building where they receive electrical power from the branch circuits. An outlet consists of a *receptacle* attached to a metal box. The electrical wires of the circuit are connected to the receptacle. A *plug* at the end of an electrical cord is inserted into the receptacle. There are convenience, special-purpose, split-wire, and lighting outlets.

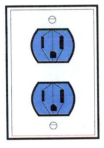

CONVENIENCE OUTLET
PROVIDES RECEPTACLE FOR
LAMPS AND APPLIANCES

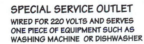

SPECIAL SERVICE OUTLET
WIRED FOR 220 VOLTS AND SERVES
ONE PIECE OF EQUIPMENT SUCH AS
WASHING MACHINE OR DISHWASHER

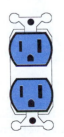

SPLIT WIRE RECEPTACLE
TOP PART OF DUPLEX RECEPTACLE IS CONTROLLED
BY LIGHT SWITCH AND BOTTOM IS CONNECTED TO
A HOT CIRCUIT

Figure 15–6.
Some more commonly used outlets and receptacles are the convenience, special-service, and split-wire outlets.

Convenience, Special-Purpose, and Split-Wire Outlets

Most wall outlets are convenience and special-purpose outlets. **Convenience outlets** provide receptacles for plugging in lamps and small appliances. The receptacles of convenience outlets are designed to receive two plugs *(duplex)*, three plugs *(threeplex)*, or four plugs *(fourplex)*. **Special-service outlets** serve one piece of equipment, such as a washing machine or dishwasher, and often have one receptacle, wired for 220 volts. A **split-wire outlet** features a duplex receptacle on the top, controlled by a wall switch, and hot circuit on the bottom. **(See Figure 15–6.)**

The locations of wall outlets are planned according to their intended use and code requirements. Convenience outlets should be placed close to the ends of wall areas so as not to be hidden by large pieces of furniture. The National Electric Code states that no point along a usable wall space, except in bathrooms and utility rooms, should be more than 6′ from a convenience outlet. The code also states that each room should have no less than three outlets.

Generally, convenience wall outlets are placed 12″ to 18″ inches above the floor. However, many outlets in kitchens and bathrooms are placed above the countertops so that small appliances can be plugged into the receptacles. Special-purpose outlets are frequently placed a few feet above the floor.

Appliances that plug into outlets near water sources present a greater shock hazard and should be equipped with a **ground fault interrupter (GFI).** This safety device is part of the receptacle and disconnects the current to the outlet faster than a circuit breaker in a service panel can.

Lighting Outlets

Lighting outlets are provided to connect light fixtures and are placed in the ceiling and upper wall sections. **(See Figure 15–7.)** The electrical current going to the lighting outlet is usually controlled by a wall switch. Some lighting fixtures are turned on and off with a pull chain.

Metal outlet boxes are required wherever receptacles or lighting outlets are placed. *Square* and *rectangular* boxes are primarily used for plug receptacles and switches. *Octagon*-shaped (eight sides) boxes provide a fastening base for lighting fixtures.

Outlet boxes are placed before both sides of the wall or ceiling surfaces are covered. They can be attached to metal bar hangers nailed to the joists. Wall outlet boxes are often nailed to the side of a stud. **(See Figure 15–8.)**

Symbols are used on the electrical plan to show the

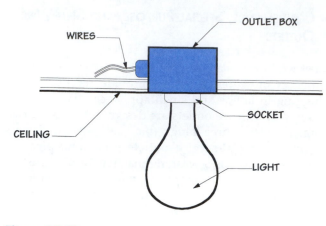

Figure 15–7.
Lighting outlets are provided to connect bulbs and lighting fixtures.

locations and types of plug receptacles and lighting outlets. Some commonly used symbols are shown in **Figure 15–9.**

SWITCHES

A **switch** is a device to open or close an electric circuit. A **circuit** begins from a point of origin such as a distribution panel. It travels through conductors (wires) to the receptacles and/or lights and back to the point of origin. **(See Figure 15–10.)** Switches are mainly used to control lighting sources such as permanent overhead and wall light fixtures. If only lamps provide lighting in a room, a switch should control at least one of the receptacle into which a lamp can be plugged.

Light switches are placed about 48″ above the floor. They should always be positioned on the *latch side* of the door entrance to a room. If a light switch is placed on the hinged side of a door, it will be blocked from use when the door is open. Bathroom switches, for safety's sake, should be out of reach of bathtubs and showers. From any point of entrance, switches should be placed so that the way can be illuminated when a person is moving through the building.

Toggle switches are used more frequently than other kinds. They operate with a snap lever and are available in single-pole, three-way, or four-way types. *The single-pole switch* opens and closes a circuit from one location. *Three-way switches* control a circuit from two locations. They are usually found in rooms with more than one entrance and at the top and bottoms of stairways, garages, and main entrances. A *four-way switch* is used with two three-way switches to control a circuit from three locations. **(See Figure 15–11.)**

Mercury switches, also manufactured in single-pole,

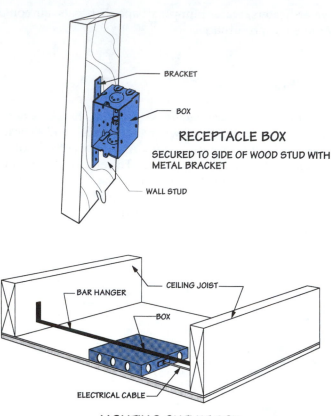

Figure 15–8.
Outlet boxes are secured to ceiling joists and wall studs.

three-way, and four-way types, contain a sealed glass tube of mercury that provides a contact when the switch is turned on. Mercury switches have the advantage of being silent and shockproof. Another type of switch is the *dimmer switch,* which makes it possible to adjust the light to a desired brightness. Also available are *push-button switches* turned on and off by pressure on button. **(See Figure 15–12.)** Switches are identified by different symbols on the plans. **(See Figure 15–13.)**

READING THE ELECTRICAL PLANS

Commercial and larger residential projects will usually have an electrical plan for each story of the building. Several types of electrical plans may be required for the different levels of electrical work. However, constant reference must be made to the basic floor plan. **(See Figure 15–14.)** The three electrical drawings that follow are for the first floor of an office building featuring a reflected ceiling.

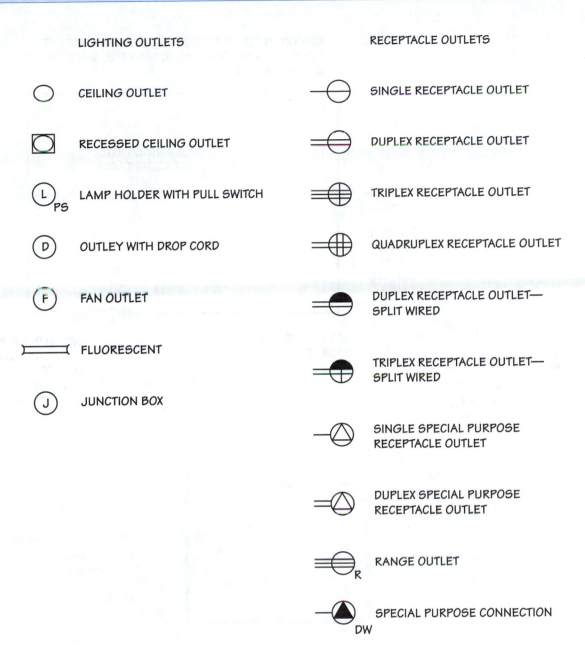

LIGHTING OUTLETS

○ CEILING OUTLET

▢ RECESSED CEILING OUTLET

Ⓛ PS LAMP HOLDER WITH PULL SWITCH

Ⓓ OUTLEY WITH DROP CORD

Ⓕ FAN OUTLET

▭ FLUORESCENT

Ⓙ JUNCTION BOX

RECEPTACLE OUTLETS

SINGLE RECEPTACLE OUTLET

DUPLEX RECEPTACLE OUTLET

TRIPLEX RECEPTACLE OUTLET

QUADRUPLEX RECEPTACLE OUTLET

DUPLEX RECEPTACLE OUTLET— SPLIT WIRED

TRIPLEX RECEPTACLE OUTLET— SPLIT WIRED

SINGLE SPECIAL PURPOSE RECEPTACLE OUTLET

DUPLEX SPECIAL PURPOSE RECEPTACLE OUTLET

RANGE OUTLET

SPECIAL PURPOSE CONNECTION

Figure 15–9.
Here are some frequently used receptacle and lighting symbols.

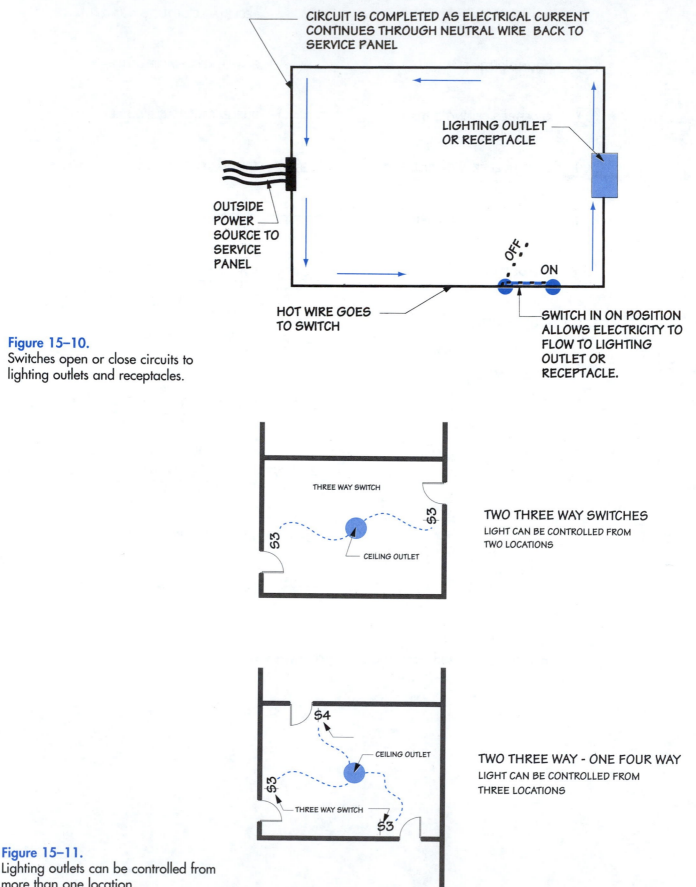

CIRCUIT IS COMPLETED AS ELECTRICAL CURRENT CONTINUES THROUGH NEUTRAL WIRE BACK TO SERVICE PANEL

LIGHTING OUTLET OR RECEPTACLE

OUTSIDE POWER SOURCE TO SERVICE PANEL

OFF

ON

HOT WIRE GOES TO SWITCH

SWITCH IN ON POSITION ALLOWS ELECTRICITY TO FLOW TO LIGHTING OUTLET OR RECEPTACLE.

Figure 15–10.
Switches open or close circuits to lighting outlets and receptacles.

THREE WAY SWITCH

$3

$3

CEILING OUTLET

TWO THREE WAY SWITCHES
LIGHT CAN BE CONTROLLED FROM TWO LOCATIONS

$4

CEILING OUTLET

$3

THREE WAY SWITCH

$3

TWO THREE WAY - ONE FOUR WAY
LIGHT CAN BE CONTROLLED FROM THREE LOCATIONS

Figure 15–11.
Lighting outlets can be controlled from more than one location.

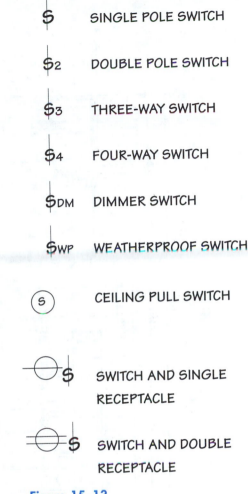

SINGLE POLE SWITCH

DOUBLE POLE SWITCH

THREE-WAY SWITCH

FOUR-WAY SWITCH

DIMMER SWITCH

WEATHERPROOF SWITCH

CEILING PULL SWITCH

SWITCH AND SINGLE RECEPTACLE

SWITCH AND DOUBLE RECEPTACLE

TOGGLE SWITCH
OPERATES WITH
SNAP LEVER

DIMMER SWITCH
ALLOWS ADJUSTING
LIGHT TO DESIRED
BRIGHTNESS

PUSH BUTTON SWITCH
PRESSURE ON BUTTON
TURNS SWITCH ON AND OFF

Figure 15–12.
Several types of switches are available for lighting circuits.

Figure 15–13.
Switches are identified by symbols on the electrical plans.

A **wiring plan** or **power plan** shows the placement of all the electrical wall outlets and the wires leading to the outlets. Circuits are labeled by letters and numbers (for example, "A-2") identifying the circuit and panel box controlling the circuit. Telephone jacks and special-purpose jacks are also shown. **(See Figure 15-15.)** A **branch circuit schedule** for each circuit panel is provided for the power plan. This schedule identifies and provides additional information about each circuit. **(See Figure 15–16.)**

A lighting plan is also included for the first floor of the office building showing the overhead lighting system. Lines show the wiring to the light outlets and the switches controlling the light outlets. **(See Figure 15–17.)** A **lighting fixture schedule** provides the code numbers of the fixtures as they are identified on the plans. The name of the manufacturer of the fixture and the model or catalog numbers are usually given. Information regarding wattage and/or amperage may also be shown. A "remarks" column is included for additional information. **(See Figure 15–18.)**

Figure 15–19 is a reflected ceiling plan showing the finished ceiling, which includes the rectangular light fixtures, the individual recessed lights, and the surrounding acoustical tile finish. The reflected ceiling is constructed of a suspended metal grid held in place by wires. The lighting fixtures and tiles fit into the metal grid. Warm-air supply ducts originating from the furnaces are in the space above the suspended ceiling. The warm-air outlets are at the ceiling surface.

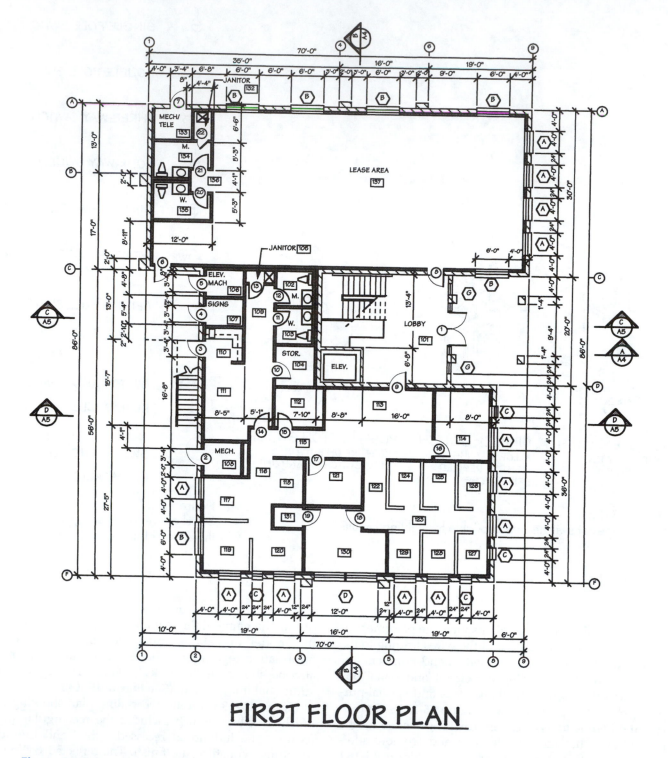

FIRST FLOOR PLAN

Figure 15–14.
The floor plan must also be studied when reading the electrical plans. (Courtesy of Josh Barclay Architects Ltd., Flagstaff, Arizona)

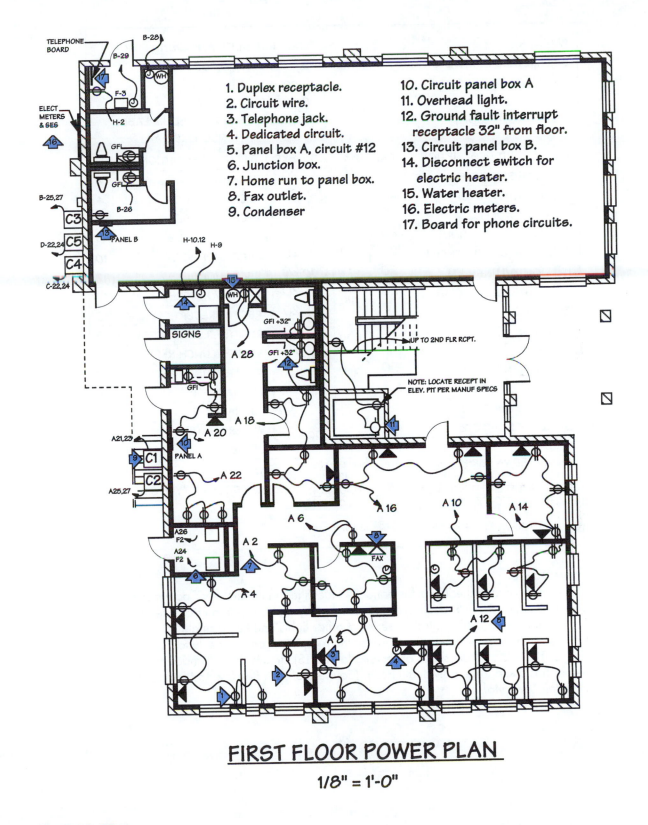

1. Duplex receptacle.
2. Circuit wire.
3. Telephone jack.
4. Dedicated circuit.
5. Panel box A, circuit #12
6. Junction box.
7. Home run to panel box.
8. Fax outlet.
9. Condenser
10. Circuit panel box A
11. Overhead light.
12. Ground fault interrupt receptacle 32" from floor.
13. Circuit panel box B.
14. Disconnect switch for electric heater.
15. Water heater.
16. Electric meters.
17. Board for phone circuits.

FIRST FLOOR POWER PLAN
1/8" = 1'-0"

Figure 15–15.
This wiring plan shows all the circuits for the first-floor receptacles. (*Courtesy of Josh Barclay Architects Ltd., Flagstaff, Arizona*)

PANEL: A
LOCATION: SUITE A, GROUND FLOOR

SERVICE: 200 A, 120/240 V: 3/4 PH - 3 WIRE
S.C. RATING: 10,000 A/C

MAINS: (MLO): 200 AMP
INCOMINGLINE: (TOP)
(3) - #3/0 CU, XHHW

MOUNTING: (FLUSH)
OTHER: ☐ SERVICE EQUIP LABEL
☐ EQUIP GRD BAR

CKT	CIRCUIT DESCRIPTION	BKR	ØA	ØB	BKR	CIRCUIT DESCRIPTION	CKT
1	LIGHTS/OFFICES	20 / 1	1320 / 900		20 / 1	RECPTS/OFFICES	2
3	LIGHTS/OFFICES	20 / 1		1410 / 900	20 / 1	RECPTS/OFFICES	4
5	LIGHTS/OFFICES	20 / 1	1410 / 900		20 / 1	RECPTS/OFFICES	6
7	LIGHTS/RECEPTION	20 / 1		1430 / 900	20 / 1	RECPTS/OFFICES	8
9	LIGHTS/CONFERENCE RM	20 / 1	680 / 900		20 / 1	RECPTS/OFFICES	10
11	LIGHTS/LUNCH RM	20 / 1		1020 / 1060	20 / 1	RECPTS/OFFICES	12
13	LIGHTS/RESTROOMS	20 / 1	660 / 900		20 / 1	RECPTS/OFFICES	14
15	SPACE				20 / 1	RECPTS/RECEPTION	16
17	SPACE			1060	20 / 1	RECPTS/CLSING RESTRMS	18
19	SPACE		1080		20 / 1	RECPTS/LUNCH RM	20
21	CONDENSER C-1*	50 /		1500	20 / 1	RECPTS/LUNCH RM	22
23	CONDENSER C-1*	2	3750 / 1500		20 / 1	FURNACE F-1	24
25	CONDENSER C-2*	50 /	3750 / 1500	3750 / 1500	20 / 1	FURNACE F-2	26
27	CONDENSER C-2*	2	3750 / 1500	3750 / 1500	20 / 1	WATER HEATER	28
29	SPACE					SPACE	30

		ØA	ØB		
CONNECTED VA PER PHASE		19.250	19.820	NOTES:	
CONNECTED AMPS PER PHASE		160.42	165.17	*PROVIDE HACR TYPE C.B.	
CONTINUOUS LIGHTING @ 25%		21.40	19.40		
LARGEST MOTOR @ 25%		6.00	6.00		
CONNECTED VA PER PHASE		187.82	190.57		

Figure 15–16.
This branch circuit schedule is for panel "A" identified in the first-floor power plan shown in Figure 15–15. Each circuit is identified by number with a circuit description. For example, circuit #1 controls a series of office lights and has a single 20-amp circuit breaker. Other circuit descriptions are for the reception, conference, lunch, and rest rooms. There is a whole series of receptacle (RCPTS) circuits. Each furnace and water heater has a separate individual circuit. (*Courtesy of Josh Barclay Architects Ltd., Flagstaff, Arizona*)

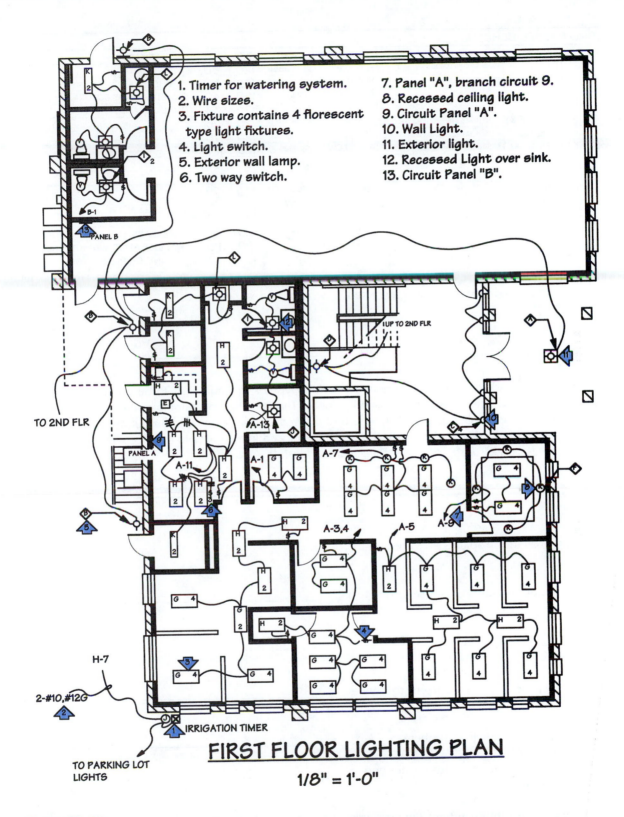

1. Timer for watering system.
2. Wire sizes.
3. Fixture contains 4 florescent type light fixtures.
4. Light switch.
5. Exterior wall lamp.
6. Two way switch.
7. Panel "A", branch circuit 9.
8. Recessed ceiling light.
9. Circuit Panel "A".
10. Wall Light.
11. Exterior light.
12. Recessed Light over sink.
13. Circuit Panel "B".

FIRST FLOOR LIGHTING PLAN

1/8" = 1'-0"

Figure 15–17.
The lighting plan shows the wiring circuits to all the lighting fixtures. The letter identifying each fixture corresponds to the letter in the mark (MK) column of the lighting fixture schedule shown in Figure 15–18. The number of lamps in each fixture is also given. For example, 4-G shows that the fixture type "G" has four fluorescent lamps. *(Courtesy of Josh Barclay Architects Ltd., Flagstaff, Arizona)*

LIGHTING FIXTURE SCHEDULE

MK	QNTY	LOCATION	MFG. NO.	LAMPS	MOUNTS	VOLTS	WATTS	REMARKS
A	1	ENTRY	PRESOLITE HD LOCOS	50W HPS	RECESSED	120	50	
B	4	REAR DR.	PRESOLITE HD 10 WCB	50W HPS	WALL	120	50	WT PHOTOCELL
C	4	LOBBY	LITECONTROL MOD S-D-10213-WA-CSO-WL-120	1-40W CW	WALL	120	40	OLD BURGANDY COLOR
D	1	LOBBY	LITECONTROL OPALESE S-D-10213-WA-CSO-WL-120	2-13W.TT	WALL	120	36	OLD BURGANDY COLOR
E	4	2ND FLR LOBBY	HALO H-44T-446(SPECULAR CLR REFL)	75W P30	RECESSED	120	75	
F	4	CONFERENCE	HALO H-43T-436(COILEX BAFFLE)	75W P30	RECESSED	120	75	
G	47	OFFICE	LITHONA PM4-2 LAMP 2X4-G	4-40W CW	GRID CLG	120	160	
H	20	OFFICE	LITHONA PM3-2 LAMP 2X4-G	2040W CW	GRID CLG	120	80	
I	6	TOILET	HALO H-2T-21G-P	150W A	RECESSED	120	150	
J	1	STORAGE	HALO H-46T-41	75W A19	RECESSED	120	75	
K	4	MECH RM	LITHONA DM240	2-40W CW	SURFACE	120	40	
L	3	JANITOR	HALO 2932T	80W A19	RECESSED	120	60	
M	2	PARKING LOT	SPAULDING OKI-55-LPS X 20' STL POLE	55W LPS	POLE	120	55	SINGLE ARM LIGHT W/PHOTOCELL
N	2	PARKING LOT	(2)SPAULDING OKI-55-LPS FIXTURES ON 20° POLE	HEAT TRACE	POLE	120	110	TWIN 150° APART W/PHOTOCELL
O	4	HEAT TRACE	EASY HEAT		GUTTER	120	1200	W/INLINE THERMOSTATS

KEY NOTE:

(1) WIRE FIXTURES SO THAT EACH SWITCH CONTROLS TWO LAMPS IN EACH FIXTURE.

Figure 15–18.
The lighting fixture schedule provides information about the type of lighting outlets to be installed. For example, mark (MK) "C" identifies four lamps in the lobby. The manufacturer's name and item number are given. The lamps are wall-mounted, 120 volts and 40 watts, and are an "Old Burgundy" color. Mark "O" shows a heat trace system that consists of a heat tape placed in the gutter to melt ice when water begins to freeze. *(Courtesy Josh Barclay Architects Ltd., Flagstaff, Arizona)*

REVIEW QUESTIONS

Enter the missing words in the spaces provided on the right.

1. The _____(use abbreviation) is used by most localities as a model electric code.

2. The _____ _____is the system bringing electricity into the building.

3. A unit of force causing electric current to flow is called a _____.

4. Volts times amperes equals the number of _____.

5. A _____reduces the voltage from the power source to the building.

6. A _____ _____is the point where feeder wires attach to the building.

7. The electricity must pass through a ___ ___before reaching the service panel.

(Review questions continued on page 206)

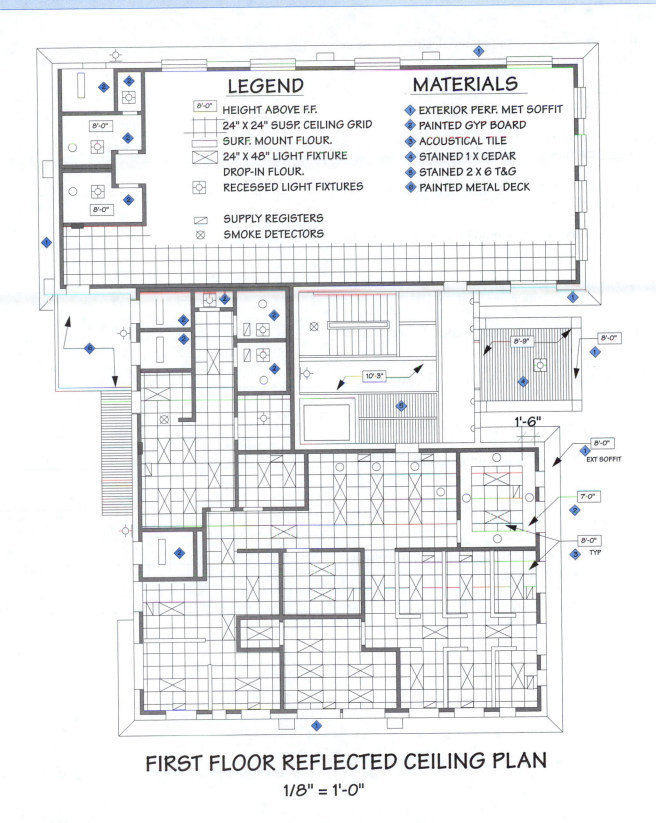

FIRST FLOOR REFLECTED CEILING PLAN

1/8" = 1'-0"

Figure 15–19.
The reflected ceiling plan shows the finished ceiling. The legend above provides the symbols for the lights, ceiling grid, warm air supply registers, and smoke detectors. A list of finish materials is also included. *(Courtesy of Josh Barclay Architects Ltd., Flagstaff, Arizona)*

8. Fuses or _____ _____ open an electric current in the event of an overload.

_____

9. Electricity is distributed through _____ _____ to the entire building.

_____

10. Convenience outlets are controlled by _____ _____ circuits.

_____

11. A washing machine may have to plug into a _____ _____ outlet.

_____

12. Convenience outlets should be placed close to the _____ of wall areas.

_____

13. Any point of usable wall space should be at least _____ from a convenience outlet

_____

14. The most frequently used type of switch is a _____ switch.

_____

15. A circuit can be controlled from two locations by a _____ _____ switch.

_____

16. _____ switches are silent and shockproof.

_____

17. The placement of wall outlets is shown on the _____ plan.

_____

18. _____ _____ schedules supplement information on the power plan.

_____

19. Switches and wires of overhead lighting systems may be shown in a _____ plan.

_____

20. Code numbers of lights on a lighting plan are also found in the _____ schedule.

_____

UNIT 16

Plumbing Plans

Separate plumbing drawings are added to a set of construction plans for commercial and larger residential buildings. These often include plan views and on-line isometric schematic diagrams of water supply, waste removal, and gas supply systems.

All plumbing work in a building must be done according to strict plumbing code requirements. Most cities, counties, or parishes adapt their local codes from a number of national plumbing codes. Two national codes widely used are the *Uniform Plumbing Code* and the *BOCA Plumbing Code*.

Plumbing plans are used by the plumbing contractor to estimate cost and supervise the work. Plumbers read the plans to install the water and waste systems and set in place the plumbing fixtures. To do this properly, they need to have an understanding of the general construction of the building.

THE WATER SUPPLY

The water supply of a building usually originates from a public municipal water source such as a lake or reservoir. In more remote areas a water supply may come from a private well. The large pipes that convey the water from its source to the community where it is being used are called **water mains.** The location of a water main below the street surface is identified on the plot plan.

A **water supply line** (also called a **water service line**) taps into the water main and conveys water to the building. The water supply line should be below the frost line to avoid the possibility of water freezing in the line during severe winter weather. A **shutoff valve** is required in a curb box close to a property line and at the point of entry next to the water meter. Other shutoff valves may be placed within the system. The water meter records how much water is used by the property owner or tenant. **(See Figure 16–1.)**

The water supply line branches into **distribution lines** that direct cold water to **fixture branch lines.** The branch lines lead to **fixture supply lines** that connect to the fixtures. Distribution lines also go from the water supply line to the hot water heaters. A hot water distribution line then moves hot water from the heater to the hot water fixture branch lines and then to the fixture

supply lines. Hot water and cold water shutoff valves should be installed at the point of connection to the fixtures. **(See Figure 16–2.)**

Materials used for piping are *galvanized steel, copper,* or *plastic.* Supply lines are often 1″ in diameter. Distribution lines are usually 3/4″ or 1/2″ in diameter. Some more frequently used fittings for water supply systems are *elbows, tees, couplings,* and *reducers.*

WASTE REMOVAL SYSTEMS

The waste removal system discharges solid and liquid wastes from the building. Its main components are vertical soil and waste stacks, and vertical vent stacks that run between the studs and joists of the building. Waste material is discharged by gravity drainage. Waste lines are much greater in diameter than water lines, and all horizontal runs should have a minimum slope of 1/4″ per foot.

Pipes 3″ and 4″ in diameter are commonly used for stack material, which may be *cast iron, rustproof copper,* or *brass alloy.* Some more frequently used fittings are *elbows, bends, tees,* and *clean-outs.*

A curved device called a **P-trap** is found beneath each fixture. Water remains in the bend of the trap, preventing the escape of gases through the drain opening in the fixture. This does not affect the flow of discharge. **(See Figure 16–3.)** A P-trap is not required beneath a toilet as the water remaining in the bowl prevents gas from escaping.

A **fixture drain pipe** is connected to the fixture trap and carries the discharge to a **horizontal branch** or **soil stack.** The horizontal branch carries the discharge from the floor above and connects with a vertical soil stack or waste stack. A **soil stack** receives the discharge from toilets and urinals and may also receive discharge from other fixtures. A **waste stack** receives only the discharge from sinks, bathtubs, and shower stalls. **Clean-outs** are found at the base of all soil and waste stacks, making it possible to eliminate congestion and backup in the stacks.

Vertical pipes called **vent stacks** extend up from the waste and soil stacks and project a minimum of 12″ through the roof to allow for air circulation and the escape of gas accumulated in the drainage system.

A **house sewer,** also called a **building drain,** con-

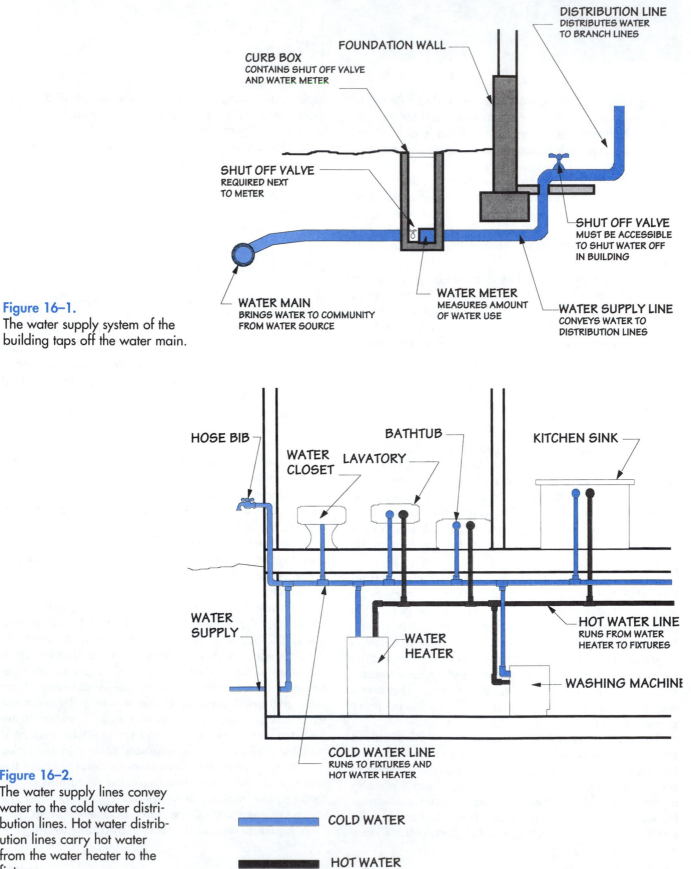

DISTRIBUTION LINE
DISTRIBUTES WATER
TO BRANCH LINES

FOUNDATION WALL

CURB BOX
CONTAINS SHUT OFF VALVE
AND WATER METER

SHUT OFF VALVE
REQUIRED NEXT
TO METER

SHUT OFF VALVE
MUST BE ACCESSIBLE
TO SHUT WATER OFF
IN BUILDING

WATER MAIN
BRINGS WATER TO COMMUNITY
FROM WATER SOURCE

WATER METER
MEASURES AMOUNT
OF WATER USE

WATER SUPPLY LINE
CONVEYS WATER TO
DISTRIBUTION LINES

Figure 16–1.
The water supply system of the
building taps off the water main.

HOSE BIB

WATER CLOSET

LAVATORY

BATHTUB

KITCHEN SINK

WATER SUPPLY

WATER HEATER

HOT WATER LINE
RUNS FROM WATER
HEATER TO FIXTURES

WASHING MACHINE

COLD WATER LINE
RUNS TO FIXTURES AND
HOT WATER HEATER

COLD WATER

HOT WATER

Figure 16–2.
The water supply lines convey
water to the cold water distribution lines. Hot water distribution lines carry hot water
from the water heater to the
fixtures.

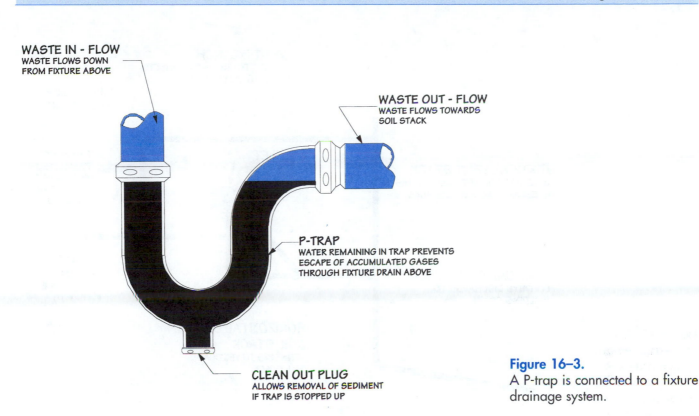

WASTE IN - FLOW
WASTE FLOWS DOWN
FROM FIXTURE ABOVE

WASTE OUT - FLOW
WASTE FLOWS TOWARDS
SOIL STACK

P-TRAP
WATER REMAINING IN TRAP PREVENTS
ESCAPE OF ACCUMULATED GASES
THROUGH FIXTURE DRAIN ABOVE

CLEAN OUT PLUG
ALLOWS REMOVAL OF SEDIMENT
IF TRAP IS STOPPED UP

Figure 16–3.
A P-trap is connected to a fixture drainage system.

nects to the lowest points of the drainage system of the building and conveys the waste to the municipal sewer system identified on the plot plan. A separate building **storm sewer** may be included to carry rainwater coming from downspouts connected to the roof gutters. Surface water and groundwater around the building can also be directed to the house sewer through drain tiles. The house sewer than carries the water to the **municipal storm sewer** that is also shown on the plot plan. **(See Figure 16–4.)**

GAS LINES

Gas lines are installed in a building where they can tap into underground gas utility lines placed beneath the street surface adjoining the construction. Gas is used to fuel furnaces, boilers, stoves, hot water heaters, and dryers.

The gas utility company installs the gas line running from the gas utility line to the building. The utility company will also install a **gas meter** to measure how much gas is used. The plumber will then put in place the gas lines where required. Gas lines below the ground are often made of *black iron*, although *plastic* is also used. Common pipe diameter sizes are 1", 3/4", or 1/2". The plot plan will show the location of the below-

surface gas utility line and the connecting gas line running to the building.

SEPTIC SYSTEMS

Septic systems are used in rural or isolated areas that do not have a public sewer system. A complete septic system consists of a *septic tank* and a *disposal field* (both described here). Septic systems must be carefully planned and properly constructed. They function best in dry and porous soils, preferably composed of sand and gravel. Doing a *percolation soil test* before installing a septic system is advisable to make sure the soil will absorb the liquid waste properly. Continuous maintenance of the system is necessary to prevent the improper disposal of sewage that can produce unpleasant odors in the surrounding area.

SEPTIC TANKS

A **septic tank** is a watertight boxlike structure built of reinforced concrete or concrete blocks. **(See Figure 16–5.)** It is placed below the ground surface, and its dimensions are generally determined by the building size and number of occupants. If automatic washers or garbage disposal units will be used, those are addition-

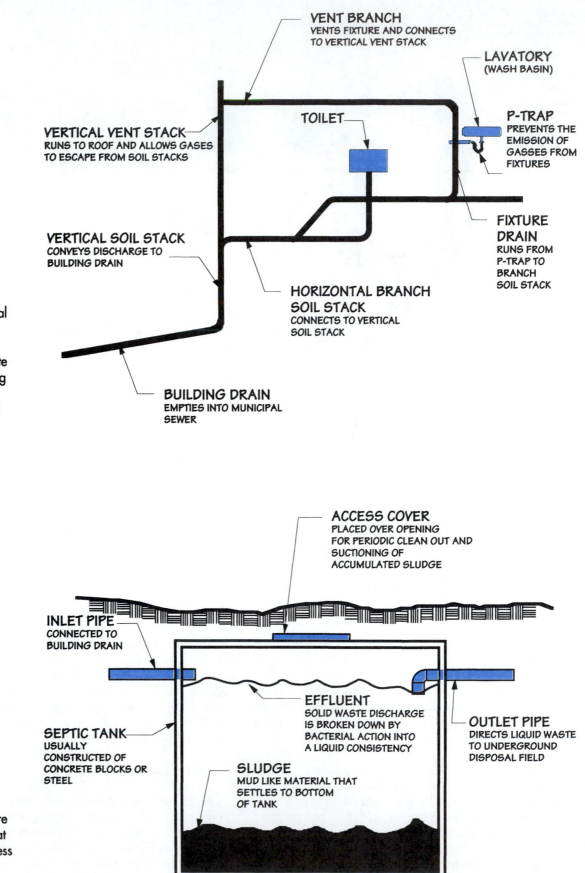

VENT BRANCH
VENTS FIXTURE AND CONNECTS
TO VERTICAL VENT STACK

LAVATORY
(WASH BASIN)

P-TRAP
PREVENTS THE
EMISSION OF
GASSES FROM
FIXTURES

TOILET

VERTICAL VENT STACK
RUNS TO ROOF AND ALLOWS GASES
TO ESCAPE FROM SOIL STACKS

**FIXTURE
DRAIN**
RUNS FROM
P-TRAP TO
BRANCH
SOIL STACK

VERTICAL SOIL STACK
CONVEYS DISCHARGE TO
BUILDING DRAIN

**HORIZONTAL BRANCH
SOIL STACK**
CONNECTS TO VERTICAL
SOIL STACK

BUILDING DRAIN
EMPTIES INTO MUNICIPAL
SEWER

Figure 16–4.
The waste removal system drains waste water and water-borne waste from the plumbing fixtures such as sinks, toilets, and tubs.

ACCESS COVER
PLACED OVER OPENING
FOR PERIODIC CLEAN OUT AND
SUCTIONING OF
ACCUMULATED SLUDGE

INLET PIPE
CONNECTED TO
BUILDING DRAIN

EFFLUENT
SOLID WASTE DISCHARGE
IS BROKEN DOWN BY
BACTERIAL ACTION INTO
A LIQUID CONSISTENCY

OUTLET PIPE
DIRECTS LIQUID WASTE
TO UNDERGROUND
DISPOSAL FIELD

SEPTIC TANK
USUALLY
CONSTRUCTED OF
CONCRETE BLOCKS OR
STEEL

SLUDGE
MUD LIKE MATERIAL THAT
SETTLES TO BOTTOM
OF TANK

Figure 16–5.
Septic systems are used in areas that do not have access to public sewer systems.

al factors in determining the appropriate size for the septic tank.

Waste discharge from the building enters the septic tank through an inlet pipe. Once in the tank, solid waste is broken down by bacterial action into a liquid or a mud like consistency called *sludge*. The sludge settles to the bottom of the tank and must periodically be suctioned out through the access holes at the top of the tank.

DISPOSAL FIELDS

The outlet pipe from the septic tank directs the liquid waste to the underground **disposal field** where it enters a network of **drain lines.** Also called **leach lines,** these drain lines are made of clay or perforated plastic or fiber. The liquid waste flows from the drain lines into the surrounding soil. Drain lines should be placed almost level, with a recommended slight slope of 1" per 50'. They should follow the contour of the land and be perpendicular with the slope.

Drain lines should be positioned so that surface water is diverted away from the disposal field. If the building water supply comes from an underground well, the drain lines should be placed downhill from the well. Local building codes usually specify the minimum distance from the house to the disposal field. The location of a septic tank and its disposal field is shown on the plot plan. **(See Figure 16–6).**

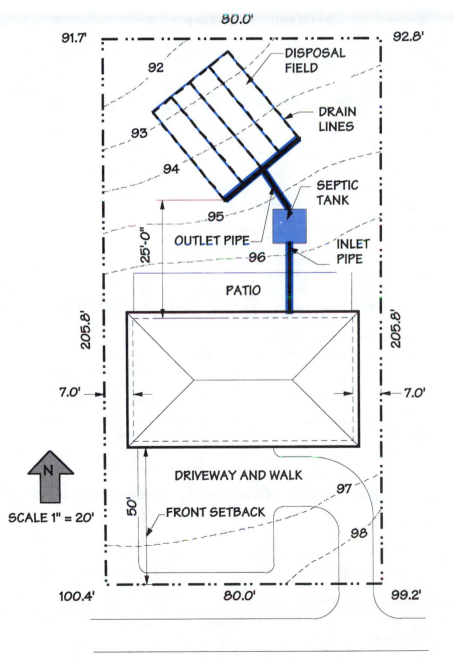

Figure 16–6.
The septic tank and disposal field are shown on the plot (site) plan.

READING THE PLUMBING PLANS

Plumbing plans are drawn by the architect who will consult with the plumbing contractor or a mechanical engineer when necessary. Regular floor and elevation plans identify fixtures. Full plumbing plans add lines showing pipes carrying water, waste, and gas. Symbols and abbreviations identify the pipes and other plumbing components such as fittings, valves, clean-outs, and drains. **(See Figure 16–7.)**

Important plumbing information is also contained in the plan specifications. Here the Plumbing division will explain the scope of plumbing work to be done and any special instructions to be followed. A description of the pipe and stack material is included, and all fixtures are identified with the names of manufacturers, colors, and model numbers.

PLUMBING DESIGN FACTORS

Several factors should be considered when laying out

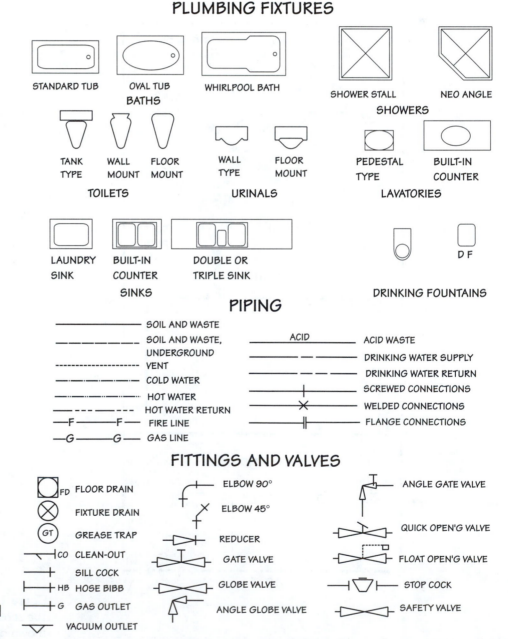

Figure 16–7.
Here are some commonly used plumbing symbols.

and designing a plumbing system. All runs of the water pipes should be as straight and short as possible with the least amount of fittings. A **run** is a section of pipe continuing in a straight line in the direction of the water flow. A straight run conserves material and reduces internal friction against flow in the pipe. Internal friction can cause pressure loss from the water main to the fixture.

If possible, rooms with plumbing fixtures (bathrooms, kitchens, laundry rooms) should be placed back to back. This makes it possible for pipes, stacks, and vents for fixtures in adjoining rooms to be shared and placed in the same wall. As a result, installation cost is reduced and plumbing noises are isolated from other living areas of the building. If feasible, rooms at different floor levels should be placed so that pipes, stacks, and vents are found in vertically aligned walls. This shortens the pipe runs and simplifies the framing of the building. **(See Figure 16–8.)**

Walls containing shared plumbing are called chase **walls** or **wet walls.** To allow more space for these components, the walls are constructed of 2 × 6s or two side-by-side 2 × 4 walls.

PLUMBING PLANS AND SCHEMATIC DRAWINGS

A plan view plumbing drawing shows the room layout and the locations of fixtures and horizontal pipe and waste runs. **(See Figure 16–9.)** Compare this plumbing plan with the floor plan of the first floor of the building (Figure 15-14, page 200). Separate plan view plumbing drawings may be provided for kitchen, bathroom, and laundry rooms, because most of the building's plumbing system occurs in those areas.

Single-line isometric **schematic drawings** provide detailed elevation views of the plumbing system that cannot be shown in a plan view. A *waste schematic drawing* gives the diameters and also the horizontal and vertical locations of the soil, waste, and vent stacks and the clean-out openings. **(See Figure 16–10).** A *water schematic drawing* provides information about the pipe runs of the hot and cold water systems, water heaters, and shutoff valves. **(See Figure 16–11.)** A *gas schematic drawing* may also be included to show the gas pipes and hookups to furnaces, hot water heaters, stoves, and dryers. **(See Figure 16–12.)**

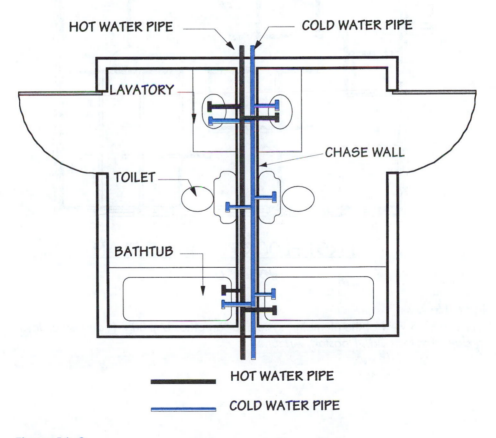

Figure 16–8.
Back-to-back bathrooms are separated by a chase wall. The same hot and cold distribution pipes service the fixtures on the opposite side of the wall.

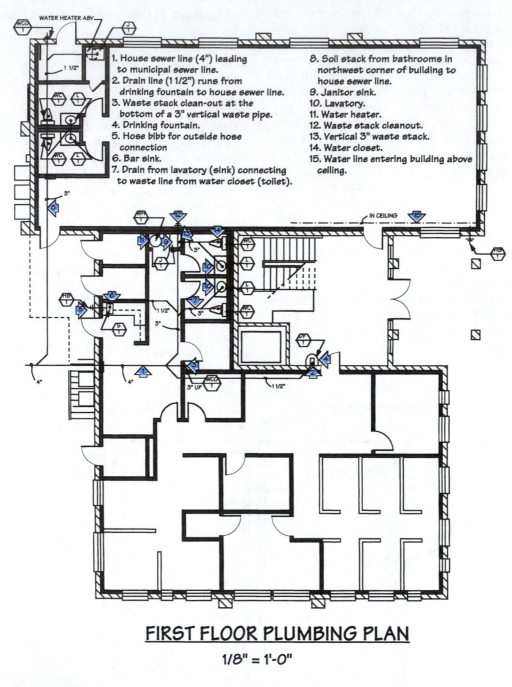

1. House sewer line (4") leading to municipal sewer line.
2. Drain line (1 1/2") runs from drinking fountain to house sewer line.
3. Waste stack clean-out at the bottom of a 3" vertical waste pipe.
4. Drinking fountain.
5. Hose bibb for outside hose connection
6. Bar sink.
7. Drain from lavatory (sink) connecting to waste line from water closet (toilet).
8. Soil stack from bathrooms in northwest corner of building to house sewer line.
9. Janitor sink.
10. Lavatory.
11. Water heater.
12. Waste stack cleanout.
13. Vertical 3" waste stack.
14. Water closet.
15. Water line entering building above ceiling.

FIRST FLOOR PLUMBING PLAN
1/8" = 1'-0"

Figure 16–9.
Here is a plan view of the plumbing system for an office building. *(Courtesy of Josh Barclay Architects Ltd., Flagstaff, Arizona)*

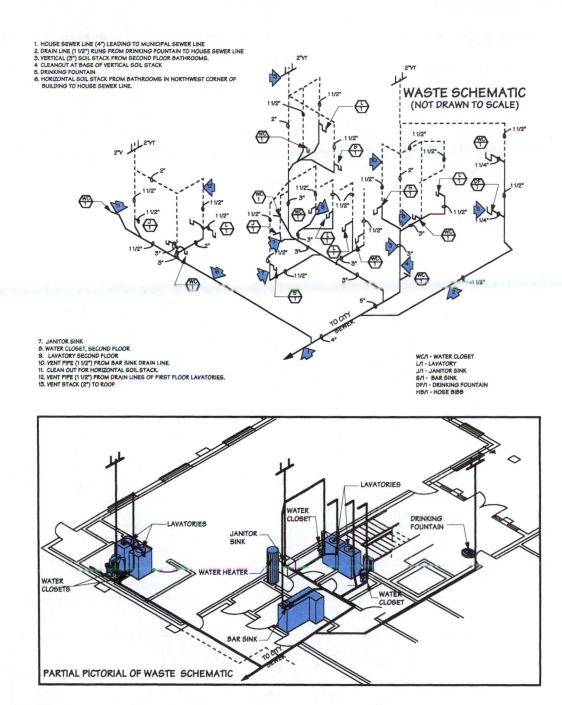

1. HOUSE SEWER LINE (4") LEADING TO MUNICIPAL SEWER LINE
2. DRAIN LINE (1 1/2") RUNS FROM DRINKING FOUNTAIN TO HOUSE SEWER LINE
3. VERTICAL (3") SOIL STACK FROM SECOND FLOOR BATHROOMS.
4. CLEANOUT AT BASE OF VERTICAL SOIL STACK
5. DRINKING FOUNTAIN
6. HORIZONTAL SOIL STACK FROM BATHROOMS IN NORTHWEST CORNER OF BUILDING TO HOUSE SEWER LINE.

7. JANITOR SINK
8. WATER CLOSET, SECOND FLOOR
9. LAVATORY SECOND FLOOR
10. VENT PIPE (1 1/2") FROM BAR SINK DRAIN LINE.
11. CLEAN OUT FOR HORIZONTAL SOIL STACK.
12. VENT PIPE (1 1/2") FROM DRAIN LINES OF FIRST FLOOR LAVATORIES.
13. VENT STACK (2") TO ROOF

WC/1 - WATER CLOSET
L/1 - LAVATORY
J/1 - JANITOR SINK
S/1 - BAR SINK
DF/1 - DRINKING FOUNTAIN
HB/1 - HOSE BIBB

WASTE SCHEMATIC
(NOT DRAWN TO SCALE)

PARTIAL PICTORIAL OF WASTE SCHEMATIC

Figure 16–10.
The schematic drawing of the waste system shows the placement of soil and vent stacks. A pictorial drawing is shown below. *(Courtesy of Josh Barclay Architects Ltd., Flagstaff, Arizona)*

PLUMBING FIXTURE SCHEDULES

Fixture schedules are often found with the plumbing drawings. They contain information that space does not allow for on the plan or elevation view drawings. The manner of presenting the information contained in the schedule may vary slightly with different architects. Most, however, will identify the fixture by symbol or location and give the number of fixtures required. The manufacturer's name and the order and/or model number are given. There is usually a "remarks" column for special information. **(See Figure 16–13.).**

More specialized schedules may be included to provide information about mechanical equipment such as water heaters, furnaces, condensers, and other plumbing-related machinery.

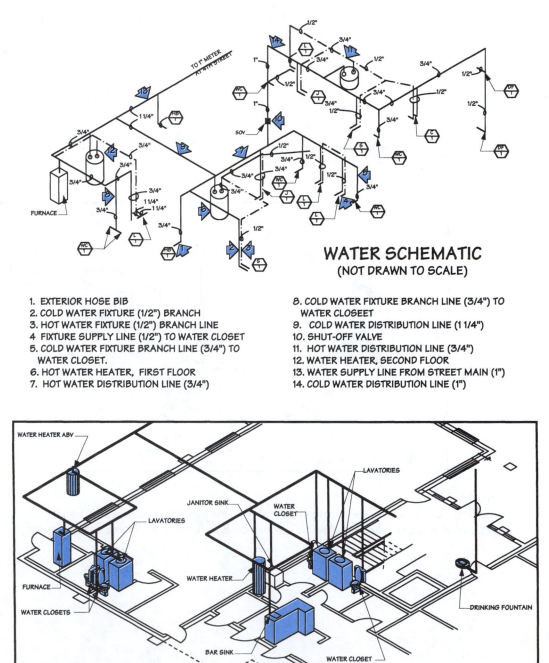

WATER SCHEMATIC
(NOT DRAWN TO SCALE)

1. EXTERIOR HOSE BIB
2. COLD WATER FIXTURE (1/2") BRANCH
3. HOT WATER FIXTURE (1/2") BRANCH LINE
4. FIXTURE SUPPLY LINE (1/2") TO WATER CLOSET
5. COLD WATER FIXTURE BRANCH LINE (3/4") TO WATER CLOSET.
6. HOT WATER HEATER, FIRST FLOOR
7. HOT WATER DISTRIBUTION LINE (3/4")
8. COLD WATER FIXTURE BRANCH LINE (3/4") TO WATER CLOSEET
9. COLD WATER DISTRIBUTION LINE (1 1/4")
10. SHUT-OFF VALVE
11. HOT WATER DISTRIBUTION LINE (3/4")
12. WATER HEATER, SECOND FLOOR
13. WATER SUPPLY LINE FROM STREET MAIN (1")
14. COLD WATER DISTRIBUTION LINE (1")

Figure 16–11.
The schematic drawing of the water system shows the placement of water pipes. A pictorial drawing is shown below. *(Courtesy of Josh Barclay Architects Ltd., Flagstaff, Arizona)*

PARTIAL PICTORIAL OF WATER SCHEMATIC

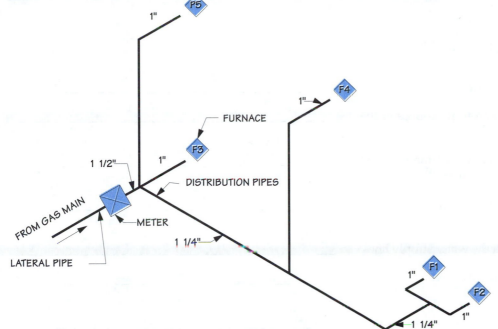

GAS SCHEMATIC

(NOT DRAWN TO SCALE)

Figure 16–12.
This gas schematic drawing shows a lateral (1 1/2") pipe coming from the gas main to the gas meter. From there the gas is carried by distribution pipes (1 1/4") to the gas furnaces and other gas outlets.

PLUMBING FIXTURE SCHEDULE				
CODE	TYPE	NO. REQ'D	MANUFACTURER & CATALOG NO.	REMARKS
WC/1	WATER CLOSET	4	CRANE ECONOMIZER	1 1/2 GAL. FLUSH, HANDICAP ELONGATED BOWLS 18" HIGH
L/1	LAVATORY	4	LAUFEN #BV 4011-1341.8	MUST COMPLY WITH ADA REQUIREMENTS AS DETAILED
WH/1	WATER HEATER	2	BRADFORD WHITE M16055513	ELECTRIC 7 GALLON
J/1	JANITOR SINK	2	FIAT #MSB 2424	WITH SERVICE FAUCET, HOSE BRACKET AND RUBBER TOOL GRIP
DF/1	DRINKING FOUNTAIN	1	HALSEY TAYLOR ##BFC - 8F	
S/1	PARTY SINK	1	ELKA 16 X 16	S.S. WITH GOOSE NECK
WCO/1	WALL CLEAN OUT	4	JAY R. SMITH 4402	S.S. COVER
HB/1	HOSE BIBB	2	JAY R. SMITH 5611	SELF DRAINING, ANTI SIPHON DEVICE

Figure 16–13.
This plumbing fixture schedule provides information about the fixtures shown in the first-floor plumbing plan (Figure 16–9).

REVIEW QUESTIONS

Enter the missing words in the spaces provided on the right.

1. BOCA is a national _____c ode.

   _____

2. _____ _____ convey water from its source to the locality where it is to be used.

   _____

3. The pipe that goes from the water main to the building is called the ____ ____ line.

   _____

4. The amount of water use for a building is recorded by a ____ ____.

   _____

5. _____ lines take water from the water supply lines.

   _____

6. Pipes that connect to fixtures are called ____ ____ lines.

   _____

7. Soil, waste, and vent _____ are the main components of a waste removal system.

   _____

8. The horizontal slopes of waste lines should be a minimum of ____ per foot.

   _____

9. The escape of gases from fixture drain openings is prevented by a ____ ____.

   _____

10. A projection of _____ above the roof is the minimum required for a vent stack.

    _____

11. Gas accumulating in a drainage system escapes through vertical ____ ____.

    _____

12. The purpose of a _____ is to eliminate congestion and backup in a stack.

    _____

13. The pipe running from the gas utility line to the building is placed by the utility _____.

    _____

14. Areas that don't have a public sewer system can use a _____ system for waste disposal.

    _____

15. Cost is reduced by ____ __ ____ fixture installation.

    _____

16. _____ walls contain shared plumbing.

    _____

17. Horizontal pipe and waste runs are shown in a ____ ____ plumbing drawing.

    _____

18. Elevation views of the plumbing system are shown in _____ drawings.

    _____

19. Single-line _____ drawings are used for schematic plans.

    _____

20. Additional plumbing fixture information is often provided by a fixture _____.

    _____

Heating, ventilation, and air conditioning **(HVAC) plans** provide information for the climate control system of a building's interior. The purpose of HVAC is to maintain a comfortable and healthy year-round interior temperature. The HVAC system produces heat when the outside temperature drops during the winter months, and cool air when the outside temperature rises during the hot summer season. A properly functioning HVAC system will maintain interior temperatures of 70 to 75 degrees Fahrenheit throughout the year.

For smaller residential projects, including the HVAC information on the floor plan may be sufficient. Such information will identify the location of the furnace and the warm-air (WA) and return-air (RA) openings. A qualified sheet metal worker will know how to connect the ducts from the furnace to the openings. Larger and more complex structures will require separate HVAC plans that show the duct runs and other additional information. The architect may require the assistance of a mechanical engineer to prepare these plans.

Besides heating and cooling systems, other factors have an important bearing on interior temperature control. Insulation materials placed in the outside walls, floors, ceilings, and roofs help prevent the transfer of hot and cold air through the shell of the building. Proper ventilation prevents the condensation of moisture in the attic and foundation crawl space. Correct solar orientation of walls and windows reduces the flow of heat or cold into the building. Insulation, ventilation, and solar orientation are discussed in Section 4.

HEAT TRANSMISSION

The three methods of heat transmission are conduction, convection, and radiation. They can be used individually or in combination with each other to generate and distribute heat within an interior space. *Conduction* moves heat through solid materials. Therefore, dense materials are the best conductors. For example, steel is a better conductor than wood. In *convection*, heat is distributed in a circular motion. This is caused by the constant rise of warm air toward the top of a heated space and its replacement by cold air at the bottom of the space. *Radiation* is the direct movement of heat through

space from a warm source. The radiated heat warms the air, which warms the walls and ceilings that reflect heat back into the living space. **(See Figure 17–1.)**

HEATING SYSTEMS

Many heating methods are used today. They are generally divided into forced-air, radiant, hot-water, heat-pump, and plenum systems. The cost and structural design of the building determine which system is best to use. All heating (and cooling) systems are thermostatically controlled.

THERMOSTATS

A **thermostat** is a device that responds to temperature changes. **(See Figure 17–2.)** When set to a desired heating level, it will control the heating system of a building to maintain that level. In **central heating,** one thermostat controls the heating for the entire building. In **zoned heating,** thermostats are placed in different areas of a building. Larger structures may require zoned heating. Some thermostats are combined with timers. These conserve fuel by lowering the heat setting at different times of the day.

The location of the thermostat is very important for proper operation of the heating system. A thermostat should not be placed against an exterior wall. The cooler surface of the wall will cause the thermostat to register a temperature lower than the actual temperature inside the building. This will cause the heating system to circulate more warm air than is necessary. Thermostats should also be placed away from heat sources such as warm air registers, fireplaces, and where direct sunlight enters from windows. They should be placed centrally in the building so that an average temperature reading is more likely.

FORCED-AIR HEATING

The heated air of a **forced-air heating system** originates from a furnace fueled by natural gas, LP (liquid petroleum) gas, or electricity. Natural gas is the best choice where municipal hookups are available. Heated air is propelled out of the furnace by a motor-driven blower and travels through ducts leading to **warm-air (WA) outlets** (also called **supply outlets**) in different rooms of the building. WA outlets are covered with louvered **registers** that can be open or shut. WA registers are most effective when they are in the lower part of a wall or in the floor. However, this placement is not possible with some structures, such as buildings with slab-on-grade foundations.

Return-Air Systems

A **return-air (RA) system** operates as follows:

1. Warm air from the furnace is distributed in the building through the supply outlets emerging from the warm air supply ducts.
2. A convection loop takes place as the warm air rises toward the ceiling and is replaced by cooler air below.
3. The cooler air is then returned to the furnace.
4. The cool air passes through a filter and then over a heating coil or combustion chamber that reheats the air.
5. The blower propels the warm air into the supply ducts, and the entire heating process is repeated.

A properly designed return-air system will assure even temperature levels throughout a building.

Air may be returned to the furnace either through ducts or grills. When **ducts** are used, the cool air is returned to the furnace from openings placed at various locations in the building. **(See Figure 17–3.)** *Non-ducted*

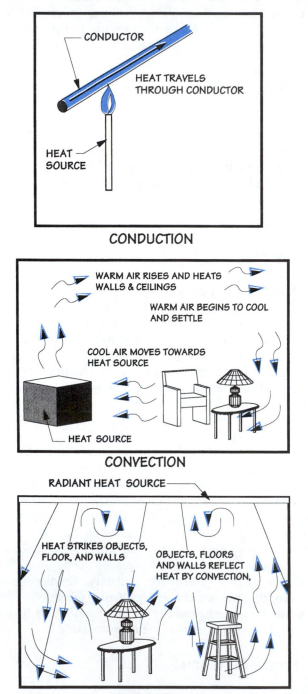

Figure 17–1.
The three methods of heat transmission are conduction, convection, and radiation.

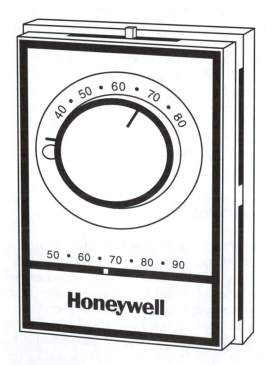

Figure 17–2.
A thermostat is a device that responds to temperature changes. It can be set to the desired heating or cooling level of a building. *(Courtesy of Honeywell, Inc.)*

systems permit air to return directly to the furnace through *grills* or *louvers*. **(See Figure 17–4.)**

Furnace Types

Three of the commonly used forced-air heating designs are the upflow, downflow, and horizontal systems. A key element in each system is the placement of the furnace, and this is determined by the structural design of the building. For best heat distribution, the furnace should be as close as possible to the center of the building.

An **upflow furnace** (also called an **updraft furnace**) is installed in a full basement. The ducts are placed between or below the floor joists. A **downflow furnace** (also called a **downdraft furnace**) is placed in the garage or elsewhere on the main-floor level of the building. This system is used in buildings over crawl-space foundations. A **horizontal furnace** hangs below the floor over a crawl-space foundation or is placed in the attic of a house built over a slab foundation. **(See Figure 17–5.)**

Perimeter Heating

Warm air in the interior of a building is drawn to the outside cold air through the outside walls and windows of a building. In **perimeter heating** all the warm-air (WA) outlets are placed against the outside walls and under the windows. This placement warms the interior surfaces of the walls and windows and lessens the heat loss. WA registers are frequently used for this purpose. **(See Figure 17–6.)** Baseboard panels may also be used, offering the advantage of better heat distribution along the wall surface.

HYDRONIC (HOT WATER) HEATING

In **hydronic heating systems,** hot water is pumped from a boiler and carried by pipe to the heat outlets or distributors. As the water cools, it flows back to the boiler. Hydronic heating systems operate quietly and produce a comfortable, clean heat. Zone control, in which each room has its own thermostat, is facilitated by hydronic

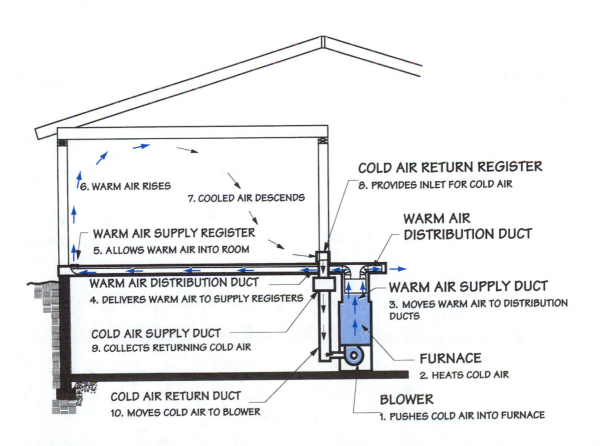

Figure 17–3.
In the forced-air heating method, warm air is supplied by a furnace and the air is returned to the furnace. In the above example a duct system returns the cold air to the furnace.

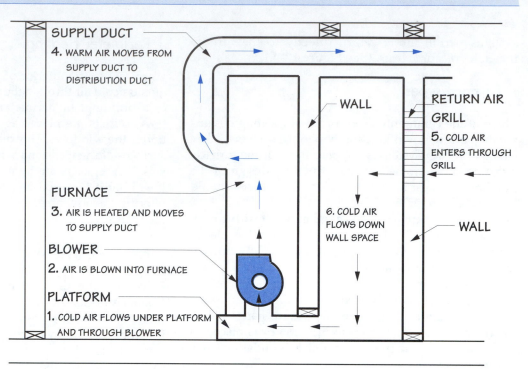

SUPPLY DUCT

4. WARM AIR MOVES FROM SUPPLY DUCT TO DISTRIBUTION DUCT

WALL

RETURN AIR GRILL

5. COLD AIR ENTERS THROUGH GRILL

FURNACE

3. AIR IS HEATED AND MOVES TO SUPPLY DUCT

6. COLD AIR FLOWS DOWN WALL SPACE

WALL

BLOWER

2. AIR IS BLOWN INTO FURNACE

PLATFORM

1. COLD AIR FLOWS UNDER PLATFORM AND THROUGH BLOWER

Figure 17–4.
Cold air moves through a grill in this non-ducted return-air system. This method is used in the HVAC plans shown in Figure 17–13.

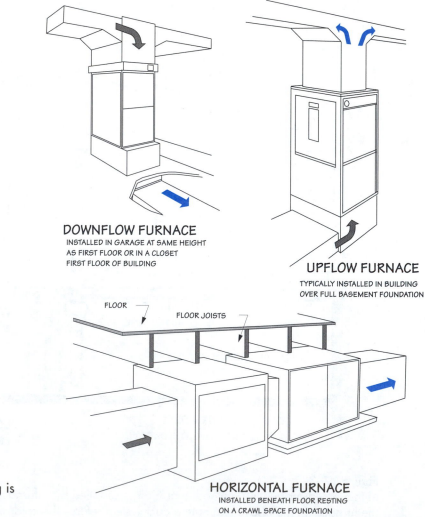

DOWNFLOW FURNACE
INSTALLED IN GARAGE AT SAME HEIGHT AS FIRST FLOOR OR IN A CLOSET FIRST FLOOR OF BUILDING

UPFLOW FURNACE
TYPICALLY INSTALLED IN BUILDING OVER FULL BASEMENT FOUNDATION

FLOOR

FLOOR JOISTS

HORIZONTAL FURNACE
INSTALLED BENEATH FLOOR RESTING ON A CRAWL SPACE FOUNDATION

Figure 17–5.
The type of furnace used to heat the building is determined by the building design.

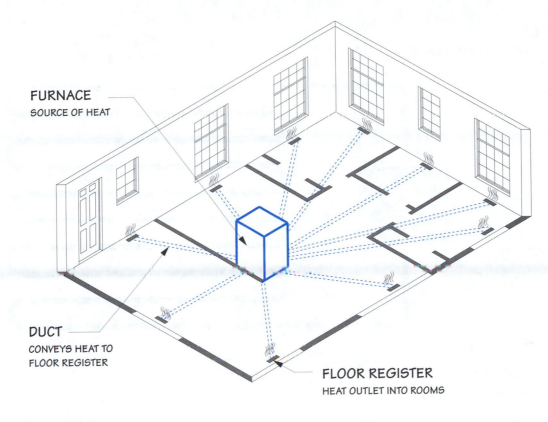

FURNACE
SOURCE OF HEAT

DUCT
CONVEYS HEAT TO
FLOOR REGISTER

FLOOR REGISTER
HEAT OUTLET INTO ROOMS

Figure 17–6.
In perimeter heating all the warm-air outlets are placed against the outside walls and under the windows.

systems. Different room temperatures can be maintained by regulating the flow of water to that room.

Unlike a forced-air system, hydronic heating cannot be combined with an air conditioning system. Therefore, hydronic heating is most practical in areas where summer temperatures do not require air-conditioning. The two major hydronic methods are the *radiant* and *loop* systems.

Hydronic Radiant Systems

A **hydronic radiant system** pumps heated water from the boiler to copper tubing placed at the floor or ceiling levels. The warm surfaces of the floor or ceiling then radiate heat to the enclosed room area.

Several types of radiant heating systems are available. Copper tubing can be embedded in a concrete floor slab or fastened to the ceiling to form heating panels covered with plaster. **(See Figure 17–7.)** Thermoplastic tubing placed on top of the floor and encased in gypsum is another effective method. This type of installation is shown in the plans for the **One-Story Residence** analyzed in Section 2.

Hydronic Loop Systems

A **hydronic loop system** conveys water from a boiler to radiators or convectors in the rooms to be heated. One-pipe or two-pipe methods are used. In *the one-pipe method* the radiators or convectors are connected in series. A single pipe circulates the water through the convectors, and back to the boiler. In the *two-pipe system* one pipe supplies hot water to the convectors, and the second pipe carries the water back to the boiler for reheating. **(See Figure 17–8.)** The two-pipe system supplies more consistent heat for larger spaces.

ELECTRIC HEATING SYSTEMS

Electric heating systems provide a very even distribution of heat and are clean and quiet to operate. Zone control, where the heat in each room can be controlled by a separate thermostat, is facilitated by electric heating systems. These systems have a slower response time than other types and cannot be combined with an air-conditioning system. Another disadvantage in some areas is the much higher cost of electricity than other

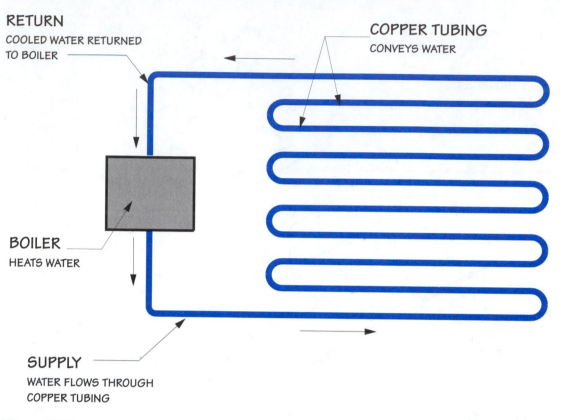

RETURN
COOLED WATER RETURNED
TO BOILER

COPPER TUBING
CONVEYS WATER

BOILER
HEATS WATER

SUPPLY
WATER FLOWS THROUGH
COPPER TUBING

Figure 17–7.
In the hydronic radiant system, copper tubing, embedded in a concrete floor or plastered ceiling, carries heated water.

fuels. Most electric heating systems are either radiant-wire, recessed wall, or baseboard types. Most of the information about an electric heating system is found in the electrical plans.

Radiant-Wire Heating

In **radiant-wire heating,** a grid pattern of resistance wires is placed in the floor or ceiling. Resistance wires produce heat that radiates into the room. Wires placed in the ceiling are stapled to lath and then covered with a 3/8″ minimum thickness of plaster. Floor systems are frequently placed in concrete floor slabs. Heavier resistance cable must be used and should be at least 1 1/2″ below the surface. **(See Figure 17–9.)**

Recessed Wall and Baseboard Heaters

Recessed wall heaters are either the radiant or convection type. The *radiant* type projects direct radiant heat in a space such as a small bathroom area. The *convection* type uses a small fan to help distribute the heat and works better than radiant heaters in larger areas.

Baseboard heaters are more effective than recessed wall heaters for larger spaces and are operated by either radiant or convection methods.

Heat Pumps

A **heat pump** is a very efficient year-round heating and cooling method for buildings in milder climates. It is considered one of the most economical types of heating. Rather than *producing* the heat, the heat pump *moves* the heat. Properly installed heat pumps can be three times as efficient as conventional resistance electric heating.

Even during cooler weather, the outside air and water contain some heat. A heat pump operates on the principle of refrigeration by removing this heat and bringing it into the building. During the warmer seasons, the heat pump acts as an air conditioner by pumping warm air from the interior to the outside of the building. The heat pump system does not need a flue, but it does require a blower and ducts. The system consists of an outdoor and indoor unit. The *outdoor unit* has a compressor, coils, and fans. The *indoor unit* has a blower and ducts to distribute or remove heat. (See **Figure 17–10.**)

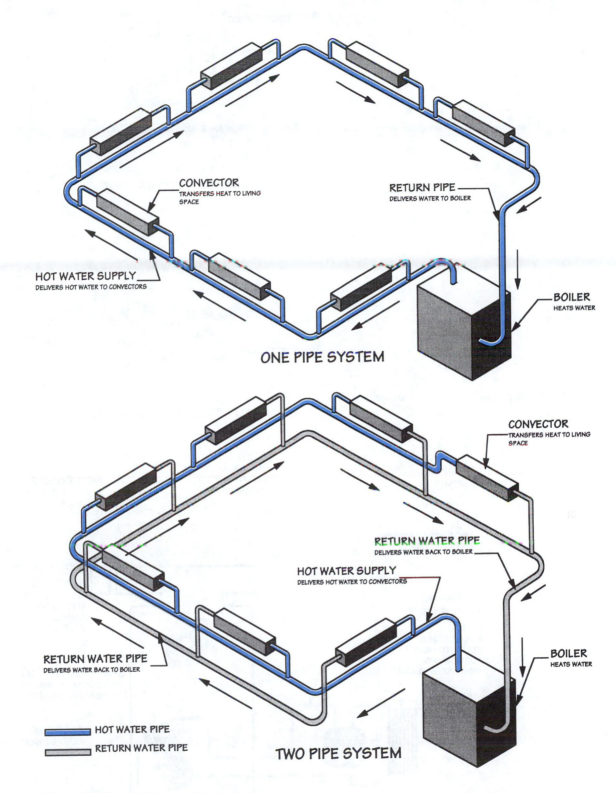

ONE PIPE SYSTEM

CONVECTOR
TRANSFERS HEAT TO LIVING
SPACE

RETURN PIPE
DELIVERS WATER TO BOILER

HOT WATER SUPPLY
DELIVERS HOT WATER TO CONVECTORS

BOILER
HEATS WATER

TWO PIPE SYSTEM

CONVECTOR
TRANSFERS HEAT TO LIVING
SPACE

RETURN WATER PIPE
DELIVERS WATER BACK TO BOILER

HOT WATER SUPPLY
DELIVERS HOT WATER TO CONVECTORS

RETURN WATER PIPE
DELIVERS WATER BACK TO BOILER

BOILER
HEATS WATER

HOT WATER PIPE
RETURN WATER PIPE

Figure 17–8.
A hydronic loop system carries hot water to the convectors that transmit heat to the living area.

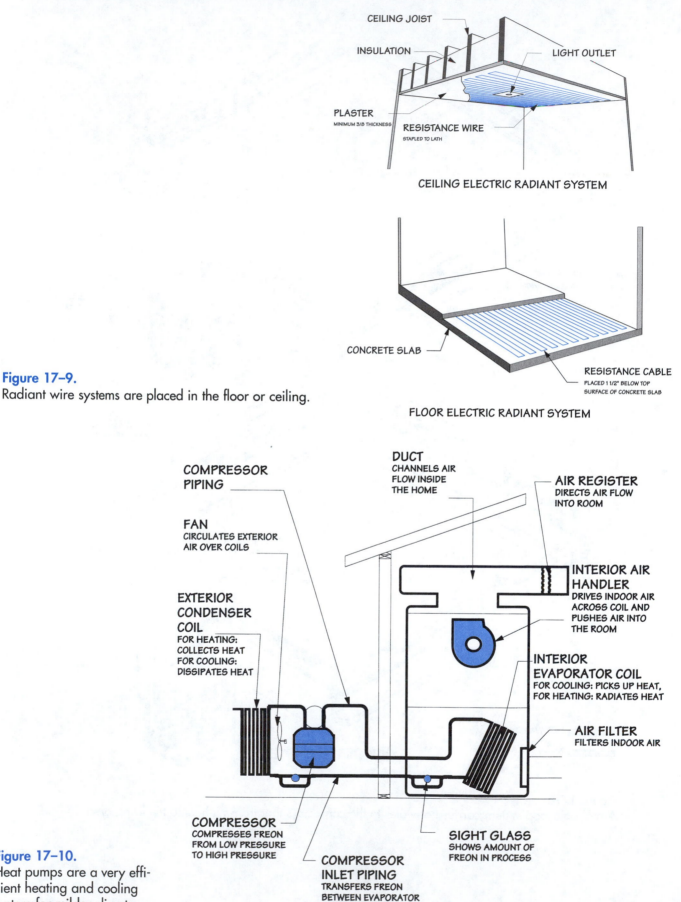

CEILING JOIST

INSULATION

LIGHT OUTLET

PLASTER
MINIMUM 3/8 THICKNESS

RESISTANCE WIRE
STAPLED TO LATH

CEILING ELECTRIC RADIANT SYSTEM

CONCRETE SLAB

RESISTANCE CABLE
PLACED 1 1/2" BELOW TOP
SURFACE OF CONCRETE SLAB

FLOOR ELECTRIC RADIANT SYSTEM

Figure 17–9.
Radiant wire systems are placed in the floor or ceiling.

COMPRESSOR
PIPING

DUCT
CHANNELS AIR
FLOW INSIDE
THE HOME

AIR REGISTER
DIRECTS AIR FLOW
INTO ROOM

FAN
CIRCULATES EXTERIOR
AIR OVER COILS

INTERIOR AIR
HANDLER
DRIVES INDOOR AIR
ACROSS COIL AND
PUSHES AIR INTO
THE ROOM

EXTERIOR
CONDENSER
COIL
FOR HEATING:
COLLECTS HEAT
FOR COOLING:
DISSIPATES HEAT

INTERIOR
EVAPORATOR COIL
FOR COOLING: PICKS UP HEAT,
FOR HEATING: RADIATES HEAT

AIR FILTER
FILTERS INDOOR AIR

COMPRESSOR
COMPRESSES FREON
FROM LOW PRESSURE
TO HIGH PRESSURE

SIGHT GLASS
SHOWS AMOUNT OF
FREON IN PROCESS

COMPRESSOR
INLET PIPING
TRANSFERS FREON
BETWEEN EVAPORATOR
AND CONDENSER

Figure 17–10.
Heat pumps are a very efficient heating and cooling system for milder climates.

COOLING SYSTEMS

A variety of cooling systems are available today, and some are more effective than others under different conditions. They all operate on the principle of removing warm air from the interior of a building. The warm air passes through a cooling mechanism and then is discharged back into the building. The three more widely used types are forced-air coolers, evaporative coolers, and individual room-cooling units.

FORCED-AIR COOLERS

A **forced-air cooler** may be a separate unit or combined with a forced-air heating unit. In either case, the cold-air system normally uses the same ducts and openings as the heating system to direct cool air into the building. In its basic operation, the warm air in the building is drawn into the air conditioning unit and passes through a filter that traps dust and other pollutants. The warm air then passes through cooling coils containing a refrigerant. A blower pushes the cooled air through the ducts and supply outlets. **(See Figure 17–11.)**

EVAPORATIVE COOLERS

Evaporative coolers are most practical for dry climate areas with low humidity. They are widely used in the American Southwest. The cooling unit is usually placed on the roof or on the ground next to the building. The cooled air can be moved through the same ducts used by the heating system. Warm air flows into the unit and is then propelled by fan through special pads moistened by a continuous stream of cold water. Finally, the cooled air is discharged into the ducts and through the supply outlets.

ROOM-COOLING UNITS

Electrically powered room cooling units are placed in window openings. Air is pulled in from the outside and passed through refrigerated coils before entering the interior of the building. These units are often combined with a room heater. They may still be found in some older houses but are usually used to provide cooling and heating for individual motel rooms.

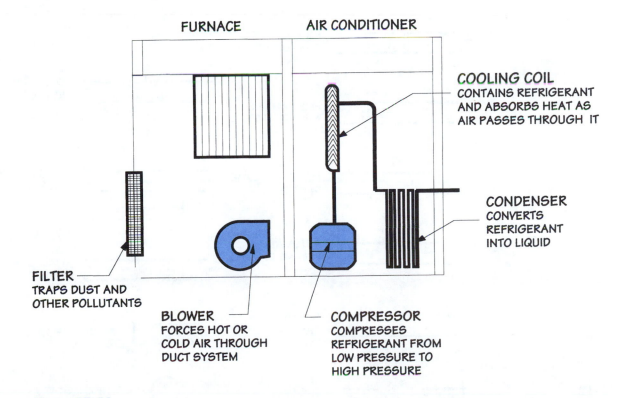

Figure 17–11.
In the combination heating and air-conditioning unit, hot or cold air is discharged into the same duct system.

READING THE HVAC PLANS

HVAC plans may be prepared by the architect, a mechanical engineer, or heating contractor. They may vary in how much information they provide. Examples of some typical HVAC symbols are shown in **Figure 17–12.**

Figure 17–13 shows a forced-air HVAC plan with a non-ducted return-air (RA) system. Compare this HVAC plan with the floor plan shown in Figure 15–14,

page 200. Two furnaces are toward the front of the building, and one is at the rear. Return-air grills (RAG) are shown leading directly to the furnaces. Because the building is constructed over a slab-on-grade foundation system, the supply ducts run above the suspended ceiling and the supply outlets open from the suspended ceiling. The plan also shows separate condenser units for the cooling system. **Figure 17–14** shows a pictorial drawing based on Figure 17–13.

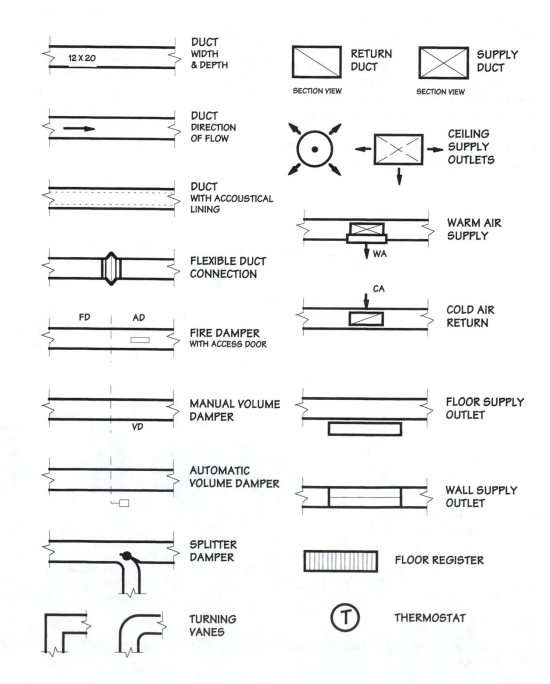

Figure 17–12.
Here are some symbols used in HVAC plans.

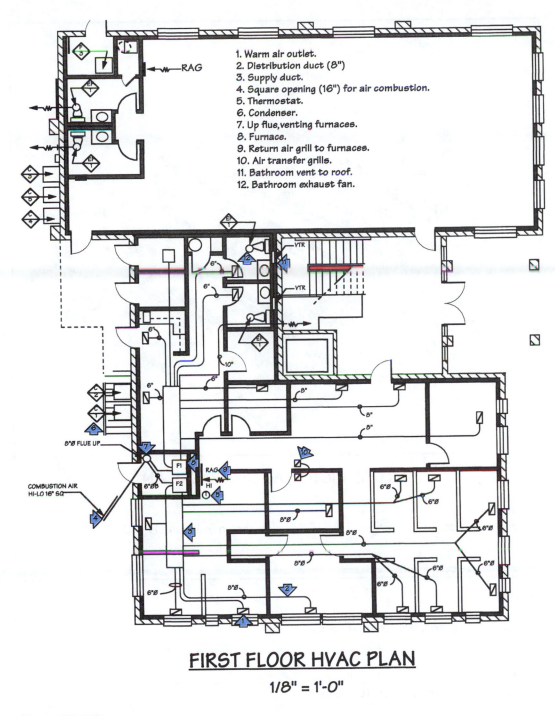

1. Warm air outlet.
2. Distribution duct (8").
3. Supply duct.
4. Square opening (16") for air combustion.
5. Thermostat.
6. Condenser.
7. Up flue, venting furnaces.
8. Furnace.
9. Return air grill to furnaces.
10. Air transfer grills.
11. Bathroom vent to roof.
12. Bathroom exhaust fan.

FIRST FLOOR HVAC PLAN

1/8" = 1'-0"

Figure 17–13.
This forced-air HVAC plan shows a non-ducted return-air heating system. For cooling purposes, Freon lines run from the condensers to the furnace cooling units. *(Courtesy of Josh Barclay Architects Ltd., Flagstaff, Arizona)*

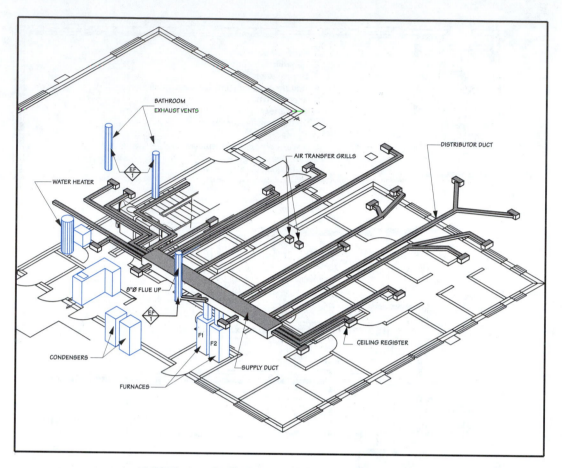

PARTIAL PICTORIAL OF HVAC PLAN

Figure 17–14.
This pictorial describes a section of the HVAC plan shown in Figure 17–13.

REVIEW QUESTIONS

Enter the missing words in the spaces provided on the right.

1. An interior temperature of __ to __degrees should be maintained by a proper HVAC system.

   _____

2. A device that controls the temperature level of a heating system is called a _____.

   _____

3. Thermostats are placed in individual rooms for _____ heating.

   _____

4. A _____ is the origin of heat in a forced-air system.

   _____

5. In a forced-air system, the cooled air is pulled into _____ _____ registers.

   _____

6. An upflow furnace is installed in buildings with a
   _____ _____ foundation.

   _____

7. In _____ heating, warm-air outlets are placed
   against outside walls and windows.

   _____

8. In _____ heating, hot water is pumped from
   a boiler to the heat outlets.

   _____

9. In a hydronic loop system, water is conveyed by
   pipe to _____ or _____.

   _____

10. A grid pattern of resistance wires is in the floor or
    ceiling of a _____ heating system.

    _____

11. A ____ ____ system operates on the principle of
    moving rather than producing heat.

    _____

12. A _____ _____ in a heating supply system is an
    enclosed area warmed by a furnace.

    _____

13. _____ foundations are best for floor plenum
    heating systems.

    _____

14. A forced-air cooling system often uses the same
    ____ and ____ as the heating system.

    _____

15. Dry climate areas with low humidity are best for
    _____ coolers.

    _____

16. Room-cooling units are placed in _____ openings.

    _____

17. A heated space from which warm air is distrib-
    uted is called a _____ _____ system.

    _____

18. Larger _____ _____ ducts begin at the furnace
    or plenum chamber.

    _____

19. A HVAC plan will give the diameters of the
    _____.

    _____

20. ____ and _____ (use abbreviations) outlets are
    identified in a HVAC plan.

    _____

SECTION 3 TEST

Electrical and Mechanical Plans

The following questions pertain to the electrical and mechanical plans included in Section 3. Many of the questions will also require studying the floor plan (Figure 15-14, page 200) that identifies the room names and numbers.

First-floor power plan in Figure 15–15, page 201, and the branch circuit schedule in Figure 15–16, page 202.

1. Scale. _____

2. Number of duplex receptacles in Room 124. _____

3. Room number of rooms with dedicated circuits. _____

4. Room number of room with a special fax outlet. _____

5. Number of receptacles in circuit A-2. _____

6. Number of furnaces in Room 108. _____

7. Number of telephone jacks in room #113. _____

8. Number of condensers along the left side of the building. _____

9. Number of receptacles in circuit A-12. _____

10. Room number of room with panel box A located at an exterior wall. _____

11. Height of GFI plugs in bathrooms. _____

12. Room number of room with panel box B located at an interior wall. _____

13. Branch circuit number for conference room lights. _____

14. Amperage of circuit #13. _____

15. Number of circuits in panel box A. _____

First-floor lighting plan in Figure 15–17, page 203, and the lighting fixture schedule in Figure 15–18, page 204.

16. Number of lamps in all G fixtures. _____

17. Number of G and H fixtures controlled by circuit A-5. _____

18. Hot and neutral wire sizes in circuit H-7. _____

19. Number of wall lights in the lobby. _____

20. Manufacturer's name of recessed ceiling light in front entry. _____

21. Room number of room with tthrr way switches. _____

22. Volts and watts of outdoor wall lamp at rear stairway. _____

23. Code letter for light in storage room. _____

24. Type of fixture mountings in toilets. _____

25. Number of required G Lithonia lightsin the office. _____

First-floor reflected ceiling plan in Figure 15–19, page 205.

26. Number of drop-in fluorescent light fixtures in Room 113. _____

27. Material used for exterior perforated overhang soffit. _____

28. Number of supply registers in Room 113. _____

29. Room numbers of rooms with surface-mounted fluorescent lights. _____

30. Material placed around light fixtures in most rooms and hallways. _____

First-floor plumbing plan in Figure 16–9, page 214.

31. Diameter of house sewer line. _____

32. Number of hose bibbs shown on the plan. _____

33. Name the room with a drinking fountain. _____

34. Name the room with a water heater. _____

35. Diameter of waste line connecting to sink and toilet waste lines. _____

36. Diameter of vertical waste stacks. _____

Waste schematic drawing in Figure 16–10, page 215.

37. Diameter of vertical soil stack connecting second-floor toilets. _____

38. Code mark for outside hose bibs. _____

39. Diameter of vent stacks to roof. _____

40. Name of 4" diameter pipe connecting to city sewer line. _____

Water schematic drawing in Figure 16–11, page 216.

41. Number of identified hose bibs. _____

42. Diameter of cold water distribution line.

43. Diameter of pipe with shutoff valve.

44. Name of 1/2" pipe connecting to water closet.

45. Number of water heaters visible in drawing.

46. Diameter of hot water distribution line.

47. Diameter of pipe connecting to meter.

Gas schematic in Figure 16–12, page 217.

48. Diameter of pipe coming from gas main.

49. Name of pipe connecting to furnaces.

50. Number of furnaces serviced by gas lines.

Forced-air HVAC plan shown in Figure 17–13, page 229.

51. Number of warm-air supply outlets in Room 114.

52. Size of duct running to warm-air outlet in Room 119.

53. Type of duct extending from furnaces to distribution ducts.

54. Size of duct running to Room 130.

55. Abbreviation for return-air grill.

56. Size of furnace flues.

57. Room number of room with thermostat.

58. Size of combustion-air opening.

59. Number of condensers.

60. Type of rooms containing exhaust fans.

SECTION

4

Plan Development

The development of the construction plan begins with the ideas agreed to by the architect and client. These ideas are subject to considerations of the natural construction site and surrounding physical environment. In addition, energy-saving features have become increasingly important elements in modern construction. These involve building orientation, insulation, and solar applications. All these factors affect both the method of construction and the appearance of a building. ■

UNIT 18

Drawing The Plans

The most familiar process of plan development begins with an architect. The architect prepares plans for an individual client or contractor. In some areas residential plans are produced by design/build contractors or owners. Stock plans, originally drawn by architects, are also available. A final set of plans must conform to all local building codes and zoning regulations and be approved by local building officials.

THE ARCHITECT

In the United States **architects** must be licensed and registered by the state in which they practice. This requires taking a series of examinations proving that the individual has the background and knowledge to design buildings that are structurally sound and meet all required health and safety standards. Canada uses a similar system, with architects registered by each province. Larger architectural firms may employ several architects. A **draftsperson** may do the actual drawings under the direction of an architect. An **illustrator** might be involved in preparing presentation drawings for a client. A **specifications writer** might also be required to supply the needed written information about materials, equipment, fixtures, and structural data. Some larger firms also employ **estimators** to calculate the approximate cost of the proposed project.

Some aspects of drawing styles may differ among architects. However, standards developed by professional organizations such as the American Institute of Architects (AIA) provide uniform recommendations for drawing the key elements of a plan. Therefore, lines, dimensioning, component and material symbols, and abbreviations can always be easily interpreted.

ARCHITECT AND ENGINEER

On larger residential and commercial projects, the architect often has to consult with and obtain drawings from structural, mechanical, electrical and soil engineers. A **structural (civil) engineer** designs the main structural components of a building (walls, floors, beams, columns) to withstand the maximum stress and loads that may be imposed on these components. **Mechanical engineers** design the more complex heat-ing and air-conditioning methods and other mechanical systems. **Electrical engineers** become involved in projects with more complicated power and electrical systems. **Soil engineers** are usually employed to test for unstable ground conditions and soil beneath buildings requiring deep excavations. The recommendations of the soil engineer strongly influence the design of the foundation.

ARCHITECT AND OWNER

For a custom-designed building, the owner and architect first decide on the general general design features of the building and whether it is to be traditional or contemporary in appearance. Other factors to be determined are:

- ✓ Lot size and shape and where structure is to be positioned on the property.
- ✓ Number and size of rooms and their arrangement.
- ✓ Style of interior and exterior trim and finish hardware.
- ✓ Types and brands of kitchen and bathroom fixtures.

A major factor is how much money the owner wishes to spend on the building. Once this is established, the architect and owner can decide whether there will have to be modifications or changes to the plan to make it fall within the proposed price range.

The next step is for the architect to make preliminary drawings and rough sketches. Usually several meetings and drawings are necessary before final agreement is reached. At that point the architect can complete the final working drawings to be signed off and approved by the owner.

If the owner desires, the architect will continue to help the owner in accepting bids for the job and selecting a building contractor. (Contracts and bids are discussed in Unit 25.) The architect may also represent the owner during construction of the building by periodically inspecting the work in progress.

TYPES OF PLANS

A major part of residential construction today is *speculative*. Here the architect creates plans requested by a building contractor who erects homes and then places them on the market. These may be individual homes on a single lot or whole subdivisions containing many houses. Several basic designs may be used for larger housing tracts, offering the potential owners some choice of price and design. In some areas this function is often performed by design/build contractors or lumber yards that directly employ their own draftspersons to produce the plans.

MANUFACTURED HOMES

Plans are also created for **manufactured homes,** a growing branch of the home-building industry. These are factory-produced homes constructed in sections. The sections are delivered by truck to the job site, where they are joined.

STOCK PLANS

Stock plan publications have multiplied in recent years. They contain simple, reduced floor plans along with pictorials and descriptions of a variety of residential plans. A set of stock plans for any of these selections can be purchased for a fee. The fee is much less than the cost of a set of plans created by an individual architect. With some adaptations to local building and zoning codes, the plans can be used by a building contractor or owner to construct a building.

PLAN REPRODUCTION

Blueprints describe an older process of plan reproduction that created white lines against a blue background. Although this method is seldom employed today, the word *blueprint* is still frequently used to refer to construction drawings. More appropriate terms are **prints** or **construction plans**.

In the traditional method of creating plans, the draftsperson uses hand instruments to make the original drawings. The completed drawings are then mechanically reproduced. Diazo and electrostatic prints are the two most common methods. In the **diazo procedure,** the drawing to be reproduced is placed together with a sheet of chemically sensitized diazo paper and fed into a white-print machine. After going through an exposure and developing process, the two sheets are separated and emerge from the printer. The result is a drawing with blue or black lines against a white background. In the **electrostatic method** the drawings to be reproduced are fed into a machine that projects an image on a selenium-coated plate. A series of chemical processes then reproduces the drawings in black lines

on a white background. This is the same process used in photocopy machines that make enlarged or reduced copies.

COMPUTER-AIDED DRAFTING AND DESIGN (CADD)

The most significant developments in recent years are the computer-generated programs for creating and reproducing construction drawings through a **computer-aided design and drafting (CADD)** system. The entire process of plan development, from presentation art to schedules, and then to the complete and final set of construction plans, can be done through a CADD program. CADD eliminates the traditional tools used in hand-drawn plans. Within a few years, CADD may very well become the prevailing system used by architects and draftspersons.

Stored in some CADD programs are many basic plan elements such as foundation sections, wall sections, and many other components of construction drawings. CADD programs also may contain complete libraries of standard architectural symbols. Therefore, CADD makes it possible to produce clear and neat drawings much faster than the traditional hand-drawn method. **(See Figure 18–1.)** Another decided advantage is the ability to make changes without redrawing the entire plan. Reproductions of the plan are produced by copying machines activated by the computer, **See Figure 18–2.**

Figure 18–1.

A growing number of architects and designers produce plans through computer-generated programs such as CADD. *(Courtesy of The CAD Shop, Coconino Blueprint Company, Inc., Flagstaff, Arizona. Photo by Mary McMacken.)*

Figure 18–2.
An ink jet plotter is activated by the computer to produce full-sized construction plans. *(Courtesy of The CAD Shop,Coconino Blueprint Company, Inc., Flagstaff, Arizona. Photo by Mary McMacken.)*

REVIEW QUESTIONS

Enter the missing words in the spaces provided on the right.

1. The _____ is the person mainly responsible for creating the design of a building.

   _____

2. An architect may request structural data from a _____ _____.

   _____

3. The abbreviated name of the architects' professional organization is _____.

   _____

4. _____-designed buildings are planned at the request of an individual owner.

   _____

5. In _____ construction, a building is placed on the market after it has been completed.

   _____

6. A residential construction plan sold through a publication is called a _____ plan.

   _____

7. A _____ is an older kind of plan reproduction with white lines against a blue background.

   _____

8. Black lines against a white background are produced by the _____ and _____ methods.

   _____

9. _____ is the abbreviated term for computer programs used to create construction drawings.

   _____

10. Computer drawings can be reproduced on a _____-_____ machine..

   _____

The Construction Site and Topography

A construction site is a plot of ground with legally defined boundaries on which a building project is to be located. The general area and topography of the construction site (lot) have a major effect on the type, design, and placement of the building. Local zoning regulations that apply to the area of the construction site directly influence building types and design. (Zoning regulations are discussed in Unit 22.) An investigation should be made of possible hazards that could affect the health of future occupants. Some examples are chemical pollution and the presence of radon in the soil. **Radon** is a gas produced by the radioactive decay of radium given off by some types of soil and rock.

TOPOGRAPHY

Topography is the graphic description of physical land surface features. This description includes the elevations, contours, shape, and size of the lot, plus its trees, plants, large boulders, streams, and creeks. Whenever possible, the design of a building should take advantage of the natural features of the construction site and blend in with them. Topography not only includes these natural features, but also the changes that can be made by grading, fill, and the planting of additional trees and shrubbery.

Contours are shown on site plans as lines connecting equal points of elevations. The different heights on the lot are identified by the elevations. These contours and elevations are major factors influencing the type of building best suited for the lot. For example, a split-level home is well suited for a sloping lot. Ranch-style and two-story houses work best on flat ground. **(See Figure 19–1.)**

Local zoning codes regulate the distance a building must be from the lot boundaries (property lines). Therefore, lot shape and size have a direct bearing on the maximum square footage of a building. A long and narrow lot may be large in its number of square feet, yet limit the shape of a building.

Existing trees can serve many useful purposes and be a factor in positioning the building on the lot.

Coniferous (evergreen) **trees** that do not lose their leaves during the winter can act as protection against strong winds. **Deciduous** (leaf-dropping) **trees** can shade windows from the sun during the summer. Then, in the fall, the leaves drop, allowing the sun to penetrate and warm the building during the winter. Trees shading paved areas such as driveways and walks cut down on reflected heat and channel breezes. Large masses of trees reduce air pollution by filtering and deodorizing the surrounding air. Noise pollution from the surrounding area can also be reduced by denser tree and plant formations.

The terrain surrounding the construction site is another factor to be considered in building design. Depending on their position, surrounding hills may funnel winter winds through the area or block them. During warmer months, air flowing over ponds or lakes within the proximity may lower temperatures around the site. Dead-air basins in warmer climates can have higher temperatures than the surrounding areas. Existing structures on adjoining lots must also be considered. Some local zoning laws restrict the height or placement of a new building so as not to shade an existing structure from the sun.

ORIENTATION

Orientation is the placement of a building on a lot. It should make the best use of prevailing climate conditions and the surrounding physical environment. One of the most important considerations is the relationship of the building's walls to the sun. The walls should be able to absorb the greatest amount of the sun's heat during the winter and shield the house from the sun during the summer. A poorly oriented house can be uncomfortably warm during the summer and not able to take full advantage of the warming rays of the sun during the winter. The views from the building's windows and protection from surrounding noise sources are other factors in orientation. In more crowded areas finding a building site that is desirable in regard to all these factors is usually difficult. In this case, it may be necessary to compromise and choose what is most important to the occupant.

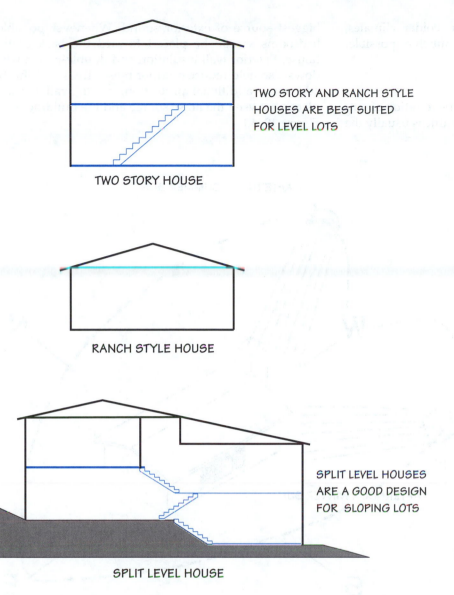

TWO STORY AND RANCH STYLE HOUSES ARE BEST SUITED FOR LEVEL LOTS

TWO STORY HOUSE

RANCH STYLE HOUSE

SPLIT LEVEL HOUSES ARE A GOOD DESIGN FOR SLOPING LOTS

SPLIT LEVEL HOUSE

Figure 19–1.
Lot contours influence building design.

CLIMATE CONTROL

Although the sun always rises in the east and sets in the west, its angle changes from summer to winter. In North America the more direct rays of the sun originate much farther south and are at a lower angle during the winter. **(See Figure 19–2.)** In cold regions, buildings should be designed with maximum window area in the walls facing south, southeast, or southwest to get the most benefits of the sun's rays during the winter. If possible, front entries should also face south, as north-facing entries maintain ice for longer periods of time. The area of north-facing windows should be limited to cut down on heat loss. In hot-climate regions, window areas in the west wall should be limited because of the strong afternoon summer sun. **(See Figure 19–3.)** Because of the higher angle of the sun's rays during the

summer, a roof overhang is effective in shielding a building from heat, yet it will not block the lower-angled rays of the winter sun.

SLOPING LOTS

Several factors should be considered when building on a sloping lot. A slope facing south gets more heat from sun exposure than one facing east or west. A north-facing slope gets the least amount of sun exposure. Therefore, building on a south-facing slope is preferable in cold climates since it takes advantage of the sun's rays. Building on a north-facing slope is preferable in warm climates since it protects from the sun's rays. Higher elevations, such as hilltops and ridges, have more wind exposure than lower levels. In warmer climates, placing the building where it will get more wind exposure may be advantageous. In colder climates, wind exposure should be avoided as much as possible.

NOISE CONTROL

Building orientation can help reduce penetration of outdoor background noises. Vehicular traffic is usually the biggest source of exterior sound. Whenever possible, bedrooms should be placed toward the back of the house. Exterior wall insulation and double-paned windows also help reduce outdoor noise. Trees and shrubs add some additional protection, as do wall barriers placed between the noise source and the building. (See Figure 19–4.)

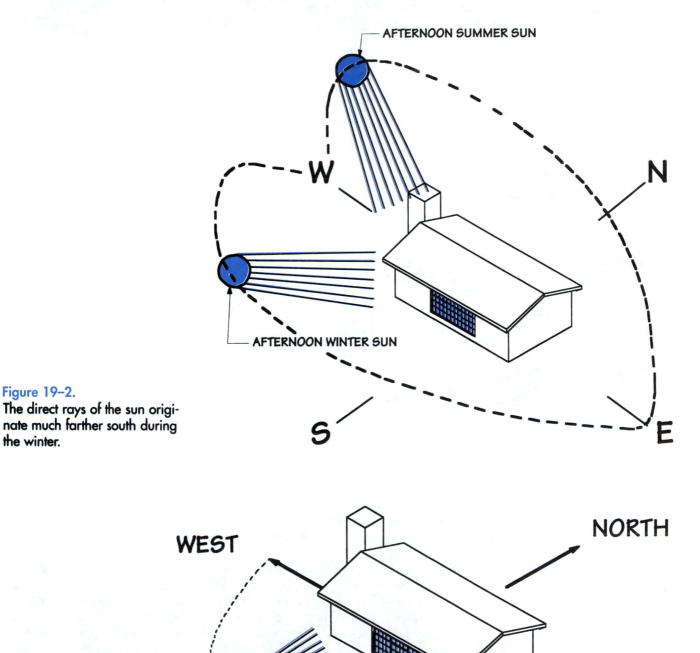

Figure 19–2.
The direct rays of the sun originate much farther south during the winter.

Figure 19–3.
In cold-winter regions, large window areas should face south, southeast, or southwest to take advantage of the winter sun.

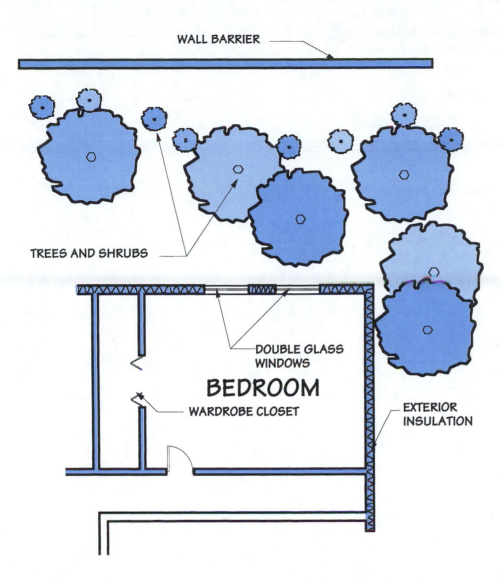

WALL BARRIER

TREES AND SHRUBS

DOUBLE GLASS
WINDOWS

BEDROOM

WARDROBE CLOSET

EXTERIOR
INSULATION

Figure 19–4.
Insulated exterior walls, double-paned glass, trees and shrubs, and a wall barrier combine to reduce exterior noise levels.

REVIEW QUESTIONS

Enter the missing words in the spaces provided on the right.

1. Lines connecting equal points of elevation on a site plan define the _____ of the building site.

   _____

2. A _____ style home is not well suited for a sloping lot.

   _____

3. Regulations regarding the distance of a building from property lines are usually found in the local _____ codes.

   _____

4. Properly positioned _____ trees help shade windows from the summer sun.

   _____

5. Proper _____ of a building means positioning it to make the best use of prevailing climate
   and physical surroundings.

   _____

6. During the winter the direct rays of the sun are at
   a much _____ angle.

   _____

7. _____ _____ houses are well suited for
   sloping lots.

   _____

8. In warmer climates, it is advisable to build on a
   _____ facing slope.

   _____

9. Outdoor noise can be reduced inside the building
   by _____ the walls.

   _____

10. _____ windows help insulate the interior
    from noise and cold.

    _____

UNIT 20

Basic Residential Design

The four basic house designs are the one-story, one-and-one-half-story, two-story, and split-level. Some of the more important considerations in choosing among these designs are the climate, number of inhabitants, room layout, space, cost, and shape and contours of the lot.

ONE-STORY (RANCH) HOUSE

The **ranch-style house** is one of the most popular one-

story house designs. It originated in the United States during a period of rapid subdivision development. Because the living space is all at one level, a medium to large ranch house covers a greater area and requires a larger lot than houses of more than one story.

Most ranch houses are constructed over a crawl-space or slab-on-grade foundation, but a full-basement foundation may also be used. The garage may be part of the main structure or detached. Ranch houses usually have sloping roofs with closed attic areas or exposed beams. An attractive feature is the easy access to patios, terraces, and porches. **(See Figure 20–1.)**

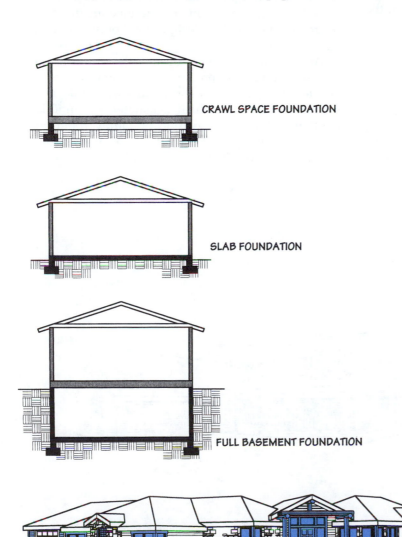

CRAWL SPACE FOUNDATION

SLAB FOUNDATION

FULL BASEMENT FOUNDATION

Figure 20–1.
The ranch style one-story house can be built with a crawl-space, slab, or full-basement foundation.

ONE-AND-ONE-HALF-STORY HOUSE

The **one-and-one-half-story house,** in addition to the main floor, has an open area under a steeply pitched roof. This area can be used for bedrooms, a bathroom, or other additional living space. **Dormers,** pitched structures that extend from the roof, are often built to provide extra light and ventilation in the attic area. Roof windows, built into the slope of the roof, may be installed for the same purpose. The one-and-one-half-story house with dormers is sometimes called a **Cape Cod house** because the design originated in that part of the country. If a building contains a second story beneath the attic area, it is called a **two-and-one-half-story house.**

Full-basement foundations are most practical for one-and-one-half-story houses. They provide a needed area for storage and/or recreation. Stairs are constructed from the basement to the main floor and from the main floor to the attic. **(See Figure 20–2.)**

TWO-STORY HOUSE

The **two-story house** consists of two complete floor levels. The house can take less space on the lot because it has two levels of living area. For this reason, it may be the preferred design when building on a smaller lot.

The two-story house can be built over either a full-basement, crawl-space, or slab-on-grade foundation. **(See Figure 20–3.)** It usually features a sloping roof with attic space. A flat roof may be used for a more contemporary design in areas where snow accumulation is not a problem. Living, dining, and family rooms, a kitchen, and often a bathroom, are usually placed on the first-floor level, and bedrooms and additional bathrooms on the second level.

SPLIT-LEVEL HOUSE

Split-level houses are most practical on steeply sloped lots. The slope may be from the front to back of the house, from the back to the front, or from side to side. The different levels are connected by short flights of stairs. Some split-level houses have as many as four levels, consisting of a basement, intermediate, living, and bedroom levels. Other designs have only a living and bedroom level.

With the full-basement design, the *basement level* is the lowest and may contain a furnace and water heater besides storage area. A short stairway would lead to the *intermediate level,* which is usually at ground level. It often includes a recreation or family room and a garage. The *living level* contains the dining room, living room, and kitchen. The highest *bedroom level*

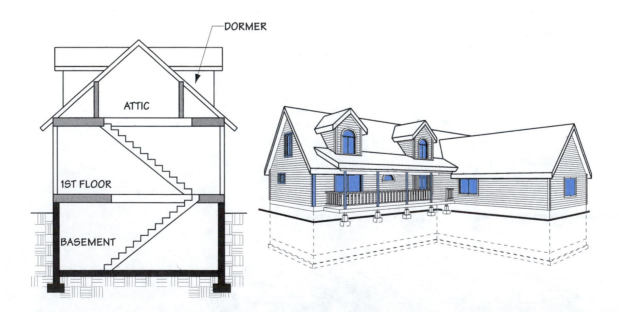

Figure 20–2.
A one-and-one-half-story house includes an attic space for an additional living area or storage space. Dormers with windows frequently project from the sloping roof.

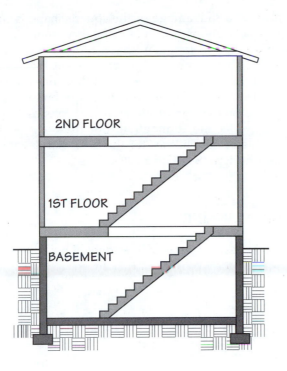

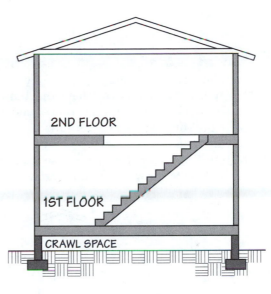

Figure 20–3.
Two-story buildings can be constructed over full-basement and crawl-space foundations. The above pictorial shows a two-story section of house combined with a post-and-beam design.

includes bedrooms and bathrooms. A simpler split-level design may feature just a ground-floor living area with a stairway leading up to the bedrooms and bathrooms. **(See Figure 20–4.)**

HOUSE SHAPES

The more usual house shapes are the square, rectangle, T-shape, U-shape, and L-shape. The shape of a building affects its room layout, outside appearance, and cost. The shape of a building will also determine the square footage of the floor area.

A **square building** is less costly because it requires fewer linear feet of exterior walls, foundation walls, and footings. A disadvantage of the square shape is that it limits the options for a satisfactory room layout. Therefore, the rectangular shape is used more often, even though this means the perimeter of the building will add up to more linear feet and thus increased cost.

The **L-shape, T-shape,** and **U-shape** buildings further increase the cost of materials for constructing the exterior walls and foundation. But these shapes offer more diverse possibilities for the floor plan layout and exterior design. **(See Figure 20–5.)**

ROOM PLANNING

The three basic sections of a residential building are the

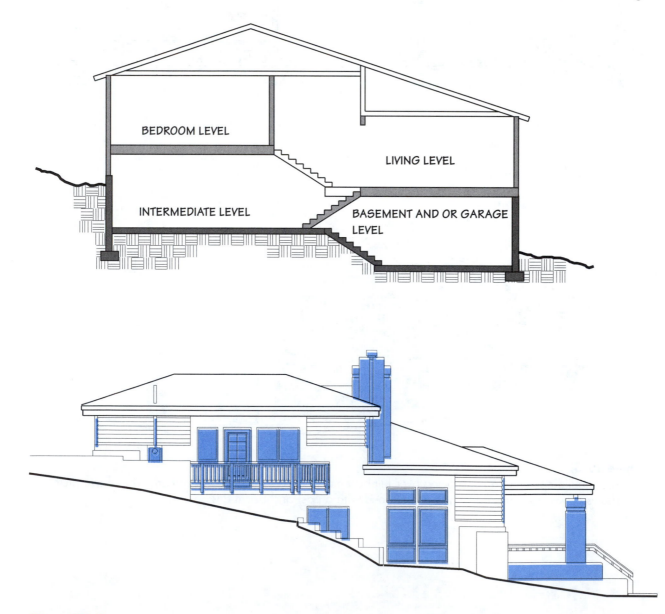

Figure 20–4.
A split-level house conforms very well to a sloping lot.

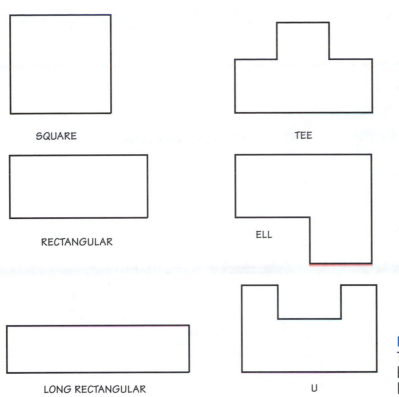

SQUARE

TEE

RECTANGULAR

ELL

LONG RECTANGULAR

U

Figure 20–5.
The more typical plan view outlines of residential buildings are square, rectangular, long rectangular, T-shape, L-shape, and U-shape.

sleeping area, living area, and the service area. **Room planning** relates to the most efficient arrangement of rooms in the different sections of the building. **(See Figure 20–6.)** Some main factors to be considered are the size and general design of the structure, the function of each individual room, and the relationship of one room to another.

SLEEPING AREAS

The **sleeping area** consists of the bedrooms, bathroom(s), and closets. Sleeping areas should be in the quietest section of the building. Bedrooms should be placed as far as possible from traffic disturbance and household noises. The number of bedrooms needed is determined by how many persons will be in the building. Each bedroom should have adequate closet space. The depth of a clothing closet should be no less than 24″. Closets should be placed along interior walls if possible. They will cut down on sound penetration between adjoining rooms.

The number of bathrooms varies with the size of the building. A **full bathroom** contains a water closet (toilet), a lavatory (sink), and a bathtub and shower, with the possible addition of a whirlpool and bidet. A **half bathroom** contains only a water closet and lavatory. The addition of a shower turns a half bathroom into a **three-quarter bathroom.** Full and three-quarter bathrooms should be near the bedrooms. Half bathrooms

are commonly placed in the living area. One full bathroom may be adequate for a small home. Two or more bathrooms are desirable for larger structures. Two-story and split-level houses should have bathroom facilities at each floor level.

LIVING AREA

The main spaces of the living area are the living room, dining room, and foyer. The living area may also include a family or recreation room and, in addition, a **special-purpose room** such as a sun porch **(atrium),** home office, or den.

The **living room** can be considered the main social center for the family and often sets the tone for the entire home. It should be large enough to accommodate the various activities that may take place there, such as TV viewing, family conversations, and entertaining guests. If possible, the living room should be placed so there is a minimum of through-traffic from different sections of the house.

A **dining room** is designed for eating, although many households use eating facilities in the kitchen for most of their meals and reserve the dining room for more formal occasions. The most efficient location for a dining room is between the kitchen and living rooms. A dining room may be a clearly defined separate room, or it may merge with the living room.

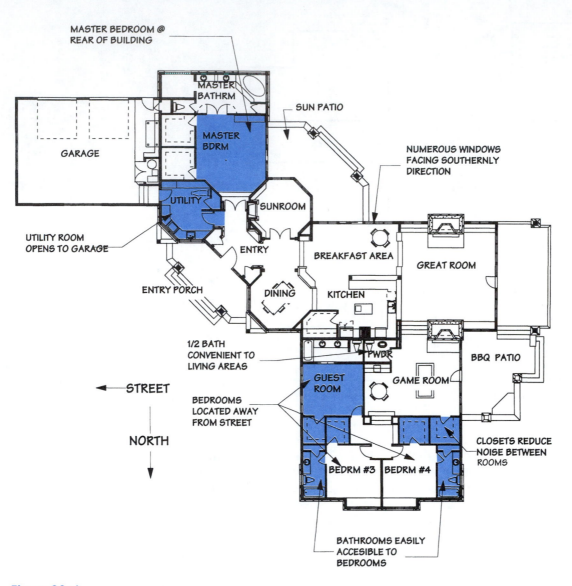

MASTER BEDROOM @
REAR OF BUILDING

MASTER
BATHRM

SUN PATIO

MASTER
BDRM

GARAGE

NUMEROUS WINDOWS
FACING SOUTHERNLY
DIRECTION

UTILITY

SUNROOM

UTILITY ROOM
OPENS TO GARAGE

ENTRY

BREAKFAST AREA

GREAT ROOM

ENTRY PORCH

DINING

KITCHEN

1/2 BATH
CONVENIENT TO
LIVING AREAS

PWDR

BBQ PATIO

GUEST
ROOM

GAME ROOM

◄— STREET

BEDROOMS
LOCATED AWAY
FROM STREET

CLOSETS REDUCE
NOISE BETWEEN
ROOMS

NORTH

BEDRM #3

BEDRM #4

BATHROOMS EASILY
ACCESIBLE TO
BEDROOMS

Figure 20–6.
Good room planning is based on the most efficient and comfortable arrangement of the
rooms. This building is located on a large lot, and the sleeping and living areas are placed in
a comfortable distance from the street. North is shown facing the bottom of the drawing
because of the placement of the house.

The **foyer** is found inside the main entry of the house. One important purpose of a foyer is to provide a buffer space between the front entry and the living room. In colder climates, foyers should be designed to keep cold air from entering the living areas of the house. Hallways may also lead from the foyer to other areas of the house. Foyers commonly include a coat closet for family members and visitors.

The main reason for a **recreation room,** also known as a **family room,** is to provide an area for family entertainment. Some families prefer to confine TV viewing to the recreation room. Other activities might include games and hobbies. Recreation rooms often adjoin other rooms in the main living area. They are often found next to the living room or dining room. Sometimes, though, they are located next to the kitchen or in the basement.

SERVICE AREA

The **service area** includes the kitchen, utility and/or mudrooms, and a garage or carport. Service areas may also provide storage and space for building maintenance equipment.

The primary purpose of a **kitchen** is food preparation. It contains appliances for this purpose, and cabinets and counters. A section of a larger kitchens is often used as an eating area. Laundry appliances are sometimes included in a nook off the kitchen. In some homes, particularly in farm areas, kitchens are the center of household social activity..

Utility rooms conveniently hold a washer and dryer, an ironing board, and counter for folding clothing.

Therefore, utility rooms are often called **laundry rooms.** They also provide storage for cleaning equipment and supplies. Utility rooms should be next to or close to the kitchen. They frequently open to the garage area. A **mudroom** is a variation of a utility room. It is usually placed inside a rear entrance. In colder climates it serves as an area for taking off wet boots and clothing before entering the other rooms.

REVIEW QUESTIONS

Enter the missing words in the spaces provided on the right.

1. The living space is all on one level in a _____-style house.

    _____

2. The Cape Cod house has _____ stories.

    _____

3. _____ or roof windows are used to provide light and ventilation to livable attic areas.

    _____

4. A one-story house requires a larger _____ than a two-story house with the same amount of living space.

    _____

5. Steeply sloped lots lend themselves to _____-_____ homes.

    _____

6. Less options for satisfactory room layout is one drawback of a _____-shaped house.

    _____

7. A room containing only a water closet and lavatory is called a _____.

    _____

8. The _____ is a space leading from the main entrance into the living area.

    _____

9. Kitchens are considered part of the _____ area of house.

    _____

10. Laundry rooms are also called _____ rooms.

    _____

UNIT 21

Energy and Building Design

Energy consumption is major concern in all of North America. As much as 70 percent of the energy used in an average home is used for heating and cooling. Therefore, interior climate control is a fundamental consideration in building design. Until recent years, buildings relied primarily on mechanical means for heating and cooling. Now there is increasing emphasis on building design orientation, and insulation to lower the use of mechanical heating and cooling systems. The result is lower costs for the homeowner and less of an energy drain on our natural resources.

Solar construction offers the most energy-saving advantages, but it has not yet become a major factor in new construction. However, very often some solar principles are incorporated in the design of conventionally designed buildings.

EXTERNAL DESIGN FEATURES

The color of the roof and exterior siding affects the temperature inside the building. Dark colors absorb the sun's heat and light colors reflect it. Therefore, darker roof covering and siding are recommended for colder climates and lighter colors for warmer climates.

Roofs should also be designed with adequate roof overhangs. These will shade the windows from the high summer sun, yet during the cold season, the low winter sun will still be able to shine into the windows. Roof louvers and vents are important to circulate air in the attic areas of a building. This air circulation helps cut down on heat buildup in the attic and heat transfer to the rooms below.

Planting trees in the proper locations can also be helpful in controlling interior house temperatures. Deciduous trees shade the sides of the building during spring and summer. After they lose their leaves in the fall, they allow the walls to be warmed by the winter sun. The strategic planting of evergreen (coniferous) trees that retain their leaves the entire year can help shelter the house from strong winds.

The greater the amount of exposed walls, the greater the heat absorption will be. Accordingly, buildings designed with less wall exposure are more comfortable during warmer weather and reduce the cost of air conditioning. For example, a square-shaped house will have less exterior wall exposure than a house of the same square footage that is L-shaped. **(See Figure 21–1.)** The most energy-efficient design is the two-story square-shaped building. It provides the greatest amount of floor space with the least amount of exposed wall area.

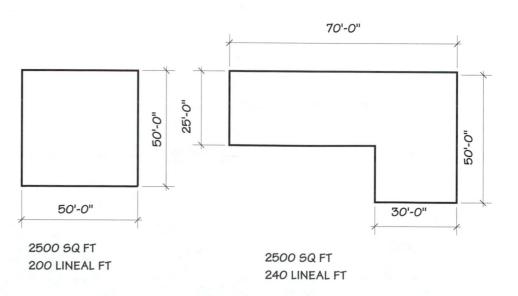

Figure 21–1.
The square and L-shaped buildings include the same number of square feet, but the L-shaped one exposes more wall space for heat absorption.

2500 SQ FT
200 LINEAL FT

2500 SQ FT
240 LINEAL FT

WINDOWS AND DOORS

Glass windows are very poor insulators, thus creating a great deal of energy loss. A single pane of glass is about 5 percent as effective as a properly insulated wall. Double-glass panes with an air space between the panes will double the insulative effectiveness, and triple panes will triple it. In severely cold areas **storm sashes** (also called **storm windows**) should be installed during the winter months. They consist of a single glazed panel that is usually attached outside the window frame. **(See Figure 21–2.)**

Metal-clad doors with foam cores are the best-insulated exterior doors. They have almost the same thermal ratings as insulated walls. Conventional wood doors have poor resistance to heat flow. Even worse is the performance of sliding glass doors.

The installation of storm doors in severely cold areas is recommended. They are hung on the door frames opposite the entrance doors. A storm door is shown in Figure 21–2.

Caulking and weatherstripping are advisable for buildings in all climates. **Caulk** is a sealing material that usually comes in a cartridge and is applied with a caulking gun. It should be placed wherever the door and window frames are nailed to the wall. **Weatherstripping** is attached directly to the doors and windows to prevent air flow. **(See Figure 21–3.)**

INSULATION

Insulation is building material that controls heat flow through the outer shell of the building. The natural movement of heat is toward colder temperatures.

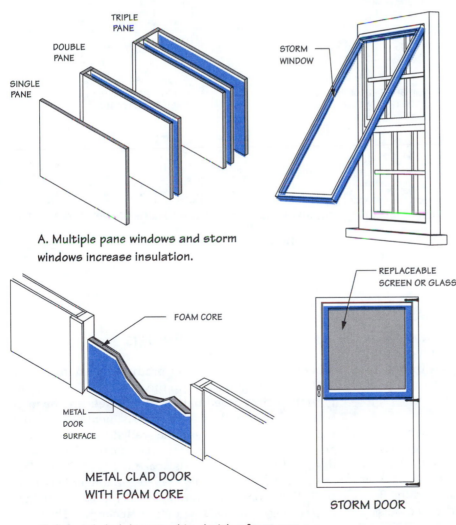

A. Multiple pane windows and storm windows increase insulation.

B. A metal clad door combined with a foam core provides the best insulation.

STORM DOOR

Figure 21–2.
Windows and doors require additional insulation to more effectively maintain comfortable interior temperatures.

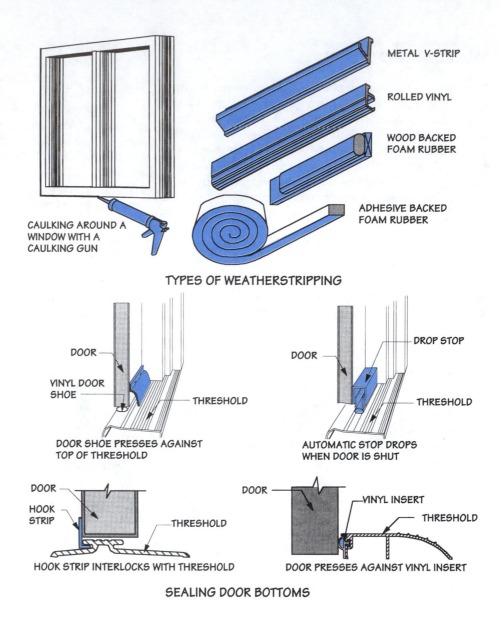

METAL V-STRIP

ROLLED VINYL

WOOD BACKED
FOAM RUBBER

ADHESIVE BACKED
FOAM RUBBER

CAULKING AROUND A
WINDOW WITH A
CAULKING GUN

TYPES OF WEATHERSTRIPPING

DOOR

VINYL DOOR
SHOE

THRESHOLD

DOOR SHOE PRESSES AGAINST
TOP OF THRESHOLD

DOOR

DROP STOP

THRESHOLD

AUTOMATIC STOP DROPS
WHEN DOOR IS SHUT

DOOR

HOOK
STRIP

THRESHOLD

HOOK STRIP INTERLOCKS WITH THRESHOLD

DOOR

VINYL INSERT

THRESHOLD

DOOR PRESSES AGAINST VINYL INSERT

SEALING DOOR BOTTOMS

Figure 21–3.
Caulking, weatherstripping,
and door bottom sealers are
installed to prevent air leakage
around doors and windows.

During the winter, warm air inside the building attempts to move through the shell toward the cold air outside the building. The opposite occurs during the warm summer months, when the warm outside air attempts to move through the shell into the cooler interior of the building.

Many types of insulating materials are available today. The three most widely used are fiberglass, cellulose, and plastic foam. They come in the form of loose fill, batts or blankets, and rigid boards. **(See Figure 21–4.)** Batts or blankets are placed between the wall studs and the floor and ceiling joists. Rigid panels are nailed to the outside face of exterior wall studs. They are also attached to concrete, and concrete-block foundation walls and placed under slab-on-grade concrete floors.

FIBERGLASS BATTS AND BLANKETS

Fiberglass is a mineral fiber produced from molten silica that is highly resistant to fire, moisture, and vermin. Flexible paper-backed **batts** and **blankets** are the more widely used fiberglass products. As mentioned earlier, they are fastened between the wall studs and the floor and ceiling joists. They are manufactured in widths designed for studs or joists spaced 16" or 24" oc. Batts are 48" long. Blankets are available in longer lengths as well. Both come in a range of thicknesses, and their effectiveness increases with the thickness. Therefore, many homes built today use 2 × 6 studs in place of 2 × 4s for the exterior walls because they allow the use of thicker insulation.

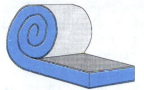

FIBROUS BLANKET INSULATION
IS PLACED BETWEEN WALL STUDS
AND CEILING AND FLOOR JOISTS

RIGID INSULATING PANELS ARE
USED AS SHEATHING AND ARE ALSO
ATTACHED TO MASONRY WALLS

CELLULOSE INSULATION IS POURED
INTO WALLS AND CEILINGS

PLASTIC FOAM INSULATION IS SPRAYED
BETWEEN WALL STUDS

Figure 21–4.
The more frequently used insulation materials
are fiberglass blankets or batts, rigid panels,
cellulose, and plastic foam.

LOOSE-FILL FIBERGLASS AND CELLULOSE.

Loose-fill fiberglass is blown into an area by pressurized hose. It is a cheaper method than using batts or blankets and is primarily applied in attic areas, directly over the ceiling below. **Cellulose** insulation consists of ground-up or shredded paper that has been chemically treated to make it resistant to fire and vermin. Cellulose insulation can also be poured or blown into walls and ceilings.

PLASTIC FOAMS

The more frequently used **plastic foams** are *polystyrene, polyurethane,* and *urea formaldehyde.* (Urea formaldehyde is not permitted in Canada and some parts of the United States.) **Foaming** is a process of pouring or spraying a material when it is in a liquid state. The material can be placed this way in an open wall before the interior finish is applied. A chemical reaction causes the foam mixture to expand and completely fill the wall cavities.

RIGID INSULATION

Polystyrene and polyurethane are used not only in foams but also in the manufacture of **rigid insulation panels.** These panels serve as sheathing as well as insulation when placed over framed exterior walls. Plus, they are often attached with mastic to the walls of full-basement concrete or concrete-block foundations and placed below soil-supported concrete slabs.

VAPOR BARRIERS

Modern construction methods and improved insulation allow very little air through the exterior walls of the building. Therefore, interior air moisture and condensation are more of a problem than they used to be. All air contains moisture, and warm air holds more moisture than cold air. During cold weather, then, there is a higher moisture content inside the building than on the outside. If nothing to prevents it, the warm and moist interior air will migrate through the walls toward the colder outside air. This process results in condensation, resulting in water accumulation inside the walls. Over time, wall studs, sheathing, insulating materials, and wallboard will rot from the water.

To prevent this problem of condensation, moisture-retarding **vapor barriers,** also called **vapor retarders,** are strongly recommended on all exterior walls. **Polyethylene plastic film** is one of the more commonly used sheet materials. For buildings in most parts of North America, the vapor barrier is placed at the inside of the insulation. It is fastened inside the wall studs directly beneath the wallboard. In the very hot and humid areas of the southern states, the procedure is reversed, and the vapor barrier is placed outside the wall. Polyethylene vapor barriers are also placed beneath concrete floor slabs to help retard moisture infiltration from the soil below.

R-VALUE

The **R (resistance) value** of a material is the rating of its thermal resistance to heat flow. The total R-value of a wall, roof, or floor is based on the combined materials used in its construction. Materials such as concrete, brick, stone, and wood have low R-values. Therefore, thermal insulating materials with high R-values are added to the building shell to make the building energy-efficient. Local building codes will often designate the required insulation R-values for ceilings, walls, and floors. General recommendations have been established for different climate areas. **(See Figure 21–5.)**

The total R-value for a ceiling, wall, or floor is found by adding the R-values of all the materials used in its construction and should be equal to or more than the required R-value for that climate zone. Tables are available that list the R-values of many commonly used building materials. **(See Table 21–1.)**

Figure 21–6 gives an example of the R-value (13) for an exterior wall located in AREA 5 as shown on the map in Figure 21–5. The materials used without added insulation, along with their thicknesses, and their R-values, are as follows:

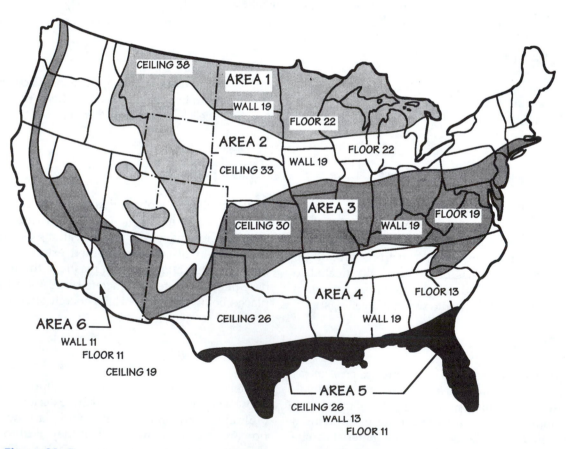

Figure 21–5.

The map shows ceiling, floor, and wall insulation R-values required in different climate areas of the United States.

TABLE OF R-VALUES
FOR VARIOUS BUILDING MATERIALS

CODE	MATERIAL DESCRIPTION	R-Value / " thickness	R-Value @ thickness listed
	INTERIOR WALL FINISH		
A	1/2" SAND AGGREGATE GYPSUM PLASTER		0.09
A	1/2" GYPSUM WALLBOARD		0.45
A	3/8" GYPSUM WALLBOARD		0.32
A	3/4" THICK WOOD PANELING		0.93
A	1/4" WOOD PANELING		0.31
	FINISH FLOOR MATERIALS		
B	CARPET & FIBROUS PAD		2.08
B	CARPET & RUBBER PAD		1.23
B	CORK TILE		0.28
B	ASPHALT OR VINYL TILE		0.05
B	HARDWOOD 3/4" (OAK, MAPLE, ETC.)		0.68
	BUILDING MEMBRANE		
C	15LB. PERMEABLE FELT		0.06
D	PLASTIC FILM		0.00
	SUB FLOORING BOARDS		
E	3/4" WOOD SUBFLOOR		0.94
E	1/2" PLYWOOD SUBFLOOR		0.62
	INSULATING BLANKET & BATTS MINERAL FIBER (ROCK, SLAG OR GLASS)		
F	2-2.75 IN		7
F	3-3.5 IN		11
F	3.50-6.5 IN		19
G	8.5 IN		30
	LOOSE FILL INSULATION		
H	CELLULOSE	3.70	
H	EXPANDED PERLITE	2.70	
	MINERAL FIBER (ROCK SLAG OR GLASS)		
H	3.75-5 IN		11
J	6.5-8.75 IN		19
J	7.5-10 IN		22
J	10.25-13.75 IN		30
J	VERMICULITE	2.13	
	FOAMED IN INSULATION		
K	FOAM-UREA FORMALDEHYDE	4.80	
K	FOAM-POLYURETHANE	6.25	

CODE	MATERIAL DESCRIPTION	R-Value / " thickness	R-Value @ thickness listed
	FOUNDATION INSULATION		
L	CELLULAR GLASS	2.63	
L	POLYSTYRENE 1" TO 2" THICK	3.57	
L	POLYURETHANE 1" TO 2" THICK	6.25	
	SHEATHING BOARDS		
M	EXPANDED POLYSTYRENE	4.00	
M	EXPANDED POLYURETHANE	6.25	
M	1/2" GYPSUM WALLBOARD		0.45
	(USE ONLY IN WEATHER PROTECTED AREAS)		
M	PLYWOOD	1.25	
M	1/2" PLYWOOD		0.62
M	3/4" PLYWOOD		0.93
M	1/2" FIBERBOARD SHEATHING		1.32
M	MEDIUM DENSITY HARDBOARD		1.37
M	MEDIUM DENSITY PARTICLEBOARD		1.06
	SIDING MATERIALS		
N	WOOD SHINGLES		1.19
N	BEVELWOOD SIDING		0.81
N	ALUMINUM OR STEEL SIDING		0.61
N	5/8" PLYWOOD SIDING		0.77
N	STUCCO		0.20
N	COMON BRICK		0.20
N	FACE BRICK		0.11
	MASONRY MATERIALS		
P	CONCRETE	0.08	
P	8" CONCRETE BLOCKS		1.11
P	12" CONCRETE BLOCKS		1.28
	MISC.		
	FIR, PINE & SIMILAR SOFTWOODS	1.25	
	SINGLE PANE GLASS		0.96
	DOUBLE PANE GLASS		1.92
	TRIPLE GLAZING		3.13

Table 21–1.
This table gives the R-values of the more commonly used building materials.

MATERIAL	R-VALUE
1/2" gypsum wallboard	0.45
.004" polyethylene vapor barrier	0.00
1" polystyrene panel sheathing	4.00
1/2" wood siding	.81
TOTAL	5.26

The total R-value of the materials (without insulation) adds up to 5.26. This means that the added insulation will have to make up a difference of 7.74 (13 minus 5.26 = 7.74). The fiberglass material shown in Table 21–1 that is closest to the 7.74 R-value is 3.5" (3 1/2") thick, which has an R-value of 11. The addition of this thickness of fiberglass results in a total R-value of 16.26 (5.26 + 11 = 16.26), more than the required amount for an AREA 5 wall.

FOUNDATION INSULATION

The outside surfaces of a full-basement foundation should first be coated with a waterproofing material.

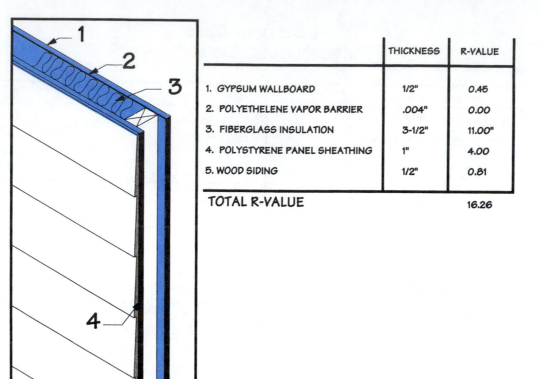

	THICKNESS	R-VALUE
1. GYPSUM WALLBOARD	1/2"	0.45
2. POLYETHELENE VAPOR BARRIER	.004"	0.00
3. FIBERGLASS INSULATION	3-1/2"	11.00"
4. POLYSTYRENE PANEL SHEATHING	1"	4.00
5. WOOD SIDING	1/2"	0.81
TOTAL R-VALUE		16.26

Figure 21–6.
The R-values of the different materials of this exterior wall add up to the total R-value of the wall.

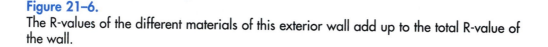

Many asphaltic, plastic, and synthetic rubber products are available for this purpose. Rigid insulating panels are then applied to the wall surface with mastic or another type of adhesive.

Full basement foundation walls can be insulated with rigid insulating panels. After the panels have been attached to the walls, a vapor barrier is placed over them. **Furring strips** (1 × 3 boards) are nailed over the insulation and vapor barrier. Finally, finish paneling or gypsum wallboard is nailed to the furring strips. Insulation is not usually placed beneath the floor slab of a full-basement foundation. However, a vapor barrier should be placed to retard water infiltration. **(See Figure 21–7.)**

Unheated crawl-space foundations require insulation only in the floor unit above the foundation walls. One procedure is to tack chicken wire beneath the joists and then lay fiberglass blankets on top of the wire. A vapor barrier is placed on top of the joists before nailing down the subfloor. **(See Figure 21–8.)**

Ground-supported slab foundations present the problems of moisture and cold. These problems are caused by water rising from the water table to the ground surface beneath the slab. Therefore, a vapor barrier should be placed over the ground before the slab is poured. Polyethylene film of 6 mil nominal thickness is often used for this purpose. Another concern is heat loss through the concrete slab. Rigid insulation placed over the vapor barrier will help prevent heat loss. **(See Figure 21–9.)**

ATTICS AND ROOFS

Figure 21–10 illustrates some insulating procedures for attics and roofs. In uninhabited attic space, insulation is placed between the ceiling joists. Fiberglass blankets or loose-fill insulation can be used. A vapor barrier should be attached to the bottom of the ceiling joists. **Vents** are provided in the attic space to enable air to circulate. This helps prevent the accumulation of moisture and lessens heat buildup during warm weather.

Higher pitched roofs, such as in one-and-one-half-story construction, allow for living space in the attic area. In this case, insulation is placed in the walls and ceiling of the attic. It is also placed in the section of roof rafters directly over the living area. Vapor barriers are fastened against all interior wall and ceiling areas.

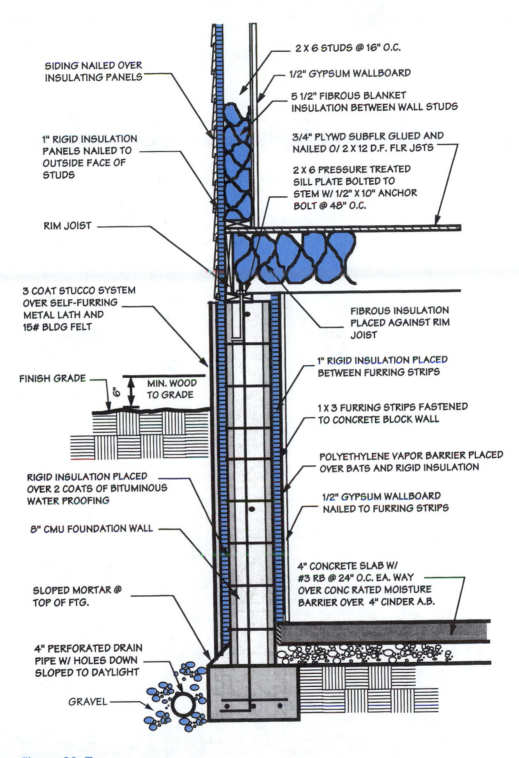

SIDING NAILED OVER
INSULATING PANELS

1" RIGID INSULATION
PANELS NAILED TO
OUTSIDE FACE OF
STUDS

RIM JOIST

3 COAT STUCCO SYSTEM
OVER SELF-FURRING
METAL LATH AND
15# BLDG FELT

FINISH GRADE

6"

MIN. WOOD
TO GRADE

RIGID INSULATION PLACED
OVER 2 COATS OF BITUMINOUS
WATER PROOFING

8" CMU FOUNDATION WALL

SLOPED MORTAR @
TOP OF FTG.

4" PERFORATED DRAIN
PIPE W/ HOLES DOWN
SLOPED TO DAYLIGHT

GRAVEL

2 X 6 STUDS @ 16" O.C.

1/2" GYPSUM WALLBOARD

5 1/2" FIBROUS BLANKET
INSULATION BETWEEN WALL STUDS

3/4" PLYWD SUBFLR GLUED AND
NAILED O/ 2 X 12 D.F. FLR JSTS

2 X 6 PRESSURE TREATED
SILL PLATE BOLTED TO
STEM W/ 1/2" X 10" ANCHOR
BOLT @ 48" O.C.

FIBROUS INSULATION
PLACED AGAINST RIM
JOIST

1" RIGID INSULATION PLACED
BETWEEN FURRING STRIPS

1 X 3 FURRING STRIPS FASTENED
TO CONCRETE BLOCK WALL

POLYETHYLENE VAPOR BARRIER PLACED
OVER BATS AND RIGID INSULATION

1/2" GYPSUM WALLBOARD
NAILED TO FURRING STRIPS

4" CONCRETE SLAB W/
#3 RB @ 24" O.C. EA. WAY
OVER CONC RATED MOISTURE
BARRIER OVER 4" CINDER A.B.

Figure 21–7.
Rigid insulation is applied to full-basement foundation walls.

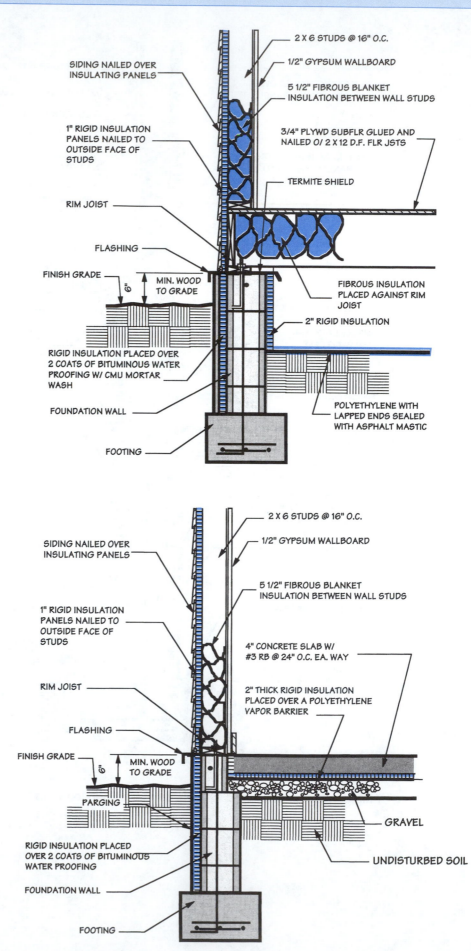

Figure 21–8.
Rigid insulation boards are placed against the walls of this crawl-space foundation. Fiberglass blankets are installed between the floor joists.

SIDING NAILED OVER INSULATING PANELS

1" RIGID INSULATION PANELS NAILED TO OUTSIDE FACE OF STUDS

RIM JOIST

FLASHING

FINISH GRADE

6"

MIN. WOOD TO GRADE

RIGID INSULATION PLACED OVER 2 COATS OF BITUMINOUS WATER PROOFING W/ CMU MORTAR WASH

FOUNDATION WALL

FOOTING

2 X 6 STUDS @ 16" O.C.

1/2" GYPSUM WALLBOARD

5 1/2" FIBROUS BLANKET INSULATION BETWEEN WALL STUDS

3/4" PLYWD SUBFLR GLUED AND NAILED O/ 2 X 12 D.F. FLR JSTS

TERMITE SHIELD

FIBROUS INSULATION PLACED AGAINST RIM JOIST

2" RIGID INSULATION

POLYETHYLENE WITH LAPPED ENDS SEALED WITH ASPHALT MASTIC

Figure 21–9.
Rigid insulation over a vapor barrier is placed below the concrete floor slab of a slab foundation. Slabs lose heat most readily at their perimeters. Therefore, in colder climates it is advisable to place insulation at the outside or inside edge of the footing wall.

SIDING NAILED OVER INSULATING PANELS

1" RIGID INSULATION PANELS NAILED TO OUTSIDE FACE OF STUDS

RIM JOIST

FLASHING

FINISH GRADE

6"

MIN. WOOD TO GRADE

PARGING

RIGID INSULATION PLACED OVER 2 COATS OF BITUMINOUS WATER PROOFING

FOUNDATION WALL

FOOTING

2 X 6 STUDS @ 16" O.C.

1/2" GYPSUM WALLBOARD

5 1/2" FIBROUS BLANKET INSULATION BETWEEN WALL STUDS

4" CONCRETE SLAB W/ #3 RB @ 24" O.C. EA. WAY

2" THICK RIGID INSULATION PLACED OVER A POLYETHYLENE VAPOR BARRIER

GRAVEL

UNDISTURBED SOIL

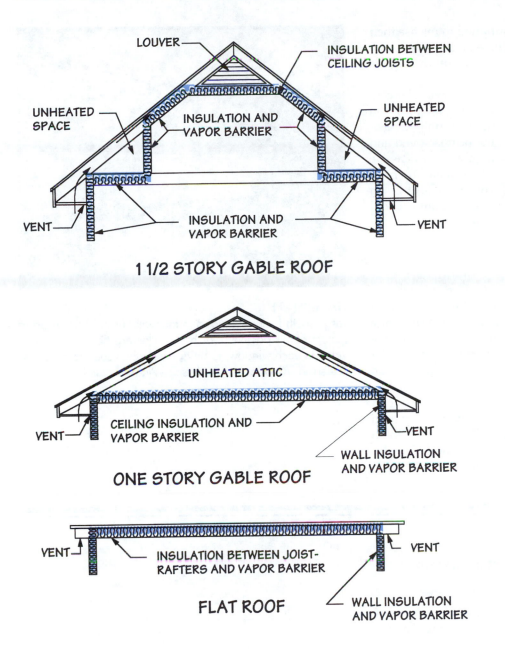

LOUVER

INSULATION BETWEEN CEILING JOISTS

UNHEATED SPACE

INSULATION AND VAPOR BARRIER

UNHEATED SPACE

VENT

INSULATION AND VAPOR BARRIER

VENT

1 1/2 STORY GABLE ROOF

UNHEATED ATTIC

VENT

CEILING INSULATION AND VAPOR BARRIER

VENT

WALL INSULATION AND VAPOR BARRIER

ONE STORY GABLE ROOF

VENT

INSULATION BETWEEN JOIST-RAFTERS AND VAPOR BARRIER

VENT

FLAT ROOF

WALL INSULATION AND VAPOR BARRIER

Figure 21–10.

These are examples of typical ceiling and roof areas requiring insulation.

In flat roofs the ceiling joists also serve as roof rafters. Blanket insulation is placed between the joist-rafters. For added protection, rigid insulation can be placed over the joists before the finish roofing material is applied.

SOLAR ENERGY AND BUILDING DESIGN

Solar energy is used to convert the natural heat of the sun into thermal energy. This free, non-polluting energy source effectively creates interior heat in buildings that experience a sufficient amount of winter sun. Solar methods are also safe and very inexpensive to operate. It is hoped that the increased use of solar energy will significantly help in the conservation of our natural resources and reduce the nation's dependency on fossil fuels. Today's solar systems are either *active* or *passive*.

PASSIVE SOLAR SYSTEMS

Passive solar systems strongly influence the design of a building and its appearance. They are based on the principle of absorbing and storing the heat energy of the sun and actually represent an idea that has been around for thousands of years. The ancient Greeks were aware of the advantage of constructing dwellings with tall walls facing south and lower walls facing north to keep out winter winds. Native Americans of the American Southwest built adobe houses in the deep alcoves of south-facing cliffs. As the winter sun rose, the adobe material of the dwellings absorbed and stored heat. At night the adobe walls continued to release their heat into the interior of the dwellings. During the summer the cliffs provided shade from the sun during the hot afternoon hours.

In the contemporary applications of passive solar construction, large areas of glass or clear plastic are placed in the south-facing walls of the building. The sunlight passes through these glazed areas and the heat is stored and absorbed within the building. **(See Figure 21–11.)** Passive solar buildings usually require some type of backup heating system during cloudy periods when there is little sun exposure. Conventional forced-air methods and wood-burning stoves are used for this purpose.

ACTIVE SOLAR SYSTEMS

Active solar systems, in contrast to passive systems, have less of an effect on the general building design. An active solar system requires a *collector,* a *thermal storage area,* and *pumps* or *fans* to distribute the heat to the desired locations of the building. The south-facing **solar collectors** are rectangular and can be mounted on an existing roof. In new construction, collectors can be placed between the roof rafters and the south gable areas of a gable roof. A collector can also be placed away from the building if the roof does not lend itself to a southern exposure for the collectors. **Figure 21–12** shows the different modes of a warm-air active solar system.

Figure 21–11.
Large south-facing windows in the wall and roof areas are a feature of this passive solar house. Pleated shades are provided in each window, to be opened when sun protection is preferred. *(Courtesy of Andersen Windows, Inc.)*

REVIEW QUESTIONS

Enter the missing words in the spaces provided on the right.

1. Heating and cooling accounts for _____ of the energy consumption of an average North American home.

   _____

2. Light colors on the exterior of a building _____ heat.

   _____

3. Roof _____ will shade windows from the summer sun.

   _____

4. Attic air circulation is provided by _____ and _____.

   _____

5. _____ sashes should be installed outside the windows in areas that experience severe winter conditions.

   _____

6. Air flow around the edges of doors and windows can be prevented by installing _____.

   _____

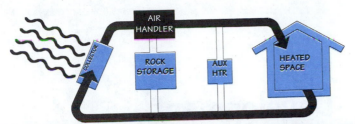

Mode 1. A centrifugal fan in the air handler moves the heated air from the collector directly to the space to be heated. Automatically controlled backdraft dampers prevent backward air leakage.

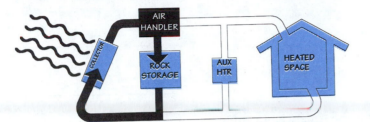

Mode 2. When heat is not needed in the building it can be directed into the rock storage area for later use. Dampers prevent the warm air from going to the building's interior.

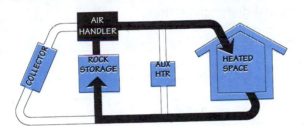

Mode 3. If stored heat is still sufficient to heat the house, the controller activates a blower and resets the dampers so that warm air is moved into the space to be heated.

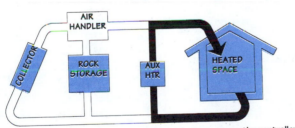

Mode 4. If there is no air available from the collector or rock storage area, the controller will automatically turn on the auxiliary heater.

Figure 21–12.
The different modes of a warm-air active solar system can be set in motion through automatic controls linked to the thermostat.

7. _____, _____ , and _____ are the most widely used insulation materials.

8. During the cold winter months, vapor barriers help prevent _____ that could result in water accumulation inside the walls.

9. The rating of a material's thermal resistance to heat flow is called its _____.

10. Solar collectors of active solar systems should always face _____ .

SECTION 4 TEST

Plan Development

Answer T (true) or F (false).

1. A civil engineer creates the basic design of a building.

   _____

2. Professional estimators calculate the approximate cost of a construction project.

   _____

3. The main professional organization for architects is the American Institute of Architects.

   _____

4. CADD is the name of a computer-generated drafting and design program.

   _____

5. Contour lines describe the soil conditions on the job site.

   _____

6. House design should not be influenced by the natural features of a construction site.

   _____

7. Orientation includes the positioning of the building.

   _____

8. The direct rays of the sun are at a lower angle during the summer.

   _____

9. Ranch-style houses are best suited for sloping lots.

   _____

10. The attic of a one-and-one-half-story house provides a usable living or storage area.

    _____

11. A square-shaped building is most costly to construct.

    _____

12. A foyer is considered part of the living area of a building.

    _____

13. If possible, front entries should face north.

    _____

14. Caulking is a sealing material placed around door and window frames.

    _____

15. The most widely used insulation materials are fiberglass, cellulose, and plastic foam.

    _____

16. Condensation is considerably reduced with the use of wall vapor barriers.

    _____

17. Passive solar systems absorb heat through roof collectors.

    _____

18. Solar energy converts the natural heat of the sun into thermal energy.

    _____

19. North-facing collectors are a feature of the active solar system.

    _____

20. Active solar systems have little effect on building design.

    _____

Choose one or more correct ways to finish each statement.

21. A final set of plans must
    a. be a design of sound construction.
    b. conform to local building codes.
    c. meet health and safety standards.
    d. be all of the above.

22. A person designing more complex power-driven equipment is called a
    a. power engineer.
    b. electrical engineer.
    c. architect.
    d. mechanical engineer.

23. Stock plans are
    a. designs for custom-built houses.
    b. used in speculative construction.
    c. found in home-design publications.
    d. stocked in bookstores.

24. More accurate terms for construction drawings today are
    a. construction plans.
    b. blueprints.
    c. dark-line prints.
    d. prints.

25. The physical and graphic description of the land surface is called
    a. geography.
    b. topography.
    c. a land survey.
    d. none of the above.

26. Elevations are the
    a. different heights identified on the lot.
    b. slope of the lot.
    c. surface features of the lot.
    d. drainage considerations.

27. Building design is influenced by
    a. access to utilities (electricity, gas, etc.).
    b. the shape and dimensions of a lot.
    c. water drainage conditions.
    d. road access to the building.

28. A building on a hilly slope
    a. facing south gets less heat from sun exposure.
    b. facing north gets the most amount of sun exposure.
    c. facing west can take best advantage of all climate conditions.
    d. facing south gets the most heat from sun exposure.

29. A ranch-style house is most often constructed over a
    a. slab-on-grade foundation.
    b. crawl-space foundation.
    c. full-basement foundation.
    d. stepped foundation.

30. Split-level homes
    a. can slope only from front to back.
    b. consist of two or more stories.
    c. are very practical for sloping lots.
    d. always have usable attic space.
    _____

31. L-shaped and U-shaped buildings
    a. are cheaper to construct.
    b. offer more diverse possibilities for room layout.
    c. have less total linear feet in their perimeters than rectangular shapes.
    d. are more limited in their exterior design.
    _____

32. The basic divisions of a residential building are the
    a. bedrooms, bathrooms, and kitchen.
    b. different floor levels.
    c. garage, basement, and living room.
    d. sleeping, living, and service areas.
    _____

33. To take better advantage of the winter sun in colder climates,
    a. dark roof covering and siding should be used.
    b. lighter exterior colors should be used.
    c. the size of roof overhangs should be minimized.
    d. the amount of exterior wall area should be maximized.
    _____

34. Batt and blanket fiberglass insulation is designed for studs and joists spaced
    a. 12" OC.
    b. 16" OC.
    c. 24" OC.
    d. 48" OC.
    _____

35. The R-value of insulation is
    a. its thickness.
    b. the type of material used.
    c. its quality.
    d. its thermal resistance.
    _____

36. Rigid insulation is often placed
    a. on the outside of full-basement foundations walls.
    b. on the interior of full-basement foundation walls.
    c. against the outside face of exterior framed walls.
    d. all of the above.
    _____

37. The direct movement of heat through space from a warm source is called
    a. radiation.
    b. convection.
    c. conduction.
    d. reflection.
    _____

38. The main heat source of a passive solar system comes through
    a. collectors on the roof.
    b. large areas of south-facing glass or plastic windows.
    c. collectors placed away from the building.
    d. none of the above.
    _____

39. The collector of an active solar system
    a. accumulates the heat.
    b. is equally effective when placed in any direction.
    c. absorbs and distributes the heat from the sun's rays.
    d. is placed only on the roof of a house.

40. Of the two types of solar systems, the building design is most affected
    a. by a passive solar system.
    b. by an active solar system.
    c. equally by both types.
    d. all of the above.

Enter the missing words in the spaces provided on the right.

41. For deeper excavations, _____ engineers are employed to test ground conditions.

42. The architect may assist the owner in selecting a building _____.

43. _____ is a term used to describe an older process of plan reproduction.

44. The entire process of plan development can be done with a _____ computer program.

45. The boundaries of a building site are also called the _____ lines.

46. _____ trees can help protect a building from prevailing winds.

47. In colder regions, windows facing _____ should be limited to reduce heat loss.

48. Dead-air _____ in warm climates have higher temperatures than surrounding areas.

49. Dormers are pitched structures that _____ from a roof.

50. _____ placed along interior bedroom walls help cut down noise penetration.

51. The main purpose of a _____ room is to provide an area for family entertainment.

52. Mudrooms are usually located inside a rear _____.

53. Modern buildings are designed to _____ the use of mechanical heating and cooling.

54. _____ colors absorb less heat from the sun's rays.

55. Metal-clad doors with foam _____ are the best insulated exterior doors.

56. Plastic _____ insulation is poured or sprayed into place.

57. To prevent moisture problems, a _____ _____ should be placed below a concrete slab.

_____

58. _____ energy converts the natural heat of the sun into thermal energy.

_____

59. Large south-facing windows are required for _____ solar systems.

_____

60. In the _____ solar system, south-facing collectors absorb energy from the sun.

_____

SECTION 5

Code Regulations and Legal Documents

The design of a building must conform with structural standards established by code regulations. Zoning codes, for example, determine what type of construction is permitted in a particular area of the community. Building codes ensure proper structural design, a high standard of workmanship, and the use of proper materials. Plans must be approved by the local government building agency, which issues a building permit. During construction, the project will be subject to a series of inspections. After the final inspection and approval, a Certificate of Occupancy can be issued.

The other legal documents in construction are the contracts between owner and architect, owner and general contractor, and general contractor and subcontractors. These contracts derive most of their information from the construction plans and specifications. ∎

UNIT 22

Zoning codes and Regulations

Most cities, counties, and states in the United States have **zoning codes** and regulations governing the type of construction permitted within different areas of communities. Developed by local planning commissions or other municipal government agencies, these codes are subject to change by the same agencies. The architect or designer of a building must be thoroughly familiar with the existing local zoning code. Frequently the codes are adapted from model zoning codes developed by several national building code agencies. An example is the **Uniform Zoning Code (UZC),** which states the following:

> The purpose of this code is to safeguard the health, property, and public welfare by controlling the design, location, use, or occupancy of all buildings through the regulated and orderly development of land and land uses within this jurisdiction.

ZONING CODE APPLICATIONS

Most zoning codes include a map showing the boundaries of each zone. The zones are usually designated as residential (R), commercial (C), and manufacturing (M). Rural areas may also identify an agricultural (A) zone.

A zoning code will contain regulations for the following:

✓ Square footage.
✓ Required distances of the building's front, rear, and side *setbacks* (also called the front, rear and side *yards*) from the property lines.
✓ Maximum building heights.
✓ Maximum density, based on the number of dwelling units per acre.

Some zoning codes also give the maximum square feet the building can occupy on the property. This is based on a percentage of the square footage of the property.

R (RESIDENTIAL) ZONES

The Uniform Zoning Code (UZC) classifies the R (residential) zones into three divisions:

Division 1. Any single-family dwelling, including private garages, accessory living quarters, recreation rooms, or private stables accessory to the main use.
Division 2. Any use permitted in the R, Division 1, zones. Added are two-family dwellings.
Division 3. Any use permitted in the R, Division 2, zones. Added are multiple-family dwellings, apartment houses, and condominiums.

Zoning codes usually include tables that clarify many regulations. **(See Table 22–1.)** (This table appears in the UZC as *Table 5-A—R (Zone Bulk Regulations)* A pictorial based on the examples used in Table 22–1 is shown in **Figure 22–1.**

C (COMMERCIAL) AND CM (COMMERCIAL-MANUFACTURING) ZONES

The UZC classifies C (commercial) zones into the following divisions:

Division 1. Any office, retail, automobile service station, restaurant, day-care center, residential, or other small-scale retail or personal service use servicing the day-to-day needs of the residents of the surrounding area.
Division 2. Any use permitted in the C, Division 1, zones. Added are nurseries, physical fitness centers, community and fraternal clubs, churches, libraries, and automated public utility facilities. Also, any uses permitted in the C, Division 2, zones.
Division 3. Any use permitted in the C, Division 2, zones. Added are supermarkets, department stores, hospitals, automobile sales, lumber yards, wholesale outlets, printing plants, cocktail lounges and taverns, self-storage centers, theaters, bowling alleys, museums, and hotels and motels with the exception of other residential use.
Division 4. Any uses permitted on the C, Division 3, zones. Added are large-scale retail, commercial and wholesale uses, bakeries, storage yards, distributors, major automobile repair shops, machine shops, and woodworking shops.

The UCZ classifies CM (commercial-manufacturing) zones into the following divisions:

TABLE 5-A—R ZONE BULK REGULATIONS
(in feet, unless noted otherwise)

DIVISION		MIN. LOT AREA/SITE (square feet)	MAXIMUM DENSITY (D.U./acre)	LOT DIMENSIONS		SETBACK REQUIREMENTS			MAXIMUM BUILDING HEIGHT
				Minimum Lot Width	Minimum Lot Depth	Minimum Front Yard	Minimum Side Yard	Minimum Rear Yard	
		× 0.0929 for m²	D.U./4047 m²	× 304.8 for mm					
1	a	35,000	1	125	150	25	10	30	35
	b	20,000	2	100	125	20	10	25	35
	c	10,000	4	75	100	20	5	25	30
	d	6,000	6	60	90	15	5	20	30
2	a	10,000	4	60	70	20	5	20	30
	b	6,000	6	60	70	15	5	20	30
3	a	6,000	8	60	70	15	5	20	30
	b	6,000	12	60	70	15	5	20	30

Division: The three divisions in this table are subdivided into a, b, c, and d categories.

Minimum Lot Area/Site: the minimum square footage permitted for lots in this area. To change square feet to meters, multiply the square feet by .0929. For example, a 20,000 square foot lot of Division 1-B equals 1858 square meters (20,000 × .0929 = 18.58).

Maximum Density: The number of dwelling units (D.U.) allowed on a one acre area of land including dedicated streets within the development. For example, an acre is 43,560 square feet and a Division 1-B lot of 20,000 square feet will go into an acre two times. Therefore, only two dwelling units are allowed in the general area of one acre.

Lot Dimension: The minimum width and length of a lot based on its division category. For example, a lot in Division 1-B must be a minimum of 100' - 0" wide and 125' - 0" deep.

Setback Requirements: Minimum distance of building from property lines. For example, a building constructed on a lot in Division 1-B must be a minimum distance of 20' - 0" from the front property line, 10' - 0" from the side property lines, and 35' - 0" from the rear property line.

Maximum Building Height: Height of building allowed in division. For example, the building height in Division 1-B can be no higher than 35' - 0".

Table 22–1.
This table gives information about minimum areas, setbacks, densities, and maximum heights for residential (R) zones. It is identified as Table 5–A in the Uniform Zoning Code.)

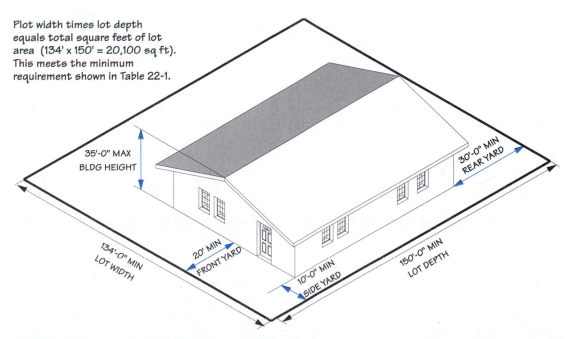

Plot width times lot depth equals total square feet of lot area (134' x 150' = 20,100 sq ft). This meets the minimum requirement shown in Table 22-1.

35'-0" MAX BLDG HEIGHT

30'-0" MIN REAR YARD

134'-0" MIN LOT WIDTH

20' MIN FRONT YARD

10'-0" MIN SIDE YARD

150'-0" MIN LOT DEPTH

Figure 22–1.
This pictorial example is based on the information provided in Table 22–1 for a Division 1-B building.

TABLE 6-A—C AND CM ZONES BULK REGULATIONS
(in feet, unless noted otherwise)

| DIVISION | MINIMUM LOT AREA (square feet) × 0.0929 for m² | MAXIMUM DENSITY (units/acre) units/4047 m² | LOT DIMENSIONS | | SETBACK REQUIREMENTS | | | MAXIMUM BUILDING HEIGHT |
| | | | Minimum Lot Width | Minimum Lot Depth | Minimum Front Yard | Minimum Side Yard | Minimum Rear Yard | |
			× 304.8 for mm					
1	6,000	12	30	70	0	0	0	30
2	N/A	N/A	30	70	0	0	0	40
3	N/A	N/A	75	100	0	0	0	50
4	N/A	N/A	75	100	0	0	0	50

Division: there are four divisions described in this table (1, 2,3, and 4).

Minimum Lot Area: The square footage permitted for a number of lots in this area. For example, the minimum lot area allowed in Division 1 is 6000 square feet, however minimum lot area requirement is *not applicable* (N/A) for Divisions 2, 3, and 4.

Maximum Density: The number of building units allowed on one acre of land including dedicated streets within the division. For example, there are 12 units allowed in Division 1, however there is no maximum number of units per acre for Divisions 2, 3, and 4.

Lot Dimensions: The minimum width and length of a lot in each division. For example, a lot size in Division 3 must be at least 75' - 0" wide and 100' - 0" deep.

Setback Requirements: Minimum distance of building from property lines. Note there are no minimum distance requirements for buildings in any division of the C and CM Zones.

Table 22–2.
This table gives information about commercial (C and CM) zones. (It is identified as Table 6–A in the Uniform Zoning Code.)

Division 1. Any C, Division 1, use. Residential use allowed except in a story or basement abutting a street grade.

Division 2. Any C, Division 2, use. Residential use allowed except in a story or basement abutting street grade.

Table 22–2 provides information about area, density, lot dimensions, and setbacks for C and CM zones. (This table appears in the UZC as *Table 6-A—C and CM (Zones Bulk Regulations.)*

MANUFACTURING (M) ZONES

The UZC classifies the M (manufacturing) zones into the following divisions:

Division 1. Any light manufacturing or industrial use, such as warehouses, research or testing laboratories, product-distribution centers, woodworking shops, auto body shops, furniture-assembly centers, dry-cleaning plants, machine shops, and boat-building storage yards.

Division 2. Any use permitted in the M, Division 1, zone. Added are stadiums and arenas, indoor swap meets, breweries, liquid-fertilizer manufacturing, carpet manufacturing, monument works, and regional recycling centers.

Division 3. Any use permitted in the M, Division 2, zone. Added are auto-dismantling yards, alcohol manufacturing, cotton gin, paper manufacturing, quarries, salt works, petroleum refining, and other similar uses

Table 22–3 provides information about area, density, lot dimensions and setbacks for M zones. (This table appears in the UBC as Table 7-A—M (Zone Bulk Regulations.)

Some additional topics and regulations included in the UZC follow:

Off-street parking
Fence heights
Locations of accessory buildings
Allowable projection into yards
Landscaping requirements
Loading spaces
Grading and excavation

TABLE 7-A—M ZONE BULK REGULATIONS
(in feet, unless noted otherwise)

DIVISION	MINIMUM LOT AREA (square feet)	MAXIMUM DENSITY (units/acre)	LOT DIMENSIONS		SETBACK REQUIREMENTS			MAXIMUM BUILDING HEIGHT
			Minimum Lot Width	Minimum Lot Depth	Minimum Front Yard	Minimum Side Yard	Minimum Rear Yard	
	× 0.0929 for m²	units/4047 m²	× 304.8 for mm					
1	N/A	N/A	50	75	0	0	0	60
2	N/A	N/A	75	100	0	0	0	80
3	N/A	N/A	100	150	0	0	0	80

Division: There are three divisions in this table (1, 2, and 3).

Minimum Lot Area: In this case the minimum lot area requirement is not applicable (N/A0 in all three divisions.

Maximum Density: Ruling regarding maximum number of units per acre is not applicable (N/A) in all three divisions.

Lot Dimensions: Minimum lot width and depth. For example, the minimum lot dimensions for Division 3 is 100′ - 0″ wide and 150′ - 0″ deep..

Setback Requirements: There are no setback requirements.

Maximum Building Height: Height of building based in division location. For example, the maximum height of a building is Division 2 is 80′ - 0″.

Table 22–3.
This table gives information about manufacturing (M) zones; it is identified as Table 7–A in the Uniform Zoning Code.

REVIEW QUESTIONS

Enter the missing words in the spaces provided on the right.

1. An A zone is a(n) _____ area.

   _____

2. A residential area is in a(n) _____ zone.

   _____

3. DU is the abbreviation for _____ _____.

   _____

4. A restaurant is permitted in a(n) ____ zone.

   _____

5. An auto body shop is permitted in a(n) _____ zone.

   _____

6. Department stores are permitted in an _____ zone.

   _____

7. The Uniform Zoning Code classification M menas _____ .

   _____

8. A one family dwelling would be located in a(n) _____ zone.

   _____

9. The minimum front yard setback in Figure 22–1, page 271 is _____ .

   _____

10. The maximum building height in division 2-A of an R zone is _____.

   _____

UNIT 23

Building Codes

Building codes are a set of legally binding regulations that cover every aspect of a building's construction from its structural procedures to its finished appearance. Therefore, building codes are one of the single biggest factors affecting the overall design of a building. When creating a set of plans, the architect or designer must be certain that every aspect of the drawing conforms to local code practices. The building contractors and subcontractors, as well as the workers on the job, should also be familiar with the code regulations that apply to their particular construction project.

Building codes decree the minimum standards to protect the safety, welfare, and health of individuals occupying or working in a building. They ensure a high quality of construction, the use of proper materials, and the use of sound construction methods. Building officials periodically inspect a job during construction and at its completion to make sure building codes are followed.

TYPES OF BUILDING CODES

A building code covers all the structural work of a building except the plumbing and electrical installation. Separate electrical and plumbing code books contain the regulations that apply to this work. Many states in the United States have codes that can be generally applied in the entire state. However, most major cities have their own code books, which are usually stricter than the state codes. In more rural areas, the different counties or parishes may use the state code or pattern their regulations on several different national **model code books.**

MODEL CODES

National model codes have been developed by building officials and industry representatives in different regions of the nation. The emergence of national codes arose out of a need to help communities in developing workable local building codes. They set uniform standards where applicable and are continuously updated to reflect industry research and the development of new methods and materials. However, they are not official codes and do not become law unless they are directly adopted by the area or locality where they are being used. Besides the model building codes, there are model codes for plumbing, electrical work, and heating. The organizations that produce these codes also publish many other pamphlets that supplement much of the material in the code books.

Most local codes in the United States are closely based on one of three model codes. The **Uniform Building Code (UBC)** is most influential in the western and some midwestern states. In the east and some midwestern states, the **Building Officials & Code Administrators National Building Code (BOCA)** is the primary model. The **Standard Building Code (SBC)** has been accepted as the model code for the southern states. Canada has a model code called the **National Building Code of Canada,** and most Canadian provinces have "provincial codes."

Over the years, three of the more widely used model plumbing codes have been the **National Plumbing Code,** the **Uniform Plumbing Code,** and the **BOCA National Plumbing Code.** More recently the BOCA National Plumbing Code has been superseded by the **BOCA International Plumbing Code.** Most local electrical codes are patterned after the **National Electric Code (NEC).** A national **Uniform Heating & Comfort Cooling Code** is available for the installation of heating and air conditioning.

CONTENTS OF THE CODES

The three national code books are models for different regions of the United States. However, they are very similar in their approach and contain much of the same information. The following Table of Contents is from the BOCA National Building Code. The Table of Contents of the Standard Building Code and the first volume of the Uniform Building Code are almost identical to that of BOCA.

BOCA NATIONAL BUILDING CODE
TABLE OF CONTENTS

Chapter 1.	Administration
Chapter 2.	Definitions
Chapter 3.	Use or Occupancy

DIVISION OF CHAPTERS

Each chapter represents a major topic. The chapters are divided into sections, and if necessary the sections are further divided into subsections. The following is the first page taken from the Standard Building Code:

CHAPTER 1
ADMINISTRATION
GENERAL
101.1 Scope
The provisions of this chapter shall govern the administration and enforcement of the Standard Building, Gas, Mechanical and Plumbing Codes, and the National Electrical Code, hereinafter referred to as the "technical codes," as may be adopted by the state or local jurisdiction.

101.2 Title
The provisions of the following chapters will constitute and be known and be cited as the "Standard Building Code," hereinafter known as "this code."

101.3 Code Remedial
101.3.1 General. This code is hereby declared to be remedial and shall be construed to secure the beneficial interests and purposes thereof, which are public safety, health, and general welfare through structural strength, stability, sanitation, adequate light and ventilation, and safety to life and property from fire and other hazards attributed to the building environment including alteration, repair, removal, demolition, use and occupancy of buildings, structures, or premises, and by regulating the installation and maintenance of all electrical, gas, mechanical and plumbing systems, which may be referred to as service systems.

101.3.2 Quality Control. Quality control of materials and workmanship is not within the purview of this code except as it relates to the purposes stated herein.

101.3.3 Permitting and Inspection. The inspection or permitting of any building system or plan by any jurisdiction under the requirements of this code, shall not be construed in any court as a warranty of the physical condition of such building, system or plan or their adequacy. No jurisdiction nor any employee thereof shall be liable in tort for damages for any defect or hazardous or illegal condition or inadequacy in such building, system or plan, nor for any failure of any component of such, which may occur subsequent to such inspection or permitting.

101.4 Applicability
101.4.1 General. Where, in any specific case, different sections of this code specify different materials, methods of construction or other requirements, the most restrictive shall govern. Where there is a conflict between a general requirement and a specific requirement, the specific requirement shall be applicable.

Standard Building Code(o)) 994

USING THE CODES

Building codes contain a vast quantity of information, including many tables and graphs and some formulas and drawings. A person develops code-reading skills by studying and working with the codes. This unit presents only an introduction to interpreting code information. The order of information required at the beginning stages of designing a building is as follows:

✓ Occupancy classification
✓ Type of building
✓ Allowable floor areas
✓ Allowable building height
✓ Fire-resistant materials and construction
✓ Means of egress
✓ Exterior envelope
✓ Building service system

In the following discussion of these topics, alternating examples of tables from three model codes will be included.

OCCUPANCY CLASSIFICATION

All three model code books use identical **occupancy classifications,** and their descriptions are very much the same. **Occupancy** is defined as "the purpose for which a building, or portion thereof is used." For example, one building might be used for public assembly, another for family living (residential), and another for mercantile purposes. Buildings may also be used for more than one occupancy purpose, and the codes explain the requirements for mixed use. **Table 23–1** provides detailed descriptions of occupancy classifications. (This table appears in the Uniform Building Code as *Table 3–A— Description of Occupancies by Group and Division.*)

TABLE 3-A—DESCRIPTION OF OCCUPANCIES BY GROUP AND DIVISION[1]

GROUP AND DIVISION	SECTION	DESCRIPTION OF OCCUPANCY
A-1	303.1.1	A building or portion of a building having an assembly room with an occupant load of 1,000 or more and a legitimate stage.
A-2		A building or portion of a building having an assembly room with an occupant load of less than 1,000 and a legitimate stage.
A-2.1		A building or portion of a building having an assembly room with an occupant load of 300 or more without a legitimate stage, including such buildings used for educational purposes and not classed as a Group E or Group B Occupancy.
A-3		Any building or portion of a building having an assembly room with an occupant load of less than 300 without a legitimate stage, including such buildings used for educational purposes and not classed as a Group E or Group B Occupancy.
A-4		Stadiums, reviewing stands and amusement park structures not included within other Group A Occupancies.
B	304.1	A building or structure, or a portion thereof, for office, professional or service-type transactions, including storage of records and accounts; eating and drinking establishments with an occupant load of less than 50.
E-1	305.1	Any building used for educational purposes through the 12th grade by 50 or more persons for more than 12 hours per week or four hours in any one day.
E-2		Any building used for educational purposes through the 12th grade by less than 50 persons for more than 12 hours per week or four hours in any one day.
E-3		Any building or portion thereof used for day-care purposes for more than six persons.
F-1	306.1	Moderate-hazard factory and industrial occupancies include factory and industrial uses not classified as Group F, Division 2 Occupancies.
F-2		Low-hazard factory and industrial occupancies include facilities producing noncombustible or nonexplosive materials that during finishing, packing or processing do not involve a significant fire hazard.
H-1	307.1	Occupancies with a quantity of material in the building in excess of those listed in Table 3-D that present a high explosion hazard as listed in Section 307.1.1.
H-2		Occupancies with a quantity of material in the building in excess of those listed in Table 3-D that present a moderate explosion hazard or a hazard from accelerated burning as listed in Section 307.1.1.
H-3		Occupancies with a quantity of material in the building in excess of those listed in Table 3-D that present a high fire or physical hazard as listed in Section 307.1.1.
H-4		Repair garages not classified as Group S, Division 3 Occupancies.
H-5		Aircraft repair hangars not classified as Group S, Division 5 Occupancies and heliports.
H-6	307.1 and 307.11	Semiconductor fabrication facilities and comparable research and development areas when the facilities in which hazardous production materials are used, and the aggregate quantity of material is in excess of those listed in Table 3-D or 3-E.
H-7	307.1	Occupancies having quantities of materials in excess of those listed in Table 3-E that are health hazards as listed in Section 307.1.1.

Group and Division: The group is the letter (A, B, etc.) defining the general use of the building. For example, A identifies assembly buildings, and E identifies buildings associated with education. The groups are subdivided into numbered divisions based on building size. For example, an A-1 assembly building has an occupant load of 1000 or more and also contains a legitimate stage.

Section: The section number is the Uniform Building code where a detailed description of the building Group can be found. For example, information regarding an A-1 building can be found in Section 303.1.1 of the UBC.

Description: These are general descriptions of each Group and Division. More detailed descriptions are given in the Sections.

Table 23–1.
This table helps clarify occupancy classification descriptions. Only a portion of the table is reproduced here. (It is identified as Table 3–A in the Uniform Building Code.)

TYPES OF CONSTRUCTION

The "types of construction" chapters in all three model code books concern themselves with a building's resistance to fire and other aspects of public safety. The chapters discuss in general terms the fire-resistive ratings of materials and assembling methods used in the construction of the building. (The code book chapters on fire protection provide much greater detail.) . A **fire-resistive rating**, also referred to as **fire resistance**, is determined by the *number of hours* a building or a building component sustains its capability to contain a fire and continues to function as a structural member. These ratings are based on tests conducted by recognized testing agencies such as the **American Society for Testing and Materials (ASTM).**

The Uniform Building Code and the BOCA National Code divide building types into classifications of **I, II, III, IV,** and **V.** The Standard Building Code also includes a **VI** classification. The codes indicate the required minimum fire resistance for the outside walls, interior partitions, columns, beams, girders, floors, ceilings, roofs, trusses, arches, and construction around the exits. All three code books contain tables in the chapters on "types of construction." For example, **Table 23–2** of the Standard Building Code provides fire resistance ratings for all building type classifications. (This table appears in the Standard Building Code as *Table 600—Fire Resistance Ratings.*)

LOCATION OF BUILDING

The location of the building concerns its distances from property lines and other buildings. These distances are measured at a right angle from each property line and must conform to the local zoning code. The location of the building is a main factor affecting the fire-resistance ratings of the exterior walls and the materials surrounding the wall openings. Tables are also provided in some code books in the chapters dealing with building location. An example is shown in **Table 23–3.** (This table appears in the Uniform Building Code as *Table 5-A— Exterior Wall and Opening Protection Based on Location on Property for All Construction Types.*)

ALLOWABLE FLOOR AREA AND BUILDING HEIGHTS

Floor area, is defined in all three model codes as the number of square feet within surrounding exterior walls of a building, excluding vent shafts and courts. The area of a building, or portion of it, that is not surrounded by exterior walls will be the usable area under the horizontal projection of the floor or roof above. The allowable floor area is the maximum floor area permit-

Table 600
Fire Resistance Ratings
Required Fire Resistance in Hours

STRUCTURAL ELEMENT	Type I	Type II	Type III	Type IV		Type V		Type VI	
				1-Hour Protected	Unprotected	1-Hour Protected	Unprotected	1-Hour Protected	Unprotected
PARTY AND FIRE WALLS (a)	4	4	4	4	4	4	4	4	4
INTERIOR BEARING WALLS Supporting columns, other bearing walls or more than one floor	(l) 4	3	2	1	NC	1 (h)	0 (h)	1	0
Supporting one floor only	3	2	1	1	NC	1	0	1	0
Supporting one roof only	3	2	1	1	NC	1	0	1	0
INTERIOR NONBEARING PARTITIONS			See 704.1, 704.2 and 705.2						
COLUMNS Supporting other columns or more than one floor	(l) 4	3	See 605 H(d)	1	NC	1	0	1	0
Supporting one floor only	3	2	H(d)	1	NC	1	0	1	0
Supporting one roof only	3	2	H(d)	1	NC	1	0	1	0
BEAMS, GIRDERS, TRUSSES & ARCHES Supporting columns or more than one floor	(l) 4	3	See 605 H(d)	1	NC	1	0	1	0
Supporting one floor only	3	2	H(d)	1	NC	1	0	1	0
Supporting one roof only	1 1/2(e,p)	1(e,f,p)	H(d)	1(e,p)	NC(e)	1	0	1	0
FLOORS & FLOOR/CEILING CONSTRUCTIONS	(l) 3	2	See 605 H (o)	(n) 1	(n,o) NC	(n) 1	(m,n,o) 0	1	(o) 0

Table 23–2.
This table gives fire-resistance ratings for different building type categories. (It is identified as Table 600 in the Standard Building Code .) *(Continued.)*

Table 600
Fire Resistance Ratings *(continued)*

STRUCTURAL ELEMENT	Type I	Type II	Type III	Type IV 1-Hour Protected	Type IV Unprotected	Type V 1-Hour Protected	Type V Unprotected	Type VI 1-Hour Protected	Type VI Unprotected
ROOFS & ROOF/CEILING CONSTRUCTIONS (g)	1 1/2(e,p)	1 (e,f,p)	See 605 H(d)	1(e,p)	NC(e)	1	0	1	0
EXTERIOR BEARING WALLS and gable ends of roof (g, i, j)	*(% indicates percent of protected and unprotected wall openings permitted. See 705.1.1 for protection requirements.)*								
Horizontal separation (distance from common property line or assumed property line).									
0 ft to 3 ft (c)	4(0%)	3(0%)	3(0%)(b)	2(0%)	1(0%)	3(0%)(b)	3(0%)(b)	1(0%)	1(0%)
over 3 ft to 10 ft (c)	4(10%)	3(10%)	2(10%)(b)	1(10%)	1(10%)	2(10%)(b)	2(10%)(b)	1(20%)	0(20%)
over 10 ft to 20 ft (c)	4(20%)	3(20%)	2(20%)(b)	1(20%)	NC(20%)	2(20%)(b)	2(20%)(b)	1(40%)	0(40%)
over 20 ft to 30 ft	4(40%)	3(40%)	1(40%)	1(40%)	NC(40%)	1(40%)	1(40%)	1(60%)	0(60%)
over 30 ft	4(NL)	3(NL)	1(NL)	1(NL)	NC(NL)	1(NL)	1(NL)	1(NL)	0(NL)
EXTERIOR NONBEARING WALLS and gable ends of roof (g, i, j)	*(% indicates percent of protected and unprotected wall openings permitted. See 705.1.1 for protection requirements.)*								
Horizontal separation (distance from common property line or assumed property line).									
0 ft to 3 ft (c)	3(0%)	3(0%)	3(0%)(b)	2(0%)	1(0%)	3(0%)(b)	3(0%)(b)	1(0%)	1(0%)
over 3 ft to 10 ft (c)	2(10%)	2(10%)	2(10%)(b)	1(10%)	1(10%)	2(10%)(b)	2(10%)(b)	1(20%)	0(20%)
over 10 ft to 20 ft (c)	2(20%)	2(20%)	2(20%)(b)	1(20%)	NC(20%)	2(20%)(b)	2(20%)(b)	1(40%)	0(40%)
over 20 ft to 30 ft	1(40%)	1(40%)	1(40%)	NC(40%)	NC(40%)	1(40%)	1(40%)	0(60%)	0(60%)
over 30 ft (k)	NC (NL)	NC(NL)	NC(NL)	NC(NL)	NC(NL)	NC(NL)	NC(NL)	0(NL)	0(NL)

1 ft = 0.305 m

NC = Noncombustible
NL = No Limits
H = Heavy Timber Sizes

Standard Building Code © 1994
144

Structural Element: The main structural elements of buildings are listed on basic categories such as *Interior Bearing Walls, Columns* and *Exterior Building Walls.* The basic categories are further divided into subcategories. For example, the columns category is divided into columns which are *supporting other columns or more than one floor, supporting one floor only,* and *supporting one roof only.*

Type: Separate columns define the fire resistance rating for the construction Types I, II, III, IV, V, VI. The construction *Type* is defined by the materials used to construct the wall. For example, an *interior bearing wall* which *supports one floor only,* and is of *Type I* construction, will have a fire resistive rating of *3 hours.* A Type II wall has a rating of *2 hours.*

For *Exterior Bearing Walls* and *Exterior Nonbearing Walls* the percentage of openings allowed in the wall is given in parentheses to the right of the first (Structural Element) column. For example, the *Exterior Nonbearing Wall* with a horizontal separation of over *3 ft to 10 ft* (more than 3'-0" to 10'-0") is required to meet *2 hour* fire resistive rating standards and is allowed to have openings that make up a maximum of 10% of the total wall area [2 (10%)].

Table 23–2.

This table gives fire-resistance ratings for different building type categories. (It is identified as Table 600 in the Standard Building Code.)

ted in a building based on the *occupancy classification* and the *type of construction.* All three model codes contain tables providing this information.

Building height is the vertical distance from a grade or datum point to the finished surface of a flat roof or to the average height of the highest roof section of a pitched roof. A **story** is the portion of a building between the upper surface of a floor to the upper surface of the floor or roof above. Again, the maximum height of a building is determined by the *occupancy classification* of the building (A, B, E, F, etc.) and the *type of construction* category (I, II, III, IV, V). Tables are provided in the relevant chapters of all three model code books that include allowable areas and height An example is shown in **Table 23–4.** (This table appears in the BOCA National Building Code as *Table 503—Height and Area Limitations of Buildings.*)

FIRE-RESISTANT MATERIAL AND CONSTRUCTION

"Fire-resistant materials and construction" chapters in all three model code books define the materials, design, and construction methods used to inhibit the spread of fire and smoke within a building. Measures to prevent the spread of fire between buildings are also discussed. Therefore, all materials used, along with their method of installation, must have the required fire- and flame-resistance ratings. These are determined by the possible fire hazards posed by the intended use and/or occupancy of the building. For example, the materials and construction methods for a factory manufacturing product that employs the use of hazardous materials will require the maximum fire-resistance ratings. So will an assembly hall that may contain many people at once.

As mentioned earlier, fire-resistance ratings are

1997 UNIFORM BUILDING CODE TABLE 5-A

TABLE 5-A—EXTERIOR WALL AND OPENING PROTECTION BASED ON LOCATION ON PROPERTY FOR ALL CONSTRUCTION TYPES[1,2,3]
For exceptions, see Section 503.4.

OCCUPANCY GROUP[4]	CONSTRUCTION TYPE	EXTERIOR WALLS Bearing	EXTERIOR WALLS Nonbearing	OPENINGS[5]
		Distances are measured to property lines (see Section 503).		
		× 304.8 for mm		
A-1	I-F.R. II-F.R.	Four-hour N/C	Four-hour N/C less than 5 feet Two-hour N/C less than 20 feet One-hour N/C less than 40 feet NR, N/C elsewhere	Not permitted less than 5 feet Protected less than 20 feet
A-1	II One-hour II-N III One-hour III-N IV-H.T. V One-hour V-N	Group A, Division 1 Occupancies are not allowed in these construction types.		
A-2 A-2.1 A-3 A-4	I-F.R. II-F.R. III One-hour IV-H.T.	Four-hour N/C	Four-hour N/C less than 5 feet Two-hour N/C less than 20 feet One-hour N/C less than 40 feet NR, N/C elsewhere	Not permitted less than 5 feet Protected less than 20 feet
A-2 A-2.1[2]	II One-hour	Two-hour N/C less than 10 feet One-hour N/C elsewhere	Same as bearing except NR, N/C 40 feet or greater	Not permitted less than 5 feet Protected less than 10 feet
A-2 A-2.1[2]	II-N III-N V-N	Group A, Divisions 2 and 2.1 Occupancies are not allowed in these construction types.		
A-2 A-2.1[2]	V One-hour	Two-hour less than 10 feet One-hour elsewhere	Same as bearing	Not permitted less than 5 feet Protected less than 10 feet
A-3	II One-hour	Two-hour N/C less than 5 feet One-hour N/C elsewhere	Same as bearing except NR, N/C 40 feet or greater	Not permitted less than 5 feet Protected less than 10 feet
A-3	II-N	Two-hour N/C less than 5 feet One-hour N/C less than 20 feet NR, N/C elsewhere	Same as bearing	Not permitted less than 5 feet Protected less than 10 feet
A-3	III-N	Four-hour N/C	Four-hour N/C less than 5 feet Two-hour N/C less than 20 feet One-hour N/C less than 40 feet NR, N/C elsewhere	Not permitted less than 5 feet Protected less than 20 feet
A-3	V One-hour	Two-hour less than 5 feet One-hour elsewhere	Same as bearing	Not permitted less than 5 feet Protected less than 10 feet
A-3	V-N	Two-hour less than 5 feet One-hour less than 20 feet NR elsewhere	Same as bearing	Not permitted less than 5 feet Protected less than 10 feet

Occupancy Group: The Occupancy Group and Division to which the requirements in the table apply are identified. In some cases, buildings of different groups are listed together because the requirements are the same. For example, occupancy groups A-2, A-2.1, A-3, and A-4 have the same requirements for the type of construction described in the table.

Construction Type: These are code identifications pertaining to minimum requirements for materials and construction methods as defined in different sections of the UBC. Some examples are *I-F.R.* (Type I Fire Resistive), *II One-Hour* (Type II One Hour), and *II-N* (Type II Noncombustible).

Exterior Walls: The requirements in the two columns are for exterior walls only.

Bearing: requirements for fire resistance ratings are given for the walls which carry a load from above. For example, the Four-Hour Noncombustible)Four-hour N/C) requirement for Occupancy Group A-ll,

Construction Type L-F.R. means that the nonbearing exterior wall system must meet the UBC standards for the Four-hour Noncombustible rating (fire resistance for four hours).

Nonbearing: Requirements for fire resistance ratings are given for the walls which do not carry a load from above. The requirements change with the distance form the property lines. For example, an Occupancy Group A-1, Construction Type L-F.R., nonbearing exterior wall must meet UBC standards for a Four-hour Noncombustible rating if it is less than 5' - 0" from the property line, but need only meet Two-hour Noncombustible standards if it is between 5' - 0" and 10' - 0" from the property line.

Openings: Allowance for doors, windows, and other openings in the exterior walls is described and is usually based on the distance of the wall from the property line. Protected openings must be constructed of a system with at least a 3/4 - hour fore protection rating.

Table 23–3.

This table provides information about fire protection for exterior walls and openings based on where the building is on the property. (It is identified as Table 5–A in the Uniform Building Code.) The table consists of seven pages in the code book; only the first page is reproduced here.

THE BOCA NATIONAL BUILDING CODE/1993

Table 503
HEIGHT AND AREA LIMITATIONS OF BUILDINGS
Height limitations of buildings (shown in upper figure as stories and feet above grade plane)[m], and area limitations of one- or two-story buildings facing on one street or public space not less than 30 feet wide (shown in lower figure as area in square feet per floor[m]). See Note a.

Use Group		Note a	Type 1 Protected Note b — 1A	Type 1 Protected Note b — 1B	Type 2 Protected — 2A	Type 2 Protected — 2B	Type 2 Unprotected — 2C	Type 3 Protected — 3A	Type 3 Unprotected — 3B	Type 4 Heavy timber — 4	Type 5 Protected — 5A	Type 5 Unprotected — 5B
A-1	Assembly, theaters		Not limited	Not limited	5 St. 65' 19,950	3 St. 40' 13,125	2 St. 30' 8,400	3 St. 40' 11,550	2 St. 30' 8,400	3 St. 40' 12,600	1 St. 20' 8,925	1 St. 20' 4,200
A-2	Assembly, nightclubs and similar uses		Not limited	Not limited 7,200	3 St. 40' 5,700	2 St. 30' 3,750	1 St. 20' 2,400	2 St. 30' 3,300	1 St. 20' 2,400	2 St. 30' 3,600	1 St. 20' 2,550	1 St. 20' 1,200
A-3	Assembly — Lecture halls, recreation centers, terminals, restaurants other than nightclubs		Not limited	Not limited	5 St. 65' 19,950	3 St. 40' 13,125	2 St. 30' 8,400	3 St. 40' 11,550	2 St. 30' 8,400	3 St. 40' 12,600	1 St. 20' 8,925	1 St. 20' 4,200
A-4	Assembly, churches	Note c	Not limited	Not limited	5 St. 65' 34,200	3 St. 40' 22,500	2 St. 30' 14,400	3 St. 40' 19,800	2 St. 30' 14,400	3 St. 40' 21,600	1 St. 20' 15,300	1 St. 20' 7,200
B	Business		Not limited	Not limited	7 St. 85' 44,700	5 St. 65' 22,500	3 St. 40' 14,400	4 St. 50' 10,800	3 St. 40' 11,400	5 St. 65' 21,600	3 St. 40' 15,300	2 St. 30' 7,000
E	Educational	Note c	Not limited	Not limited	5 St. 65' 34,200	3 St. 40' 22,500	2 St. 30' 14,400	3 St. 40' 19,800	2 St. 30' 14,400	3 St. 40' 21,600	1 St. 20' 15,300 Note d	1 St. 20' 7,200 Note d
F-1	Factory and industrial, moderate		Not limited	Not limited	6 St. 75' 22,800	4 St. 50' 15,000	2 St. 30' 9,600	3 St. 40' 13,200	2 St. 30' 9,600	4 St. 50' 14,400	2 St. 30' 10,200	1 St. 20' 4,800
F-2	Factory and industrial, low	Note h	Not limited	Not limited	7 St. 85' 34,200	5 St. 65' 22,500	3 St. 40' 14,400	4 St. 50' 19,800	3 St. 40' 14,400	5 St. 65' 21,600	3 St. 40' 15,300	2 St. 30' 7,200
H-1	High hazard, detonation hazards	Notes e, i, k, l	1 St. 20' 16,800	1 St. 20' 14,400	1 St. 20' 11,400	1 St. 20' 7,500	1 St. 20' 4,800	1 St. 20' 6,600	1 St. 20' 4,800	1 St. 20' 7,200	1 St. 20' 5,100	Not permitted
H-2	High hazard, deflagration hazards	Notes e, i, j, l	5 St. 65' 16,800	3 St. 40' 14,400	3 St. 40' 11,400	2 St. 30' 7,500	1 St. 20' 4,800	2 St. 30' 6,600	1 St. 20' 4,800	2 St. 30' 7,200	1 St. 20' 5,100	Not permitted
H-3	High hazard, physical hazards	Notes e, l	7 St. 85' 33,600	7 St. 85' 28,800	6 St. 75' 22,800	4 St. 50' 15,000	2 St. 30' 9,600	3 St. 40' 13,200	2 St. 30' 9,600	4 St. 50' 14,400	2 St. 30' 10,200	1 St 20' 4,800
H-4	High hazard, health hazards	Notes e, l	7 St. 85' Not limited	7 St. 85' Not limited	7 St. 85' 34,200	5 St. 65' 22,500	3 St. 40' 14,400	4 St. 50' 19,800	3 St. 40' 14,400	5 St. 65' 21,600	3 St. 40' 15,300	2 St. 30' 7,200
I-1	Institutional, residential care		Not limited	Not limited	9 St. 100' 19,950	4 St. 50' 13,125	3 St. 40' 8,400	4 St. 50' 11,550	3 St. 40' 8,400	4 St. 50' 12,600	3 St. 40' 8,925	2 St. 35' 4,200
I-2	Institutional, incapacitated		Not limited	Not limited	4 St. 50' 17,100	2 St. 30' 11,250	1 St. 20' 7,200	1 St. 20' 9,900	Not permitted	1 St. 20' 10,800	1 St. 20' 7,650	Not permitted
I-3	Institutional, restrained		Not limited	Not limited	4 St. 50' 14,250	2 St. 30' 9,375	1 St. 20' 6,000	2 St. 30' 8,250	1 St. 20' 6,000	2 St. 30' 9,000	1 St. 20' 6,375	Not permitted
M	Mercantile		Not limited	Not limited	6 St. 75' 22,800	4 St. 50' 15,000	2 St. 30' 9,600	3 St. 40' 13,200	2 St. 30' 9,600	4 St. 50' 14,400	2 St. 30' 10,200	1 St. 20' 4,800
R-1	Residential, hotels		Not limited	Not limited	9 St. 100' 22,800	4 St. 50' 15,000	3 St. 40' 9,600	4 St. 50' 13,200	3 St. 40' 9,600	4 St. 50' 14,400	3 St. 40' 10,200	2 St. 35' 4,800
R-2	Residential, multiple-family		Not limited	Not limited	9 St. 100' 22,800	4 St. 50' 15,000 Note f	3 St. 40' 9,600	4 St. 50' 13,200 Note f	3 St. 40' 9,600	4 St. 50' 14,400	3 St. 40' 10,200	2 St. 35' 4,800
R-3	Residential, one- and two-family and multiple single-family		Not limited	Not limited	4 St. 50' 22,800	4 St. 50' 15,000	3 St. 40' 9,600	4 St. 50' 13,200	3 St. 40' 9,600	4 St. 50' 14,400	3 St. 40' 10,200	2 St. 35' 4,800
S-1	Storage, moderate		Not limited	Not limited	5 St. 65' 19,950	4 St. 50' 13,125	2 St. 30' 8,400	3 St. 40' 11,550	2 St. 30' 8,400	4 St. 50' 12,600	2 St. 30' 8,925	1 St. 20' 4,200
S-2	Storage, low	Note g	Not limited	Not limited	7 St. 85' 34,200	5 St. 65' 22,500	3 St. 40' 14,400	4 St. 50' 19,800	3 St. 40' 14,400	5 St. 65' 21,600	3 St. 40' 15,300	2 St. 30' 7,200
U	Utility, miscellaneous		Not limited	Not limited								

Use Group: As defined in BOCA code, the *Use Group* categories give a brief description of their purpose. For example, *Use Group A-1* is the listing *for Assembly theaters.*

Type of Construction: The building materials used determine the construction type. Types 1 and 2 are defined as *Noncombustible* construction using materials such as steel and concrete. Types 3 and 4 are either *Noncombustible* or *Combustible* depending on the amount of protection offered by materials such as wall board or concrete surrounding the framing member. Type 5 is defined as *combustible* construction such as unprotected wood. In some cases, the types are further divided in the code. For example, *Type 1* construction is divided into *1A* and *1B,* depending on the building material used.

For each *Use Group* and *Type of Construction,* the table gives the limitations for the number of stories, height, and maximum allowable area of the building. For example, the table entry for an *A-1 Use Group of Type 2A* construction is 5 St. 19,950. This means that the building may be up to 5 stories high, 65'0" above grade, with an allowable area of 19,950 square feet. This area is based on the measurements between outside walls to identify the amount of space the building takes up to the lot. It is not the sum of the areas of each floor.

Table 23–4.

This table provides information about allowable building height and area. (It is identified as Table 503 in the BOCA National Building Code.)

established by industry-recognized testing agencies such as the American Society for Testing and Materials (ASTM). The ratings are based on the nature of the material and the protective covering placed over and around the material.

OPENINGS

Door openings are very vulnerable to the spread of fire within a building. For this reason, building codes require special-rated **fire doors** in all building types, with the exception of some residential structures. These are doors constructed of materials that conform to the standards established by an approved testing agency such as the ASTM. A typical fire door is covered with metal and contains a solid core of fire-treated wood or other fire-resistant material. Fire doors are rated according to the time of protection they offer—from 3/4 of an hour to 1 1/2 hours. **Table 23–5** provides ratings for fire door assemblies. (This table appears in the BOCA National Code as *Table 716.l—Opening Protective Fire Rating*.)

Rated fire-resistant windows are also installed where required. They consist of wired and tempered glass in a metal sash placed in metal frames. For example, metal-framed wired glass 1/4" thick meets the requirements of a 3/4-hour fire-resistant assembly.

Means of Egress

Building codes provide very specific details and requirements to ensure the greatest possible safety when occupants must rapidly leave the building in case of fire and other emergencies. Chapters on "means of egress" are found in all three model code books. **Means of egress** describes the way out of a building. The Standard Building Code defines it as follows:

A continuous and unobstructed path of travel from any point in a building or structure to a *public way*. A *means of egress* consists of three separate and distinct parts: the *exit access*; the *exit*, and the *exit discharge*. A means of egress comprises the vertical and horizontal means of travel and shall include *intervening room spaces, doors, hallway, corridors, passageways, balconies, ramps, stairs, enclosures, lobbies, horizontal exits, courts* and *yards*.

A *public way* is any street, alley, or other parcel of land that is open from the ground to the sky and is permanently allocated for the use of the public with a minimum clear width of not less than 10′. An *exit access* is the part of a means of egress between the completion point of an exit and the beginning of a public way. The *exit* is the part of a means-of-access that provides a protected

Table 716.1
OPENING PROTECTIVE FIRE PROTECTION RATING

Type of assembly	Required assembly rating (hour)	Minimum opening protection assembly (hour)
Fire walls and fire separation assemblies having a required fireresistance rating greater than 1 hour	4 3 2 1½	3 3 1½ 1½
Fire separation assemblies: Shaft and exit enclosure walls Other fire separation assemblies	1 1	1 ¾
Fire partitions: Exit access corridor enclosure wall Other fire partitions	1 ½ 1	⅓[a] ⅓[a] ¾

Note a. For testing requirements, see Section 716.1.1.

Type of Assembly: This pertains to the combination of materials and methods used to construct a wall.

Required Assembly Rating: The fire protection requirements explained in other parts of BOCA are shown in hours. For example, in the type of assembly shown at the top of the table the required assembly fire protection rating is 4 hours.

Minimum Opening Protection Assembly: The minimum fire protection rating for a door or window opening in the wall assembly described above is given in hours. For example, if a fire resistive wall if required to provide 4 hours of fire protection, any window or door in that wall must be rated to provide 3 hours of fire protection.

Table 23–5.

This table gives fire-resistive ratings for walls and wall openings. (It is identified as Table 10–A in the BOCA Code.)

way of travel to the exit discharge. The *exit discharge* is the means-of-access section between the end of an exit and the public way. **(See Figure 23–1.)** Elevators, escalators, and moving sidewalks are not considered means of egress, because they are subject to the possibilities of mechanical breakdowns during emergencies.

Many factors must be considered when deciding the means of egress and the number of exits for a building. A key element is the character of the *group occupancy*. For example, the *Group B* (business) occupancy will obviously have a different means of egress arrangement than a *Group R* (residential) occupancy. Related to this is the *occupant load*, the calculated minimum number of people for whom the means of egress is designed. All three model codes provide related tables. An example is shown in **Table 23–6**. (This table appears in the Uniform Building Code as *Table 10-A—Minimum Egress Requirements*.)

Horizontal travel is the level plane of travel of a

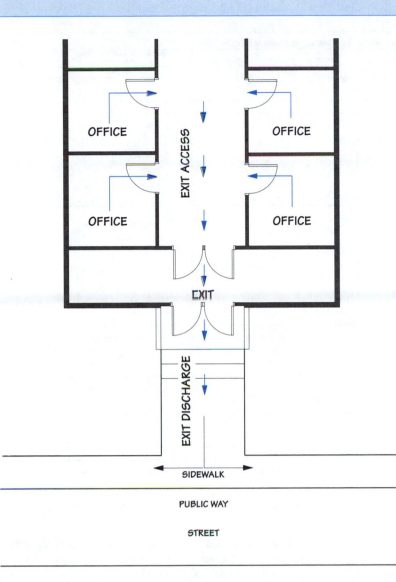

OFFICE

OFFICE

EXIT ACCESS

OFFICE

OFFICE

EXIT

EXIT DISCHARGE

SIDEWALK

PUBLIC WAY

STREET

Figure 23–1.
This means of egress is a clear path of travel to a public way from any point in the building.

means of egress. An important consideration is the **travel distance.** This is the total length of an exit path a person must travel from any location within the occupied part of a building to exit entirely from the building. This line of travel may have to pass through intervening room spaces, passageways, and balconies. All passageways must conform to fire-resistant requirements, and their minimum widths are given in the code books. Building codes usually include the maximum allowable travel distance.

Passageways of public buildings must be well lighted, and exit signs (illuminated if called for) should be provided. Doors must be of the pivoted and side-hinged type. Locks on public-access exit doors must be openable from the inside without the use of a key. Panic hardware, when used, must be of the push-bar type.

Stairways

Building codes go into a great deal of detail regarding stairway construction and protection. A properly

designed stair system is basic to the safety of any means of egress system. Following are some items generally discussed in building codes.

✓ Fire-resistant enclosures for interior stairways. (Not all interior stairways in one- and two-family dwelling units need to be enclosed.)
✓ Exterior stairways.
✓ Stairways to roofs.
✓ Minimum widths of stairways based on usage.
✓ Permissible handrail projection.
✓ Rise and run.
✓ Headroom.
✓ Winding stairways.
✓ Circular stairways.
✓ Spiral stairways.
✓ Landings.
✓ Guardrails on open sides.

TABLE 10-A—MINIMUM EGRESS REQUIREMENTS[1]

USE[2]	MINIMUM OF TWO MEANS OF EGRESS ARE REQUIRED WHERE NUMBER OF OCCUPANTS IS AT LEAST	OCCUPANT LOAD FACTOR[3] (square feet) × 0.0929 for m²
1. Aircraft hangars (no repair)	10	500
2. Auction rooms	30	7
3. Assembly areas, concentrated use (without fixed seats) Auditoriums Churches and chapels Dance floors Lobby accessory to assembly occupancy Lodge rooms Reviewing stands Stadiums Waiting area	50 50	7 3
4. Assembly areas, less-concentrated use Conference rooms Dining rooms Drinking establishments Exhibit rooms Gymnasiums Lounges Stages Gaming: keno, slot machine and live games area	50 50	15 11
5. Bowling alley (assume no occupant load for bowling lanes)	50	4
6. Children's homes and homes for the aged	6	80
7. Classrooms	50	20
8. Congregate residences	10	200
9. Courtrooms	50	40
10. Dormitories	10	50
11. Dwellings	10	300
12. Exercising rooms	50	50
13. Garage, parking	30	200
14. Health care facilities— Sleeping rooms Treatment rooms	 8 10	 120 240
15. Hotels and apartments	10	200
16. Kitchen—commercial	30	200
17. Library— Reading rooms Stack areas	 50 30	 50 100
18. Locker rooms	30	50

Use: The particular building uses are listed.

Minimum of Two Exits other than Elevators are Required where Number of Occupants is at least: If the number of occupants of a building or building area for the described use meets or exceeds the number in this column, at least two exits from the area are required. For example, in Classrooms (use item #7) the number of occupants is 50 or more at least two exits are required.

Occupant Load Factor: This determined how many occupants are permitted in a building or portion of a building. It is found by dividing the total area by the Occupant Load Factor that is shown in the table. The Occupant Load Factor for classrooms is 20. For example, to find the number of occupants permitted in a classroom of 1200 square feet, divide 1200 square feet by the Occupant Load Factor 20 (1200 ÷ 20 = 60). Therefore, the classroom will require two exits as the table shows that a classroom with 50 or more occupants will require two exits.

Table 23–6.
This table provides information for determining whether two exits are required for a designated area. (It is identified as Table 10–A in the Uniform Building Code.) The table consists of two pages in the code book; only the first page is reproduced here.

STRUCTURAL CODE REQUIREMENTS

Thus far this unit has discussed eight major areas of building code information essential to the general design of a building. But a set of construction plans also contains many structural details showing how the building components are put together, and these structural details must also conform to building code requirements. Tables and schedules are provided in the codes to give required component sizes (walls, beams, posts, etc.) based on span and anticipated loads, proper fastening methods, and maximum and minimum measurements. Proper framing methods of floor and wall openings are also discussed. This information is often identical or very similar in all three model code books. However, there are some regional differences due to climatic and seismic conditions. Every few years sections or subsections of a code book may contain some revisions. The references that follow are taken from the latest printed codes at the time of this book's publication. They are the Uniform Building Code (UBC) (1967), BOCA National Building Code (1966), and the Standard Building Code (SBC) (1997). Listed below are a few examples of how the three model codes define the requirements for the same structural procedures:

GENERAL FOOTING DESIGN

UBC

SECTION 1806- FOOTINGS

1806.1 General. Footings and foundations shall be constructed of masonry, concrete or treated wood in conformance with Division II and shall extend below the frost line. Footings of concrete and masonry shall be of solid material. Foundations supporting wood shall extend at least 6 inches (152 mm) above the adjacent finish grade. Footings will have a minimum depth as indicated in Table 18-I-D unless depth is recommended by some foundation inspection.

BOCA

SECTION 1806.0 DEPTH OF FOOTINGS

1806.1 Frost protection: Except where erected on solid rock or otherwise protected from frost, foundation walls, piers and other permanent supports of all buildings and structures larger than 100 square feet (9.30 m squared) in area or 10 feet (3048 mm) in height shall extend to the frost line of the locality and spread footings of adequate size shall be provided where necessary to distribute properly the load withing the allowable load bearing value of the soil. Alternatively, such structures shall be supported on piles where solid earth or rock is not available. Footings shall not bear on frozen soils unless such frozen condition is of a permanent character.

SECTION 1804 FOOTINGS AND FOUNDATIONS

1804.1 General
1804.1.1 Foundations will be built on undisturbed soil or properly compacted fill material. Foundations will be constructed of materials described in this chapter.
1804.1.2 Pile foundations will be designed and constructed by 1805.
1804.1.3 The bottom of foundations shall extend below the depth of frost penetration shown in

Figure 1804.1, but no less than 12 inches (305 mm) below finish grade.

WALL FRAMING

UBC

2326.11 WALL FRAMING

2326.11.1 Size, height and spacing. The size, height and spacing of studs shall be in accordance with Table 23-I-R except that Utility Grade studs shall not be spaced more than 16 inches (406 mm) on center, or support more than a roof and ceiling, or exceed 8 feet (2438 mm) in height for walls and load bearing walls or 10 feet (3048 mm) for interior non load bearing walls.

2326.11.2 Framing details. Studs shall be placed with their wide dimension perpendicular to the wall. Not less than three studs shall be installed at each corner of an exterior wall.

Exception: At corners, a third stud may be omitted through the use of wood spacers or backup cleats of 3/8 inch (9.5 mm) wood structural panel. 3/8 inch (9.5 mm) Type M "Exterior Glue" particle board, 1 inch thick (25 mm) lumber or other approved devices that will serve as an adequate backing for the attachment of facing materials. Where fire resistance ratings or shear values are involved, wood spacers, backup cleats or other devices shall not be used unless specifically approved for such use.

Bearing and exterior wall studs shall be capped with double top plates installed to provide overlapping at corners and intersection with other partitions. End joints in double top plates shall be offset at least 48 inches (2438 mm).

Exception: A single top plate may be used, provided the plate is adequately tied at joints, corners and intersecting walls by at least the equivalent of 3-inch by 6-inch (76 mm by 152 mm) by 0.036-inch thick (0.9 mm) galvanized steel that is nailed to each wall or segment of wall by six 8d nails or equivalent, provided the rafters, joists, or trusses are centered over the studs with a tolerance of no more than 1 inch (25 mm).

When bearing studs are spaced at 24 inch (610 mm) intervals and top plates are less than two 2 inch by 6 inch (51 mm by 152 mm) or two 3 inch by 4 inch (76 mm by 102 mm) members and when the floor joists, floor trusses roof trusses which they support are spaced at more than 16 inch (406 mm) intervals, such joists or trusses shall bear within 5 inches (127 mm) of the studs beneath or a third plate shall be installed.

Interior non bearing partitions may be capped

with a single top plate installed to provide over-lapping at corners and intersections with other walls and partitions. The plate shall be continuously tied at joints by solid blocking at least 16 inches (406 mm) in length and equal in size to the plate or by 1/8 inch by 1 1/2 inches (3.2 mm by 38 mm) metal ties with spliced sections fastened with two 16d nails on each side of the joint.

Studs shall have full bearing on a plate or sill not less than 2 inches (51 mm) in thickness having a width not less than that of the wall studs.

BOCA

2305.4.1: Wall framing: Studs shall be placed with the wide dimension perpendicular to the wall. Not less than three studs shall be installed at each corner of an exterior wall

Exception: At corners, a third stud is not required where wood spacers or back up cleats of 3/8 inch thick wood structural panel is used.

2305.4.2 Double top plates: Stud walls shall be capped with double top plates installed to provide overlapping corners at wall intersections. Top plate joints shall be offset not less than 48 inches (1219 mm).

2305.4.3 Bottom plates: Studs shall have full support by a plate or sill. The sill shall have not less that a 2 inch nominal thickness and a width at least equal to the width of the studs.

SBC

2308 VERTICAL FRAMING
2308.1 Exterior Wall Framing
2308.1.1 Studs in one and two story buildings shall be not less than 2×4 with the wide face perpendicular to the wall. In three story buildings, studs in the first story shall be not less than 3×4 or 2×6. Studs shall be spaced in accordance with Table 2308.1A.
2308.1.2 Utility grade studs shall not be spaced more than 16 inches (406 mm) on centers, support more than a roof and ceiling and exceed 8 ft (2438 mm) in height for exterior load bearing walls.

RISE AND RUN OF STAIRWAYS

UBC

1003.3.3.3 Rise and Run. The rise of steps shall not be less than 4 inches (102 mm) or greater than 7 inches (178 mm). Except as permitted in Sections 1006.4 and 1006.6, the run shall not be less than 11 inches (279 mm) as measured horizontally between the vertical planes of the furthermost

projection of adjacent treads. Except as permitted in Sections 1006.4, 1006.5 and 1006.6, the largest tread run within any flight of stairs shall not exceed the smallest by more than 3/8 inch (9.5 mm). The greatest riser height within any flight of stairs shall not exceed the smallest by more than 3/8 inch (9.5 mm).

Exceptions: Private steps and stairways serving an occupant load of less than 10 and stairways to unoccupied roofs may be constructed with an 8-inch maximum (203 mm) rise and a 9 inches minimum (229 mm) run.

Where the bottom or top riser adjoins a sloping public way, walk or driveway having an established grade and serving as a landing, the bottom or top riser may be reduced along the slope to less than 4 inches (102 mm) in height with the variation in height of the bottom or top riser not to exceed 3 inches (76 mm) in every 3 feet (914 mm) of stairway.

BOCA

1014.6 Treads and Risers: Maximum riser height shall be 7 inches (178 mm) and minimum riser height shall be 4 inches (102 mm). Minimum tread depth shall be 11 inches (279 mm) measured horizontally between the vertical planes of the foremost projection of adjacent treads and at a right angle to the tread's leading edge.

Exceptions:
1. Winders in accordance with Section 1014.6.3.
2. Spiral stairways in accordance with Section 1014.6.4.
3. Circular stairways in accordance with Section 1014.6.5.
4. Alternating tread stairways in accordance with Section 1014.6.6.
5. Stairways serving as aisles in assembly seating areas where the stairway pitch or slope is set, for sight line reasons, by the slope of the adjacent seating area.
6. Any stairway replacing an existing stairway with a space where, because of existing construction, the pitch or slope cannot be reduced.
7. Existing stairways.
8. In occupancies in Use Group R-3 and within dwelling units in occupancies in Use Group R-2, the maximum riser height shall be 8-1/4 inches (210 mm) and the minimum tread depth shall be 9 inches (229 mm). A 1 inch (25 mm) nosing shall be provided on stairways with solid risers.
9. Stairways in penal facilities serving guard

towers, observation stations and control rooms not more than 250 square feet in areas shall be permitted to have risers not exceeding 8 inches (203 mm) in height and treads not less than 9 inches (229 mm) in depth.

SBC

1007.3 Treads and Risers

1007.3.1 Treads and risers of stairs shall be so proportioned that the sum of two risers and a tread, exclusive of projection of nosing, is not less than 24 inches (610 mm) nor more than 25 inches (635 mm). The height of risers shall not exceed 7-3/4 inches (197 mm), and treads, exclusive of nosing, shall be not less than 9 inches (229 mm) wide.

Exception: Special stairs in 1007.8.

1007.3.2 Every tread less than 10 inches (254 mm) wide shall have a nosing or effective projection, of approximately 1 inch (25.4 mm) over the level immediately below that tread.

1007.3.3 Tread depth shall be measured horizontally between the vertical planes of the foremost projection of adjacent treads and at a right angle to the tread's leading edge.

Exception: Tread depth of special stairs in 1007.8 shall be measured on a line perpendicular to the centerline of tread.

1007.3.4 Treads shall be of uniform depth and risers of uniform height in any stairway between the two floors. There shall be no variation exceeding 3/16 inch (4.8 mm) in the depth of adjacent treads or in the height of adjacent risers and the tolerance between the largest and smallest tread shall not exceed 3/8 inch (9.5 mm) in any flight. The uniformity of winders and other tapered treads, complying with 1007.8.1, 1007.8.2, and 1007.8.3 shall be measured at consistent distances from the narrower end of the treads.

Exception: Where the bottom or top riser adjoins a sloping public way, walk or driveway having an established grade and serving as a landing, a variation in height of the riser of not more than 3 inches (76 mm) for every 3 ft (914 mm) of stairway width is permitted.

REVIEW QUESTIONS

Enter the missing words in the spaces provided on the right.

1. Many state and local codes are patterned after a regional _____ code book.

2. The most widely used model code in the southern states is the _____ _____ _____ .

3. The _____ model code (use abbreviation) is widely used in the east and midwest. .

4. Western and some midwestern states model their codes on the _____ _____ _____

5. The Canadian National Uniform _____ and ____ code is available for heating and air conditioning work.

6. Chapter _____ of the BOCA code covers duties and powers of code officials..

7. Chapter _____ of the BOCA code covers foundations and retaining walls.

8. In Table 3–8 of the UBC, a(n) _____ building group and division includes an assembly hall with an occupant load greater than 1,000.

9. Quality control is discussed in subsection _____ of the Standard Building Code.

   _____

10. The number of _____ _____ in the floor area is one of the initial things to know when designing a building.

    _____

11. The _____(use abbreviation) is an industry-recognized testing agency for fire-resistant ratings.

    _____

12. A metal-clad door containing a solid fireproof core is called a(n) _____ door.

    _____

13. A method of exiting a building is called a(n) _____ of _____.

    _____

14. The _____ travel is the level plane of travel of a means of egress.

    _____

15. Building codes provide information regarding _____ on open sides of stairways.

    _____

16. The UBC states foundations supporting wooden floor units must extend at least _____ inches above grade.

    _____

17. BOCA states that bottom plates must be at least _____ inch nominal thickness.

    _____

18. According to the SBC code, the bottoms of foundations shall be no less than _____ below finish grade.

    _____

19. The UBC states that utility grade studs shall not be spaced more than _____ OC.

    _____

20. The maximum rise of steps shall be _____ or no less than _____ inches, according to the BOCA code.

    _____

All city and county governments have a bureau or department to make certain that local zoning and building codes are strictly enforced. These agencies go by such names as Department of Building Inspection, Building Department, or Building Division. All three model code books offer general organizational guidelines for structuring local government building departments. This unit deals with the permit and inspection process for residential and light commercial work. Larger commercial projects entail more detailed procedures and applications.

BUILDING DEPARTMENT

Building Departments are staffed with personnel qualified to judge submitted building applications and carry out the necessary inspections that guarantee code compliance. The key positions usually require a number of years of experience as an architect, civil engineer, contractor, or superintendent of construction. The primary purposes of a building department are to

1. Review applications along with building plans for permission to begin construction.
2. Examine the plans for zoning and code conformity and impose any changes if necessary.
3. Grant a permit to begin construction.
4. Provide inspections during construction to ensure code conformity.
5. Grant a certificate of occupancy following the final inspection after completion of construction.

PERMIT APPLICATION PROCESS

A **building permit** is an official document or certificate issued by a local Building Department authorizing the beginning of a construction project. Most construction projects require a building permit. A project begun without a permit is considered illegal. If apprehended, the owner or contractor may have to tear down the work done, pay a higher fee for a permit, and possibly pay a fine. The following requirements for obtaining a permit appear in the BOCA National Building Code.

SECTION 107.0 APPLICATION FOR PERMIT
107.1 Permit application: An application shall be submitted to the code official for the following activities, and these activities shall not commence without a permit being issued in accordance with Section 108.0.

1. Construct or alter a structure.
2. Construct an *addition*.
3. Demolish or move a structure.
4. Make a *change of occupancy*.
5. Install or alter any equipment that is regulated by this code.
6. Move a *lot line* that affects an existing structure.

107.1.1 Repairs: Application or notice to the code official is *not* required for ordinary repairs to structures. Such repairs shall not include the cutting away of any wall, partition or portion thereof, the removal or cutting of any structural beam or loadbearing support, or the removal or change of any required *means of egress,* or rearrangement of parts of a structure affecting the egress requirements; nor shall ordinary repairs include *addition* to, *alterations* of, replacement or relocation of any standpipe, water supply, sewer, drainage, drain leader, gas, soil, waste, vent or similar piping, electric wiring or mechanical or other work affecting public health or general safety.

Some examples of work that may not require a permit are

1. One-story detached buildings used for tool and storage sheds or playhouses, providing the projected roof area does not exceed 120 square feet.
2. Fences less than 6'-0" high.
3. Movable partitions, counters, or cases less than 9'-5" in height.
4. Retaining walls less than 4'-0" from the bottom of the footing to the top of the wall.
5. Painting, papering, and similar finish work.

APPLYING FOR THE PERMIT

Application for a permit can be made by the owner or a *registered design professional.* The latter is often called the *architect* or *engineer of record*. In some regions the contractor applies for the permit. Application forms may differ a little from one locality to another. **(See Figure 24–1.)** However, some frequently required information is as follows:

1. Location and legal description of job site.
2. General description of work.

APPLICATION FOR BUILDING/GRADING PERMIT

Residential submittals require 2 sets of plans; Commercial submittals require 3 sets. For construction with a value of $8-$12,000, a $50 deposit is required; a value exceeding $12,000 requires a $100 deposit. Review and approval of any off-site permits or grading permits may delay issuance of building permits. For residential **and** grading permit applications, please complete the grading questions on the reverse side of this form.

SITE ADDRESS:_____

SUBDIVISION: _____ **LOT:**____ **PARCEL NO.** _____

OWNER: _____ **TENANT:** _____

ADDRESS: _____

CITY/STATE/ZIP: _____

CONTACT NAME: _____ **DAY PHONE:**_____

ESTIMATE OF CONSTRUCTION VALUE: _____

Contact the Registrar of Contractors Office (526-2325) regarding licensed contractor requirements. If you are using a licensed General Contractor and/or subcontractors, you must provide that information prior to the permit being issued.

GENERAL CONTRACTOR: _____

ADDRESS: _____

CITY/STATE/ZIP: _____

CONTACT NAME: _____ **DAY PHONE:** _____

CONTRACTOR'S LICENSE NO: _____ **STATE SALES TAX NO.** _____

    PLUMBING SUB: _____ **LICENSE NO.** _____

    ELECTRICAL SUB: _____ **LICENSE NO.** _____

    MECHANICAL SUB: _____ **LICENSE NO.** _____

OFFICE USE ONLY

PERMIT NO._____

THIS ROUTING SLIP MUST BE GIVEN PRIORITY.

REC'D BY: _____

DATE: _____

Route to the next Division or Department with either your approval or reasons for denial on Screen 3 of Build Tech. Please list any conditions on Screen 1 of Build Tech. In order to save the contractor or owner any avoidable delay, please have the courtesy to notify them immediately should there be any problems in approving this application.

	Project Manager	Initials		Project Manager	Initials
PUBLIC WORKS	_____	_____	**ENGINEERING**	_____	_____
UTILITIES	_____	_____	**PLANNING**	_____	_____
FIRE	_____	_____	**BUILDING**	_____	_____

DRB No. _____ **DATE:** _____ **P & Z NO.** _____

COMMENTS: _____

Figure 24–1.
An application for a building permit can be made by the owner, contractor, or a registered design professional.

3. Type of occupancy.
4. Estimated construction value.
5. Portion of lot not covered by structure.
6. Name and license number of general contractor.
7. Names and license numbers of plumbing, electrical, and mechanical subcontractors.

At the time of application, most Building Departments will require at least two sets of complete construction plans including specifications. The site plans must show the size and location of all new construction and any structures already existing on the property. Distances from property lines must be included, as well as existing and finish grades. Any required excavation and fill replacement must be described. The rest of the drawings define the general design of the building and must contain adequate details of structural, mechanical, and electrical work.

Computations, stress diagrams, and other important technical data are included if necessary. These data must be signed by the architect or engineer responsible for this information.

Some Building Departments will require a $50 to $100 deposit at the time of application. The balance must then be paid when the permit is granted. The amount of a permit fee is based on the total value of the construction work. An example of a fee schedule for residential construction work is shown in **Table 24–1.** (This table appears in the Uniform Building Code as *Table 1-A—Building Permit Fees.*)

A copy of the permit should be kept in a protected location on the job site. Most Building Departments will declare a permit invalid if the authorized work does not begin within six months after issuance of the permit, or if the authorized work is suspended or abandoned for six months after the work has begun.

TABLE 1-A—BUILDING PERMIT FEES

TOTAL VALUATION	FEE
$1.00 to $500.00	$22.00
$501.00 to $2,000.00	$22.00 for the first $500.00 plus $2.75 for each additional $100.00, or fraction thereof, to and including $2,000.00
$2,001.00 to $25,000.00	$63.00 for the first $2,000.00 plus $12.50 for each additional $1,000.00, or fraction thereof, to and including $25,000.00
$25,001.00 to $50,000.00	$352.00 for the first $25,000.00 plus $9.00 for each additional $1,000.00, or fraction thereof, to and including $50,000.00
$50,001.00 to $100,000.00	$580.00 for the first $50,000.00 plus $6.25 for each additional $1,000.00, or fraction thereof, to and including $100,000.00
$100,001.00 to $500,000.00	$895.00 for the first $100,000.00 plus $5.00 for each additional $1,000.00, or fraction thereof
$500,001.00 to $1,000,000.00	$2,855.00 for the first $500,000.00 plus $4.25 for each additional $1,000.00, or fraction thereof, to and including $1,000.00
$1,000,001.00 and up	$4,955.00 for the first $1,000,000.00 plus $2.75 for each additional $1,000,000.00, or fraction thereof

Other Inspections and Fees:
1. Inspections outside of normal business hours .... $42.00 per hour*
   (minimum charge—two hours)
2. Reinspection fees assessed under provisions of
   Section 108.8 .... $42.00 per hour*
3. Inspections for which no fee is specifically indicated .... $42.00 per hour*
   (minimum charge—one-half hour)
4. Additional plan review required by changes, additions
   or revisions to plans .... $42.00 per hour*
   (minimum charge—one-half hour)
5. For use of outside consultants for plan checking and
   inspections, or both .... Actual costs**

*Or the total hourly cost to the jurisdiction, whichever is the greatest. This cost shall include supervision, overhead, equipment, hourly wages and fringe benefits of the employees involved.

**Actual costs include administrative and overhead costs.

Table 24–1.
This table suggests permit fees based on the total value of the construction work. (It is identified as Table 1–A in the Uniform Building Code.)

REVIEWING THE PLANS

Once the plans are submitted, they are examined by the various branches of the Building Department. Different localities may have some minor differences in the organization of their review process. However, the process is always divided into the general areas of *site, planning,* and *structural review.*

Besides the data provided in the site plan, other features, such as flood plains, easements, utilities, and other topographical features, are examined. The plans are checked to see if they conform to the zoning code requirements where the construction takes place. The structural review examines every aspect of the building's design.

If there are any unacceptable features on the plans, the structural plan examiner will make the corrections or return the plans to the owner, architect, or engineer of record for the corrections.

When the plans are approved, the designated building official will stamp the plans accordingly. One set of the approved plans will be retained by the Building Department. The other set will be returned to the applicant and kept on the job site for the duration of construction.

INSPECTION

A *building inspector* will carry out periodic inspections during the construction of the building. An *inspection record card* must be posted in a protected location on the job. Entries are made on the card at the time of each inspection. **(See Figure 24–2.)** Requests for inspections should be made by the person doing the work authorized by the permit. Most Building Departments request 24-hour notification.

The major areas of inspection are building, electrical, plumbing, mechanical, and gas. *Building inspection* includes the excavation and forms constructed for the foundation, the rough frame, and exterior and interior finish work. *Electrical, plumbing, mechanical,* and *gas inspections* occur when underground conduit, cable, piping, and ducts have been installed after the trenches or ditches have been excavated. Also included in these inspections is the rough-in work placed in the building frame. **Rough-in** work is the concealed work of the plumbing and electrical systems. Main items of plumbing rough-in work include water, drain, and vent pipes. Main items of electrical rough-in work include conduits and/or wiring, and electrical boxes.

The sequence of inspection is usually listed on the posted inspection card. Generally, the framing inspection cannot be made until all the electrical, plumbing, and mechanical rough-in work has been completed and "signed off." Framing inspection approval at this time means that the walls and ceilings are ready for drywall or plaster. Adherence to a proper order of inspection will avoid delay and inconvenience for all parties involved. Covering work that has not been inspected, for example, means that the wall covering will have to be removed and then reapplied.

Following is an example of the sequence of residential inspection:

1. **Footing trenches.** Required for earth-formed footing trenches, formed footings, and drilled piers.
2. **Foundation walls.** Required when forms are completed and reinforcing steel is set in place.
3. **Damp-proofing.** Usually required for full-basement foundations. Sometimes required with crawl-space and slab foundations.
4. **Rough plumbing (underground).** Required when concrete or fill is to be placed over underground sewer lines and water pipes.
5. **Rough mechanical (underground).** Required when concrete or fill is to be placed over mechanical components such as underground ducts and fuel piping.
6. **Rough electrical (underground).** Required when concrete or fill is to be placed over underground cable or conduit.
7. **Rough plumbing.** Required when rough-in plumbing installation is completed and water and air is in the lines for testing according to code specifications.
8. **Rough mechanical (duct and pipe).** Required when all duct work and/or piping has been completely installed in the framework that is ready for drywall. The air pressure tests described above are also applicable.
9. **Rough mechanical (flue and vent).** Required after roofing is completed to check whether there is proper separation of the flue from combustible materials.
10. **Rough electrical.** Required when rough-in electrical work has been installed in the framework and the walls and ceilings are ready for drywall. The rough-in work includes all wiring for the circuits, boxes, plaster rings, panel boxes set, as well as the made-up neutral and ground wires.
11. **Fireplace.** First inspection is required when firebox is completed and first flue liner is in place. Surrounding framing must also be visible. Second

SPECIAL INSPECTION REQUIRED   ☐ ☐
NOTED WITH * BELOW   Yes No

CITY OF FLAGSTAFF
INSPECTION CERTIFICATE

INSPECTOR _____

Date Issued

OWNER _____   BLDG. PERMIT NO. _____ _____

PLMB. PERMIT NO. _____ _____

ADDRESS _____   MECH. PERMIT NO. _____ _____

ELEC. PERMIT NO. _____ _____

LOT NO./SUBDIV. _____   CONTRACTOR _____

DESCRIPTION OF WORK _____

DATE

LOCATION ON LOT _____ _____

EXCAVATION/STEEL _____ UFER GROUND _____ FOOTING _____ _____

STEM WALL/PIERS _____ _____

SEWER/WATER UNDER FLOOR _____ _____

FILL UNDER FLOOR _____ _____

EXTERIOR WALLS _____ _____

FLOORS _____ _____

PARTITIONS/FIREBLOCK/FIREWALL _____ _____

ROOF STRUCTURE _____ _____

ROOF COVERING _____ _____

ROUGH MECHANICAL _____ _____

COMBUSTION AIR/VENTS _____ _____

ROUGH-IN ELECTRIC _____ _____

ROUGH-IN PLUMBING _____ _____

STOVE/FIREPLACE (PIPE/ARRESTOR) _____ _____

INSULATION _____ _____

DRYWALL _____ _____

CEILING INSULATION _____ _____

SEWER LINE _____ _____

GAS YARD LINE _____ _____

WATER SUPPLY LINE _____ _____

ELECTRIC METER _____ TEMPORARY ELECTRIC _____ _____

FINAL INSPECTION:

PLUMBING FIXTURES _____ _____

RANGE-DRYER VENT _____ _____

ELECTRICAL FIXTURES _____ _____

WOODSTOVE _____ _____

GAS APPLIANCE FIXTURES _____ _____

SMOKE DETECTOR _____ _____

ENTIRE STRUCTURE _____ ADDRESS NUMBERS _____ _____

I HEREBY CERTIFY THAT THE ABOVE STRUCTURE CONFORMS WITH THE BUILDING CODE FOR WORK DESCRIBED ABOVE:

_____ / _____

BUILDING INSPECTOR   /DATE

Revised 4/20/95

Figure 24–2.
An inspection record card should be posted in a protected location on the job site.

inspection is required when chimney has been completed except for capping.

12. **Framing.** Required when all framing is completed and roof is in place. All plumbing, heating, and electrical rough-ins must be completed and approved.

13. **Insulation.** Required when the entire building is insulated and before drywall is installed.

14. **Lathing or drywall.** Required before taping drywall or applying plaster to lath work.

15. **Final inspections:**
    a. All structural and finish work must be completed.
    b. All surface plumbing, mechanical, and electrical fixtures and equipment must be in place.
    c. If the project includes a landscape plan,

parking requirements, Fire Department regulations, drainage and/or public improvements, all these must be inspected and approved.

CERTIFICATE OF OCCUPANCY

In most localities a building cannot be inhabited until the owner has been issued a **Certificate of Occupancy**. This document is granted by the Building Department after the final inspection and approval of the building. A Certificate of Occupancy contains such information as the use and group classification, type of construction, use zone, name and address of the owner, locality, and the signature of the building official. **(See Figure 24–3.)**

Certificate of Occupancy

City of Flagstaff
Building and Safety Division

This Certificate issued pursuant to the requirements of Section 109 of the Uniform Building Code certifying that at the time of issuance this structure was in compliance with the various ordinances of the City regulating building construction or use. For the following:

Use Classification _____ Bldg. Permit No. _____

Group _____ Type Construction _____ Use Zone _____

Owner of Building _____ Address _____

Building Address _____ Locality _____

_____ By: _____
Building Official

Date: _____

POST IN A CONSPICUOUS PLACE

Figure 24–3.
A Certificate of Occupancy is granted after the final inspection and approval of the building.

REVIEW QUESTIONS

Enter the missing words in the spaces provided on the right.

1. A certificate authorizing that construction can begin is called a_____ _____.

   _____

2. The BOCA code states a permit is required to construct or ____ a structure.

   _____

3. A permit is required to make a _____ of occupancy, according to the BOCA code.

   _____

4. Moving counters or cases less than ___ may not require a permit.

   _____

5. A permit is not needed for a____ ___ less than 4′ from the bottom of footing.

   _____

6. An application for a permit must specify the _____ of the job site and provide a legal description of it.

   _____

7. ___ sets of construction plans are usually required with a building application.

   _____

8. The amount of a permit fee is based on the total _____ of the construction work.

   _____

9. A permit can be declared invalid if work does not begin within __ ___ of issuance.

   _____

10. The plan review process is generally divided into ____, _____, and _____ reviews.

    _____

11. Every aspect of the building's design is examined in the _____ review.

    _____

12. The concealed plumbing and electrical installation is called the _____ work.

    _____

13. The first inspection is usually for the _____ trenches or forms.

    _____

14. At the time of the _____ inspection, all structural and finish work must be completed.

    _____

15. A_____ of _____ must be issued before anyone can inhabit the building.

    _____

UNIT 25

Contracts and Bids

A contract is a legal, enforceable agreement between two or more parties. Several important contracts must be drawn up and signed by the persons and firms primarily responsible for the construction project. The major ones are between (1) the owner and architect, (2) the owner and general contractor, and (3) the general contractor and subcontractors.

The *architect* produces the construction plans for the building. A **general contractor,** also called a **building contractor,** is in overall charge of the construction project and employs office personnel, estimators, carpenters, and laborers. **Subcontractors** are directly responsible to the general contractor. They supervise and provide workers to do the installation of plumbing, electrical, sheet metal, roofing, painting, and other specialized functions. **(See Figure 25–1.)**

The bidding process precedes the awarding of contracts based on competitive cost proposals. Bids are based on information in the construction plans and specifications.

The information provided in this unit is based on documents prepared by the **American Institute of Architects (AIA).** These documents are intended only as guidelines; an attorney should be consulted about specific agreements. Local laws may vary on some items, such as contractors' licensing laws, arbitration, compensation for loss, and other legal questions. Copies of AIA documents are available from local and state offices and from the national office of the AIA. In Canada a similar set of standard documents has been prepared by the Canadian Construction Association (CCA).

OWNER-ARCHITECT CONTRACTS

AIA Document B151 is entitled "Abbreviated Form of Agreement Between Owner and Architect for Construction Projects of Limited Scope." This type of contract would most often apply to residential and smaller commercial buildings. Some major areas covered are architect services, owner responsibilities, construction costs, fees, and use of architects' construction documents.

The *architect services* include functions done by the architect, employees of the architect, and consultants. The architect is committed to work with the owner in the initial design process and produce construction documents (plans, specifications, etc.) based on projected cost estimates subject to the owner's approval. The architect will also help the owner in obtaining bids or negotiated proposals that lead to preparing and awarding the construction contract to the chosen general contractor. The architect will be deemed the author of the construction documents and retains all common law and statutory rights as well as copyrights of these documents.

Owner responsibilities include providing the architect with full information regarding the owner's goals, schedules, limitations, and budget considerations. It is also the obligation of the owner to furnish surveys

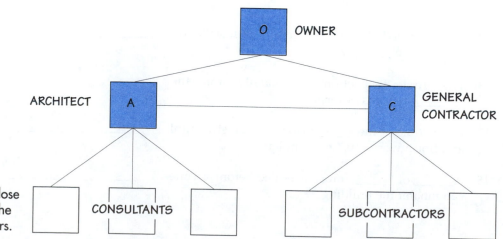

Figure 25–1.
A building project requires a close working relationship between the owner, architect, and contractors.

regarding physical and topographical characteristics of the job site, utility locations, and legal descriptions of the site. Qualified sources and/or consultants must be used to obtain this information. Other owner obligations include any necessary legal, accounting, and insurance services during construction.

A clear definition of construction costs applicable to the building project is included in an owner-architect contract. The construction cost is the total cost or estimated cost of the building designed by the architect. The cost is based upon current market rates of labor and materials.

The owner-architect contract should also include arbitration procedures to be followed if claims or disputes occur between the parties to the agreement. Conditions for termination or suspension of the agreement must be explained. The agreement should define the method of payments to the architect for basic services or additional reimbursable expenses.

OWNER–GENERAL CONTRACTOR AGREEMENTS

The contract between the owner and the general contractor is the most important agreement of the construction process. It defines the cost, scope, time frame, legal considerations, and all other functions related to the project.

The owner–general contractor agreements used most often are "stipulated sum" and the "cost of work plus fee" contracts. The first kind of contract establishes an agreed upon price for the construction of a building. The second kind of contract, often called a cost-plus contract, bases its fee on the total cost of the labor and materials plus a percentage of the labor cost.

"STIPULATED SUM" CONTRACT

Following is a duplication of the "stipulated sum" contract.

AIA Document A205
General Conditions of the Contract for Construction of a Small Project
1993 SMALL PROJECTS EDITION

Article 1
GENERAL PROVISIONS
1.1 THE CONTRACT
The Contract represents the entire and integrated agreement between the parties and supersedes prior negotiations, representations or agreements, either written or oral. The Contract may be amended or modified only by a written modification.
1.2 THE WORK

The term "Work" means the construction and services required by the Contract Documents, and includes all other labor, materials, equipment and services provided by the Contractor to fulfill the Contractor's obligations.
1.3 INTENT
The intent of the Contract Documents is to include all items necessary for the proper execution and completion of the Work by the Contractor. The Contract Documents are complementary, and what is required by one shall be as binding as if required by all.
1.4 OWNERSHIP AND USE OF ARCHITECT'S DRAWINGS, SPECIFICATIONS AND OTHER DOCUMENTS
Documents prepared by the Architect are instruments of the Architect's service for use solely with respect to this project. The Architect shall retain all common law, statutory and other reserved rights, including the copyright. They are not to be used by the Contractor or any Subcontractor, Sub-subcontractor or material or equipment supplier for other projects or for additions to this project outside the scope of the Work without the specific written consent of the Owner and Architect.

ARTICLE 2
OWNER
2.1 INFORMATION AND SERVICES REQUIRED OF THE OWNER
2.1.1 If requested by the Contractor, the Owner shall furnish and pay for a survey and a legal description of the site.
2.1.2 Except for permits and fees which are the responsibility of the Contractor under the Contract Documents, the Owner shall obtain and pay for other necessary approvals, easements, assessments and charges.
2.2 OWNER'S RIGHTS TO STOP THE WORK
If the Contractor fails to correct Work which is not in accordance with the Contract Documents, the Owner may direct the Contractor in writing to stop the Work until the correction is made.
2.3 OWNER'S RIGHT TO CARRY OUT THE WORK
If the Contractor defaults or neglects to carry out the Work in accordance with the Contract Documents and fails within a seven day period after receipt of written notice from the Owner to correct such default or neglect with diligence and promptness, the Owner may, without prejudice to other remedies, correct such deficiencies. In such case, a Change Order shall be issued deducting the cost of correction from payments due the Contractor.

2.4 OWNER'S RIGHTS TO PERFORM CONSTRUCTION AND TO AWARD SEPARATE CONTRACTS

2.4.1 The Owner reserves the right to perform construction or operations related to the project with the Owner's own forces, and to award separate contracts in connection with other portions of the project.

2.4.2 The Contractor shall coordinate and cooperate with separate contractors employed by the Owner.

2.4.3 Costs caused by delays or by improperly timed activities or defective construction shall be borne by the party responsible therefor.

ARTICLE 3
CONTRACTOR
3.1 EXECUTION OF THE CONTRACT

Execution of the Contract by the Contractor is a representation that the Contractor has visited the site, become familiar with local conditions under which the Work is to be performed and correlated personal observations with requirements of the Contract Documents.

3.2 REVIEW OF CONTRACT DOCUMENTS AND FIELD CONDITIONS BY CONTRACTOR

The Contractor shall carefully study and compare the Contract Documents with each other and with information furnished by the Owner. Before commencing activities, the Contractor shall: (1) take field measurements and verify field conditions; (2) carefully compare this and other information known to the Contractor with the Contract Documents; and (3) promptly report errors, inconsistencies or omissions discovered to the Architect.

3.3 SUPERVISION AND CONSTRUCTION PROCEDURES

3.3.1 The Contractor shall supervise and direct the Work, using the Contractor's best skill and attention. The Contractor shall be solely responsible for and have control over construction means, methods, techniques, sequences and procedures, and for coordinating all portions of the Work.

3.3.2 The Contractor, as soon as practicable after award of the Contract, shall furnish in writing to the Owner through the Architect the names of subcontractors or suppliers for each portion of the Work. The Architect will promptly reply to the Contractor in writing if the Owner or the Architect, after due investigation, has reasonable objection to the subcontractors or suppliers listed.

3.4 LABOR AND MATERIALS

3.4.1 Unless otherwise provided in the Contract Documents, the Contractor shall provide and pay for labor, materials, equipment, tools, utilities, transportation, and other facilities and services necessary for proper execution and completion of the Work.

3.4.2 The Contractor shall deliver, handle, store and install materials in accordance with manufacturers' instructions.

3.5 WARRANTY

The Contractor warrants to the Owner and Architect that: (1) materials and equipment furnished under the Contract will be new and of good quality unless otherwise required or permitted by the Contract Documents; (2) the Work will be free from defects not inherent in the quality required or permitted, and (3) the Work will conform to the requirements of the Contract Documents.

3.6 TAXES

The Contractor shall pay sales, consumers, use and similar taxes that are legally required when the Contract is executed.

3.7 PERMITS, FEES AND NOTICES

3.7.1 The Contractor shall obtain and pay for the building permit and other permits and governmental fees, licenses and inspections necessary for proper execution and completion of the Work.

3.7.2 The Contractor shall comply with and give notices required by agencies having jurisdiction over the Work. If the Contractor performs Work knowing it to be contrary to laws, statutes, ordinances, building codes, and rules and regulations without notice to the Architect and Owner, the Contractor shall assume full responsibility for such Work and shall bear the attributable costs. The Contractor shall promptly notify the Architect in writing of any known inconsistencies in the Contract Documents with such governmental laws, rules and regulations.

3.8 SUBMITTALS

The Contractor shall promptly review, approve in writing and submit to the Architect Shop Drawings, Product Data, Samples and similar submittals required by the Contract Documents, Shop Drawings, Product Data, Samples and similar submittals are not Contract Documents.

3.9 USE OF SITE

The Contractor shall confine operations at the site to areas permitted by law, ordinances, permits, the Contract Documents and the Owner.

3.10 CUTTING AND PATCHING

The Contractor shall be responsible for cutting, fitting or patching required to complete the Work or to make its parts fit together properly.

3.11 CLEANING UP

The Contractor shall keep the premises and surrounding area free from accumulation of debris and trash related to the Work.

3.12 INDEMNIFICATION

To the fullest extent permitted by law, the Contractor shall indemnify and hold harmless the Owner, Architect, Architect's consultants and agents and employees of any of them from and against claims, damages, losses and expenses, including but not limited to attorneys' fees, arising out of or resulting from performance of the Work, provided that such claim, damage, loss or expense is attributable to bodily injury, sickness, disease or death, or to injury to or destruction of tangible property (other than the Work itself) including loss of use resulting therefrom, but only to the extent caused in whole or in part by negligent acts or omissions of the Contractor, a Subcontractor, anyone directly or indirectly employed by them or anyone for whose acts they may be liable, regardless of whether or not such claim, damage, loss or expense is caused in part by a party indemnified hereunder.

ARTICLE 4
ARCHITECT'S ADMINISTRATION OF THE CONTRACT

4.1 The Architect will provide administration of the Contract as described in the Contract Documents. The Architect will have authority to act on behalf of the Owner only to the extent provided in the Contract Documents.

4.2 The Architect will visit the site at intervals appropriate to the stage of construction to become generally familiar with the progress and quality of the Work.

4.3 The Architect will not have control over or charge of and will not be responsible for construction means, methods, techniques, sequences or procedures, or for safety precautions and programs in connection with the Work, since these are solely the Contractor's responsibility. The Architect will not be responsible for the Contractor's failure to carry out the Work in accordance with the Contract Documents.

4.4 Based on the Architect's observations and evaluations of the Contractor's Applications for Payment, the Architect will review and certify the amounts due to Contractor.

4.5 The Architect will have authority to reject Work that does not conform to the Contract Documents.

4.6 The Architect will promptly review and approve or take appropriate action upon Contractor's submittals such as Shop Drawings, Product Data and Samples, but only for the limited purpose of checking for conformance with information given and the design concept expressed in the Contract Documents.

4.7 The Architect will promptly interpret and decide matters concerning performance under and requirements of the Contract Documents on written request for either the Owner or Contractor.

4.8 Interpretations and decisions of the Architect will be consistent with the intent of and reasonably inferable from the Contract Documents and will be in writing or in the form of drawings. When making such interpretations and decisions, the Architect will endeavor to secure performance by both Owner and Contractor, will not show partiality to either and will not be liable for results of interpretations or decisions so rendered in good faith.

4.9 The Architect's duties, responsibilities and limits of authority as described in the Contract Documents will not be changed without written consent of the Owner, Contractor and Architect. Consent shall not be unreasonably withheld.

ARTICLE 5
CHANGES IN THE WORK

5.1 After execution of the Contract, changes in the Work may be accomplished by Change Order or by order for a minor change in the Work. The Owner, without invalidating the Contract, may order changes in the Work within the general scope of the Contract consisting of additions, deletions or other revisions, the Contract Sum and Contract Time being adjusted accordingly.

5.2 A Change Order shall be a written order to the Contractor signed by the Owner and Architect to change the Work, Contract Sum or Contract Time.

5.3 The Architect will have authority to order minor changes in the Work not involving changes in the Contract Sum or the Contract Time and not inconsistent with the intent of the Contract Documents. Such changes shall be written orders and shall be binding on the Owner and Contractor. The Contractor shall carry out such written orders promptly.

5.4 If concealed or unknown physical conditions are encountered at the site that differ materially from those indicated in the Contract Documents or from those conditions ordinarily found to exist, the Contract Sum and Contract Time shall be subject to equitable adjustment.

ARTICLE 6
TIME

6.1 Time limits stated in the Contract Documents are of the essence of the Contract.

6.2 If the Contractor is delayed at any time in progress of the Work by changes ordered in the Work, or by labor disputes, fire, unusual delay in deliveries, unavoidable casualties or other causes beyond the Contractor's control, the Contract Time shall be extended by Change Order for such reasonable time as the Architect may determine.

ARTICLE 7
PAYMENTS AND COMPLETION
7.1 CONTRACT SUM

The Contract Sum stated in the Agreement, including authorized adjustments, is the total amount payable by the Owner to the Contractor for performance of the Work under the Contract Documents.

7.2 APPLICATIONS FOR PAYMENT

7.2.1 At least ten days before the date established for each progress payment, the Contractor shall submit to the Architect an itemized Application for Payment for operations completed in accordance with the values stated in the Agreement. Such application shall be supported by such data substantiating the Contractor's right to payment as the Owner or Architect may reasonably require and reflecting retainage if provided for elsewhere in the Contract Documents.

7.2.2 The Contractor warrants that title to all Work covered by an Application for Payment will pass to the Owner no later than the time of payment. The Contractor further warrants that upon submittal of an Application for Payment, all Work for which Certificates for Payment have been previously issued and payments received from the Owner shall, to the best of the Contractor's knowledge, information and belief, be free and clear of liens, claims, security interests or other encumbrances adverse to the Owner's interests.

7.3 CERTIFICATES FOR PAYMENT

The Architect will, within seven days after receipt of the Contractor's Application for Payment, either issue to the Owner a Certificate for Payment, with a copy to the Contractor, for such amount as the Architect determines is properly due, or notify the Contractor and Owner in writing of the Architect's reasons for withholding certification in whole or in part.

7.4 PROGRESS PAYMENTS

7.4.1 After the Architect has issued a Certificate for Payment, the Owner shall make payment in the manner provided in the Contract Documents.

7.4.2 The Contractor shall promptly pay each Subcontractor and material supplier, upon receipt of payment from the Owner, out of the amount paid to the Contractor on account of such entities' portion of the Work.

7.4.3 Neither the Owner nor the Architect shall have responsibility for the payment of money to a Subcontractor or material supplier.

7.4.4 A Certificate for Payment, a progress payment, or partial or entire use or occupancy of the project by the Owner shall not constitute acceptance of Work not in accordance with the requirements of the Contract Documents.

7.5 SUBSTANTIAL COMPLETION

7.5.1 Substantial Completion is the stage in the progress of the Work when the Work or designated portion thereof is sufficiently complete in accordance with the Contract Documents so that the Owner can occupy or utilize the Work for its intended use.

7.5.2 When the Work or designated portion thereof is substantially complete, the Architect will prepare a Certificate of Substantial Completion which shall establish the date of Substantial Completion, shall establish the responsibilities of the Owner and Contractor, and shall fix the time within which the Contractor shall finish all items on the list accompanying the Certificate. Warranties required by the Contract Documents shall commence on the date of Substantial Completion of the Work or designated portion thereof unless otherwise provided in the Certificate of Substantial Completion.

7.6 FINAL COMPLETION AND FINAL PAYMENT

7.6.1 Upon receipt of a final Application for Payment, the Architect will inspect the Work. When the Architect finds the Work acceptable and the Contract fully performed, the Architect will promptly issue a final Certificate for Payment.

7.6.2 Final payment shall not become due until the Contractor submits to the Architect releases and waivers of liens, and data establishing payment or satisfaction of obligations, such as receipts, claims, security interests or encumbrances arising out of the Contract.

7.6.3 Acceptance of final payment by the Contractor, a Subcontractor or material supplier shall constitute a waiver of claims by that payee except those previously made in writing and identified by that payee as unsettled at the time of final Application for Payment.

ARTICLE 8
PROTECTION OF PERSONS AND PROPERTY
8.1 SAFETY PRECAUTIONS AND PROGRAMS

The Contractor shall be responsible for initiating, maintaining and supervising all safety precautions and programs, including all those required by law in connection with performance of the Contract. The Contractor shall promptly remedy damage and loss to property caused in whole or in part by the Contractor, or by anyone for whose acts the Contractor may be liable.

ARTICLE 9
CORRECTION OF WORK

9.1 The Contractor shall promptly correct Work rejected by the Architect as failing to conform to the requirements of the Contract Documents. The Contractor shall bear the cost of correcting such rejected Work.

9.2 In addition to the Contractor's other obligations including warranties under the Contract, the Contractor shall, for a period of one year after Substantial Completion, correct work not conforming to the requirements of the Contract Documents.

9.3 If the Contractor fails to correct nonconforming Work within a reasonable time, the Owner may correct it and the Contractor shall reimburse the Owner for the cost of correction.

ARTICLE 10
MISCELLANEOUS PROVISIONS
10.1 ASSIGNMENT OF CONTRACT

Neither party to the Contract shall assign the Contract as a whole without written consent of the other.

10.2 TESTS AND INSPECTIONS

10.2.1 Test, inspections and approvals of portions of the Work required by the Contract Documents or by laws, ordinances, rules, regulations or orders of public authorities having jurisdiction shall be made at an appropriate time.

10.2.2 If the Architect requires additional testing, the Contractor shall perform these tests.

10.2.3 The Owner shall pay for tests except for testing Work found to be defective for which the Contractor shall pay.

10.3 GOVERNING LAW

The Contract shall be governed by the law of the place where the project is located.

ARTICLE 11
TERMINATION OF THE CONTRACT
11.1 TERMINATION BY THE CONTRACTOR

If the Owner fails to make payment when due or substantially breaches any other obligation of this Contract, following seven days' written notice to the Owner, the Contractor may terminate the Contract and recover from the Owner payment for Work executed and for proven loss with respect to materials, equipment, tools, construction equipment and machinery, including reasonable overhead, profit and damages.

11.2 TERMINATION BY THE OWNER

11.2.1 The Owner may terminate the Contract if the Contractor:

.1 persistently or repeatedly refuses or fails to supply enough properly skilled workers or proper materials;

.2 fails to make payment to Subcontractors for materials or labor in accordance with the respective agreements between the Contractor and the Subcontractors;

.3 persistently disregards laws, ordinances, or rules, regulations or orders of a public authority having jurisdiction; or

.4 is otherwise guilty of substantial breach of a provision of the Contract Documents.

11.2.2 When any of the above reasons exist, the Owner, after consultation with the Architect, may without prejudice to any other rights or remedies of the Owner and after giving the Contractor and the Contractor's surety, if any, seven days' written notice, terminate employment of the Contractor and may:

.1 take possession of the site and of all materials thereon owned by the Contractor;

.2 finish the Work by whatever reasonable method the Owner may deem expedient.

11.2.3 When the Owner terminates the Contract for one of the reasons stated in Subparagraph 11.2.1, the Contractor shall not be entitled to receive further payment until the Work is finished.

11.2.4 If the unpaid balance of the Contract Sum exceeds costs of finishing the Work, such excess shall be paid to the Contractor. If such costs exceed the unpaid balance, the Contractor shall pay the difference to the Owner. This obligation for payment shall survive termination of the Contract.

"COST OF WORK PLUS FEE" CONTRACT

"Cost of work plus fee" contracts are often drawn up for alteration (remodeling) work. Alteration work is much more difficult to estimate than new building and presents a greater risk to the contractor. These contracts contain provisions similar to those of stipulated-sum contracts except cost and payment provisions. They are based on labor and material plus a percentage added to the labor and material price. For example, a contract stipulates that the contractor's fee will be 10% added to the cost of labor and material. The final cost of labor and material is $5,350.00. Ten percent of $5,350.00 is $535.00. Therefore, the total cost of the construction is $5,885.00 ($5,350.00 + $535.00 = $5,855.00). The contract may also contain a *guaranteed maximum price* provision. If the guaranteed maximum price is exceeded, it is paid by the contractor.

GENERAL CONTRACTOR–SUBCONTRACTOR AGREEMENTS

The general contractor is responsible for making contracts with various subcontractors doing work on the construction project. Obviously, the general contractor must work closely with the subcontractors in coordinating the work of the different crafts. This is very important in maintaining good working relations and progress on the job. Guidelines for drawing up a general contractor–subcontractor agreement are contained in AIA Document A406, Standard Form of Agreement Between Contractor and Subcontractor.

The agreement between the two parties must not only clearly define the scope and quality of the work but also their mutual rights and responsibilities. The general contractor must cooperate with the subcontractors in scheduling work to avoid conflict and interference among them. The general contractor must promptly inform a subcontractor of any information that affects the agreement. Conditions and methods for filing claims against a subcontractor should be included.

The agreement should also define the subcontractor's responsibilities as they relate to the progress of the work. Some of these responsibilities include cooperation with the general contractor in scheduling work and the submission of new shop drawings to already existing construction plans. The subcontractor is also responsible for periodic progress reports and taking precautions to protect the work and cooperate with the general contractor, owner, and other subcontractors to avoid potential conflicts in the flow and organization of the work.

Other conditions of an agreement should deal with the responsibility for obtaining and paying for permits and government fees, licenses, and the inspections for the subcontractor's completed work. Safety precautions, cleaning up, and proof and warranty of good-quality materials and equipment are additional areas that should be defined in the contract. Clauses should be included concerning procedures to be followed in case of changes

in the work and the arbitration of controversies or claims between the general contractor and subcontractor. The conditions and procedure for terminating or suspending a subcontractor and assigning a replacement should be clearly defined in the agreement.

BIDDING AND AWARDING CONTRACTS

When the drawings for a building have been completed, the owner can begin to accept bids from general contractors. A **bid,** as it applies to the building industry, is an offer to contract the performance of work described in the construction documents at a specified cost. The owner can thus compare bids in making a selection.

Public bidding is sometimes required by law, such as for public work projects. This requires placing legal notices in one or more local newspapers. However, this method is not used often for bidding on private work. A frequent procedure selecting private bidders is for the owner to enlist the architect's aid in drawing up a list of general contractors who have established a reputation for quality construction and financial integrity. Each contractor asked to bid is provided, without cost, a full set of plans and any other documents covering all the work that must be included in the bid. (Sometimes a refundable deposit may be required of the contractor.) The owner and architect must notify the bidders of the time, date, and place where the bids will be opened, allowing a reasonable amount of time for bidders to draw up their proposals.

After the contract has been awarded to the lowest bidder, the remaining bidders must return their documents to the owner or architect. It is advisable that the bids be opened when all the bidders are present. If this is not done, each bidder is entitled to a tabulation of all the bids within ten days.

The owner reserves the right to decline any or all of the submitted bids. However, this cannot be done to accept a different bid that was not offered before the prices of the other bids were publicly announced. In other words, rejection of all bids should not be a subterfuge for getting estimates of the cost of work and then awarding the agreement to a contractor selected before the open-bidding process.

Usually the qualified bidder offering the lowest price is awarded the contract. Upon signing it, the bidder is legally obligated to do all the work named in the contract for the price of the bid. However, the low bidder may request the withdrawal of his or her bid if a reasonable claim can be made that a serious mistake was made in calculating the bid. It then becomes the decision of the owner and architect whether to permit withdrawal of the bid. Errors in mathematical calculations or duplication of documents can be considered justification for bid withdrawal. Incorrect judgment of the cost of labor and material, though, is not considered a valid excuse. If a bid is withdrawn by the contractor or rejected by the owner, the lowest of the remaining bids is then considered for acceptance.

INSURANCE AND BONDS

Insurance and bond coverage are very important provisions in any agreement between the contractor and owner. The contractor is responsible for obtaining **insurance** for the protection of the owner and the contractor. **Bonds** are financial guarantees by a surety company assuring payment for labor, material, and all other obligations related to the performance of work described in the contract.

REVIEW QUESTIONS

Enter the missing words in the spaces provided on the right.

1. The _____ _____is in overall charge of the construction project.

　　　_____

2. Guidelines for drawing up agreements are available from the_____(use abbreviation).

　　　_____

3. In a stipulated-sum contract between owner and general contractor, a _____ price is agreed upon.

　　　_____

4. AIA Document A–205 states that the _____shall pay for the building permit.

　　　_____

5. AIA Document A–205 states the _____ has the authority to reject unsatisfactory work.

　　　_____

6. Time _____ are included in owner–general contractor agreements.

　　　_____

7. Cost-plus contracts are based on material and labor cost plus a _____of that cost.

　　　_____

8. Subcontractors sign agreements with the_____for the scope and cost of their work.

　　　_____

9. Public works projects require public _____.

　　　_____

10. A financial guarantee by a surety company is called a _____.

　　　_____

SECTION 5 TEST

Code Regulations and Legal Documents

Answer T (true) or F (false).

1. Zoning codes give requirements for the structural components of a building. _____

2. Zoning codes never limit the number of square feet of building allowed on a lot. _____

3. An R—Div. 1 zone allows apartment buildings, according to the UZC. _____

4. Supermarkets are allowed in **C** zones. _____

5. An **M** zone is reserved for mobile homes. _____

6. Building codes ensure the use of high-quality construction methods. _____

7. There are model codes for plumbing, electrical, and heating and air conditioning work. _____

8. The UBC model code is mainly directed toward the southern states. _____

9. Local codes in the eastern and some midwestern states are based on the BOCA model code. _____

10. The table of contents of the UBC, BOCA, and SBC model codes are almost identical. _____

11. Local governments have agencies for enforcing building codes. _____

12. An addition to an existing building does not require a building permit. _____

13. The estimated value of a building project must be included with a permit application. _____

14. Floodplains are not a concern of building-agency plan reviewers. _____

15. Inspections are carried out only at the beginning and end of a construction project. _____

16. Some examples of subcontracted areas are painting, roofing, and sheet metal work. _____

17. A job bid is a submitted offer to do the construction for the amount set forth in the bid. _____

18. A cost-plus contract is the same as a stipulated-bid contract. _____

19. An owner is legally mandated to accept the lowest bid on the job.   _____

20. Job insurance is the responsibility of the general contractor.   _____

Choose one or more correct ways to finish each statement.

21. Zoning codes contain regulations about   _____
     a. depths and widths of property.
     b. front, rear, and side setbacks.
     c. maximum density.
     d. all of the above.

22. An R—Div. 3 (residential) zone includes   _____
     a. supermarkets.
     b. single-family homes.
     c. wholesale outlets.
     d. condominiums.

23. An **M** (manufacture) zone includes   _____
     a. playgrounds.
     b. churches.
     c. woodworking shops.
     d. community centers.

24. The number of dwelling units allowed on an area of land has to do with the _____
     a. setback requirements
     b. minimum lot area.
     c. permitted maximum density.
     d. building height.

25. Additional topics included in the Uniform Zoning Code are: _____
     a. off street parking.
     b. locations of accessory buildings.
     c. fence heights
     d. none of the above.

26. A building code concentrates most of its information on   _____
     a. plumbing regulations.
     b. safety procedures.
     c. electrical and mechanical work.
     d. the structural work of the building.

27. The names of the national model codes are the   _____
     a. Uniform Building Code.
     b. National Building Code.
     c. Building Officials and Code Administrators (BOCA) National Building Code.
     d. Standard Building Code.

28. National model codes are   _____
     a. government codes.
     b. advisory codes only.
     c. enforceable by law.
     d. codes superseding any local codes.

29. The term *means of egress* in a building code book has to do with _____
    a. the method of entry into the building.
    b. fire-escape provisions.
    c. a path of travel from any point in a building to the outside ground level.
    d. passageways between rooms in the interior of the building.

30. Occupancy classification as defined in the code books identifies the _____
    a. type of building.
    b. type of fire-resistant materials.
    c. purpose for which a building, or portion thereof, is used.
    d. all of the above.

31. The level plane of travel of a means of egress is called the _____
    a. horizontal travel.
    b. vertical travel.
    c. travel distance.
    d. parallel travel.

32. A permit to begin new construction is usually issued by the local government's _____
    a. Maintenance Department.
    b. Streets and Highway Division.
    c. Zoning Commission.
    d. Building Department.

33. Projects that require a building permit include _____
    a. constructing and altering a structure.
    b. building a one-story detached tool shed.
    c. making a change of occupancy.
    d. painting the interior of the building.

34. Application for a permit can be made by the _____
    a. institution making the construction loan.
    b. the owner.
    c. any of the subcontractors.
    d. the architect.

35. When applying for a building permit, you will find that most building
    departments require _____
    a. one set of plans and specifications.
    b. a materials list for the entire project.
    c. two sets of plans and specifications.
    d. none of the above.

36. The inspection of rough-in plumbing and electrical work can be done _____
    a. before the walls are erected.
    b. prior to drywall application.
    c. at any time during construction.
    c. after drywall application.

37. Contracts usually required for construction projects are between the _____
    a. owner and general contractor.
    b. architect and general contractor.
    c. owner and subcontractors.
    d. general contractor and subcontractors.

38. A contract between the owner and general contractor includes   _____
    a. payment for wear and tear on the contractor's equipment.
    b. a preliminary land survey.
    c. the cost of the entire project.
    d. the scope of the work.

39. The labor and material of a cost plus project is $35,180.00, and the agreed
    upon profit margin is 10 percent. What is the total cost of the project?   _____
    a. $35,500.00.
    b. $38,698.00.
    c. $36,351.00.
    d. $38,689.00.

40. Public bidding is usually required for   _____
    a. public works projects.
    b. private projects.
    c. cost-plus projects.
    d. all of the above.

Enter the missing words in the spaces provided on the right.

41. Regulations governing property depth and width are found in the _____code.   _____

42. Zoning codes provide information regarding _____building heights.   _____

43. Setback requirements for buildings give the minimum distances from ____ ____.   _____

44. Maximum density has to do with number of _____ _____ per acre.   _____

45. _____ zones allow light-manufacture and industrial-use buildings.   _____

46. Most local electrical codes are patterned after the _____ _____ code.   _____

47. The _____building code is the model code for most southern states.   _____

48. The portion of a building between floors is called a _____.   _____

49. The _____(use abbreviation) is an industry-recognized testing agency that
    establishes fire resistance ratings.   _____

50. The discussion of headroom in a code book pertains to _____.   _____

51. An application for a building permit usually includes the _____ construction
    value.   _____

52. A permit may be declared invalid if work does not begin within ___ _____.   _____

53. The major areas of inspection are ____, electrical, plumbing, mechanical, and gas.   _____

54. The first residential inspection is for the _____forms or trenches.   _____

55. The insulation inspection occurs before the _____is installed.   _____

56. All structural and finish work must be completed for the _____inspection.   _____

57. A _____ is a legal and enforceable agreement between two or more parties.   _____

58. The major agreement in the construction process is the _____ and _____ contract.   _____

59. The abbreviation AIA stands for the _____ _____ of _____.   _____

60. The owner reserves the right to _____ a construction bid.   _____

SECTION

6 | Related Mathematics

In reading a construction plan, it is often necessary to do mathematical calculations in order to interpret various types of dimensions. For example, decimal and common fractions must frequently be interchanged. Combined feet-and-inch measurements may have to be added or subtracted from each other, multiplied or divided. Although the U.S. construction industry has not yet adopted the metric system of measurement, it may do so in the future. Therefore, an understanding of how to change English (customary) linear units to equivalent metric units is recommended.

This section consists of mathematical problems and exercises. Units 26 and 27 should be studied before proceeding to the remaining units. ■

UNIT 26

When reading construction plans, you may need to convert linear quantities. Following are some of the more commonly required conversions:

✓ Inches to feet
(example: 40" = 3'-4").
✓ Feet to Inches
(example: 5'-9" = 69").
✓ Common fractions to decimal fractions
(example: 5/8 = .625).
✓ Decimal fractions of an inch to common fractions.
(example: .75" = 3/4").
✓ Mixed numbers to decimal fractions
(example: 9 3/8 = 9.375).
✓ Decimal inch fractions to common fractions of an inch
(example: .1875" = 3/16").
✓ Inches to decimal fractions of a foot
(example: 5" = .4166').
✓ Decimal fractions of a foot to inches
(example: .9166' = 11").

CONVERTING INCHES TO FEET

To change inches to feet, divide the number of inches by 12. **(See Example 1.)** The answer is expressed as combined feet and inches if there is a remainder. **(See Example 2.)**

Example 1: Change 60" to feet.

Step 1. Divide 60 by 12.

$$
\begin{array}{r}
5 \\
12 \overline{)\ 60} \\
\underline{60} \\
00
\end{array}
$$

Final Answer: 60" = 5'

Example 2: Change 41" to a feet-and-inches measurement.

Step 1. Divide 41 by 12.

$$
\begin{array}{r}
3 \\
12 \overline{)\ 41} \\
\underline{36} \\
5 \leftarrow \textbf{inches remainder}
\end{array}
$$

The remainder (5) is the number of inches left over after dividing 41 inches by 12.

Step 2. Bring up the 5 and place a hyphen between the 3 and 5.

$$
\begin{array}{r}
3\text{ - }5 \\
12 \overline{)\ 41} \\
\underline{36} \\
5
\end{array}
$$

Final Answer: 41" = 3'-5"

QUIZ 26-1

Convert the following inches to feet.

	Inches	Feet
1.	48"	_____
2.	72"	_____
3.	60"	_____
4.	96"	_____
5.	29"	_____
6.	81"	_____
7.	52"	_____
8.	16"	_____
9.	115"	_____
10.	173"	_____

CONVERTING FEET TO INCHES

To change feet to inches, multiply the number of feet by 12. **(See Example 3.)** To change combined feet-and-inches measurements entirely to inches, multiply the number of feet by 12 and then add the inches. **(See Example 4.)**

Example 3: Change 9' to inches.

Step 1. Multiply 9 times 12.

$$
\begin{array}{r}
12 \\
\times\ 9 \\
\hline
108
\end{array}
$$

Final Answer: 9' = 108"

Example 4: Change 5'-7" to inches.

Step 1. Multiply 5 feet times 12.

$$
\begin{array}{r}
12 \\
\times\ 5 \\
\hline
60
\end{array}
$$

Step 2. Add 7 inches to 60 inches.

$$
\begin{array}{r}
60 \\
+\ 7 \\
\hline
67
\end{array}
$$

Final Answer: 5'-7" = 67"

QUIZ 26-2

Change the following feet and feet-and-inches measurements to all inches.

Feet	Inches
1. 3'-0"	_____
2. 7'-0"	_____
3. 11'-0"	_____
4. 21'-0"	_____
5. 1'-8"	_____
6. 4'-5"	_____

7. 9'-3"	_____
8. 15'-4"	_____
9. 18'-2"	_____
10. 24'-7"	_____

CONVERTING COMMON FRACTIONS TO DECIMAL FRACTIONS

To change a common fraction to a decimal fraction, divide the numerator by the denominator of the common fraction. **(See Example 5.)** To change a mixed number (such as 7 1/2 or 14 7/8) to a whole number with a decimal remainder (resulting in 7.5 and 14.88, respectively, for the preceding example), divide the numerator by the denominator of the fraction, and then combine the decimal fraction with the whole number. **(See Example 6.)**

Example 5: Change the common fraction 3/16 to a decimal fraction. Work the remainder to three places and round off to two places.

Step 1.
  A. Place the numerator (3) in a frame, and place the denominator (16) to the left of the frame.
  B. Place a decimal point to the right of the number (3) in the frame.
  C. Work the problem to a remainder of three places; you would put three zeros to the right of the decimal point in the frame.
  D. Place a decimal point above the frame directly over the decimal position below.

　　　　· ← **place decimal point above frame**
16) 3.000 ← **place three zeroes**

Step 2. Divide the number in the frame (3.000) by the divisor (16). Round the remainder to two places.

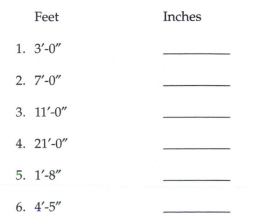

Final Answer: 3/16 = .19

Example 6: Change the mixed number 5 1/3 to a whole number with a decimal remainder.

Step 1. Separate the whole number, 5, from the common fraction, 1/3, and change the 1/3 to a decimal fraction. Work the decimal remainder to three places and round off to two places.

```
          . 333 = .33
    3 ) 1.000
         9
        10
         9
        10
         9
```

Step 2. Combine the whole number, 5, with the decimal, .33 (5 and .33 = 5.33)

Final Answer: 5 1/3 = 5.33

QUIZ 26-3

Change the following common fractions to decimal fractions, and mixed numbers to whole numbers with a decimal remainder. Round off the decimal remainders to the number of places indicated.

	Common Fraction	Decimal Fraction
1.	1/4	(2 places) _____
2.	5/8	(3 places) _____
3.	1/3	(2 places) _____
4.	7/16	(3 places) _____
5.	5/6	(4 places) _____
6.	6 3/4	(2 places) _____
7.	9 3/5	(1 place) _____
8.	3 7/8	(2 places) _____
9.	5 1/16	(4 places) _____
10	7 11/32	(5 places) _____

CONVERTING DECIMAL FRACTIONS OF AN INCH TO NEAREST COMMON FRACTIONS

To change a decimal fraction of an inch to the nearest common fraction, multiply the decimal by 16. The number to the left of the decimal point in the answer is the number of sixteenths. The number to the right of the decimal is a fraction of a sixteenth. **(See Example 7.)** If the number to the right of the decimal is .5 or more, add 1/16 to the final answer. **(See Example 8.)**

Example 7: Change .63″ to the nearest inch fraction.

Step 1. Multiply 16 times .63. Note that the number to the right of the decimal is less than 5.

```
     .63
   × .16
    378
    63
  10.08  sixteenths (.08 is less than .5)
```

Step 2. Place the number 10 over 16 and reduce the fraction.

$$\frac{10}{16} = \frac{5}{8}$$

Final Answer: .63″ = 5/8″

Example 8: Change .85″ to the nearest inch fraction.

Step 1. Multiply 16 times .85.

```
     .85
   ×  16
    510
    85
  13.60  sixteenths (.60 is more than .5)
```

Step 2. The number .6 to the right of the decimal is .5 or more. Add 1/16 to the number at the left of the decimal.

```
    13
   + 1
    14
```

Step 3. Place the number 14 over 16 and reduce the fraction.

$$\frac{14}{16} = \frac{7}{8}$$

Final Answer : 85″ = 7/8″

QUIZ 26-4

Change the following decimal fractions of an inch to the nearest common fractions.

	Decimal Fraction	Common Fraction
1.	.25″	_____
2.	.41″	_____
3	.57″	_____
4.	.097″	_____
5.	.94″	_____
6.	.208″	_____
7.	.384″	_____
8.	.795″	_____
9.	.4209″	_____
10.	.6758″	_____

CONVERTING INCHES TO A DECIMAL FRACTION OF A FOOT

To change inches to a decimal fraction of a foot, first change the inches to a common fraction of a foot. Then change the common fraction to a decimal fraction. **(See Example 9.)** To change a combined inch-and-inch-fraction measurement to a decimal fraction of a foot, convert the inch fraction to a decimal. **(See Example 10.)**

Example 9: Change 5″ to a decimal fraction of a foot.

Step 1. Place 5 over 12.

$$\frac{5}{12}$$

Step 2. Divide the numerator (5) by the denominator (12). Work the problem to four places.

```
       .4166
  12 ) 5.0000
       4 8
        20
        12
        80
        72
        80
        72
```

Final Answer : 5″ = .4166′

Example 10: Change 6 3/4″ to a decimal fraction of a foot.

Step 1. Change the 3/4 common fraction to a decimal.

```
      .75
  4 ) 3.00
      2 8
       20
       20
```

3/4 = .75

Step 2. Divide 6.75 by 12. Work the problem to four decimal places.

```
        .5625
  12 ) 6.7500
        6 0
        75
        72
        30
        24
         60
         60
```

Final Answer: 6 3/4″ = .5625′

QUIZ 26-5

Change the following inches to a decimal fraction of a foot. Work the decimal answer to four places, if necessary.

	Inches	Decimal
1.	6″	_____
2.	7″	_____

Inches	Decimal
3. 9″	_____
4. 11″	_____
5. 8″	_____
6. 4 1/8″	_____
7. 8 3/4″	_____
8. 5 1/2″	_____
9. 3 5/16″	_____
10. 7 7/8″	_____

CONVERTING DECIMAL FRACTIONS OF A FOOT TO INCHES

To change a decimal fraction of a foot to inches, multiply the decimal fraction by 12. **(See Example 11.)** If there is a decimal remainder in the answer, multiply the remainder by 16 in order to change it to an inch fraction. **(See Example12.)**

Example 11: Change .75′ to inches.

Step 1. Multiply 12 times .75.

```
  .75
× 12
  150
   75
 9.00
```

Final Answer: .75′ = 9″

Example 12: Change .684′ to inches.

Step 1. Multiply 12 times .684.

```
  .684
×  12
 1368
  684
8.2 0 8″
```

Step 2. Multiply 16 times the decimal remainder (.208).

```
  .208
×  16
 1248
  208
 3.328 = 3/16″
```

Step 3. Add together 8″ and 3/16″.

```
  8″
+   3/16″
 8  3/16″
```

Final Answer: .684′ = 8 3/16″

QUIZ 26-6

Change the following decimal fractions of a foot to inches:

Decimal	Inches
1. .25′	_____
2. .5′	_____
3. .54′	_____
4. .39′	_____
5. .83′	_____
6. .704′	_____
7. .492′	_____
8. .568′	_____
9. .8305′	_____
10. .7067′	_____

UNIT 27

Calculating English Measurements

Plan reading requires the ability to mathematically calculate linear measurements. Some of these calculations are as follows:

- ✓ Adding combined feet-and-inches measurements **(examples: 5'-7" + 9'-4" = 14'-11").**
- ✓ Adding separate feet and inches to combined feet-and-inches measurements **(example: 11" + 5' = 71" or 5'-11").**
- ✓ Subtracting combined feet-and-inches measurements **(example: 53'-8" − 17'-10" = 35'-10").**
- ✓ Subtracting inches from feet and feet from inches **(example: 87" − 5'-8" = 19" or 1'-7").**
- ✓ Multiplying combined feet-and-inches measurements **(example: 3'-11" × 8 = 31'-4").**
- ✓ Dividing combined feet-and-inches measurements **(example: 36'-9" ÷ 3 = 12'-3").**

ADDING COMBINED FEET-AND-INCHES MEASUREMENTS

To add combined feet-and-inches measurements, place them in a vertical column with the feet in line with feet and the inches in line with inches. Add the feet and inches separately. **(See Example 1.)** If the sum of the inches is 12 or more, change it to a feet-and-inches sum and add this to the sum of the feet. **(See Example 2.)**

Example 1: Add 7'-5" and 9'-3".

Step 1. Place 7'-5" and 9'-3" in a vertical column and add the feet and inches and separately.

```
  7'-5"
+ 9'-3"
 16'-8"
```

Final Answer: 16'-8"

Example 2: Add 5'-9" and 21'-8".

Step 1. Place 5'-9" and 21'-8" in a vertical column, and add the feet and inches separately.

```
   5'-9"
+ 21'-8"
 26'-17"
```

Step 2. Change the 17" total to a feet-and-inches total.

$$12\,)\overline{\,17\,}^{\;1\text{-}5} \quad \frac{12}{5}$$

17" = 1'-5"

Step 3. Add 1'-5" to the feet total (26').

```
 26'-0"
+ 1'-5"  ← changed from 17"
 27'-5"
```

Final Answer: 27'-5"

QUIZ 27-1

Add the following feet-and-inches measurements..

1. 2'-3" + 5'-6" _____
2. 4'-5" + 6'-2" _____
3. 13'-2" + 21'-7" _____
4. 11'-8" + 9'-9" _____
5. 18'-5" + 6'-11" _____
6. 7'-9" + 4'-6" + 8'-4" _____
7. 15'-7" + 19'-5" + 31'-9" _____
8. 17'-2" + 4'-10" + 23'-8" + 33'-6" _____
9. 12'-3" + 8'-0" + 13'-11" _____
10. 98'-7" + 256'-10" + 5'-9" + 324'-11" _____

ADDING SEPARATE FEET OR INCHES TO COMBINED FEET-AND-INCHES MEASUREMENTS

To add feet to inches, when the sum is to be expressed as inches, change the feet measurement to inches first. **(See Example 3.)** If the sum is to be expressed as feet, change the inch measurement to feet. **(See Example 4.)**

Example 3: Add 5'-8" and 36". Express the sum in inches.

Step 1. Change 5'-8" to inches.

```
    12
  × 5
    60
  + 8
    68
```

5'-8" = 68"

Step 2. Place 68" and 36" in a vertical column and add them together.

```
    68"  ←  changed from 5'-8"
  + 36"
   104"
```

Final Answer: 104"

Example 4: Add 52" and 12'-3". Express the sum in feet and inches.

Step 1. Change 52" to feet.

```
        4 - 4
  12 ) 52
        48
         4
```

52" = 4'-4"

Step 2. Place 4'-4" and 12'-3" in a vertical column, and add the feet and inches separately.

```
    4'-4"   ←  changed from 52"
  + 12'-3"
   16'-7"
```

Final Answer: 16'-7"

QUIZ 27-2.

Add the following feet-and-inches measurements to feet or inches. Express the answers as indicated.

1. 3'-6" + 25"   _____ (inches)

2. 18'-4" + 76"   _____ (inches)

3. 4'-7" + 33" + 7'-2"   _____ (inches)

4. 11'-9" + 41" + 7'-3" + 52"   _____ (inches)

5. 25" + 7'-8" + 9'-3" + 13"   _____ (inches)

6. 27" + 6'-8"   _____ (inches)

7. 81" + 15'-6"   _____ (ft. - in.)

8. 54" + 19'-2" + 30"   _____ (ft. - in.)

9. 21" + 8'-11" + 21'-7" + 91"   _____ (ft. - in.)

10. 142" + 3'-5" + 17" + 87"   _____ (ft. - in.)

SUBTRACTING COMBINED FEET-AND-INCHES MEASUREMENT

Computed measurement may necessitate subtracting separate quantities of feet or inches that will only require following the established procedure for subtracting whole numbers. Very often subtracting combined feet and inches is also necessary.

To subtract combined feet-and-inches measurements, place them in a vertical column with the feet in line with the feet and the inches in line with the inches. Subtract the feet and inches separately. **(See Example 5.)** If the number of inches above is less than the number of inches below, borrow 1' (12") from the number of feet above. Add the borrowed 12" to the number of inches above. **(See Example 6.)**

Example 5: Subtract 5'-7" from 12'-10".

Step 1. Place 5'-7" below 12'-10", and subtract the feet and inches separately.

12'-10"
− 5'- 7"
7'- 3"

Final Answer: 7'-3"

Example 6: Subtract 13'-10" from 21'-5".

Step 1. Place 13'-10" below 21'-5".

21'- 5"
− 13'-10"

Step 2.
A. The 10" in the *lower* number is greater than the 5" above. Borrow 1' (12") from the 21' in the *top* dimension. This will reduce 21' to 20'.
B. Add the borrowed 1' (12") to the 5", producing a total of 17". The quantity 21'-5" is now expressed as 20'-17".

borrowed from 21' ⌐
                    ↓
Reduced from 21' →   20'-17"   (5" + 12" = 17")

Step 3. Subtract 13'-10" from 20'-17".

20'-17"
− 13'-10"
7'- 7"

Final Answer: 7'-7"

QUIZ 27-3

Subtract the following feet-and-inches measurements.

                                          Answer

1. 9'-10" – 3'-4"   _____

2. 15'-9" – 7'-7"   _____

3. 53'-11" – 36'-2"   _____

4. 13'-3" – 5'-7"   _____

5. 21'-2" – 8'-6"   _____

6. 33'-8" – 19'-9"   _____

7. 41'-6" – 12'-8"   _____

8. 72'-9" – 69'-11"   _____

9. 129'-3" – 98' - 7"   _____

10. 207'-10" – 143'-11"   _____

SUBTRACTING INCHES FROM FEET AND FEET FROM INCHES

To subtract inches from feet or feet from inches when the remainder is to be expressed as inches, change the feet quantity to inches. **(See Example 7.)** If the answer is to be expressed as feet and inches, change the inch measurement to feet. **(See Example 8.)**

Example 7: Subtract 3'-5" from 59". Express the answer in inches.

Step 1. Change 3'-5" to inches.

12
× 3
36
+ 5
41

3'-5" = 41"

Step 2. Subtract 41" from 59".

59"
− 41"
18"

Final Answer: 18"

Example 8: Subtract 37" from 5'-9". Express the answer in a feet-and-inches measurement.

Step 1. Change 37" to feet.

            3' - 1"
12) 37
      36
       1" ← **remainder**

37" = 3'-1"

Step 2. Subtract 3'-1" from 5'-9".

5'-9"
− 3'-1"
2'-8"

Final Answer: 2'-8"

QUIZ 27-4

Subtract the following inches from feet, and feet from inches. Express the answers as indicated.

Answer

1. 29″ – 2′-0″ _____ (inches)

2. 62″ – 2′-9″ _____ (inches)

3. 78″ – 5′-11″ _____ (inches)

4. 137″ – 8′-7″ _____ (inches)

5. 469″ – 15′-9″ _____ (inches)

6. 3′-10″ – 30″ _____ (feet)

7. 5′-8″ – 54″ _____ (feet)

8. 9′-3″ – 89″ _____ (feet)

9. 17′-8″ – 166″ _____ (feet)

10. 54′-3″ – 347″ _____ (feet)

MULTIPLYING COMBINED FEET-AND-INCHES MEASUREMENTS

To multiply combined feet-and-inches measurements, multiply the feet and inches separately. **(See Example 9.)** If the product of the inches is 12 or more, change the inches to a feet-and-inches total. Add this quantity to the product of the feet. **(See Example 10.)**

Example 9: Multiply 4 times 5′-2″.

Step 1. Place the multiplier (4) below the 5′-2″ dimension, and multiply the feet and inches separately.

```
  5′-2″
×   4
 20′-8″
```

Final Answer: 20′-8″

Example 10: Multiply 8 times 7′-5″.

Step 1. Place the multiplier (8) below the 7′-5″ dimen-

sion, and multiply the feet and inches separately.

```
  7′-5″
×   8
 56′-40″
```

Step 2. Change the inches product (40″) to a feet-and-inches product.

```
        3 - 4
12 ) 40
       36
        4
```

40″= 3′-4″

Step 3. Add 3′-4″ to the feet product (56′).

```
  56′-0″
+ 3′-4″
 59′-4″
```

Final Answer: 59′-4″

QUIZ 27- 5

Multiply the following combined feet-and-inches measurement.

Answer

1. 3 × 11′-3″ _____

2. 2 × 2′-5″ _____

3. 4 × 12′-5″ _____

4. 7 × 6′-2″ _____

5. 6 × 9′-3″ _____

6. 8 × 13′-7″ _____

7. 13 × 11′-4″ _____

8. 15 × 16′-8″ _____

9. 9 × 27′-6″ _____

10. 21 × 43′-3″ _____

DIVIDING COMBINED FEET-AND-INCHES MEASUREMENT

To divide inches and feet separately by a whole number, follow the established procedure for dividing whole numbers. Very often it is necessary to divide combined feet and inches by a whole number.

First divide the feet of the dimension, and then divide the inches. **(See Example 11).** If there is a remainder after dividing the feet, change the remainder to inches, and add this to the inches of the dimension. **(See Example 12.)**

Example 11: Divide 8'-4" by 2.

Step 1.
A. Place the dimension (8'-4") in a frame, and place the divisor (2) to the left of the frame.
B. Divide the feet and inches separately.

$$\begin{array}{r} 4\text{-}2 \\ 2\,\overline{)\,8'\text{-}4''} \\ \underline{8\ \ 4} \\ 0\ \ 0 \end{array}$$

Final Answer: 4'-2"

Example 12: Divide 25'-6" by 12.

Step 1.
A. Place the dimension (25'-6") in a frame, and place the divisor (12) to the left of the frame.
B. Divide the 25' of the dimension by 12. The answer is 2 with a remainder of 1'.
C. Place the 2 over the 25' part of the dimension.

$$\begin{array}{r} 2 \\ 12\,\overline{)\,25'\text{-}6''} \\ \underline{24} \\ 1' \leftarrow \textbf{ remainder} \end{array}$$

Step 2.
A. Change the 1' remainder to 12".
B. Add 12" to the 6" quantity of the dimension. producing an 18" total.

$$\begin{array}{r} 2 \\ 12\,\overline{)\,25'\text{ - }6''} \\ \underline{24'\ 12''} \leftarrow \textbf{ 1' remainder added to 6''} \\ 1'\ 18'' \leftarrow \textbf{ total} \end{array}$$
remainder ↑

Step 3. Divide the inch total (18) by 12.

$$\begin{array}{r} 1\ 6/12 = 1\,1/2 \\ 12\,\overline{)\,18} \\ \underline{12} \\ 6 \end{array}$$

Step 4. Place the 1 1/2" over the 6" part of the number being divided (25'-6").

$$\begin{array}{r} 2'\text{-}\quad 1\,1/2'' \\ 12\,\overline{)\,25'\text{-}\quad 6''} \\ \underline{24'\quad 12''} \leftarrow \text{ 1' remainder added to 6''} \\ 1'\quad 18'' \leftarrow \text{ divided by 12 = 1 1/2''} \end{array}$$
remainder ↑

Final Answer: 2'-1 1/2"

QUIZ 27-6

Divide the following combined feet-and-inches measurements by the whole numbers indicated:

		Answer
1.	45'-10" ÷ 5	_____
2.	32'-8" ÷ 4	_____
3.	24'-6" ÷ 7	_____
4.	45'-10" ÷ 11	_____
5.	73'-6" ÷ 6	_____
6.	62'-6" ÷ 30	_____
7.	72'-4" ÷ 14	_____
8.	51'-5" ÷ 8	_____
9.	83'-7" ÷ 9	_____
10.	167'-9" ÷ 22	_____

UNIT 28

A conventional ruler can be used to measure or draw a line to scale by multiplying the scale times the total dimension of the line. **(See Examples 1 and 2).**

Example 1: A drawing shows a 15'-0" line drawn to 1/2"=1'-0" scale. What will this measure on a conventional ruler?

Step 1. Multiply 1/2 times 15.

$$\frac{1}{2} \times \frac{15}{2} = 15 = 7\ 1/2$$

Final Answer: 7 1/2"

Example 2: A drawing shows a 25'-8" line drawn to 1/8" = 1'-0" scale. What will this measure on a conventional ruler?

Step 1. Change the 8" to a decimal fraction of a foot.

$$\frac{8}{12} = \frac{2}{3}$$

Step 2. Change the common fraction (2/3) to a decimal fraction.

```
      .666
3 ) 2.000
    1 8
     20
     18
     20
     18
```

2/3 = .666

Step 3. Multiply 25.666 times 1/8.

$$25.666 \times \frac{1}{8} = \frac{25.666}{8}$$

```
      3.207
8 ) 25.6660
    24
    1 6
    1 6
      06
      00
      60
      56
```

Step 4. Change the decimal remainder (.207) to a common fraction (.207 × 16).

```
   .207
 ×  16
  1242
   207
  3.312  = 3/16
```

Step 5. Combine the 3/16" common fraction with the whole number 3.

```
   3
 +  3/16
   3  3/16
```

Final Answer: 3 3/16"

QUIZ 28-1

Using the following scales and dimensions, find the measurements on a conventional ruler.

		Answer
1. 18'-0" @ 1/4" scale		_____
2. 25'-0" @ 3/4" scale		_____
3. 13'-0" @ 3/16" scale		_____
4. 1'-0" @ 3/8" scale		_____

5. 60'-0" @ 3/32" scale   _____

6. 9'-0" @ 1 1/2" scale   _____

7. 47'-0" @ 1/8" scale   _____

8. 17'-0" @ 3/4" scale   _____

9. 53'-0" @ 3/16" scale   _____

10. 29'-0" @ 1/4" scale   _____

11. 8'-6" @ 3/8" scale   _____

12. 10'-4" @ 1/2" scale   _____

13. 17'-8" @ 3/4" scale   _____

14. 23'-2" @ 1/8" scale   _____

15. 21'-5" @ 1/4" scale   _____

16. 7'-3" @ 1 1/2" scale   _____

17. 19'-7" @ 3/16" scale   _____

18. 35'-10"@ 3/32" scale   _____

19. 43'-11"@ 3/8" scale   _____

20. 18'-1" @ 1 1/2" scale   _____

UNIT 29

Rise and Run of Roofs

FINDING TOTAL RISE OF ROOFS

The total rise (height) of a sloping roof is found by multiplying the total run times the number of inches in the unit rise. In Figure 29–1 the total run is 4'-0" and the unit rise is 8". Therefore, the total rise is 32" (4 × 8' = 32"). **(See Examples 1 and 2.)**

Example 1: Find the total rise of a roof with a total span of 22'-0" and an 7" unit rise

Step 1. Divide the total span by 2 in order to find the total run.

$$2 \overline{) \underset{22'\text{-}0''}{\overset{11'\text{-}0'' \leftarrow \textbf{ total run}}{}}}$$

Step 2. Multiply the total run (11) times the unit rise (7").

$$\begin{array}{r} 11 \\ \times 7 \\ \hline 77'' \end{array}$$

Step 3. Change 77" to a feet-and-inches measurement.

$$12 \overline{) \overset{6\text{ - }5}{77}} \\ \underline{72} \\ 5$$

Final Answer: 6'-5"

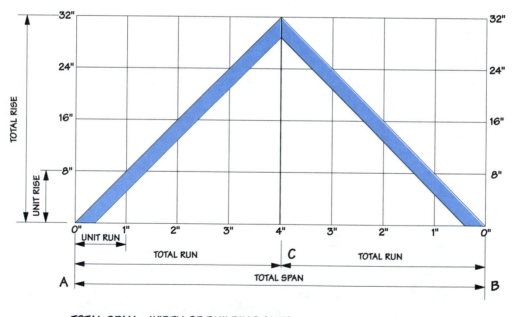

TOTAL SPAN = WIDTH OF BUILDING (A-B)
TOTAL RUN = ONE HALF OF SPAN (A-B & C-B)
UNIT RUN = 12" (ALWAYS)
UNIT RISE = NUMBER OF VERTICAL INCHES ROOF RISES PER FOOT OF RUN
TOTAL RISE = TOTAL HEIGHT OF ROOF

Figure 29–1.
Total run times unit rise of the roof equals the total rise.

EXAMPLE 2: Find the total rise of a roof with a total span of 17'-4" and an 11" unit rise.

Step 1. Divide the total span by 2.

$$2 \overline{)\ \overset{8}{17'\text{-}4''}}$$
$$\underline{16}$$
$$1 \ \leftarrow \ \textbf{1' remainder (12")}$$

Step 2.
A. Change the 1' remainder to 12".
B. Add the 12" to the 4" above, and divide by 2.

$$2 \overline{)\ \overset{8'\ \text{-}\ \ 8''}{17'\ \text{-}\ \ 4''}}$$
$$\underline{16 \text{-} 12} \ \leftarrow \ \textbf{Add 1' (12") TO 4"}$$
$$1 \text{-} 16 \ \leftarrow \ \textbf{Divide total by 2 (16 ÷ 2 = 8)}$$

8'-8" = TOTAL RUN

Step 3. Change the 8" To a decimal fraction of a foot.

$$\frac{8}{12} = \frac{2}{3}$$

$$3 \overline{)\ \overset{.666}{2.000}}$$
$$\underline{1\ 8}$$
$$20$$
$$\underline{18}$$
$$20$$
$$18$$

8" = .666'

Step 4. Multiply the total run (8.666) times the unit rise (11").

$$\begin{array}{r} 8.666 \\ \times\ \ \ \ 11 \\ \hline 8666 \\ 8666\ \ \\ \hline 95.326'' \end{array}$$

Step 5: Change .326" to a common fraction.

$$\begin{array}{r} .325 \\ \times\ \ \ 16 \\ \hline 1956 \\ \underline{326\ \ } \\ 5.216 = 5/16'' \end{array}$$

Step 6. Combine 95" with 5/16"

$$\begin{array}{r} 95 \\ +\ \ \ \ 5/16 \\ \hline 95\ 5/16'' = \text{total rise} \end{array}$$

Step 7. Change 95 5/16" to a feet-and-inches measurement.

$$12 \overline{)\ \overset{7'\text{-}11\ 5/16}{95}}$$
$$\underline{84}$$
$$11$$

Final Answer: 7'-11 5/16" total rise

QUIZ 29-1

Find the total rise of the following roofs:

	Unit Rise	Span	Total Rise
1.	8"	20'-0"	_____
2.	16"	24'-0"	_____
3.	12"	26'-0"	_____
4.	6"	18'-0"	_____
5.	18"	22'-0"	_____
6.	16"	28'-0"	_____
7.	12"	30'-0"	_____
8.	8"	42'-0"	_____
9.	4"	19'-0"	_____
10.	10"	13'-0"	_____
11.	6"	12'-0"	_____
12.	8"	24'-6"	_____
13.	4"	30'-6"	_____

	Unit Rise	Span	Total Rise
14.	16″	18′-0″	_____
15.	9″	12′-4″	_____
16.	8″	21′-6″	_____
17.	18″	33′-8″	_____
18.	7″	51′-4″	_____
19.	11″	50′-10″	_____
20.	17″	63′-6″	_____

UNIT 30

Stair Dimensions

CALCULATING RISERS AND TREADS

The calculation of stair dimensions is based on the total rise and total run of the stairway. **(See Figure 30–1.)** Preferred riser heights range from 7″ to 7 1/2″. Recommended tread widths range from 9″ to 12″ **(See Examples 1 and 2.)**

Example 1: What are the tread and riser dimensions for a stairway with a 6′-0″ total rise and an 8′-9″ total run?

Step 1. Change the total rise (6′-0″) to all inches.

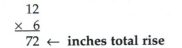

$$\begin{array}{r} 12 \\ \times\ 6 \\ \hline 72 \end{array} \leftarrow \textbf{inches total rise}$$

Step 2. Divide the total rise (72″) by 7 to see how many even times 7 will go into the total rise. The 72″ total rise divided by 7 is 10 with a remainder of 2″. Disregard the remainder.

$$\begin{array}{r} 10 \leftarrow \textbf{number of risers} \\ 7\,\overline{)\,72} \\ \underline{70} \\ 2 \leftarrow \textbf{disregard remainder} \end{array}$$

Step 3. Divide the total rise by the number of even risers (10). Work the problem to a decimal remainder.

$$\begin{array}{r} 7.2 \\ 10\,\overline{)\,72.0} \\ \underline{70} \\ 20 \\ 20 \end{array}$$

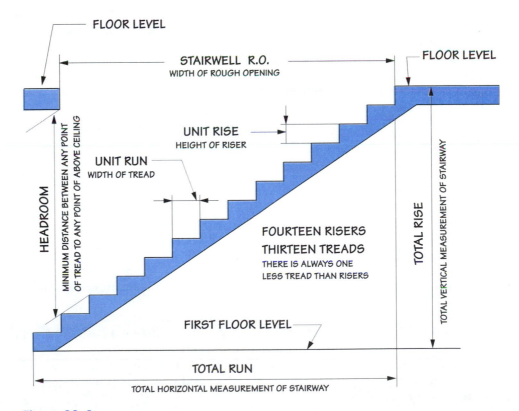

Figure 30–1.
The total rise and run of a stairway are the keys to finding the measurements of risers and treads.

Step 4.
  A. Change the decimal-fraction remainder (.2) to a common-fraction remainder.
  B. Multiply the decimal fraction by 16.

$$
\begin{array}{r}
16 \\
\times\ .2 \\
\hline
3.2 \leftarrow \textbf{ sixteenths}
\end{array}
$$

Step 5.
  A. The .2 remainder is less than 5; therefore, the decimal fraction 3.2 equals 3/16".
  B. Combine the 7" and the 3/16" fraction.

$$
\begin{array}{r}
7 \\
+\qquad 3/16 \\
\hline
7\ \ 3/16 \leftarrow \textbf{ inches riser height}
\end{array}
$$

Step 6. Change the 8'-9" total run measurement to all inches.

$$
\begin{array}{r}
12 \\
\times\ 8 \\
\hline
96 \\
+\ 9 \\
\hline
105 \leftarrow \textbf{ inches total run}
\end{array}
$$

Step 7.
  A. Divide the total run measurement (105") by the number of treads (9) . Remember, there is always one less tread than risers.
  B. Work the decimal remainder to three places and round off to two places.

$$
\begin{array}{r}
11.666 = 11.67'' \\
9\,)\overline{\,105.000} \\
\underline{9}\qquad\qquad \\
15\qquad\quad \\
\underline{9}\qquad\quad \\
60\qquad \\
\underline{54}\qquad \\
60\quad \\
\underline{54}\quad \\
60 \\
\underline{54} \\
6
\end{array}
$$

Step 8.
  A. Change the decimal-fraction remainder (.67) to a common-fraction remainder.
  B. Multiply the decimal fraction times 16.

$$
\begin{array}{r}
.67 \\
\times\ 16 \\
\hline
402 \\
67\ \ \\
\hline
10.72 \leftarrow \textbf{ sixteenths.}
\end{array}
$$

Step 9. The number .72 is more than half of a sixteenth; therefore, raise the 10/16 to 11/16.

$$
\begin{array}{r}
10/16 \\
+\ \ 1/16 \\
\hline
11/16
\end{array}
$$

Step 10. Combine the 11" whole number with the fraction 11/16.

$$
\begin{array}{r}
11 \\
+\qquad 11/16 \\
\hline
11\ \ 11/16 \leftarrow \textbf{ inches of tread width}
\end{array}
$$

Final Answer: 7 3/16" riser height and 11 11/16" tread width

Example 2. What are the tread and riser dimensions for a stairway with a 7'-4" total rise and a 9'-8 5/8" total run.

Step 1. Change the total rise (7'-4") to all inches.

$$
\begin{array}{r}
12 \\
\times\ 7 \\
\hline
84 \\
+\ 4 \\
\hline
88 \leftarrow \textbf{ inches total rise}
\end{array}
$$

Step 2. Divide the total rise (88") by 7 to see how many even times 7 will go into the total rise.

$$
\begin{array}{r}
12 \leftarrow \textbf{ number of risers} \\
7\,)\overline{\,88}\qquad\qquad\qquad\quad \\
\underline{7}\qquad\qquad\qquad\qquad\ \\
18\qquad\qquad\qquad\quad \\
\underline{14}\qquad\qquad\qquad\quad \\
4 \leftarrow \textbf{ disregard remainder}
\end{array}
$$

Step 3.
  A. Divide the total rise (88") by the number of risers (12).
  B. Work the problem to three decimal remainders and round off to two places.

```
    7.333 = 7.33"
12 ) 88.000
     84
     40
     36
     40
     36
     40
     36
      4
```

```
      10.602 = 10.6
11 ) 116.630
     11
     06
     00
     66
     66
     03
     00
     30
     24
      6
```

Step 4.
 A. Change the decimal-fraction remainder (.33) to a common-fraction remainder.
 B. Multiply the decimal fraction times 16.

```
    .33
  × 16
   198
    33
  5.28 ← sixteenths
```

Step 5.
 A. The .28 remainder is less than 5; therefore, the decimal fraction 5.28 equals 5/16".
 B Combine the 7" and 5/16 inch fraction.

```
    7
  +    5/16
    7  5/16 ← inches riser height
```

Step 6. Change the total run measurement (9'-8 5/8") to all inches.

```
     12
   ×  9
    108
  +    8 5/8
    116 5/8 ← inches total run
```

Step 7.
 A. Change the 5/8" common fraction to a decimal fraction.
 B. Work the decimal remainder to to three places and round off to two places.

```
     .625 = .63
  8 ) 5.000
     4 8
     20
     16
     40
     40
```

Step 8. Divide the numbers of inches in the total run (116.623) by the number of treads (11).

Step 9.
 A. Change the decimal-fraction remainder (.6) to a common-fraction remainder.
 B. Multiply the decimal fraction by 16.

```
     16
  ×  .6
    9.6 ← sixteenths
```

Step 10. As .6 is greater than one half of a sixteenth, raise the 9/16 to 10/16.

```
    9/16
  + 1/16
    10/16 = 5/8
```

Step 11. Combine the whole number (10) with the fraction (5/8).

```
    10
  +    5/8
    10  5/8 ← inches tread
```

Final Answer: 7 5/16" riser height and 10 5/8" tread width

QUIZ 30-1

Find the riser heights and tread widths for the following stairways. Work decimal remainders to three places if necessary and round off to two places.

	Total Rise	Total Run	Riser	Tread
1.	5'-8"	7'-0"	_____	_____
2.	6'-5"	8'-0"	_____	_____
3.	9'-8"	13'-9"	_____	_____
4.	8'-4"	11'-3"	_____	_____

Total Rise	Total Run	Riser	Tread
5. 9′-9″	12′-6″	_____	_____
6. 5′-8″	7′-0″	_____	_____
7. 9′-2″	12′-2″	_____	_____
8. 5′-1″	5′-11″	_____	_____
9. 5′-11 1/4″	8′-2″	_____	_____
10. 6′-5″	9′-10 3/4″	_____	_____
11. 7′-3 1/2″	8′-8 1/2″	_____	_____
12. 10′-4″	15′-2″	_____	_____
13. 7′-8 5/8″	12′-3″	_____	_____
14. 5′-1 1/2″	6′-0 5/8″	_____	_____
15. 6′-9 1/8″	7′-8 1/2″	_____	_____
16. 5′-7 1/2″	8′-1 1/2″	_____	_____
17. 8′-9 7/8″	12′-10 3/8″	_____	_____
18. 12′-5 5/8″	15′-10 1/8″	_____	_____
19. 9′-6 3/8″	11′-10 1/2″	_____	_____
20. 13′-10 11/16″	20′-8 3/4″	_____	_____

UNIT 31

Metric Conversion

English-metric conversion can be done by using the formulas in **Table 31-1. (See Examples 1 through 8.)**

To Convert	Multiply	
1. inches to millimeters	inches times	25.4
2. millimeters to inches	millimeters times	0.0394
3. inches to centimeters	inches times	2.54
4. centimeters to inches	centimeters times	0.394
5. inches to meters	inches times	0.0254
6. meters to inches	meters times	39.37
7. feet to meters	feet times	0.3048
8. meters to feet	meters times	3.2808

Table 31-1
Frequently Used English-Metric Linear Conversions

CONVERTING INCHES TO MILLIMETERS

To convert inches to millimeters (mm), multiply the inches times 25.4.

Example 1. Convert 39" to millimeters.

```
    25.4
 ×   39"
  2286
   762
  990.6 mm
```

QUIZ 31-1

Convert the following inches to millimeters.

Inches	Millimeters
1. 17	_____
2. 25	_____
3. 34.5	_____
4. 87.25	_____
5. 246.33	_____

CONVERTING MILLIMETERS TO INCHES

To convert millimeters to inches, multiply the millimeters by .0394.

Example 2. Convert 1130 mm to inches.

```
    .0394
 × 1130  millimeters
   0000
   1182
   394
   394
  44.522"
```

QUIZ 31-2

Convert the following millimeters to inches.

Millimeters	Inches
1. 920	_____
2. 1470	_____
3. 632	_____
4. 3804	_____
5. 16351	_____

CONVERTING INCHES TO CENTIMETERS

To convert inches to centimeters (cm), multiply the inches times 2.54.

Example 3. Convert 27" to centimeters.

```
   2.54
 × 27  inches
  1778
   508
  68.58 cm
```

Quiz 31-3

Convert the following inches to centimeters.

	Inches	Centimeters
1.	14	_____
2.	31	_____
3.	19.33	_____
4.	21.5	_____
5.	42.625	_____

Converting Centimeters to Inches

To convert centimeters to inches, multiply the centimeters times .394.

Example 4. Convert 89 cm to inches.

$$
\begin{array}{r}
.394 \\
\times\,89 \text{ cm} \\
\hline
3546 \\
3152 \\
\hline
35.066 = 35.07''
\end{array}
$$

Quiz 31-4

Convert the following centimeters to inches.

	Centimeters	Inches
1.	32	_____
2.	69	_____
3.	81	_____
4.	102	_____
5.	73	_____

Converting Inches to Meters

To convert inches to meters (m), multiply the inches times .0254.

Example 5. Convert 57″ to meters.

$$
\begin{array}{r}
.0254 \\
\times\,57 \text{ inches} \\
\hline
1778 \\
1270 \\
\hline
1.4478 = 1.45 \text{ m}
\end{array}
$$

Quiz 31-5

Convert the following inches to meters.

	Inches	Meters
1.	39	_____
2.	63	_____
3.	21	_____
4.	179	_____
5.	14	_____

Converting Meters to Inches

To convert meters to inches, multiply the meters times 39.37.

Example 6. Convert 2.5 m to inches.

$$
\begin{array}{r}
39.37 \\
\times\,2.5 \text{ m} \\
\hline
19685 \\
7874 \\
\hline
98.425 = 98.43''
\end{array}
$$

Quiz 31-6

Convert the following meters to inches.

	Meters	Inches
1.	4	_____
2.	3	_____
3.	1.75	_____
4.	5.33	_____
5.	9.25	_____

CONVERTING FEET TO METERS

To convert feet to meters, multiply the feet times .3048.

Example 7.   Convert 29′ to meters.

```
    .3048
  ×  29′
  27432
  6096
  8.8392 = 8.84 m
```

QUIZ- 31-7

Convert the following feet to meters.

Feet	Meters
1. 35	_____
2. 14	_____
3. 73	_____
4. 1.25	_____
5. 48.5	_____

CONVERTING METERS TO FEET

To convert meters to feet, multiply the numbers of meters times. 3.2808.

Example 8.   Change 17 m to feet.

```
    3.2808
  ×  17 m
  229656
  32808
  55.7736 = 55.77′
```

QUIZ 31-8

Convert the following meters to feet.

Meters	Feet
1. 13	_____
2. 25	_____

Meters	Feet
3. 43.8	_____
4. .25	_____
5. 93	_____

CONVERTING ENGLISH TO METRIC MEASUREMENT

English-metric conversion can also be done through the the use of available tables and mathematical calculations. **(See Examples 9 and 10.)**

Example 9: Change the English measurement 5′-7″ to metric measurement in millimeters.

Step 1. Place the 5′ measurement in a frame.

```
  )5
```

Step 2. Change the 7″ measurement to a decimal fraction of a foot.

```
         .5833
  12 ) 7.0000
       6 0
       1 00
         96
         40
         36
         40
         36
```

Step 3. Place the decimal fraction of a foot (.5833) next to the 5′ in the frame. Divide by the number 3.281 (There are 3.281 feet in a meter.) Work the decimal remainder to four places and round off to three places.

```
              1.7017
  3.281 ) 5.5833.00
          3 281
          2 3023
          2 2967
            5600
            3281
           23190
           22967
              23
```

Step 4. Multiply 1.7017 meters by 1000 to change the meters to millimeters, by moving the decimal three places to the right.

1.7017 meters = 1701 millimeters

Final Answer: 5'-7" = 1701 millimeters

Example 10: Change the English measurement 12'- 7 3/8" to its metric measurement in millimeters.

Step 1. Place the 12' measurement in a frame.

$$\overline{)12}$$

Step 2. Change 7 3/8" to a decimal fraction of a foot.
  A. Divide 3 by 8.

```
       .375
   8 ) 3.000
       2 4
        60
        56
         40
         40
```

  B. Divide .375 by 12. Work the problem to three decimal places.

```
         .614
   12 ) 7.375
        7 2
         17
         12
          55
          48
           7
```

Step 3. Place the decimal fraction of a foot, .614, next to the 12' in the frame. Divide by the number 3.281. (There are **3.281 feet** in **1 meter.**) Work the decimal remainder to three places and round off to two places.

```
              3.844  = 3.84
   3.281 ) 12.614.000
    ↑   ↑    ↑   ↑
           9 843
           2 7710
           2 6248
            14620
            13124
            14960
            13124
             1836
```

12.614' = 3.84 m

Step 4. Multiply the 3.84 meters by 1000 to change the meters to millimeters, moving the decimal point three places to the right.

   3.840 m = 3840 mm

Final Answer: 12'-7 3/8" = 3840 mm

Quiz 31-9

Change the following feet-and-inches measurements to millimeters.

Feet and Inches	Millimeters
1. 4'-5"	_____
2. 9'-3"	_____
3. 13'-8"	_____
4. 6'-4 1/2"	_____
5. 9'-2 1/4"	_____
6. 11'-7 1/8"	_____
7. 18'-6 5/8"	_____
8. 21'-8 3/4"	_____
9. 32'-4 3/8"	_____
10. 45'-11 7/8"	_____

SECTION 6 TEST 1

Related Mathematics

Convert the following inches to feet.

Inches	Feet
1. 36″	_____
2. 84″	_____
3. 92″	_____
4. 45″	_____
5. 138″	_____

Convert the following feet to inches.

Feet	Inches
6. 4′-0″	_____
7. 9′-0″	_____
8. 2′-11″	_____
9. 8′-5″	_____
10. 31′-7″	_____

Convert the following common fractions to decimal fractions. (Round off the decimal remainder to number of places indicated.)

Common	Decimal	
11. 3/8	_____	(3 places)
12. 3/4	_____	(2 places)
13. 13/16	_____	(4 places)
14. 1/2	_____	(1 place)
15. 1/8	_____	(3 places)

Convert the following mixed numbers to whole numbers with a decimal remainder. (Round off decimal remainders to number of places indicated.)

Common	Decimal	
16. 5 1/4	_____	(2 places)
17. 9 5/8	_____	(3 places)
18. 11 1/2	_____	(1 place)
19. 21 11/16	_____	(4 places)
20. 13 7/8	_____	(3 places)

Convert the following decimal inch fractions to the nearest common fractions.

Decimal	Common
21. .33″	_____
22. .58″	_____
23. .235″	_____
24. .87″	_____
25. .597″	_____

Convert the following inches to decimal fractions of a foot. (Work the decimal answer to four places, if necessary.)

Inches	Decimal
26. 8″	_____
27. 10″	_____
28. 4 1/2″	_____
29. 9 1/8″	_____
30. 6 9/16″	_____

Convert the following decimal fractions of a foot to inches.

Decimal of a Foot	Inches
31. .15′	_____
32. .66′	_____
33. .37′	_____

34. .093′ _____

35. .9648′ _____

Calculate the following linear measurements.

Answer

36. 3′-2″ + 4′-5″ _____

37. 23′-9″ + 9′-8″ _____

38. 18′-5″ + 8′-7″ + 32′-11″ _____

39. 4′-9″ + 35″ _____ (in.)

40. 15′-3″ + 17″ + 8′-7″ _____ (ft.)

41. 12″ + 11′-8″ + 45′-6″ + 47″ _____ (ft.)

42. 10′-9″ − 7′-5″ _____

43. 34′-2″ − 29′-8″ _____

44. 103′-7″ − 83′-11″ _____

45. 57″ − 3′-6″ _____ (in.)

46. 25′-3″ − 76″ _____ (ft.)

47. 141″ − 9′-11″ _____ (ft.)

48. 5 × 9′-2″ _____

49. 8′ × 6′-9″ _____

50. 14 × 7′-3″ _____

51. 12′-9″ ÷ 3 _____

52. 74′-4″ ÷ 8 _____

53. 133′-6″ ÷ 24 _____

54. 102′-9″ ÷ 9 _____

SECTION 6 TEST 2

Related Mathematics

Compute the following scales for rule measurements.

Answer

1. 19'-0" @ 1/2" scale _____

2. 31'-0" @ 3/16" scale _____

3. 7'-9" @ 3/4" scale _____

4. 28'- 5" @ 3/32" scale _____

5. 17'-11" @ 1 1/2" scale _____

Compute the following total roof rises.

	Unit Rise	Span	Total Rise
6.	6"	22'-0"	_____
7.	4"	18'-0"	_____
8.	9"	36'-0"	_____
9.	14"	28'-0"	_____
10.	16"	34'-0"	_____
11.	8"	19'-0"	_____
12.	12"	41'-0"	_____
13.	7"	18'-6"	_____
14.	5"	51'-4"	_____
15.	10"	75'-8"	_____

Compute the following stair riser and tread measurements.

	Total Rise	Total Run	Riser	Tread
16.	5'-0"	5'-10"	_____	_____
17.	6'-5"	8'-0"	_____	_____
18.	9'-8"	13'-9"	_____	_____
19.	8'-4"	11'-3"	_____	_____
20.	9'-2"	12'-2"	_____	_____
21.	5'-11 1/4"	8'-2"	_____	_____
22.	5'-5 3/8"	6'-10 3/4"	_____	_____
23.	5'-8 5/8"	7'-8 1/4"	_____	_____
24.	3'-7 1/8"	4'-4 1/2"	_____	_____
25.	8'-9 3/16"	12'-4 5/8"	_____	_____

Convert the following feet-and-inches measurements to millimeters. (There are 3.281 feet in 1 meter.)

	Feet and Inches	Millimeters
26.	5'-4"	_____
27.	7'-9"	_____
28.	15'-6"	_____
29.	4'-7 1/2"	_____
30.	8'-3 1/4"	_____
31.	19'-7 5/8"	_____
32.	22'-5 3/4"	_____
33.	35'-2 3/8"	_____
34.	47'-9 7/8"	_____
35.	72'-11 1/4"	_____

SECTION
7 | Materials

Building plans and specifications mention types and sizes of materials used at all stages of the construction work. The *type* of material determines its structural value and/or its appearance. The *size* of the material determines dimensions and sizes of building components. Plan reading requires a knowledge of both the types and sizes of the materials being used. Space does not allow for a description here of all the materials that can be used in construction. However, this section will discuss some of the most commonly used materials found in residential and commercial work, as well as all the materials identified in the plans featured in this book. ■

UNIT 32

Masonry

Reinforced concrete and concrete masonry units (CMUs) are used to build both foundation systems and exterior and interior walls of buildings. Foundations may be constructed entirely of concrete, or they may consist of concrete footings supporting CMU walls. Clay masonry units (brick and tile) are used for both solid and veneer walls.

CONCRETE

Concrete is used in the erection of foundation footings and walls, entire buildings, bridges, dams, and many other types of heavy construction work. It is a mixture of cement, aggregate, and water. The manufacture of cement begins with material crushed from limestone. Other materials are blended into the limestone, and then the mixture is processed into a fine powder. **Aggregate** consists of *fine* particles of sand and coarse gravel or crushed rock. When water is added to the cement and aggregates, the cement acts as a paste that binds the mixture together in a chemical process called **hydration.** The proportion of cement, fine and coarse aggregates, and water in a batch of concrete is called the **concrete mix.** Cement makes up the smallest portion of a concrete mix. Different types of structures and related conditions require different proportions in the concrete mix. Information regarding the mix for a particular project is usually found in the specifications and/or notes included with a set of plans. Except when very small quantities are required, **ready-mixed concrete** is used on construction projects. Ready-mixed concrete is manufactured in a *batch plant* and delivered by truck to the job site. **(See Figure 32-1.)**

CONCRETE MASONRY UNITS

Concrete masonry units (CMUs), used in the construction of both foundations and walls, are manufactured from a concrete mixture and are also referred to as **concrete blocks.** They come in many shapes and standard sizes. When placed, they are bonded with mortar. Although solid concrete bricks are available, most concrete blocks have a hollow core. For greater strength, vertical rebars can be placed through the cavities. Horizontal rebars or ladder-and-truss ties can be placed in the horizontal mortar joints. **Grout** is then placed in the hollow cores of the blocks. Grout is a concrete mix with a consistency that allows it to be poured easily into the hollow cores. **(See Figure 32–2** and **Figure 32–3.)**

CLAY MASONRY UNITS

Clay masonry units are usually *solid brick* or *hollow tile.* Another product, *terra cotta,* has been popular in the past for ornamental and veneer purposes but is not often used today. The manufacture of clay masonry begins with the extraction of clay from open pits or mines. The clay is broken down, pulverized, and combined with other ingredients while subjected to a heating process. Clay bricks may be solid or hollow and come in many surface textures. Clay tiles lend themselves to a greater variety of sizes and shapes. Both are available in different standard dimensions and can be used for solid or veneer wall construction. **(See Figure 32–4.)**

Figure 32–1.
Ready-mixed concrete has been delivered by truck and is being placed for a floor slab. *(Courtesy of Portland Cement Association)*

STONE MASONRY

Stone masonry is natural rock taken from the earth and broken down into sizes suitable for construction. It may be placed in mortar much like clay bricks or concrete blocks to build a solid wall or a veneer wall. The two major classifications are rubble and ashlar. **Rubble masonry** is made of random-shaped pieces of rock. **Ashlar masonry** consists of squared pieces of rock. In *coursed* stone masonry the stone is laid in a continuous horizontal line. *Uncoursed* stones are placed in an irregular pattern. **(See Figure 32–5.)**

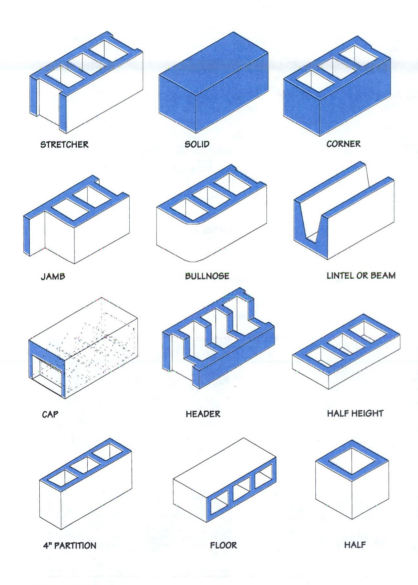

STRETCHER   SOLID   CORNER

JAMB   BULLNOSE   LINTEL OR BEAM

CAP   HEADER   HALF HEIGHT

4" PARTITION   FLOOR   HALF

Figure 32–2.
Concrete masonry units are produced in a variety of shapes and sizes.

Figure 32–3.
A partially completed CMU wall is shown above. Vertical rebars project from the hollow cores that are to be filled with grout. *(Courtesy of Portland Cement Association)*

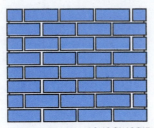

SECTION OF WALL CONSTRUCTED OF
STANDARD MODULAR 3 3/4" THICK,
2 1/4" HIGH, AND 8" LONG BRICK.

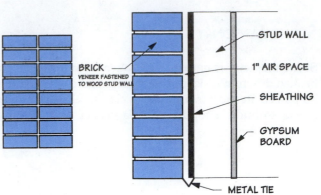

BRICK
VENEER FASTENED
TO WOOD STUD WALL

STUD WALL

1" AIR SPACE

SHEATHING

GYPSUM
BOARD

METAL TIE

SOLID CLAY BRICK

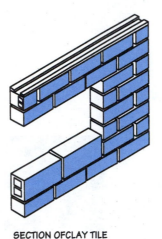

SECTION OF CLAY TILE
WALL WITH WINDOW
OPENING

12 X 12

12 X 12

8 X 12

"SHOWN HERE ARE A FEW EXAMPLES
OF STRUCTURAL CLAY TILE DESIGN.
THICKNESSES VARY FROM 6" TO 12"

STRUCTURAL CLAY TILE

Figure 32–4.
Solid clay bricks and hollow clay tiles are used as structural units or as exterior veneer.

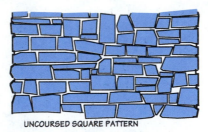

UNCOURSED SQUARE PATTERN

RANDOM BROKEN COURSED PATTERN

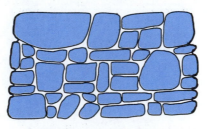

UNCOURSED FIELDSTONE PATTERN

COURSED PATTERN

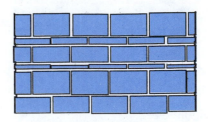

Figure 32–5.
Rubble and ashlar stone masonry can be laid in coursed and uncoursed patterns.

RUBBLE STONE MASONRY
RANDOM SHAPED PIECES OF ROCK

ASHLAR STONE MASONRY
SQUARED PIECES OF ROCK

Review Questions

Enter the missing words in the spaces provided on the right.

1. _____, _____, and _____ make up the con-   _____
   crete mixture.

2. Aggregate consists of _____ and _____.   _____

3. A commonly used trade term for steel reinforce-   _____
   ment is _____.

4. _____ (use abbreviation) is used to reinforce con-   _____
   crete slabs.

5. The technical term for concrete blocks is _____   _____
   (use abbreviation).

6. For greater strength, the hollow cores of concrete   _____
   blocks are filled with _____.

7. Terra cotta is a _____ material   _____

8. Veneer wall construction is often of clay bricks or   _____
   _____ _____.

9. _____and _____ are the two major classifica-   _____
   tions of stone masonry construction.

10. _____ stones are placed in an irregular pattern.   _____

UNIT 33

Lumber

Lumber is identified by the species of tree from which it is produced. Softwoods and hardwoods are the two major classifications of construction lumber. **Softwood** comes from *coniferous* trees (also called *evergreens*), which have needle-like or scalelike leaves during all climate seasons. About 75 percent of the total lumber manufactured is from softwood trees, and this accounts for most of the structural lumber on construction projects. Softwoods are also used as paint-grade finish materials. Some examples of frequently used softwoods are *Douglas fir, western hemlock, red spruce,* and *southern yellow pine.*

Hardwoods come from broad-leafed *(deciduous)* trees, which shed their leaves at the end of the growing season. The cell structure of hardwoods produces attractive grain patterns for stained finishes. Therefore, hardwoods are often used for interior molding, paneling, cabinets, and furniture. Some examples of hardwoods are *ash, black walnut, mahogany,* and *oak.*

STRUCTURAL LUMBER

Structural lumber consists of boards, dimension lumber, and timbers. These are identified by their *nominal* (named) dimensions. For example, the piece of lumber

called a "two by four" (2 × 4) is approximately 2" thick and 4" wide when produced in a sawmill. After the rough surface is removed in the planing mill, and the lumber has seasoned sufficiently (dried and shrunk), the *actual* dimensions of a 2 × 4 will be 1 1/2" × 3 1/2". (See **Appendix D** for a table of actual lumber sizes. See **Appendix E** for tables of metric equivalents of standard nominal and dressed lumber sizes.)

Boards are pieces of lumber less than 2" in nominal thickness and 2" or more in nominal width. They have been traditionally used for sheathing, subfloor, and form construction, although today panel products are used instead.

Dimension lumber, also considered a category of **framing lumber**, are pieces 2" to 4" in nominal thickness and 2" or more in nominal width. This includes most materials used for studs, plates, blocking, and headers for wood framed walls. The wider sizes are used for floor and ceiling joists, floor planks, and roof rafters. **Timbers,** commonly used for posts and beams, are pieces of lumber 5" or more in nominal thickness and width.

All structural lumber is graded according to its strength and appearance. (See **Appendix B** and **Appendix C** for grading information.) Building plans and/or specifications are very exacting about what lumber grades to use on the construction projects. **(See Figure 33–1.)**

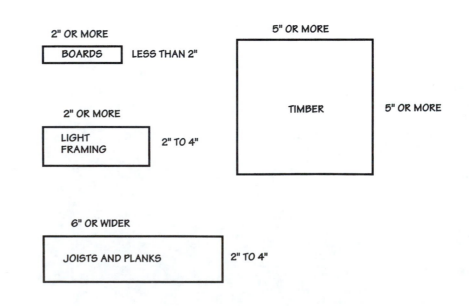

Figure 33–1.
Structural lumber includes boards and dimension lumber.

APPEARANCE LUMBER

Appearance lumber, also called *finish lumber,* is primarily determined by the look of its outer surface. Appearance lumber ranges from exposed timbers to interior and exterior molding and includes both softwood and hardwood species. **(See Figure 33–2.)** Hardwood makes a much better appearance for stained surfaces. (See **Appendix B** and **Appendix C** for appearance lumber grading systems.)

PANEL PRODUCTS

Panel products are widely used for subfloors and wall and roof sheathing. Most commonly used are plywood or reconstituted wood panels such as particleboard, waferboard, oriented strandboard, and composite panels. **Plywood** is the oldest product and is still very popular. It is made up of several wood veneer sheets placed at right angles to each other and glued together under high pressure. Structural plywood panels are also available as exterior siding featuring outside veneers of different designs and textured surfaces. Interior plywood panels with finish veneer surfaces are frequently used for interior wall paneling.

Reconstituted wood panels consist of wood particles, flakes, or strands bonded together and molded into full size sheets. **Particleboard** is made up of smaller wood chips and particles. **Waferboard** is made of large wafer-like wood flakes compressed and bonded into panels. **Oriented strandboard** is fabricated from long strand-like particles compressed and glued into

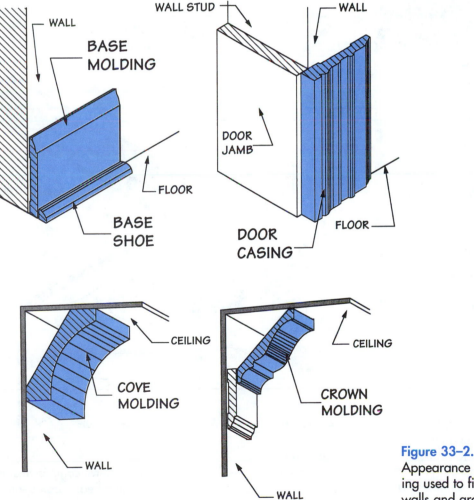

Figure 33–2.
Appearance (finish) lumber includes the molding used to finish off the tops and bottoms of walls and around door and window openings.

three to five layers. The strands in a layer are oriented in the same direction, and the layers placed at right angles to each other. **Composite** panels have plywood outside veneers glued to a core of reconstituted wood fibers. **(See Figure 33–3.)** All wood panel products are graded according to strength and appearance. **(See Appendix F** for grading information.)

PREFABRICATED COMPONENTS

Prefabricated structural wood components are built with dimension lumber, panel products, and metal fasteners. Some more frequently used components in residential and commercial construction are roof trusses, floor trusses, glued-laminated (glulam) timbers, veneered I-beams and joists, and box beams. Prefabricated components save labor and material, and make it possible to span greater distances without central support.

Roof trusses consist of top and bottom chords and webs tied with metal fasteners or plywood cleats. The *top chord* acts as a roof rafter, and the *bottom chord* serves as a ceiling joist. The *webs,* arranged in a network of triangles, tie the chords into a rigid and self-supporting unit. **Floor trusses** are similar in design to roof trusses, but both top and bottom chords are flat.

Glued-laminated (glulam) timbers are made up of several pieces of lumber joined with a very strong adhesive. Glulam timbers can be fabricated in curved and tapered shapes. **Veneered joists** are made of softwood veneers glued together. **Veneered I-beams** consist of veneer pieces glued together. The top and bottom flanges are notched to receive the edges of the veneer pieces. **Box beams** are fabricated with plywood sides nailed to top and bottom chords held together with vertical stiffeners. **(See Figure 33–4.)**

Oriented Strand Board
Composite Board
Textured Ply with Grooves
Plywood

Figure 33–3.
Besides plywood, other panel products, such as oriented strandboard and composite board, are used for wall and roof sheathing and subfloor. Structural textured panels for finish siding are also widely used. *(Courtesy of APA—An Engineered Wood Association)*

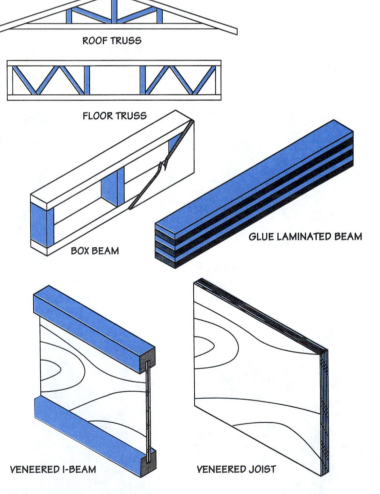

ROOF TRUSS

FLOOR TRUSS

BOX BEAM

GLUE LAMINATED BEAM

VENEERED I-BEAM

VENEERED JOIST

Figure 33–4.
Some frequently used prefabricated structural components are the roof truss, floor truss, glued laminated beam, box beam, veneered I-beam, and veneered joist.

REVIEW QUESTIONS

Enter the missing words in the spaces provided on the right.

1. Coniferous trees produce _____ lumber.

   _____

2. Hardwood lumber comes from _____ trees.

   _____

3. Softwood is used primarily as _____ lumber.

   _____

4. Surfacing and shrinkage produces the _____ dimensions of structural lumber.

   _____

5. The actual dimensions of a 2 × 4 are ___ by ____.

   _____

6. Light framing lumber comes under the category of _____ lumber.

   _____

7. Lumber 5" or more in thickness and in width is called _____.

   _____

8. The oldest available type of panel product is _____.

   _____

9. Strength and appearance determines the _____ of a piece of structural lumber.

   _____

10. Plywood is made up of _____ _____ at right angles and bonded together.

    _____

11. _____ wood panels are made of wood particles, flakes, or strands bonded together.

    _____

12. _____ panels have plywood outside veneers glued to a reconstituted wood core.

    _____

13. The basic components of a roof truss are the top and bottom _____ and _____.

    _____

14. Several pieces of lumber joined with a strong adhesive are called _____ timbers.

    _____

15. _____ _____ consist of veneered pieces fastened to notched top and bottom flanges.

    _____

UNIT 34

Metals

Steel, sheet metal, and aluminum are the metals most widely used in the construction industry. Steel is used for the basic framework of entire buildings, bridges, overpasses, and other structures. Wooden residential and commercial buildings include many metal members such as beams, columns, ductwork, and siding. And in more recent years the use of light metal materials to frame smaller buildings has increased.

BEAMS AND COLUMNS

Steel **I-beams,** also called **wide-flange beams,** are made up of top and bottom flanges separated by a vertical web. They resemble a capital letter I when viewed from one end. I-beams are often used for the internal support of wooden joists below the main floor. *Lally* pipe columns are another common feature in residential and light commercial construction. These cylindrically shaped steel columns have a hollow core that is filled with concrete for greater bearing strength. They are used to support steel or wooden beams. **(See Figure 34-1.)**

FASTENERS

The more commonly used metal fasteners are nails, staples, screws, bolts, concrete, or masonry anchors, and metal connectors. Section view and detail drawings will usually give the number, type, and size of the required fasteners.

NAILS

The nails most frequently used to fasten wooden members are generally defined as common or finish. **Common nails** have flat heads with thicker shanks and are used for rough work such as wood framing. **Finish nails** have very small heads with thinner shanks. They are used to fasten finish woodwork such as molding and can be set below the surface with a nail set. The term **penny (d)** identifies the length of a nail. Common nail lengths range from 2d (1") to 60d (6"). Nails with lengths from 10d and up are called **spikes.** Finish nails

range in size from 2d (1") to 20d (4"). **(See Figure 34–2.)**

Nails are driven by hammers or pneumatic nailing guns. **Staples** may be used in place of nails to secure building paper, roof sheathing, paneling, and roof finish materials such as shingles. They are driven by electric or pneumatic tools or by hand-operated tools.

SCREWS

Screws come with flat, round, or oval heads that have single or cross (Phillips) slots. *Wood screws* are used to fasten into wood. *Machine screws* are placed in threaded holes in metal. For fastening to softer metals, *self-driving screws,* mounted in an electric screwdriver, can drill a hole and cut threads in one operation.

In the past screws were used primarily to attach finish hardware. With the development of highly efficient cordless screwdrivers, however, screws are now often used to secure subfloors, porch decks, railings, and other wood components as well. Screws have greater holding power than either nails or staples. **(See Figure 34–3.)**

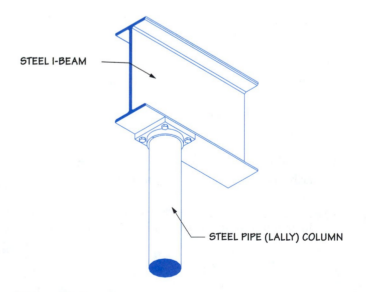

STEEL I-BEAM

STEEL PIPE (LALLY) COLUMN

Figure 34–1.
Steel I-beams and lally columns are often used in wood framed buildings to support floor and ceiling systems.

COMMON NAILS ARE USED FOR WOOD FRAMING, CONCRETE FORM CONSTRUCTION, AND OTHER ROUGH WORK.

BOX NAILS ARE LESS LIKELY TO CAUSE SPLITS WHEN NAILING THINNER MATERIALS SUCH AS EXTERIOR SHEATHING AND INSULATION BOARD. BOX NAILS DO BEND MORE EASILY THAN COMMON HAILS BECAUSE OF THEIR THINNER SHANKS.

FINISH NAILS AND BRADS ARE DESIGNED FOR NAILING FINISH MATERIALS. BRADS HAVE THINNER SHANKS THAN FINISH NAILS. THE SMALL TAPERED HEADS OF FINISH NAILS AND BRADS MAKE IT POSSIBLE TO SET THE HEADS BELOW THE WOOD SURFACE SO THAT THE HOLES CAN BE PUTTIED BEFORE PAINTING OR STAINING.

CASING NAILS HAVE THICKER SHANKS THAN OTHER FINISH NAILS AND ARE USED TO FASTEN HEAVIER PIECES OF TRIM MATERIALS.

CONCRETE NAILS ARE MADE OF SPECIAL HARDENED STEEL AND CAN BE DRIVEN SHORT DISTANCES INTO CONCRETE AND OTHER MASONRY. THEY MAY BE USED TO FASTEN THIN MATERIALS SUCH AS FURRING STRIPS.

DEFORMED SHANK NAILS HAVE SUPERIOR HOLDING POWER AND ARE OFTEN USED TO FASTEN GYPSUM WALLBOARD, FLOOR UNDERLAYMENT, AND OTHER MATERIALS THAT MUST NOT WORK LOOSE AFTER APPLICATION.

CUT NAILS ARE AN OLDER TYPED ONCE USED FOR FRAMING PURPOSES. THEY ARE SOMETIMES USED TODAY FOR NAILING FINISH FLOORING BECAUSE THEIR SQUARE ENDS MINIMIZE SPLITTING OF MORE BRITTLE TYPES OF HARDWOOD FLOORING MATERIAL.

ROOFING NAILS HAVE LARGE HEADS TO SUPPLY SUFFICIENT HOLDING POWER FOR SOFT ROOFING MATERIALS SUCH AS ASPHALT SHINGLES.

Figure 34–2.
Shown above are some nails frequently used in construction work.

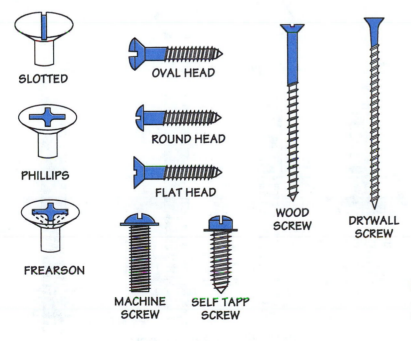

SLOTTED

PHILLIPS

FREARSON

OVAL HEAD

ROUND HEAD

FLAT HEAD

MACHINE SCREW

SELF TAPP SCREW

WOOD SCREW

DRYWALL SCREW

Figure 34–3.
Screws are available in many types and sizes to fasten objects to wood and metal.

BOLTS

Bolts connect heavier wood and metal materials. The more commonly used bolts come in all lengths and range from 3/8" to 1" in diameter. Major types include machine, carriage, stove, and lag bolts. Machine bolts, carriage bolts, and lag screws are used most frequently in construction work. **Machine bolts** have hexagonal or square heads with a section of threaded shank. Nuts tightened at the threaded section hold the bolted pieces together. Flat washers should be placed against the wood under the head and nut. **Carriage bolts** have an oval head with a square shoulder under the head. The shoulder is embedded in the wood and prevents the bolt from turning as it is being tightened with a nut at the threaded end. A flat washer is required only under the nut. **Lag bolts,** also called **lag screws,** have square or hexagonal heads with a section of coarse-pitched thread. **(See Figure 34–4.)**

CONCRETE AND MASONRY FASTENERS

Special concrete and masonry fasteners are used to attach wood or metal pieces to concrete or masonry. Some equipment and devices used are powder-actuated guns, expansion anchors and shields, and chemical anchors. **Powder-actuated tools** shoot specially hardened nails called **drive pins** or **studs** through the material being fastened and into the concrete. **Expansion anchors** are used to secure lighter objects. Often made of plastic or lead, expansion anchors are placed in predrilled holes. As a screw is driven into the anchor, it expands so that the anchor is held in place by pressure. **Expansion shields,** made of a zinc alloy, are used as medium- to heavy-duty anchors. Various types can receive machine bolts or lag screws. **Chemical anchors** consist of a bolt placed in a predrilled hole and held in place by synthetic resin and quartz aggregate filler. **(See Figure 34–5.)**

FRAMING AND TIMBER FASTENERS

Metal fasteners are extensively used to strengthen the ties between wooden framing members. This is most important in areas where buildings are subjected to earthquakes, tornadoes, or hurricanes. Zinc-coated 16 or 18 gauge metal connectors are used with light framing materials. Heavier steel devices are used for heavy timber attachments. **(See Figure 34–6a, b.)**

LIGHT-GAUGE STEEL FRAMING

In light-gauge steel framing, steel is used in place of wood for plates, studs, joists, rafters, roof trusses, and all the other basic framing components. They are made of a light-gauge galvanized steel that is cold-formed into C-shaped members. Heavier-gauge steel may also be used when required. Wood, gypsum board, and other surface materials are screwed to the steel surfaces. **(See Figure 34–7.)**

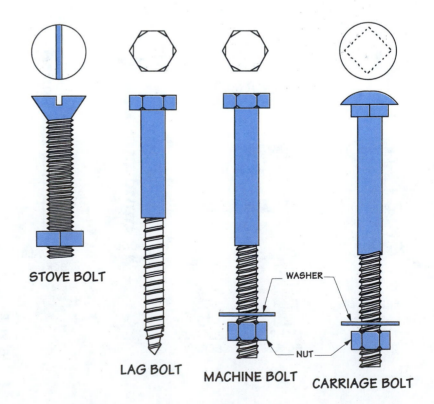

STOVE BOLT

LAG BOLT MACHINE BOLT CARRIAGE BOLT

WASHER

NUT

Figure 34–4.
Bolts are used to fasten heavier wood, plastic, and metal materials.

POWER ACTUATED FASTENERS

FASTENERS ARE SHOT INTO THE CONCRETE WITH A POWDER ACTUATED TOOL. DRIVE PINS DIRECTLY NAIL WOOD OR METAL TO THE CONCRETE. STUDS LEAVE A THREADED BOLT ABOVE THE CONCRETE AFTER BEING FASTENED.

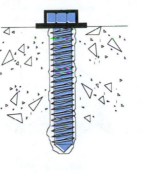

CHEMICAL ANCHOR

ANCHORS HARD HELD IN PLACE BY A SYNTHETIC RESIN AND QUARTZ AGGREGATE FILLER AFTER A HOLE HAS BEEN DRILLED.

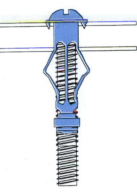

EXPANSION ANCHOR WITH WOOD SCREW

SIDES EXPAND WHEN DRIVING SCREW. THIS TYPE IS NORMALLY USED FOR LIGHTER OBJECTS

EXPANSION SHIELD WITH MACHINE BOLT

EXPANSION SHIELD WITH LAG BOLT

BOTH OF THE ABOVE EXPANSION SHIELDS ARE ADEQUATE FOR FASTENING HEAVIER OBJECTS. THE MACHINE BOLT TYPE IS USED MORE OFTEN WHEN FASTENING METAL MATERIALS. SHIELDS WITH LAG BOLTS WORK BEST WHEN FASTENING WOODEN PIECES.

HOLLOW WALL ANCHOR

AS THE SCREW IS TIGHTENED THE SHIELD EXPANDS AND FLATTENS AGAINST THE INSIDE SURFACE OF THE WALL MATERIAL.

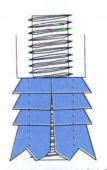

SELF DRILLING EXPANSION SHIELD

THIS SHIELD DRILLS ITS OWN HOLE WHILE DRIVEN WITH A ROTARY IMPACT HAMMER. AN EXPANDER PLUG AT THE END WEDGES THE SHIELD SECURELY IN PLACE.

Figure 34–5.
There are many types of masonry fasteners. A few examples are shown above.

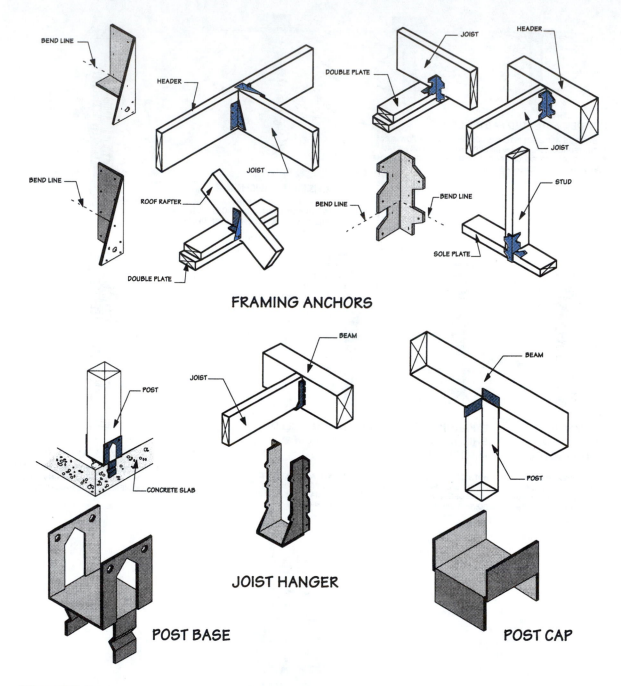

FRAMING ANCHORS

JOIST HANGER

POST BASE

POST CAP

Figure 34–6a.
Metal fasteners are used to strengthen ties between structural members.

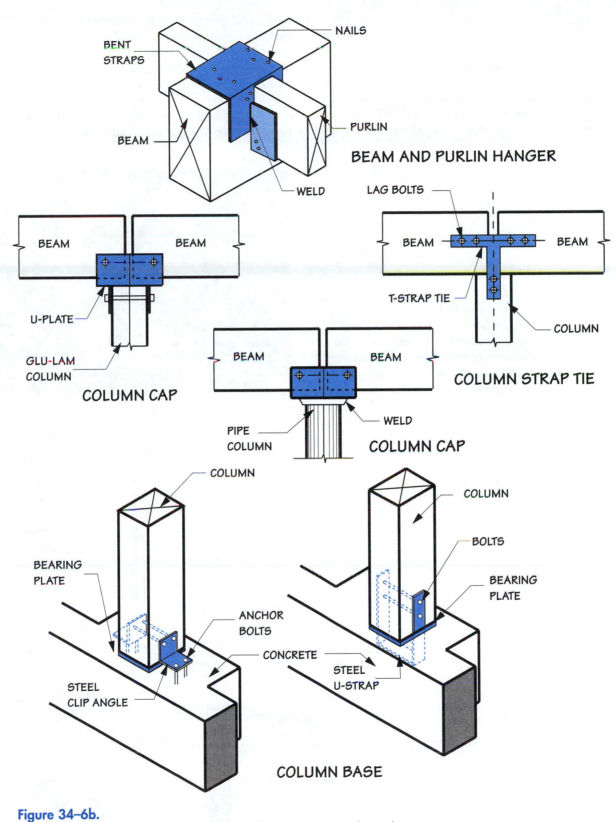

BENT STRAPS

NAILS

BEAM

PURLIN

WELD

BEAM AND PURLIN HANGER

BEAM   BEAM

U-PLATE

GLU-LAM COLUMN

COLUMN CAP

LAG BOLTS

BEAM   BEAM

T-STRAP TIE

COLUMN

COLUMN STRAP TIE

BEAM   BEAM

PIPE COLUMN

WELD

COLUMN CAP

COLUMN

BEARING PLATE

ANCHOR BOLTS

CONCRETE

STEEL CLIP ANGLE

COLUMN

BOLTS

BEARING PLATE

STEEL U-STRAP

COLUMN BASE

Figure 34–6b.
Metal fasteners are used to strengthen ties between structural members.

FINISH METALS

Steel and aluminum finish materials may be used within or outside buildings. For example, *fire-resistant* interior and exterior steel doors and frames are most often used in commercial work. Sometimes residential buildings also feature steel-clad exterior doors. *Prefabricated* steel stair ways and fireplaces are commonplace, and plumbing systems extensively use steel for fixtures, sinks, tubs, and shower enclosures. Aluminum *exterior siding* is available with different colors, simulated patterns, and designs. Aluminum is also widely used for all types of window units and sliding glass doors. Sheet metal is commonly used for gutters, downspouts, and even as a finish roof covering. **(See Figure 34–8.)**

Figure 34–7.
Light-gauge steel is used for all the framing components of this building. *(Courtesy of Ayers Construction Company, Inc., Lexington, Kentucky)*

ALUMINUM SIDING
AVAILABLE IN SIMULATED GRAIN
PATTERNS AS WELL AS PLAIN COLORS

STEEL CLAD DOOR
A RIGID FOAM CORE PROVIDES
A GOOD THERMAL BARRIER
AND SOUND RETARDATION

STEEL DOOR FRAME
COMES IN DIFFERENT
WIDTHS TO ACCOMMODATE
STANDARD WALL THICKNESSES

ALUMINUM GUTTER
LOCATED AT EDGE OF ROOF
AND CARRIES WATER FROM
RAIN AND SNOW MELT TO
DOWNSPOUT

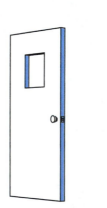

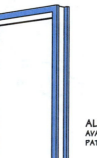

ALUMINUM ROOF SHINGLES
SHINGLES ARE PATTERNED AFTER
CEDAR SHAKES

Figure 34–8.
Many metal finish products are used in the construction of a building. A number of examples are shown above.

ALUMINUM WINDOW UNIT
SLIDING AND STATIONARY
WINDOWS SET IN A FRAME

REVIEW QUESTIONS

Enter the missing words in the spaces provided on the right.

1. Beams with bottom and top flanges separated by a vertical web are called _____.

   _____

2. Cylindrically shaped steel columns filled with concrete are called _____ columns.

   _____

3. _____and _____ nails are the two main categories of nails.

   _____

4. The range of common nail lengths is ___ to ____ inches.

   _____

5. Threaded holes in metal receive _____ screws.

   _____

6. _____ _____ screws can simultaneously drill a hole and cut threads in soft metal.

   _____

7. Bolts with a square or hexagonal head and threaded shanks are called_____ bolts.

   _____

8. Lag screws are also called _____ _____.

   _____

9. A drive pin is driven by a _____ _____ tools.

   _____

10. _____ _____ and _____ ____ can be received by heavy-duty expansion shields.

   _____

11. Bolts held in place in predrilled holes by a synthetic resin are called ____ _____ .

   _____

12. Ties between wood framing members are greatly strengthened with _____ _____ .

   _____

13. Light-gauge steel framing can be done instead of _____ framing.

   _____

14. Fire-resistant doors are often _____ _____ on both surfaces.

   _____

15. Sheet metal is used to fabricate _____ for heating and cooling systems.

   _____

UNIT 35

Plastics

Plastics are a synthetically produced material made up of combinations of various chemicals. They can be molded and formed into different solid shapes or very thin sheets and are generally divided into thermoplastics and thermosets. Recent years have seen a major growth in the use of plastics in the construction industry.

THERMOPLASTICS

Thermoplastics may be softened by reheating and are hard and rigid when cooled again. Some better known thermoplastics are acrylics, polyethylene, polypropylene, polystyrene, and vinyl. **Acrylics** are transparent materials and can be used for window glazing, roof domes, skylights, and translucent panels for ceiling lighting systems. **Polyethylene** is a sheet material used most often as vapor barriers for floors and walls and beneath concrete slabs resting on the ground. **Polypropylene** is well suited for insulation around piping and pipe fittings. **Polystyrene** is used for the manufacture of lighting fixtures and molded pieces of hardware. Foamed (expanding) polystyrene is frequently placed for wall, floor, and ceiling insulation. It is also used for duct and pipe insulation and insulation for walk-in refrigerators and freezers. **Vinyls** are one of the most widely used plastics in construction work. As an interior material, vinyl is used for sheet and tile floor finish. Roofs commonly feature vinyl for gutters, downspouts and flashing. Vinyl is also produced with simulated patterns for exterior siding. (See Figure 35–1.)

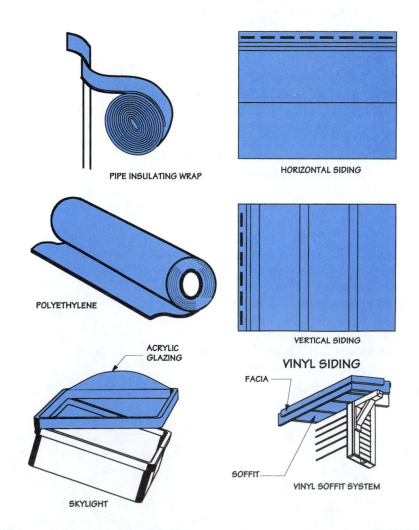

PIPE INSULATING WRAP

HORIZONTAL SIDING

POLYETHYLENE

VERTICAL SIDING

ACRYLIC GLAZING

VINYL SIDING

FACIA

SOFFIT

VINYL SOFFIT SYSTEM

SKYLIGHT

Figure 35–1.
Thermoplastic materials are used for a growing number of products. A few examples are shown.

THERMOSETS

Thermosets are plastics that cannot be softened after they have been formed. They are generally harder and more heat resistant than thermoplastic materials. The three major categories of thermosets are melamines, phenolics, and polyesters. **Melamines** are best known as plastic laminates for countertops and cabinet finishes. **Phenolics** are used for electrical parts such as sockets, switch boxes, and circuit breakers. **Polyesters** combined with a fiberglass reinforcing agent are produced as translucent sheets for roofing, interior partitions, and window units. Polyesters are also molded into sinks, bathtubs, and shower stalls. **(See Figure 35–2.)**

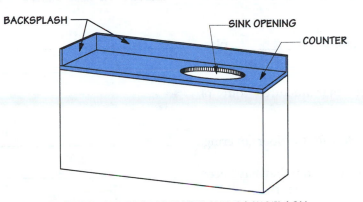

PLASTIC LAMINATE COUNTER AND BACKSPLASH

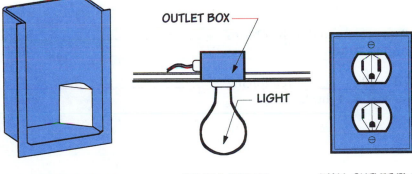

MOLDED SHOWER STALL   CEILING OUTLET   WALL OUTLET PLATE

Figure 35–2.
A very common example of thermoset plastic is melomines used for plastic laminate countertops. Electrical parts are made from phenolic plastic. Sinks and shower stalls are fabricated from molded polyesters.

REVIEW QUESTIONS

Enter the missing words in the spaces provided on the right.

1. The _____ category of plastics can be reheated and softened.

   _____

2. A transparent plastic that can be used for window glazing is called _____ .

   _____

3. Vapor barriers are often made of _____ plastic sheet material.

   _____

4. Lighting fixtures are manufactured out of _____ plastic.

   _____

5. _____ plastic is widely for tile floor covering.

   _____

6. Plastics unable to be softened after they have been formed are in the _____ category.

   _____

7. The major categories of thermosets are _____, _____, _____ .

   _____

8. The main use of melamines is for _____ _____ countertops.

   _____

9. Electrical parts such as sockets and switch boxes are made from _____ plastic.

   _____

10. Polyesters combined with a _____ reinforcing agent produce translucent roof sheets.

    _____

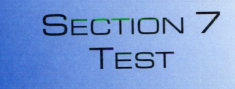

Materials

Answer T (true) or F (false).   Answers

1. Cement is a mixture of concrete, aggregate, and water.   _____

2. Concrete mixed at a batch plant and delivered to the job is called ready-
    mixed concrete.   _____

3. The main purpose of rebars in a concrete wall is to resist vertical pressures. _____

4. Terra cotta is made from a concrete mixture.   _____

5. Coursed stone is placed in an irregular pattern.   _____

6. Mahogany is considered hardwood lumber.   _____

7. Joists and planks are 2" to 6" thick and 6" or more wide.   _____

8. Both softwoods and hardwoods are used for finish work.   _____

9. Particleboard is made of large wood-flake wafers compressed and
    bonded into panels.   _____

10. Veneered joists are fabricated from softwood veneers glued together.   _____

11. The end view of a wide-flange beam resembles the letter U.   _____

12. Common nails are used for finish work.   _____

13. Carriage bolts have an oval head with a square shoulder under the head. _____

14. Expansion anchors are held in place by pressure from the inserted screw
    or bolt.   _____

15. Stairways and fireplaces can be prefabricated of steel material.   _____

16. Acrylics are transparent plastics.   _____

17. Polystyrene is used for foamed wall insulation.   _____

18. Vinyl is commonly used for plastic laminate countertops.   _____

19. The major use of polyethylene is for wall insulation.   _____

20. Thermosets are plastics that cannot be softened after they have been
    formed.   _____

Choose one or more correct ways to finish each statement.

Answers

21. Ready-mixed concrete is prepared
    a. on the job site.
    b. at a batch plant.
    c. by a stationary mixer.
    d. all of the above.

    _____

22. The proportion of cement, aggregate, and water in concrete is called the concrete
    a. batch.
    b. ratio.
    c. proportions.
    d. mix.

    _____

23. Reinforced steel bars
    a. are commonly called rebars.
    b. are placed only vertically in walls.
    c. have deformed surfaces.
    d. are placed only horizontally in walls.

    _____

24. Concrete masonry units are
    a. used in the construction of footings and walls.
    b. commonly used to construct foundation footings.
    c. only used to construct low foundation walls.
    d. unable to be reinforced.

    _____

25. Stone masonry
    a. is applied only to veneer walls.
    b. consists only of random-shaped rocks.
    c. can be placed in coursed or uncoursed patterns.
    d. none of the above.

    _____

26. The thickness and width of a piece of surfaced and dried structural lumber are its
    a. nominal dimensions.
    b. English or metric dimensions.
    c. actual dimensions.
    d. metric dimensions only.

    _____

27. Some examples of softwood lumber are
    a. ash.
    b. Douglas fir.
    c. yellow pine.
    d. oak.

    _____

28. Wood material used for studs, plates, and blocking fall under the category of
    a. framing lumber.
    b. appearance-grade lumber.
    c. timbers.
    d. dimension lumber.

    _____

29. The oldest, and a still widely used, panel product is
    a. particleboard.
    b composite panels.
    c. waferboard.
    d. plywood.

    _____

30. Some examples of wooden prefabricated components are
    a. glulam timbers.
    b. box beams.
    c. window trimmers.
    d. floor trusses.

31. Steel I-beams are used
    a. only in steel frame buildings.
    b. primarily for columns.
    c. for the internal support of wooden joists below the main floor.
    d. in light-gauge steel framing.

32. Some of the more widely used wood fasteners are
    a. contact cement.
    b. nails.
    c. screws.
    d. bolts.

33. An advantage of screws over nails is that they
    a. are cheaper.
    b. have better holding power.
    c. are easier to drive.
    d. none of the above.

34. Powder-actuated guns are used to install
    a. expansion anchors.
    b. chemical anchors.
    c. expansion shields.
    d. drive pins.

35. Metal connectors are
    a. nails
    b. fasteners used between metal items.
    c. devices used to strengthen ties between wooden framing members.
    d. bolts and screws.

36. Synthetically produced materials made up of chemical combinations are called
    a. plastics.
    b. synthetics.
    c. compounds.
    d. all of the above.

37. Polyethylene is used mainly for
    a. floor tile.
    b. vapor barrier material.
    c. window glazing.
    d. electrical parts.

38. Plastics that can be reheated and softened are called
    a. thermosets.
    b. phenolics.
    c. polyesters.
    d. thermoplastics.

39. A plastic used for floor, wall, and ceiling insulation is called
    a. polypropylene.
    b. vinyl.
    c. foamed polystyrene.
    d. polyester.

    _____

40. Vinyl plastic is used for
    a. floor tile.
    b. exterior siding.
    c. gutters and downspouts.
    d. all of the above.

    _____

Match the drawings with their correct name:

41. Bolt   _____

42. Wide-flange beam _____

43. Joist   _____

44. Expansion shield _____

45. Hollow wall anchor _____

46. CMU   _____

47. Lally column _____

48. Rebar   _____

49. Molding   _____

50. Timber   _____

51. Floor truss   _____

52. Stone masonry _____

53. Glulam   _____

54. Lag screw   _____

55. WWM   _____

56. Metal connector _____

57. Board   _____

58. Veneered I-beam _____

59. Roof Truss   _____

60. Solid brick   _____

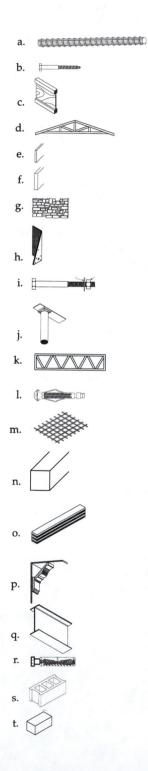

a.

b.

c.

d.

e.

f.

g.

h.

i.

j.

k.

l.

m.

n.

o.

p.

q.

r.

s.

t.

SECTION

8

Plan Assignments

The sets of plans composing the packet included with *Construction Print Reading* are the working drawings for construction projects designed by three different architects. These plans were chosen not only to show the similarities among working drawings but also some of the different methods used by architects in presenting and arranging information. Some drawings have been rearranged for text purposes but none have been altered in any way that affects content or meaning. Although most of the symbols and abbreviations in the different sets of plans are the same, you will note some variations in definition and style. For example, the terms *section drawing* and *detail drawing* may be used to define the same type of drawing.

The assignment included with each set of plans is designed to encourage a very thorough study of the plans. Read the comments provided in the assignments before attempting to answer the questions. ■

Assignment 1
Garlinghouse Plans

This set of drawings is one of a variety of hand drawn **stock plans** offered by the L.F. Garlinghouse Company. A brief description of this house, a pictorial drawing, and a very simple floor plan drawing are shown in one of the catalogs published by the company. A full set of the plans can be purchased from the company. The plans provide full and comprehensive information for the construction of the building. They may require some changes or corrections because of local building and zoning codes.

The Garlinghouse plans are for a three-bedroom, one-level home with two bathrooms and the customary kitchen, dining, and living rooms. Sloped ceilings are shown for the dining and living rooms and for the master bedroom. Sliding glass doors open from the dining room to an outside deck. This set of plans is available with drawings for a full-basement, crawl-space, or slab foundation. Only the full-basement drawings are included for this assignment.

The Garlinghouse plans include a first-page cover sheet containing a pictorial drawing of the house, General Notes, Abbreviations, and an Architectural Drawing Index. The *General Notes* serve the same purpose as specifications, and it is fairly common practice to use them in place of specifications in plans of smaller buildings. The *Architectural Drawing Index* lists the sheet numbers and their drawings. However, for purposes of this assignment, the numbering of the sheets has been changed. No compass directions are shown on the plan, as this must be determined at the building site. Therefore, the different sides of the building are called *front, rear, right,* and *left.*

Some of the dimension questions will require adding or subtracting room dimensions. Other questions will require adding or subtracting a 2 × 4 wall thickness (3 1/2″ actual thickness) or a 2 × 6 wall thickness (5 1/2″ actual thicknesses).

SHEET 1
FOUNDATION PLAN, DETAILS, AND SITE PLAN

Only a very general sketch of a **site plan** is included. Measurements, grades, and other information must be added based on the conditions of the property. The foundation is for a full basement, and the walls and piers are constructed entirely of concrete. The term *foundation detail* is used in these drawings to identify *section views* of the foundation. The Foundation Detail 1/1A describes the full-basement foundation walls. The Foundation Wall at Porch detail 2/1A focuses on the front porch construction. The Garage Foundation Detail 3/GA describes the walls and slabs beneath the garage area.

SHEET 1 QUESTIONS

Enter the missing words in the spaces provided on the right.

1. What is the length of the rear foundation wall? _____

2. Give the distance from the outside face of the left foundation wall at the rear of the building to the center of the first pier footing. _____

3. The foundation footings shown in Detail 1/1A are _____ wide and _____ high. _____

4. Beneath what section of the building is the unexcavated area? _____

5. What is the thickness of the concrete floor slab in the basement? _____

6. All the pier footings are the same size (T or F?). _____

7. How many lights are controlled by a wall switch in the basement? _____

8. There is a total of _____ pull-chain lights in the basement area. _____

9. The C mark in the plan view identifies the _____ _____ . _____

10. What is the distance between the inside faces of the left and right garage foundation walls? _____

11. Give the diameter of the perforated drain tile at the perimeters of the foundation. _____

12. How many, and what is the size of, the continuous horizontal rebars placed in the foundation footings? _____

13. According to the General Notes, the footings must bear on firm soil a minimum of _____ below grade, or below the _____ line. _____

14. The foundation walls are 8″ thick (T or F?). _____

15. Give the length and spacing of the dowels extending from the footing. _____

16. To damp-proof the exterior foundation wall below grade, a _____ coating is applied to the outside surfaces. (See General Notes.) _____

17. What is the thickness of the expansion joint between the basement floor slab and the foundation wall? _____

18. What is the minimum distance between the top of the foundation wall and grade? _____

19. Identify the size and spacing of the vertical rebars in the basement walls. _____

20. There are no horizontal rebars required in the basement walls (T or F?) _____

21. Give the size and spacing of the sill plate anchor bolts. _____

22. The minimum required soil-bearing capacity is _____. (See General Notes.) _____

23. Give the size and type of reinforcement placed in the concrete slabs. _____

24. Treated 2 × 4 material is used for the sill plates (T or F?). _____

25. What is the minimum rating of the insulation to be placed between the floor joists? (See General Notes.) _____

26. Describe the furring, and specify what its spacing is at the inside wall of the stairway. _____

27. The total slope of the garage slab is _____ from back to front. _____

28. What is the total number of risers in the stairway? _____

29. How many sump pump pits are there in the basement floor? _____

30. There is one smoke detector identified in the basement area (T or F?). _____

31. What are the diameter and type of material of the columns placed over the pier footings? _____

32. The distance from the outside face of the right foundation wall to the center of the first rear basement window is _____. _____

33. Water pumped out of the sump pit is directed into _____ perforated drain tubing of _____ diameter. _____

34. Give the center-to-center distance between the basement windows in the right side wall. _____

35. What is the door width at the top of the stairway? _____

36. The semicircular line next to each basement window identifies a(n) _____. _____

37. The outside ends of the girders rest in pockets provided in the concrete wall (T or F?). _____

38. Give the distance between the inside concrete stair wall and the center of the opposite framed wall. _____

39. What is the measurement from the outside of the rear foundation wall to the centers of the pier footings? _____

40. The furnace and water heater connect to the same vent pipe (T or F?). _____

SHEET 2
FLOOR PLAN AND WINDOW SCHEDULE

The floor plan shows three bedrooms, two bathrooms, a living room, dining area, kitchen, an outside deck accessible from the dining area, and a two-car garage. The floor plan also shows a fireplace in the living room and gives a clearer view of the stairway leading down to the basement. A **window schedule** is provided to give the finish opening (unit dimension) and the rough opening size of each window. A door schedule is not included with this set of plans. The finish door widths are noted at the door openings, and it can be assumed that all the doors are a 6'-8" standard height. Additional door and window information is provided in the General Notes. Dimensions are given from the outside of the exterior walls to one side of the adjacent partitions. Measurements to windows are to the centers of the windows.

SHEET 2 QUESTIONS

Enter the missing words in the spaces provided on the right.

41. Give the width and length of the rear (optional) deck.

_____

42. What is the distance from the outside face of the front garage wall to the front side of the deck?

_____

43. The deck of the front porch is finished off in brick (T or F?).

_____

44. _____ type doors open up to the laundry room.

_____

45. Give the width and length of the attic access opening.

_____

46. Between what rooms is a fire-rated door required?

_____

47. The time rating for the fire door is _____.

_____

48. Name the finish floor material in the kitchen.

_____

49. Give the width and height of the window rough opening in Bedroom #2.

_____

50. The stud size of the exterior walls is _____ spaced _____ OC.

_____

51. How many risers and treads are in the stairway?

_____

52. The inside face-to-face measurements of Bedroom #2 are _____ wide and _____ long.

_____

53. What is the inside width measurement (wall to wall) of Master Bedroom #1?

_____

54. Give the finish (unit dimensions) of the living room window.

_____

55. Which room has a fixed window?

_____

56. The distance from the outside face of the rear wall to the center of the small bathroom window is _____.

_____

57. The master bedroom ceiling slopes from the opposite walls to a peak at the center of the room (T or F?).

_____

58. Give the width and name of the hearth material in front of the fireplace.

_____

59. What supports the ceiling above the opening between the kitchen and dining room?

_____

60. The General Notes say that all windows must be _____ glazed.

_____

61. Name the manufacturer and give the model number of the lower window in Den/Bedroom#3.

_____

62. What is the total width of the bifold closet doors in Bedroom #2?

_____

63. The inside wall-to-wall measurement of the hall is _____.

_____

64. How high must the sill of at least one window be in the sleeping areas? (See General Notes.)

_____

65. Name the type, width, and height of the garage door.

_____

66. What wooden members must be pressure-treated at the rear deck?

_____

67. Based on information in the "Notes" provided on Sheet 2, what is the floor finish in Bedroom #1.

_____

68. A _____ wide door opens to Bedroom #2.

_____

69. According to the General Notes, all interior doors must be foam-core insulated doors (T or F?).

_____

70. What type and of what thickness are the gypsum boards required for the walls and ceiling of the garage?

_____

71. Give the size of the prefabricated steel fireplace unit.

_____

72. Name the type, width, and height of the door opening from the dining room to the deck.

_____

73. What is the measurement from the left living room wall to the center of the living room window?

_____

74. The garage wall next to the stairway is constructed of 2 × 6 studs (T or F?).

_____

75. Which room shows a bay window?

_____

76. The front section of the foyer receives a _____ floor finish.

_____

77. What is the length of the front of the kitchen sink counter?

_____

78. The flue coming up from the basement furnace is shown in the corner of a _____ in _____.

_____

79. How many steps lead down from the kitchen to the garage floor?

_____

80. Insulation is shown for the left garage wall (T or F?).

_____

SHEET 3
EXTERIOR ELEVATIONS AND INTERIOR CABINET ELEVATIONS

Exterior and **interior elevations** are also contained on Sheet 3. The exterior drawings are identified as front, rear, right side, and left side elevations. Interior elevations are provided for the kitchen, master bathroom, and bathroom #2 cabinets. The elevation drawings should be read along with the floor plan. Note that the same section reference symbols and numbers appear here as on the floor plan. They also refer to the section drawings on Sheet 4.

SHEET 3 QUESTIONS

Enter the missing words in the spaces provided on the right.

_____

81. The Front Elevation is drawn to _____ scale.

_____

82. What is the floor-to-ceiling height shown on the elevations?

_____

83. Give the thickness and width of the finish corner boards.

_____

84. Name the type and exposure of the siding at the front of the house.

_____

85. What is the unit rise for the intersecting gable roofs shown in the Front Elevation?

_____

86. Crown molding is shown at the sides of the garage door opening (T or F?).

_____

87. If the front porch stairway has three or more risers, a _____ is required.

_____

88. How wide are the soffits at the front gable overhangs?

_____

89. What material is placed over all windows, doors, and at roof valleys to prevent water leakage? (See General Notes.)

_____

90. Name the roof finish material.

_____

91. The diameter of the louvered attic openings shown in the front elevation is _____.

_____

92. What device at the roof ridges helps provide attic ventilation?

_____

93. Brick veneer finishes off the wall next to the front entrance door (T or F?).

_____

94. Give the size of the barge board at the gable ends of the roof.

_____

95. Name the type of siding required for the left and rear sides of the building.

_____

96. What is the unit rise of the main roof?

_____

97. The chimney should extend no less than 2'-0" above the point where the chimney passes through the roof (T or F?).

_____

98. Describe the material used for flashing. (See General Notes.)

_____

99. The elevations show a _____ unit rise for the short roof slope over the front porch.

_____

100. What is the spacing of the gutter fasteners? (See General Notes.)

_____

101. How many cabinet doors are shown at the left kitchen wall?

_____

102. What are the width and height of the cabinet over the refrigerator?

_____

103. The dishwasher is skewed at a(n) _____ angle to the sink counter.

_____

104. A 3'-6" space is provided between cabinets for the range (T or F?).

_____

105. Give the width and height of the space provided for the refrigerator.

_____

106. What is the total height from the floor to the top of the kitchen wall cabinets?

_____

107. The distance between the kitchen floor and bottom of the wall cabinets is _____.

_____

108. What is the total width of the bottom cabinet units at the front kitchen wall?

_____

109. How many rows of shelves are shown in the top wall cabinets?

_____

110. Find the distance from the left wall to the left side of the dishwasher.

_____

111. The total width of the cabinet units in the master bathroom is 4'-3" (T or F?).

_____

112. What is the height of the floor cabinets in both bathrooms?

_____

113. How many drawers are shown in the cabinet units in the smaller bathroom?

_____

114. The distance from the floor to the top of the bathroom mirrors is _____.

_____

115. What is the width of the mirror in the master bathroom?

_____

SHEET 4
SECTION AND DETAIL DRAWINGS

Reference symbols for the **section drawings** appear on the floor plan and the elevation drawings. Therefore, each section drawing has to be studied in relation to the floor plan and elevations. In some cases a section drawing will refer to one of the framing plans for further clarification. Sheet 4 includes a transverse building cross section drawing that is referenced by the symbol and line A/4. The arrows of the corresponding symbols on the floor plan (Sheet 4) point to the left. This pro-

vides a view from front to rear of the garage wall, front entrance, foyer, and the railing on one side of the stairwell. Also viewed is the living room, the opening from the living room to the hall in front of the laundry room, the fireplace, and the opening from the living room to the dining room. There are also section drawings for the fireplace, stairway, and a wall. There are three ceiling section drawings.

SHEET 4 QUESTIONS

Enter the missing words in the spaces provided on the right.

116. What is the highest floor-to-ceiling height in the living room?

    _____

117. Give the thickness, width, and spacing of the bracing below the rafters.

    _____

118. A 2 × 6 _____ is nailed at the top ends of the rafter braces.

    _____

119. Name the type of vapor barrier attached to the ceiling over the foyer.

    _____

120. What is the unit rise of the ceiling slope over the living room?

    _____

121. According to the General Notes, R-19 insulation is required in the floors over unheated basement foundations (T or F?).

    _____

122. Give the size and spacing of the collar ties.

    _____

123. Give the scale of the cross section drawing.

    _____

124. What is the thickness of the gypsum board to be nailed to the interior ceilings?

    _____

125. Identify the thickness and type of roof sheathing.

    _____

126. Name the material placed directly below the composition roof shingles.

    _____

127. What supports the inside ends of the ceiling joists over the foyer and the lower ends of the sloped joists in the living room?

    _____

128. _____ size subfascia board is nailed to the lower ends of the roof rafters over the living room.

    _____

129. What is the reason for furring down the roof rafters over the living room?

    _____

130. How wide is the bituthane eaves flashing?

    _____

131. The width of the roof overhang at the rear of the building is _____.

    _____

132. The width of the continuous soffit vent is 2 1/2″ (T or F?).

    _____

133. Give the thickness and width of the frieze board below the overhang soffit.

_____

134. According to the General Notes, vapor barrier material applied to the walls and ceiling must face the _____ _____.

_____

135. What is the height of the stairwell railing in the foyer?

_____

136. Name the two kinds of wind bracing that can be used on the walls.

_____

137. A 1" thick ridge is nailed to the top ends of the roof rafters (T or F?).

_____

138. Wind bracing should be placed at the corners and at every _____ linear feet.

_____

139. The overhang shown in the wall section drawing is _____ wide.

_____

140. Insulated foam _____ is placed on the exterior walls.

_____

141. Give the size and name of the framing member supporting the rafters extending over the bay in the master bedroom.

_____

142. The raised ceiling in the master bedroom is _____ above the lower ceiling.

_____

143. What two framing components are supported by the LVL beam at one side of the dining area?

_____

144. How wide is the plant shelf?

_____

145. There is a flat ceiling in the master bathroom (T or F?).

_____

146. Name the type of fireplace being used.

_____

147. The minimum distance between the outside faces of the chimney frame above the roof is _____.

_____

148. What size header is placed over the fireplace opening?

_____

149. Identify the type and total width of the hearth material.

_____

150. According to the General Notes, the fireplace must be provided with tempered metal doors (T or F?).

_____

151. The thickness and type of material used to sheath the chimney is _____ .

_____

152. Name the device placed over the top of the chimney.

_____

153. What is the total run of the stairway?

_____

154. How many stringers support the stairway?

_____

155. The railing height at the stairway is _____.

_____

156. What is the width of the stair tread?

_____

157. The riser height is _____.

_____

158. The total rise of the stairway is 8'-1 1/2" (T or F?).

_____

159. What is the minimum required headroom for the stairway?

_____

160. The width of the landing at the foot of the stairs is _____.

_____

161. The top of the stringers butt against a _____ _____.

_____

162. The space below the stairway can be used for _____.

_____

163. How many risers are there in the entire stairway?

_____

164. The bottom of the stairway rests on a _____.

_____

165. What is the diameter of the stair railing?

_____

SHEET 5
FRAMING PLANS AND ELECTRICAL FLOOR PLAN

The floor, ceiling, and roof **framing plans** are very helpful in the construction process, and should be studied together with the floor plan on Sheet 2. Framing plans show the actual placement of the major framing components.

The **electrical floor plan** of the Garlinghouse drawings shows the locations of all the plugs, lights, fan/lights, and the smoke detectors. Also shown are all the switch locations and the wiring connected to the lights. A schedule of electric symbols is conveniently provided.

According to the General Notes, plumbing diagrams and drawings will be provided by the plumbing contractor, and heating and cooling duct drawings or diagrams will be provided by the HVAC contractor. These drawings may not even be necessary, because the building requires fairly routine and simple plumbing and HVAC systems. The locations of all the plumbing fixtures are quite clear on both the regular floor plan and the electrical floor plan. Neither plan shows the heating outlets. Therefore, the HVAC contractor can establish their proper locations and place the ducts accordingly.

SHEET 5 QUESTIONS

Enter the missing words in the spaces provided on the right.

166. Give the size of the deck ledger board bolted to the house, and give the size and type of fasteners.

_____

167. The thickness and width of the floor joists are _____ and the spacing of most of the joists is _____ OC.

_____

168. What is the beam size running beneath the dining and living room areas?

_____

169. Beneath what rooms are the floor joists spaced 8" OC?

_____

170. How far do the cantilevered joists extend beyond the rear of the building?

_____

171. The purpose of the cantilevered joists mentioned in Question 170 is to support the bay section (T or F?).

_____

172. According to the General Notes, what grade of lumber is used for the floor and ceiling joists?

_____

173. What sizes are required, when laminated veneer lumber beams are shown supporting the floor joists?

_____

174. In what section of the first-floor framing plan is the framing member pressure-treated?

_____

175. The cantilevered joists extending into the garage area support a _____ and _____.

_____

176. Of what thickness and type is the plywood used for the subfloor? (See General Notes.)

_____

177. What lumber grade is required for the posts, beams, and headers? (See General Notes.)

_____

178. The span over the garage ceiling calls for 2 × 8 joists (T or F?).

_____

179. The cantilevered ceiling joists over the dining room support a _____.

_____

180. Between what rooms is there a ceiling beam supporting a knee wall?

_____

181. What is the header size over the garage door?

_____

182. Over what room is the attic access? _____

183. All the ceiling joists run in the same direction (T or F?). _____

184. A 6 × 8 flush beam is placed above _____. _____

185. Name the device required to support any abutting joists. (See General Notes.) _____

186. What is the distance of the front gable roof overhang? _____

187. Give the thickness and width of the ridge boards. _____

188. The size of the material used for the roof rafters is _____. _____

189. What is the reason for adding 2 × 6s to the bottoms of the 2 × 8 rafters over the dining room, living room, and master bathroom? _____

190. All the roof rafters are spaced 24" OC (T or F?). _____

191. What is the width of the main roof? _____

192. Over what space does the roof framing plan show an over-framed area? _____

193. What size are the valley rafters over the Bedroom #2 and #3 areas? _____

194. _____ grade lumber is required for the roof rafters. (See General Notes.) _____

195. Flat 2 × 10 valley rafters are placed over the over-framed area (T or F?). _____

ELECTRICAL FLOOR PLAN QUESTIONS

Enter the missing words in the spaces provided on the right.

196. Which rooms contain ground fault interrupter switches?
_____

197. How many ceiling light fixtures are there in the garage?
_____

198. Which rooms have overhead fan/lights?
_____

199. There are two three-way switches in Bedroom #2 (T or F?)
_____

200. A _____ light is found in the kitchen ceiling.
_____

201. How many smoke detectors are shown in the building?
_____

202. Which rooms have recessed adjustable ceiling fixtures?
_____

203. Combination ceiling lights and exhaust fans are shown in both bathrooms (T or F?).
_____

204. The exterior twin floodlights at the front of the garage are controlled by three-way switches in the _____ and _____.
_____

205. How many exterior waterproof outlets are provided for the building?
_____

206. What type of light fixture is shown above the sinks in the bathrooms?
_____

207. The ceiling plug in the garage provides power to open and close the _____ _____ _____ .
_____

208. What is the total number of ceiling light fixtures in the building, excluding the fan and lights, and fluorescent lights?
_____

209. The same switch operates the fan and light in the overhead fan/lights (T or F?).
_____

210. How many smoke detectors are shown in the electrical floor plan?
_____

UNIT 37

Assignment 2
Stafford
Residence Plans

The Stafford residence plans call for the remodeling of an U-shaped one-story residential structure into a two-story building. **(See Figure 37–1 and Figure 37–2.)** All components and features of the building in its original state are identified as *Existing (E)*. Renovation work is identified as *New (N)*. The existing house includes three bedrooms, a kitchen and breakfast room, a dining room, a living room, and two bathrooms. The new second story will add a study and two more bedrooms and bathrooms. The ceilings of the master bedroom and study slope in four directions from the center. The complete set of plans includes six sheets. Specifications on the original set of plans were printed on a seventh sheet. They are now reproduced at the end of this unit **(See page 399).**

The plan for the Stafford residence was prepared by Michael Connell, architect of San Francisco, California.

Figure 37–1.
Front view of the Stafford residence prior to the second-floor addition.

Figure 37–2.
Front view of the Stafford residence with the second-floor addition.

SHEET A1
SITE PLAN, EXISTING FIRST FLOOR PLAN, NEW FIRST FLOOR PLAN, AND NEW SECOND FLOOR PLAN

SITE PLAN

Additions are being made on a building on existing property. Therefore, the main purpose of the site plan is to give a plan view perspective of the property and the existing building.

SITE PLAN QUESTIONS

Enter the missing words in the spaces provided on the right.

1. What is the total length of the SW to NW property line? _____

2. How wide is the second-story addition (east to west)? _____

3. Name the type of zoning building occupancy. _____

4. What is the NW to NE length of the garage? _____

5. The building has a _____ zoning classification. _____

6. Who is responsible for all excavation and demolition? (See specs.) _____

7. What is the angle of the SW to SE property line? _____

8. Give the maximum height limit for buildings in this zone. _____

9. How wide is the N to S side of the existing building? _____

10. What is the distance between the north side of the house and the garage? _____

EXISTING FIRST FLOOR PLAN

The Existing First Floor Plan shows the original room layout of the first floor before renovation. Wall symbols shown below the New Second Floor Plan identify the different wall conditions. Note that the north compass direction also shown below the New Second Floor drawing, points in a different direction than it did in the site plan drawing. This is done to place the front of the building toward the bottom of all the floor plan drawings.

Existing walls that will remain are shown with solid lines. Walls to be removed are shown with dashed lines. For example, the bearing wall between the den and living room is to be removed and replaced by a beam. Windows, fixtures, and other items to be removed are also identified with dashed lines. For example, the drawing shows that an existing window in the rear bedroom is to be removed. Dashed lines and written instructions inform that the fireplace will be relocated.

EXISTING FIRST FLOOR PLAN QUESTIONS

Enter the missing words in the spaces provided on the right.

11. What type of line identifies walls and objects to be removed?

    _____

12. Give the width and length of the salvaged skylight.

    _____

13. Name the fixtures to be removed from the bathroom next to the den.

    _____

14. A _____ _____ fireplace is to be salvaged.

    _____

15. The non-bearing bookcases are to be salvaged (T or F?).

    _____

16. Name the rooms where an existing window is to be replaced.

    _____

17. To what scale is the Existing First Floor Plan drawn?

    _____

18. The _____ and _____ _____ rooms are separated by a bearing wall that is to be removed.

    _____

NEW FIRST FLOOR PLAN

This drawing shows the changes taking place during renovation of the first floor. Three types of reference symbols appear on this drawing. An arrow placed over a circle identifies an *elevation reference* symbol. The number above the horizontal line gives the number of the drawing, and the number below the line identifies the sheet number. For example, the symbol pointing to the front (west) wall identifies elevation drawing number 1 on sheet 2. A *section reference* symbol also consists of an arrow over a circle. However, note the line going from the symbol into the drawing. Also note that a line with a flag identification at its end extends from the opposite side of the drawing. A *detail reference* symbol consists of a circle divided by a straight line extending from the circle to the object on the plan detailed. Shaded walls identify new sections of walls to be constructed.

NEW FIRST FLOOR PLAN QUESTIONS

Enter the missing words in the spaces provided on the right.

19. What size studs are to be used to construct the short walls between the living and dining rooms?

    _____

20. Detail reference symbol 9/A-6 points to the _____.

    _____

21. What material covers the fireplace hearth?

    _____

22. The first-floor bathroom remains at the same location (T or F?).

    _____

23. Section line symbol 2/A-3 points (give compass direction) _____ and cuts through the _____ and _____ rooms.

    _____

24. What room is shown as being larger in the first-floor renovations than it was in the Existing First Floor Plan?

    _____

25. The original hot water heater is to be replaced (T or F?).

    _____

26. What type of flooring is to be placed in the former bathroom area?

    _____

27. What is the total run of the stairway?

    _____

28. The stair risers are _____ high, and the treads are _____ wide.

    _____

29. How many risers and treads does the stairway have?

    _____

30. What is the inside width measurement of the stairway?

    _____

31. The length of the second-floor landing is _____. _____

32. How do the new windows in the west living room wall differ from the original windows? _____

33. What are the code numbers of the dining room bay windows? _____

34. Between what rooms are new walls to be constructed? _____

35. The metal firebox is to be installed at the (give compass direction) _____ wall of the living room. _____

36. How wide is the opening in the wall separating the living room and dining room? _____

37. What is the total width of the wall between the living room and dining room, and what is the stud size? _____

38. The tile hearth in front of the fireplace is _____ wide and _____ long. _____

39. Under what division does the specifications provide more information about the fireplace? _____

40. The center of the fireplace is _____ from the east living room wall. _____

41. Give the depth and width of the new bookcases next to the fireplace. _____

42. The symbol 4/ A-2 refers to the _____ elevation drawing. _____

43. The symbol 9/A-6 refers to what detail drawing? _____

44. How far does the master bedroom bay project from the east wall? _____

NEW SECOND FLOOR PLAN

The addition of a second story to the house is described in this floor plan. This second story includes two bedrooms, two bathrooms, and a study. Note the rectangle with dashed lines and curved corners surrounding the two bathrooms. The rectangle identifies an enlarged floor plan for the bathrooms only, shown on Sheet A4.

NEW SECOND-FLOOR PLAN QUESTIONS

Enter the missing words in the spaces provided on the right.

45. What is the total length of the north wall of the second-floor addition?

   _____

46. What is the total length of the west wall of the second-floor addition?

   _____

47. _____ is the measurement from outside the west wall to the center of the bathroom wall.

   _____

48. What type of door opens from the master bedroom to the master bathroom?

   _____

49. What is the distance from the outside face of the north wall of the master bathroom to the center of the 2 × 6 wall next to the shower stall?

   _____

50. How wide (N to S) is the shower stall in the smaller bathroom?

   _____

51. How many pocket sliding doors are shown on the second-floor plan?

   _____

52. According to the specifications, the interior doors are _____ thick.

   _____

53. How wide is the second floor balcony?

   _____

54. The _____ of the bay in the master bedroom should be aligned with the window below.

   _____

55. What is the distance from the northeast corner of the building to the center of the first bedroom window in the north wall?

   _____

56. Give the diameter and name of the material of the handrail at the stairway.

   _____

57. The center of the double doors leading to the balcony is _____ from the center of the north wall of the study.

   _____

58. Swinging doors open to the bedroom closets (T or F?).

   _____

59. What items are identified inside the bedroom closets?

   _____

60. What are the angles of the side walls of the second-floor bays?

   _____

61. The exterior east wall of the new second-floor addition extends past the south wing of the existing first floor (T or F?).

    _____

62. How many swinging doors are shown for the second-floor addition?

    _____

SHEET A2
EXTERIOR ELEVATION DRAWINGS

The Exterior Elevation Drawings show the west (front), east (rear), south (side), and north (side) exterior walls. As you study each wall drawing, relate it to the same wall shown on the floor plans. Also identified is the wall siding, trim around the windows and doors, and roof finish materials. Vertical heights from floor to floor and from grade to top of roof are included.

The *west elevation* shows double-hung windows with divided lights. Fixed windows are set between the double-hung windows in the dining room wall and the first and second-floor bays. The front entrance doors are also seen.

The *north elevation* provides a view of one side of the bays extending from the first and second floors. At the first floor of the north elevation there are a pair of casement windows replacing the previous double hung windows. At the second floor, two casements are shown. The roof skylight is also visible.

The *east elevation* shows casement and double-hung windows. Sliding glass doors open from the first-floor breakfast room to the outside of the building. French doors open from the first-floor living room and upstairs study. The *south elevation* includes awning windows hinged at the top as well as casement windows.

SHEET A2 QUESTIONS

Enter the missing words in the spaces provided on the right.

63. What type of wood siding is shown on all the elevation drawings?

    _____

64. Name the finish roofing material?

    _____

65. For which rooms are windows shown at the first-floor of the west elevation drawing?.

    _____

66. Aluminum windows are shown for the second-floor addition (T or F?).

    _____

67. What is the height from the existing grade to the existing first floor as shown in the west elevation?

    _____

68. How many casement windows are shown for the entire building?

    _____

69. The top of the chimney extends a distance of _____ from the highest point of the roof.

    _____

70. What material is used for roof flashing?

    _____

71. Give the width and height of the shutters shown at the second floor of the west elevation.

72. The west elevation shows a distance of _____ from the first floor to the top of the roof.

73. Give the distance from grade to the second floor.

74. What is the unit rise of the second-story addition roof?

75. The roof of the second-floor addition shows a 1'-6" overhang (T or F?).

76. The gutters and rainwater leaders at the second floor are made of what material?

77. The elevation drawings show a total number of _____ vents for air circulation in the crawl space.

78. What type of doors open to the second-floor balcony?

79. Which elevation drawing shows two flue pipes projecting from the roof?

80. What types of windows are shown in the south elevation?

SHEET A3
SECTIONS

Four building section drawings provide structural information not found on any of the other drawings. Again, relating these views with the floor plan is necessary. Also included are views of an older type *battered foundation* around the crawl space and several rows of pier footings supporting posts and girders. A T-shaped *spread foundation* is shown beneath the north bathroom wall, suggesting that a section of building was added after the original construction.

The *Section at Front* drawing is identified by section lines 1/A-3 in the New First Floor Plan and the New Second Floor Plan that point toward the east (rear) side of the building. At the first-floor level, the section line runs right to left through a bathroom, hallway, living room, dining room, closet, and a second bathroom. At the second-floor level the section line runs through a bedroom, linen closet, bathroom, and master bathroom. A new row of pier footings and posts will be placed in the crawl space close to the battered foundation wall to help support the weight of the south wall of the second-floor addition.

The *Cross Section Facing South* drawing is identified by section lines 2/A-3 in the New First Floor Plan and New Second Floor Plan that point toward the south side of the building. At the first-floor level the section line runs front to back through the dining room and kitchen. At the second-floor level the section line runs through the master bathroom and master bedroom. Side views of the bays can be seen extending from the front and back of the building.

The *Section at Rear* drawing is identified by section lines 3/A-3 in the New First Floor Plan and New Second Floor Plan that point toward the east (rear) of the building. At the first-floor level the section line runs through a bedroom, stairs, living room, kitchen, and breakfast room. At the second-floor level the section line runs through the study and master bedroom.

The *Cross Section Facing North* drawing is identified by

lines 4/A-3 in the New First Floor Plan and the New Second Floor Plan that point toward the north side of the building. At the first-floor level the section line runs front to rear through the living room. At the second-floor level the section line runs through a bedroom and study

As previously mentioned, it is very important to refer to the floor plans when studying the different building section drawings. The arrows at the end of the section lines point in the direction of the view seen in the section drawings.

SHEET A3 QUESTIONS

Enter the missing words in the spaces provided on the right.

81. What is the floor-to-ceiling height of the second-floor addition?

    _____

82. The new position of the skylight is over the _____ _____.

    _____

83. The Section at Front drawing shows a door leading into the kitchen (T or F?).

    _____

84. In the Section at Front drawing, the door shown in a hallway next to the bathroom opens to a _____ in the first floor.

    _____

85. What species of lumber should be used for all structural work? (See specs.)

    _____

86. The existing concrete pier footings are _____ wide, _____ high, and spaced _____ OC.

    _____

87. Give the size and spacing of the wall studs.

    _____

88. According to the specifications, loose-fill insulation is used between the ceiling joists (T or F?).

    _____

89. In addition to the roofing composition shingles, the specifications also mention_____, _____, and _____.

    _____

90. What is the R-value of the exterior wall insulation?

    _____

91. Give the size and spacing of the roof rafters over the second-floor addition.

    _____

92. The Cross Section Facing North shows an opening between the _____ and _____ rooms at the first floor.

    _____

93. The top of the roof is a distance of _____ above the second floor.

    _____

94. What is the concrete slab shown in the Cross Section Facing South?

    _____

95. Which section drawing does not show the new row of pier footings?

_____

96. Combining the first and second floors, give the number of bedrooms and bathrooms in the building.

_____

97. What do the series of vertical lines identify in the master bedroom as shown in the Cross Section Facing South?

_____

98. The symbol 5/A-6 refers to a balcony drawing showing the _____ and _____ .

_____

99. How high is the railing around the stairwell?

_____

100. What is the spacing between the I-joists placed beneath the second-floor addition?

_____

101. What species and grade of lumber must be used for all horizontal framing members? (See specs.)

_____

102. What is the rating of insulation material over the upstairs bedroom as shown in the Cross Section Facing North?

_____

103. Name the rooms in the second-story addition that have sloped ceilings.

_____

104. Give the size and spacing of the of the girders supporting the first floor joists.

_____

105. In the Cross Section Facing North, a _____ is identified by parallel dashed lines from the fireplace to the chimney.

_____

106. There are no heating ducts indicated in any of the building section drawings (T or F?).

_____

SHEET A4
BATHROOM PLANS AND LIVING ROOM AND STUDY ELEVATIONS

The bathroom plans on this sheet include a plan view of the two bathrooms in the second-floor addition and interior elevation drawings of the bathroom. The floor plan drawing is an enlargement of the area enclosed in dashed lines on the New Second Floor Plan shown on Sheet A1. The larger room with two sinks and a bathtub is designated as the master bathroom and opens from the master bedroom. The smaller bathroom opens from the hallway between the bathroom and the second bedroom. Note the north compass direction shown below the plan. Compare the enlarged plan view drawing of the bathrooms with the full New Second Floor Plan. Interior elevations are provided for the north, south, east, and west walls of the two bathrooms. The four master bathroom elevations are shown at the bottom of the sheet. The second-bathroom elevations appear in the second row of the drawings. It is important to relate each elevation drawing to its position on the floor plans.

The *Living Room North Elevation* on Sheet A4 clarifies the relocation of the fireplace. New cabinets and open shelves are also described.

The main purpose of the *Living Room and Study South Elevation* drawing on this sheet is to show an elevation view of the stairway leading from the first-floor living room to the second-floor study, and the second-floor railing around the stairwell.

SHEET A4 QUESTIONS

Enter the missing words in the spaces provided on the right.

107. What is the floor finish in the bathrooms?

108. Name the material used for the shower doors.

109. A _____ medicine cabinet is shown in the wall between the two bathrooms.

110. Give the dimensions of the fiberglass shower base in the smaller bathroom.

111. What items run up the framed section next to the sink in the smaller bathroom?

112. Plastic laminate is used to finish off the sink counter in the master bathroom (T or F?).

113. Give the overall length and height of the base cabinet and counter in the master bathroom.

114. The base cabinet in the master bathroom contains a total of _____ working drawers and _____ cabinet doors.

115. Wood base is shown at the bottom of the wall behind the toilet in the master bathroom (T or F?).

116. What scale is used for the interior elevation drawings?

117. Give the width and height of the upper cabinet over the toilet in the master bathroom.

118. How high is the backsplash below the wood shelf over the counter?

119. How many towel bars are shown in the master bathroom?

120. The bathtub enclosure is _____ wide.

121. A large mirror is located above the base cabinet and counter (T or F?).

122. Name the wall material above the ceramic tile in the master bathroom.

_____

123. What should be done to protect the exposed corners and edges of the gypsum wall board? (See specs.)

_____

124. What type of window is shown in the master bathroom?

_____

125. What is the base material used at the bottoms of the walls in the smaller bathroom?

_____

126. How long is the towel bar in the master bathroom?

_____

127. The wall between the two bathrooms is constructed with _____ studs.

_____

128. The distance from the floor to the top of the window opening in the smaller bathroom is _____.

_____

129. The sink in the smaller bathroom is set in a base cabinet (T or F?).

_____

130. What is the floor finish in the shower stall of the master bathroom?

_____

NORTH ELEVATION AND SOUTH ELEVATION FOR LIVING ROOM AND STUDY

The _Living Room North Elevation_ clarifies the relocation of the fireplace. New cabinets and open shelves are also described. The main purpose of the _Living Room and Study South Elevation_ interior drawing is to show an elevation view of the stairway leading from the first-floor living room to the second-floor study, and the second-floor railing around the stairwell.

NORTH ELEVATION AND SOUTH ELEVATION FOR LIVING ROOM AND STUDY QUESTIONS

Enter the missing words in the spaces provided on the right.

131. What is the vertical height of the stair railing?

_____

132. The maximum spacing of the balusters is _____.

_____

133. What type of material is used to finish off the bottoms of the walls?

_____

134. What type of material is used at the joint between the wall and ceiling in the living room?

_____

135. The stair treads are made of oak material (T or F?).

_____

136. A _____ finish is required for the wood cabinets in the living room.

_____

137. What material is used around the face of the fireplace?

_____

138. What is the height of the base cabinets in the living room?

_____

139. The opening in the north living room wall leads to the _____ room.

_____

140. A completely new mantle is to be used over the fireplace (T or F?).

_____

SHEET A5
FINISH WINDOW AND DOOR PLAN, DOOR AND WINDOW SCHEDULES, ELECTRICAL AND MECHANICAL PLANS, PLUMBING FIXTURE SCHEDULE, AND ELECTRICAL FIXTURES AND SYMBOLS

The Stafford Residence plans include an additional *Finish Window and Floor Plan* as well as *Door and Window Schedules*. The *finish* designation includes the finish materials used for the floors, walls, ceiling, and base trim. Code numbers and letters shown for the windows and doors refer you to the schedules below the Finish Window and Door Plan.

The door schedule, besides giving the finish door size, has columns for the code letter operation, style, hardware, and comments. The style column explains that these are all stock doors. Only the finish door sizes are given. Therefore, when determining the rough width and height of the framed opening, you must include the thicknesses of side jamb materials and clearances, plus the head jamb and clearances above and below the door.

The Windows, Skylights, and Exterior Glass Doors Schedule includes columns for the code number, nominal size, operation, model number, frame, and comments. The *frame* designation gives the rough opening for the windows, so they will not have to be calculated on the job.

FINISH WINDOW AND DOOR PLAN DRAWING AND SCHEDULES QUESTIONS

Enter the missing words in the spaces provided on the right.

141. Name the floor finish material in the two bathrooms.

142. What thickness and type of finish material is placed on the walls and ceiling?

143. Name the coats of paint required for the flat finish of gypsum board. (See specs.)

144. What type of finish floor material is required in the study and master bedroom?

145. A _____ or _____ finish is required for the interior trim. (See specs.)

146. What is the code letter of the door between the master bedroom _____ and master bathroom?

147. How many No. 2 windows are there in the second-floor addition?

148. A pair of _____ doors opens from the study to the balcony and the dimensions of each door are _____ × _____.

149. Name the recommended manufacturer of the windows.

150. What is the size, operation, and style of the individual closet doors in the master bedroom?

151. All the interior doors are _____ thick.

152. The pocket sliding door opening to the smaller bedroom is a hollow-core panel door (T or F?).

153. The model number of the side bay windows opening from the dining room on the first floor is _____. (Refer to New First Floor drawing, Sheet A1.)

154. What comments are there in the window schedule regarding the skylight?

155. Which room has tile base?

156. The contractor must consult with the _____, regarding floor pile selection.

ELECTRICAL AND MECHANICAL PLAN AND PLUMBING AND ELECTRICAL FIXTURE SCHEDULES

The *Electrical and Mechanical Plan* for the Stafford Residence primarily concerns itself with the electrical and heating systems for the second-floor addition. Also shown are the bathroom and kitchen plumbing fixtures.

The *Plumbing Fixtures Schedule* identifies the manufacturer's name and model number of fixtures that are not to be selected by the owner. The *Electrical Fixtures Schedule* identifies the light fixtures and in this case also includes the electrical symbols used in the Electrical and Mechanical Plan.

ELECTRICAL AND MECHANICAL PLAN PLUMBING AND ELECTRICAL FIXTURE SCHEDULES QUESTIONS

Enter the missing words in the spaces provided on the right.

157. What does the abbreviation GFI stand for? _____

158. Which room shows wall-mounted lights? _____

159. Who provides all the electrical fixtures shown on the plans? (See specs.) _____

160. Where is a wall mounted exterior light shown on the plan? _____

161. What are the heights of the GFI plugs in the bathrooms? _____

162. Which rooms feature a ceiling-mounted light and overhead fan? _____

163. What rooms show three-way wall switches? _____

164. _____ type lights are placed above the bathroom sinks. _____

165. All the rooms have dimmer wall switches (T or F?). _____

166. Which room has a floor-mounted duplex outlet? _____

167. Which rooms have recessed ceiling lights? _____

168. There are a total of _____ heat registers in the whole building? _____

169. Which rooms show half-hot duplex outlets? _____

170. Individual _____ turn on the recessed fan/heat lamps in the bathrooms. _____

171. What are the heights of the wall switches unless otherwise noted? _____

172. Name the manufacturer and model number of the recessed fan/heat lamps. _____

173. There is an attic access shown in the ceiling above the hallway (T or F?).

_____

174. How many 3-way switches are identified in the electrical and mechanical plan?

_____

175. Unless otherwise noted, who makes the final selection of the light fixtures?

_____

176 In which rooms are heating ducts shown on the plan?

_____

177. Crossed dashed lines in the master bedroom and study identify a _____ ceiling.

_____

178. What is the finish on all plumbing fixtures and faucets unless otherwise noted?

_____

179. Where is the source of the hot air that will run through the ducts heating the second floor?

_____

180. What choices of manufacturers are given for the bathtub in the master bathroom?

_____

SHEET A6
DETAILS

All detail drawings refer to reference symbols appearing in the *New Second-Floor Plan* (A-1), *section* drawings (A-3), and the *Bathroom Plan and Interior Elevations* (A-4). Therefore, each detail drawing should be studied together with these plans. Following are the detail drawings appearing on Sheet A-6:

Detail 1—Skylight Curb: The main consideration, in installing the skylight is its proper construction to prevent future water leakage. Refer to the *Section at Front* drawing (Sheet A-3).

Detail 2—Roof Ridge: A close-up view is provided of the ridge construction, attic area, and the level and sloping ceiling joists below. Refer to the *Cross Section Facing South* drawing (Sheet A-3).

Detail 3—New Wall at Existing Roof: A large segment of the existing roof has to be dismantled to allow construction of the second-story addition. The remaining roof sections must be reattached to the new walls so there will be no water leakage. Refer to the *Section at Rear* drawing (Sheet A-3).

Detail 4—Wall Section at Roof: This drawing provides details of the typical roof overhang at the exterior walls. Refer to the *Section at Front* drawing (Sheet A3).

Detail 5—Deck Guardrail: The guardrail must be safely secured to the balcony deck. This is done by fastening short posts to the joists below and then attaching the guardrail to the top of the posts. The guardrail itself consists of vertical balusters nailed between top and bottom rails. Refer to the *Cross Section Facing North* section drawing (Sheet A-3).

Detail 6—Floor Section at Deck: The deck must be solidly fastened to the building and measures taken to prevent water leakage at the juncture of the deck and building. Support for the deck is provided by cantilevered joists. These cantilevered joists extend back into the building and project the width of the balcony. Water leakage is prevented with flashing and sealant materials. Note the 4 × 6 blocking placed between the existing ceiling joists over the first floor. A 2 X 4 plate is placed on top of the blocking and ceiling joists, and the new I-joists rest on the plate. Refer to the *Cross Section Facing North* section drawing (Sheet A3).

Detail 7—Cantilevered Floor: The bay in the rear wall of the second floor addition extends past the supporting wall of the lower floor. Refer to the *Cross Section Facing South* (Sheet A-3) and the *North Elevation* drawing (Sheet A-2). In this case the I-joists will cantilever out beneath the bay section. Note that new insulation is to be placed where the joists project from the wall. The underside of the bay is enclosed except for a continuous screened vent strip.

Detail 8—Floor Section: The purpose of this detail is to show typical floor and wall construction. Refer to the *New Second Floor Plan* (Sheet A-1).

Detail 9—Typical Stair: The drawings show a side view and frontal view of a section of the stairway. Information is provided about the treads and risers and handrails. The frontal drawing shows that wood base is to be nailed against the wall before the wall stringer is placed. Refer to the the *New First Floor Plan* (Sheet A-1) and the *Living Room and Study South Elevation* drawings (Sheet A-4).

Detail 10—Guardrail: Three different related drawings provide details about the guardrail around the stairwell. The *Living Room and Study South Elevation* drawing (Sheet A-4) shows posts and balusters. Three posts are used with balusters placed between the posts.

The posts are bolted to pieces of laminated veneer lumber (LVL) that finishes off three sides of the stairwell opening. The *plan view* at the top of Detail 10 looks down on the post and LVL and shows that metal angle brackets are also used to help secure the post. The *section* view at the right side of the drawing adds more information about the post. Refer also to the *Living Room and Study South Elevation* drawing (Sheet A-4).

Detail 11—Base Cabinet: This detail provides a side elevation drawing of the base cabinets found on either side of the fireplace and at the northwest corner of the living room wall. Refer to the *Living Room North Elevation* drawing (Sheet A4).

Detail 12—Bookshelves: Open bookshelves fill the space between the base cabinets and the ceiling. A blind panel covers a corner section of the bookshelves where the flue angles up from the fireplace. Refer to the *Living Room North Elevation* drawing (Sheet A4).

SHEET A6 QUESTIONS

Enter the missing words in the spaces provided on the right.

181. What scale is used for the detail drawings?

    _____

182. The skylight curb is constructed of _____ material.

    _____

183. What is placed at the juncture of the upper skylight curb and the roof to prevent leakage?

    _____

184. Name the inside finish material in the skylight shaft.

    _____

185. How does the drawing describe the skylight cover?

    _____

186. Give the thickness and width of the ridge rafter.

    _____

187. The Roof Ridge detail shows that Simpson metal angle clips are used to secure the _____ to the _____.

    _____

188. Name the thickness and type of material used for roof sheathing.

    _____

189. Give the thickness, width, and spacing of the ceiling joists below the roof rafters.

    _____

190. The roof rafters and ceiling joists are the same size (T or F?).

    _____

191. The short supporting wall in the enclosed attic area is constructed of _____ size studs spaced _____ OC, and the face of the wall is _____ from the center of the ridge.

_____

192. Give the type and rating of the insulation placed between the ceiling joists.

_____

193. What supports the end of the existing roof where it butts up against the new wall?

_____

194. How much does the new wall siding overlap the GSM flashing?

_____

195. Plywood 3/4″ thick is used for sheathing on the new wall (T or F?).

_____

196. What type of metal fastener ties the ends of the rafters to the top of the stud wall as shown in the Wall Section at Roof detail?

_____

197. A _____ thick by _____ wide _____ board is nailed to the ends of the roof rafters.

_____

198. What is provided between the rafters to allow air circulation in the attic area?

_____

199. The end view of a _____ is shown nailed to the fascia board.

_____

200. Give the size and lumber species of the decking material.

_____

201. The 2 × 2 balusters are nailed directly to the top and bottom rails (T or F?).

_____

202. The redwood posts are spaced _____ OC and are fastened with _____ to the 2 × 10 joists below.

_____

203. How far do the posts extend above the deck?

_____

204. What is the height of the guardrail from the top of the posts?

_____

205. The space between the 2 × 6 deck material is _____.

_____

206. The top rails must have _____ edges.

_____

207. What prevents air and water infiltration beneath the the door to the balcony?

_____

208. New insulation has been placed between the 2 × 10 ceiling joists (T or F?).

_____

209. What written description is given for the door in the Floor Section at Deck detail?

_____

210. In order to prevent water infiltration where the porch deck is fastened to the building, _____ and _____ are placed over the top row of siding.

_____

211. Name the type of finish floor material shown in the cantilevered floor detail.

_____

212. Give the size and spacing of the second-floor wall studs.

_____

213. Give the name and thickness of the subfloor material.

_____

214. _____ _____ is used as the interior wall finish.

_____

215. What rating is the wall insulation?

_____

216. Give the thickness and name of the material nailed to the underside of the bay.

_____

217. What is the rating of the floor insulation placed in the cantilevered section?

_____

218. The Floor Section shows that there is a _____ at the perimeter of the hardwood flooring.

_____

219. What is nailed between the ends of the I-floor joists?

_____

220. Describe the thickness and type of subflooring to be used.

_____

221. In addition to being nailed, the subfloor must be _____ to the joists.

_____

222. Which detail drawing contains a description of the flashing required at the juncture of the wall and lower roof?

_____

223. There is a double top plate in the framed wall supporting the floor joists (T or F?).

_____

224. What must be placed between the wall siding and sheathing?

_____

225. What material is used for the treads?

_____

226. Give the diameter and lumber species of the handrail.

_____

227. What is the vertical height of the handrail?

_____

228. The total width of the stair tread material including the nosing is _____.

_____

229. How many stringers are required, and from what size material are they made?

_____

230. Name the finishes on the tread and risers.

_____

231. What is placed in the wall to secure the handrail bracket?

_____

232. A size _____ cleat is used to hold the bottom of the stringer away from the wall.

_____

233. What size, type, and number of bolts are used to secure the guardrail post to the LVL header?

_____

234. How high is the top of the guardrail post from the finish floor?

_____

235. The cross-section dimensions of the balusters are _____.

_____

236. Give the size and description of the top rail.

_____

237. How far down is the rail from the top of the post?

_____

238. Oak trim is shown at the top and bottom of the posts (T or F?).

_____

239. Name the finish and the number of coats to be used on the interior trim and doors (See specs.)

_____

240. How high and deep is the recessed toe space at the bottom of the cabinets?

_____

241. The thickness and width of the band (molding around the face of the cabinet doors) is _____ at the top and bottom and _____ at the sides.

_____

242. How wide is the edgeband that gives a finished appearance to the front of the shelves?

_____

243. There is one adjustable shelf inside the cabinet (T or F?).

_____

244. How many hinges are required for each cabinet door?

_____

245. What is the total depth of the base cabinet?

_____

246. Give the dimensions and name of the trim material placed at the juncture of the ceiling and bookshelves.

_____

247. Describe the surface finish for the bookshelves.

_____

248. The blind panel covers a space containing the _____.

_____

249. The blind panel is recessed a distance of _____ from the front face of the bookshelves.

_____

250. The shelves are made of a single piece of 1 1/2″ plywood (T or F?).

_____

STAFFORD RESIDENCE SPECIFICATIONS

DIVISION 1: GENERAL CONDITIONS

Sec. 1. All work shall comply with the Uniform Building Code, 1991 Edition, and any other codes, rulings, or regulations having jurisdiction.

Sec. 2. The general contractor is responsible for checking contract documents, field conditions, grades, and dimensions for accuracy and confirming that work is buildable before proceeding with construction.

Sec. 3. The general contractor shall obtain all necessary construction permits required. Any discrepancies shall be called to the attention of the architect and structural engineer prior to the start of construction.

DIVISION 2: SITE WORK

Sec. 1. All excavation and demolition shall be the responsibility of the general contractor. Existing conditions shall be removed as required.

Sec. 2. Care shall be taken not to damage existing structure and finishes. Damage to existing building, adjacent properties, or utilities shall be restored, repaired, or replaced.

DIVISION 3: CONCRETE AND REINFORCEMENT

Sec. 1. Perform cast-in-place concrete work in accordance with applicable ACI standards, the Uniform Building Code, and the structural engineer's specifications.

Sec. 2. Concrete shall meet the requirements of 2500 psi at 28 days.

Sec. 3. Exposed surfaces of concrete shall be kept moist or cured by protective coatings or coverings.

Sec. 4. Reinforcing shall be in accordance with ASTM 615, grade 40, and shall be free from rust or other coatings that will reduce bond. Fabricate and place in accordance with ACI standards, the UBC and the structural engineer's specifications.

DIVISION 4: WOOD AND FASTENING

Sec. 1. All structural lumber and fastenings shall conform to the Uniform Building Code and the structural engineer's specifications.

Sec. 2. Sills on concrete shall be pressure-treated Douglas Fir bedded in 1:23 cement group 1/2" thick. Bolt to concrete with 3/4" diameter anchor bolts at 4'-0" OC, with a bolt within 9" of each end piece, and a minimum of two bolts per sill. Embed bolts 8" minimum in concrete.

Sec. 3. Provide 2× blocking between all joists over supports. In all wood stud walls and partitions provide 2× continuous horizontal blocking placed at mid-height of studs.

Sec. 4. All lumber to be Douglas Fir. Light framing to be standard or better. Posts and all horizontal framing to be Douglas Fir #1.

Sec. 5. Arrange all joists and horizontal members with crown up. Floor joists to be manufactured I-joists. Fabrication and installation shall conform to the manufacturer's and structural engineer's specifications, and the UBC.

Sec. 6. All nails shall be common wire nails, unless otherwise noted. Bolts shall be in accordance with ASTM A307. Nails and bolting shall be in accordance with the requirements of the Uniform Building Code.

Sec. 7. Timber fasteners are referred to by "Simpson Company" designation. Fasten in accordance with manufacturer's specifications.

DIVISION 5: THERMAL AND MOISTURE PROTECTION

Sec. 1. Furnish and install all labor and materials needed to provide a complete weather-tight and waterproof roofing system in compliance with all applicable building codes.

Sec. 2. Roofing shall be composition shingle roofing, to match existing. The complete installation of the roofing, including shingles, felts, flashing, and counter flashing, shall be in complete accordance with the manufacturer's and the National Roofing Contractors Association standards and recommendations, and the Uniform Building Code. Verify manufacturers and products: Johns-Manville Fire King III, Celotex Fiber Glass Asphalt Shingles/25, GAF Royal Sovereign, or equal.

Sec. 3. Provide fiberglass batt insulation at exterior wood stud walls and ceilings to R-values as shown on plans. Insulation shall conform to the California Residential Energy Conservation Standards. Caulk all penetrations.

DIVISION 6: DOORS AND WINDOWS

Sec. 1. Provide new doors and windows to dimensions given on door and window schedule. Weatherstrip all exterior doors and windows. Provide all necessary hardware for proper hanging and operation of doors and windows.

Sec. 2. Exterior French doors to be double-glazed, 1 3/4" thick. Interior doors to be 1 3/8" thick. Provide "Pemko" metal threshold at exterior door, or approved equal.

Sec. 3. Windows to be clad wood, awning, casement, double-hung, or fixed, double-glazed or equal. All glazing shall conform to the California Residential Energy Conservation Standards.

DIVISION 7: FINISHES

Sec. 1. Contractor to provide 1/2" gypsum wallboard at all new walls and ceilings, and W/R gypboard at bathroom. Gypboard to be taped, free from imperfections, with smooth texture. Provide metal edges and J-clips at all exposed corners and edges.

Sec. 2. Contractor to provide and install ceramic tile at bathrooms, thickset at floors, mastic set at walls over cement board. Install per *Tile Council of America's Handbook for Ceramic Tile Installation* recommendations. Coordinate tile with owner's preferences prior to installation.

Sec. 3. Provide new hardwood flooring to match existing at locations shown on drawings. Install per *NOFMA/OF Hardwood Flooring Installation Manual* recommendations. Provide new wood base and wood trim at interior walls, to match existing.

Sec. 4. Provide exterior wood siding to match existing, over building paper, as shown on plans. Provide new wood trim to match existing.

Sec. 5. Provide paint or painted finishes at all new and remodeled interior and exterior surfaces. Recommended manufacturers: Kelley-Moore, Sherwin Williams, Benjamin Moore, Fuller-O'Brien, Glidden.

Sec. 6. Recommended paint systems:

a. Flat finish on gypsum wallboard—one coat latex sealer, one coat flat latex. Semigloss finish on gypsum wallboard at bathrooms and kitchen: one coat latex sealer, two coats latex semi-gloss enamel.

b. Semigloss finish on interior trim and doors: one coat acrylic enamel undercoater, two coats acrylic semi-glass enamel.

c. Transparent finish on interior trim and doors: one coat stain, two coats polyurethane. Transparent finish on wood floor: three coats polyurethane.

d. Exterior wood siding and trim: one coat exterior wood primer, two coats semi-gloss acrylic latex.

DIVISION 8: PLUMBING AND MECHANICAL

Sec. 1. All mechanical and plumbing work shall be done in strict accordance with applicable code requirements. Provide all necessary connections, obtain all permits, and pay all fees required.

Sec. 2. Plumbing fixtures shall be as shown on plumbing schedule or as determined by owner.

Sec. 3. Mechanical contractor to reuse existing natural gas-fired central heating furnace. Mechanical contractor shall provide proper information on sizing, ductwork, insulation, operation, and any other pertinent information for conformance to all regulations and codes, and install accordingly.

Sec. 4. Final installation, sizing, and balancing of the entire heating system shall be the responsibility of the mechanical contractor.

Sec. 5. Provide automatic setback thermostat. Heating system shall conform to the California Residential Energy Conservation Standards.

Sec. 6. Provide new 50-gallon natural gas fired hot water heater, and install as required.

Sec. 7. Plumbing contractor to provide all bathroom fixtures as shown on plans, including sinks, faucets, toilets, tub, and shower, and install as required. Plumbing contractor shall verify all manufacturers and model numbers with owner prior to installation.

DIVISION 9: ELECTRICAL

Sec. 1. All electrical work shall be done in strict accordance with applicable code requirements, and the California Residential Energy Conservation Standards.

Sec. 2. Electrical contractor to provide all necessary electrical service, wiring, circuits, switches, duplexes, and fixtures necessary to complete the work as shown on the plans or otherwise required. Electrical contractor to provide all electrical fixtures as shown on plans.

Sec. 3. Electrical contractor shall verify all manufacturers and model numbers with owner prior to installation.

DIVISION 10: SPECIALTIES

Sec. 1. Fireplaces to be approved metal fireplaces. Fireplace must conform with all applicable code requirements.

Sec. 2. Contractor to provide and install firebox, glass doors, combustion air kit, flue termination with built-in spark arrester, flashing and storm collar, firestops, flue sections, offset elbows, and any other parts necessary for proper fireplace installation and operation.

Sec. 3. Provide and install tile hearth set in grout, wood mantle, and any other finishes at fireplace. Coordinate finishes with owner prior to installation.

Assignment 3
Pulver
House Plan

The Pulver House plan is for a two-and-one-half-story residence with an adjoining garage in a suburban area close to Atlanta, Georgia. (**See Figure 38–1** and **Figure 38–2.**) The design takes up a small area of the lot. This allows for a greater amount of landscaping and retains more of the existing vegetation. However, at the same time this design allows for room spaces equal to those of a one-story house spread out over a much greater area.

The complete set of the Pulver House plan does not include an electrical or mechanical plan. Some drawings of less complex small building design, like the Pulver House, do not include an electrical or mechanical plan, but instead leave it to the electrical and plumbing subcontractors to make shop drawings for their work. In place of specifications, a *Description of Materials* list is used, as described in Unit 14, (p. 163), also shown at end of this unit. The plan for the Pulver House has been prepared by architect Tim Pulver of Marietta, Georgia.

SHEET A1
SITE PLAN AND AREA MAP

The main property area is shaped like a partial trapezoid, with an additional area extending at a right angle from the east property line. A grass-earth driveway begins at the end of the existing paved road and leads to the driveway circle. The driveway then continues to a brick courtyard with a brick walk on the north side extending to the entrance of the building. A grass-earth driveway continues from the east side of the courtyard and leads to the garage. A right of way (ROW) area outside the Pulver property boundaries is identified. This is an easement allowing the placement of the driveway leading into the Pulver property.

In this plan the front, sides, and rear setbacks do not measure directly to the house. The lines identified show the boundaries of the yard area where the house can be placed. To establish the building lines, measurements are given to the southeast corner of the garage section of the building. At that point a transit is set up and establishes the angle of the wall extending to the northeast corner. The remaining walls are laid out at 90-degree angles to each other.

Figure 38–1.
Front view of the Pulver House as shown in the south elevation drawing.

Figure 38–2.
Partial side view of the Pulver House as shown in the east elevation drawing.

The septic absorption field is toward the front of the sloping lot. The septic design notes refer to a future additional absorption field in the rear portion of the lot if this becomes necessary. A pump system is required because of the upward slope of the lot.

The Area Map shows the subdivision in which the Pulver House is found. The main streets are shown and identified. Lot outlines are shown and the exact location of the Pulver House is shown in the shaded area.

SHEET A1 SITE PLAN AND AREA MAP QUESTIONS

Enter the missing words in the spcaes provided on the right.

1. What is the width of the driveway running from the end of the road to the circle? _____

2. The diameter of the driveway circle is _____. _____

3. How wide is the right of way (ROW) area? _____

4. Name the subdivision where the Pulver House is found. _____

5. The _____ property adjoins the west and north sides of the job site. _____

6. What is the zoning of the Pulver property? _____

7. Name the road leading to the property. _____

8. Give the elevation of the highest point of the lot. _____

9. The total distance of the front yard setback is _____. _____

10. What is the grade difference between the contour lines? _____

11. What is the total length of the east property line? _____

12. Give the total length of the west property line. _____

13. What is the difference in length between the west and east property lines? _____

14. The distance between the outer driveway circle perimeter and the interior landscaped circle is _____. _____

15. Give the length of the south property line. _____

16. The main area of the lot is shaped like a rectangle (T or F?). _____

17. What type of walk runs along one side of the circle and the driveway leading to the road? _____

18. What is the diameter of the inner landscaped circle? _____

19. Give the distance between the southeast corner of the lot to the straight-line intersection of the projecting property. _____

20. What is the exact angle of the straight property line projecting from the east property line? _____

21. The contour elevation along the existing creek is _____. _____

22. How wide is the grass-earth driveway leading from the circle to the brick court? _____

23. What are the length and width of the brick court? _____

24. Name the septic absorption system in the absorption field. _____

25. The scale of the site plan is _____. _____

26. What is the volume of the septic tanks? _____

27. How long is the straight property line of the projecting property? _____

28. What is the total length of the curved line of the projecting property? _____

29. The angle of the property line running from the northwest to northeast corners is _____. _____

30. Give the elevation of the invert pipe leading from the house to the septic tank. _____

31. The garage slab is at the same elevation as the first floor of the house (T or F?). _____

32. What is the elevation of the garage slab? _____

33. What are the west and east side yard setbacks? _____

34. The width of the walk leading from the brick court to the building is _____. _____

35. At which wall of the building (compass direction) is the electrical service panel? _____

36. How far is the southeast corner of the building from the east property line? _____

37. The SE corner of the building is 152'-7 7/8" from the center of the driveway circle (T or F?).

_____

38. What is the difference in elevation between the NW and SE corners of the lot?

_____

39. The ground elevation at the NE corner of the house is _____.

_____

40. What is the distance between the top of the garage slab and the ground elevation?

_____

SHEET A1.1
FOUNDATION PLAN

Section views are given for the typical crawl-space footing, turned-down slab, pier footing, and junction between the garage slab and house foundation. The typical wall section provides details of the outside walls, floors and ceilings, and the roof-attic area.

SHEET A1.1 FOUNDATION PLAN QUESTIONS

Enter the missing words in the spcaes provided on the right.

41. What is the total combined west-to-east dimension of the house foundation _____ and the garage?

_____

42. What are the width and height of a typical crawl-space footing?

_____

43. _____ size rebars run across the width of the footing and are spaced _____.

_____

44. Give the minimum distance allowed from the top of the crawl space footing to finish grade.

_____

45. The rebars running across the length of the crawl space footings must be lapped at least 10" (T or F?).

_____

46. How wide is the CMU wall?

_____

47. What scales are used for all the section drawings?

_____

48. The total minimum distance from the bottom of the crawl-space footing to the finish grade is _____ .

_____

49. Give the total north-to-south plan view dimension of the crawl-space foundation.

_____

50. The scale for the typical wall section is 3/8" - 1'-0" (T or F?).

   _____

51. Size # _____ vertical rebars in the CMU foundation wall are spaced _____ apart.

   _____

52. What elevation is given for the top of the garage slab?

   _____

53. What is the distance from the west foundation wall to the centers of the piers?

   _____

54. Give the size of the glulam beam over the piers.

   _____

55. The distance from the side face of the north foundation wall to the center of the first pier is
   _____.

   _____

56. What is the size of the welded wire mesh reinforcing in the garage slab?

   _____

57. Name the material used as a ground cover within
   the crawl space. (See Description of Materials.)

   _____

58. A vapor barrier is required under the garage slab
   (T or F?).

   _____

59. Excavation is required to a depth of _____below
   grade to _____ soil. (See Description of
   Materials.)

   _____

60. What proportion is being used for the concrete
   mix of the footings? (See Description of Materials.)

   _____

61. What is the psi strength of the concrete? (See
   Description of Materials.)

   _____

62. The cross measurements of the piers within the
   crawl space are _____ square.

   _____

63. The north-to-south center spacing of the piers is
   _____ .

   _____

64. Give the thickness and name of the material
   placed beneath the garage floor slab.

   _____

65. Name the species and grade of lumber used for
   the pressure-treated sill plates. (See Description of
   Materials.)

   _____

66. Give the size and the number of vertical rebars
   required for each CMU pier.

   _____

67. The pressure-treated sill plates are doubled (T or
   F?).

   _____

68. Give the thickness and width of the rim joist at the second floor.

_____

69. What type of subfloor will be used, and how thick will it be?

_____

70. The thickness and width of the external wall studs are _____ and, _____, and they are spaced ____.

_____

71. What is the rating of the wall insulation?

_____

72. Name the material used to finish off the interior walls.

_____

73. Give the first and second floor-to-floor heights.

_____

74. Plywood siding is used as exterior finish material (T or F?).

_____

75. What type, thickness, and width of material is used as the exterior sheathing? (See Description of Materials.)

_____

76. The second-floor ceiling joists are ___thick and ____wide with ___rated insulation.

_____

77. Give the size and spacing of the roof rafters.

_____

78. What is the height of the attic wall?

_____

79. What is the horizontal distance of the roof overhang?

_____

80. The unit rise of the roof is ____ .

_____

81. What is placed as underlayment under the roof sheathing?

_____

82. Wood shingles are used for the roof covering (T or F?).

SHEETS A2, A3, AND A4
FLOOR PLANS

The Pulver House plan contains a First Floor Plan, Second Floor Plan, and an Attic Floor Plan. Referring to the schedules on Sheet A6 for the sizes of door and window opening is necessary. Windows are identified by a code letter on the floor plan and the window schedule. However, the doors in this plan are identified by code room number on the floor plan and door schedule. The finish schedule giving the finish materials for the floors, walls, and ceiling is also on Sheet A6.

Note that dimensions to and between openings are to one side of the openings. As in the previous assignments, some of the dimension questions will require adding or subtracting room dimensions. Other questions will require adding or subtracting a 2 × 4 wall thickness (3 1/2″ actual thickness) or a 2 × 6 wall thickness (5 1/2″ actual thickness).

SHEET A2 FIRST FLOOR PLAN QUESTIONS

Enter the missing words in the spcaes provided on the right.

83. What is the radius for the curve of the front porch?   _____

84. The slope of the front porch is _____ per foot.   _____

85. Give the diameter of the wood columns at the porch.   _____

86. What is the distance between the (B) windows at the south wall of the living room?   _____

87. Name the stud size and spacing of the interior partitions (walls). (See Description of Materials.)   _____

88. The interior walls will require a single top and bottom plate (T or F?).   _____

89. The fireplace hearth in the living room is _____ above the finish floor.   _____

90. What is the width of the two openings leading from the living to dining room?   _____

91. What is the diameter of the wood columns on each side of the foyer opening to the living room?   _____

92. Give the code number, width, and height of the door at the north dining room wall.   _____

93. What is the total face-to-face distance between the north and south living room walls?   _____

94. What is the distance measured from the outside face of the west living room wall to the inside face of the east living room wall?   _____

95. Name the size and type of material used to surface the fireplace hearth.   _____

96. The widest west-to-east dimensions between the foyer walls, from inside face to inside face, is_____.   _____

97. Name the type of door that opens to the pantry.   _____

98. How wide is the bay at the north wall of the kitchen?   _____

99. Give the width and length dimensions for the powder room.   _____

100. What are the width and length dimensions shown for the powder room?

    _____

101. The measurement from the outside of the west wall of the house to the outside of the east wall of the garage is _____.

    _____

102. The west-to-east slope of the concrete apron in front of the garage is 1/4″ per foot (T or F?).

    _____

103. What is the reference number of the north-to-south section line?

    _____

104. The height of the bench along the east kitchen wall is _____.

    _____

105. How many winder treads are shown at the base of the stairway?

    _____

106. The plans indicate that there is a(n) _____ area under the stairway.

    _____

107. What is the width of each garage door opening?

    _____

108. Give the distance between the garage door openings.

    _____

109. The north-to-south dimension of the west house wall is _____.

    _____

110. The south powder room wall is framed with _____ studs.

    _____

SHEET A3 SECOND FLOOR PLAN QUESTIONS

Enter the missing words in the spcaes provided on the right.

111. How many balustrades are shown at the second-floor porch railing?

    _____

112. What is the spacing of the pickets at the second-floor porch railings?

    _____

113. Give the code number and width and height of the door leading to the second-floor porch.

    _____

114. There are 14 treads in the stairway going from the first to second floor (T or F?).

    _____

115. The distance from the south building wall to the center of the chimney is _____.

    _____

116. _____ _____ type windows are shown in the south wall of the master bedroom.

    _____

117. Give the north-to-south inside measurement of the second-floor linen closet.

    _____

118. What is the inside width of the stairway?

    _____

119. Between what two rooms is there a pocket sliding door?

    _____

120. The north-to-south inside measurement of the shower stall in the master bathroom is _____.

    _____

121. What is the dimension from the outside face of the north exterior wall to the side of the first window opening in the west wall?

    _____

122. Ceramic tiles are the floor finish material used in the master bathroom (T or F?).

    _____

123. What are the north-to-south and west-to-east inside measurements of Room 207?

    _____

124. The sizes of windows shown in the north master bathroom wall are _____ × _____ .

    _____

125. Name the wall and ceiling finish in the laundry room.

    _____

126. How many sinks are there in the master bathroom, and what are their shapes?

    _____

127. The height of the low wall at the top of the stairway is _____.

    _____

128. How is the top of the low wall mentioned in Question 127 finished off?

    _____

129. What are the interior width and length dimensions of the family/recreation room?

    _____

130. Name the type, width, and height of the dormer windows in the family/recreation room.

    _____

131. All window sashes in the building are wood (T or F?).

    _____

132. The dormers extend _____ from the outside face of the wall.

    _____

SHEET A4 ATTIC FLOOR PLAN QUESTIONS

Enter the missing words in the spcaes provided on the right.

133. How many treads are there in the stairway from the attic down to the second floor? _____

134. What are the face-to-face inside measurements of room #301? _____

135. What is the floor finish in the study? _____

136. The laundry chute is found below the _____ in the bathroom. _____

137. The doors to the bedroom closets are wooden, hollow-core, six-panel doors (T or F?). _____

138. What is the finish floor covering in the toilet area? _____

139. The interior length of the bathroom space is _____. _____

140. Name the type of windows in the east wall of the study and give the window dimensions. _____

141. What is the total width of the window group in the west wall of the bedroom? _____

142. The window group at the wall mentioned in Question 141 is placed exactly at the center of that wall (T or F?). _____

143. The typical roof overhang is _____. _____

144. In which room and wall is an access attic panel suggested? _____

SHEET A5
EXTERIOR ELEVATIONS

The exterior elevation drawings are identified as north, south, east, and west, and should be studied together with the floor plans. Additional information is given regarding exterior finish, floor-to-ceiling heights, and elevations.

SHEET A5 EXTERIOR ELEVATIONS QUESTIONS

Enter the missing words in the spcaes provided on the right.

145. What scale is being used for the elevation drawings? _____

146. Which elevation shows the front entrance to the house? _____

147. The first-floor elevation reading is _____. _____

148. Give the total vertical dimension from the first floor to the attic floor. _____

149. The garage door consists entirely of wooden panels (T or F?). _____

150. What is the total height from the balcony deck to the top of the railing? _____

151. Name the finish roof material. _____

152. Which elevation drawing show side projections of the bay? _____

153. What type of roofs are over the house and garage, and what is their unit rise? _____

154. Gutters and downspouts are placed at all the roof overhangs (T or F?). _____

155. The west side elevation shows a total of _____ risers going up to the back porch. _____

156. What type of window is shown for the entire building? _____

157. Give the difference in elevation between the top of the 4" thick garage slab, and the first-floor elevations. _____

158. What is the thickness and width of the barge board at the gable end of the roof? _____

159. What is the total number of lights shown in each of the dormer windows? _____

160. There is a total of _____ panels in the front entrance doors. _____

SHEET A6
SCHEDULES, SECTIONS, DETAILS, ELEVATIONS AND ROOF FRAMING PLANS

Sheet A6 provides the Door and Window Schedules that give the size, type, material, and the frame material for the doors and windows. You are already familiar with these schedules because reference to them was necessary in studying the floor plans and elevations. The Finish Schedule gives the finish materials of the floors, walls, and ceiling applied toward the end of the construction project.

The Cross Section drawing greatly adds to the understanding of the interior of the building. It should be studied with the floor plans. This Cross Section (1/A6) is a transverse cut running from north to south, and is a view facing east. In order to properly orient the cross section, refer to the First-Floor Plan. Imagine yourself standing on the section line (1/A6) looking east. The Cross Section drawing would then be facing in the direction of the section line shown in the First-Floor Plan. The Cross Section drawing will show the following:

✓ **First Floor:** (1) Front of the first step and two winder steps with a window above. (2) Section of stair railing and side view of the stair stringer continuing up to the second floor. (3) Door to a storage closet beneath the stairway. (4) Door leading to the garage. (5) Refrigerator in the kitchen, 18″ high bench-cabinet, 36″ high cabinet, stove, and sink. (6) Upper wall cabinets.

✓ **Second Floor:** (1) Front of two winder steps with a window above. (2) Section of railing and side view of the stair stringer continuing up to the attic floor. (3) Low wall section next to the stairway below. (4) Door leading to the family/recreation room.

✓ **Third-Floor Attic:** (1) Low walls from the floor to the rafters that define the living area. (2) A sink cabinet next to the bathtub. (3) A bathtub in the bathroom. (4) A window in the east wall of the study. (5) A louver in the upper portion of the attic space.

✓ **The Hearth Edge detail** relates to the fireplace hearth. A Fireplace Elevation gives a frontal view of the fireplace. The Cabinet Elevation provides additional information about the kitchen cabinets. A simple roof plan helps to explain the roof construction.

SCHEDULES QUESTIONS

Enter the missing words in the spcaes provided on the right.

161. The surface of the fire-rated door leading to the garage is covered with _____ and has a _____ fire-rating.

_____

162. The porch rails will have a stained finish (T or F?).

_____

163. How many door openings contain bifold doors?

_____

164. Name the garage door and frame material.

_____

165. In which room is there a G window?

_____

166. Which rooms will have both paint and wallpaper on their walls?

_____

167. All the bedrooms will have exposed wooden floors (T or F?).

_____

168. A _____ finish is used on the ceilings of all the rooms.

_____

169. Which rooms have a vitreous clay tile floor finish?

_____

170. What two areas show a concrete floor finish?

_____

CROSS SECTION QUESTIONS

Enter the missing words in the spaces provided on the right.

171. What is the clearance beneath the first-floor joists? _____

172. The code number of the door leading to the storage area below the stairway is _____, and its size is _____ by _____. _____

173. What is the floor-to-ceiling height in the attic area? _____

174. A microwave oven and a hood with a fan are located above the _____. _____

175. A 1/8" = 1'-0" scale is used for the Cross Section drawing (T or F?). _____

176. The doorway shown to the right of the refrigerator leads to the _____. _____

177. What are the floor-to-ceiling heights at the first and second floors? _____

178. How many stair treads, including winders, are there in the stairway going from the second floor to the attic floor? _____

179. There is a total of _____ panels in the door opening to the family/recreation room. _____

180. What is the unit rise of the roof? _____

181. Give the width and height of the foundation footing. _____

182. A side view of a _____ and _____ are shown in the third-floor attic bathroom. _____

HEARTH DETAIL— FIREPLACE ELEVATION QUESTIONS

Enter the missing words in the spcaes provided on the right.

183. What are the size and spacing of the framing members supporting the top of the hearth? _____

184. Name the type and size of the finish material at the surface of the hearth. _____

185. What is the distance from the subfloor to the top of the hearth (minus thickness of the tiles)? _____

186. Give the scale of the Hearth Edge Detail. _____

187. The total height of the fireplace is _____. _____

188. What material finishes off the front of the fireplace? _____

189. What finishes off the top of the fireplace? _____

190. The owner-builder contract calls for a plaster cast medallion above the fireplace (T or F?). _____

CABINET ELEVATION—ROOF PLAN QUESTIONS

Enter the missing words in the spcaes provided on the right.

191. The Cabinet Elevation drawing relates to the (compass direction) _____ kitchen wall. _____

192. What is the dimension from the finish floor to the top of the floor cabinets? _____

193. The distance between the top of the floor cabinets to the top of the wall cabinets is _____. _____

194. How many drawers are shown in the floor cabinets? _____

195. The distance between the top of the floor cabinet and the bottom of the wall cabinets is _____ . _____

196. How deep are the top and bottom cabinets? _____

197. What is the floor-to-ceiling height shown in the drawing? _____

198. Dormers are framed over the _____ room and the _____ . _____

199. What is the spacing of the roof rafters? _____

200. Give the type and size of the roof rafters. _____

Description of Materials

U.S. Department of Housing and Development,
Department of Veterans Affairs, and
Farmers Home Administration

HUD's OMB Approval No. 2502-192 (exp. 1/31/96)

Public Reporting Burden for this collection of information is estimated to average 0.5 hours per response, including the time for reviewing instructions, searching existing data sources, gathering and maintaining the data needed, and completing and reviewing the collection of information. Send comments regarding this burden estimate or any other aspects of this collection of information, including suggestions for reducing this burden, to the Reports Management Officer, Office of Information Policies and Systems, U.S. Department of Housing and Urban Development, Washington, D.C. 20410-3600 and to the Office of Management and Budget, Paperwork Reduction Project (2502-0192), Washington, D.C. 20503. Do not send this completed form to either of these addresses.

☒ Proposed Construction   No. __N/A__

☐   Under Construction

Property Address: ___141 KINGS ROW___   City: __MARIETTA__   State __GA__

Mortgagor or Sponsor: __REGIONS BANK OF SPAULDING COUNTY__   P.O. BOX 108, DALLAS, GA
(Name)   (Address)

Contractor or Builder __PRIME CONSTRUCTION CO.__   61 MERRITT ST. S.E. MARIETTA 30060

Instructions

1. For additional information on how this form is to be submitted, number of copies, etc., see the instructions applicable to the HUD Application for Mortgage Insurance, VA Request for Determination of Reasonable Value, or FmHA Property Information and Appraisal Report, as the case may be.

2. Describe all materials and equipment to be used, whether or not shown on the drawings, by marking an X in each appropriate check-box and entering the information called for in each space. If space is inadequate enter "See misc." and describe under item 27 or on an attached sheet.. THE USE OF PAINT CONTAINING MORE THAN THE PERCENTAGE OF LEAD BY WEIGHT PERMITTED BY LAW IS PROHIBITED.

3. Work not specifically described or shown will not be considered unless required, then the minimum acceptable will be assumed. Work exceeding minimum requirements cannot be considered unless specifically described.

4. Include no alternates, "or equal" phrases, or contradictory items. (Consideration of a request for acceptance of substitute materials or equipment is not thereby precluded.)

5. Include signatures required at the end of this form.

6. The construction shall be completed in compliance with the related drawings and specifications, as amended during processing. The specifications include this Description of Materials and the applicable Minimum Property standards.

1. EXCAVATION:

Bearing soil, type ___CLAY 16" BELOW GRADE___

2. FOUNDATIONS:

Footings: concrete mix __1-3-5__   strength psi __3,000__   Reinforcing __#4 REBAR__
Foundation wall: material __C.M.U__   Reinforcing __#4 REBAR @ GROUTED CELLS__
Interior foundation wall: material ___   Party foundation wall ___
Columns: material and sizes ___   Piers: material and reinforcing __CONCRETE W/ #4 REBAR__
Girders: material and sizes __DOUGLAS FIR, LAMINATED 6"x14'__   Sills material __#2 PRESSURE TREATED SOUTHERN YELLOW PINE__
Basement entrance areaway ___   Window areaways ___
Waterproofing ___   Footing drain ___
Termite protection __SOIL TREATED SYSTEM BY LICENCED BONDED TERMITE CONTRACTOR__
Basement less space: ground cover __6 MIL POLY__ ;   Insulation __R-19__ ;   foundation vents _5_
Special foundations ___
Additional information: ___

3. CHIMNEYS:

Material __Metal Asbestos__   Prefabricated (make and size) __36"__
Flue lining: material __88 TRIPLE WALL__   Heater Flue size ___   Fireplace flue size ___
Vents (material and size): gas or oil heater __METAL ASBESTOS 6"__ ; water heater __METAL ASBESTOS 3"__
Additional information ___

4. FIREPLACES

Type: ☒ solid fuel; ☐ gas-burning; ☐ circulator (make and size) __MAJESTIC 36"__   Ash dump and clean-out ___
Fireplace: facing; ___ ; lining __FIRE BRICK__ ; hearth __TILE__ ; mantel __WOOD__
Additional information: __GAS STARTER__

5. EXTERIOR WALLS

Wood frame; wood grade, and species __CONSTRUCTION GRADE 2X4 D.F.__ ☐ Corner bracing. Building paper or felt __NONE__

Sheathing __INSULATING__ ; thickness __½"__ ; width ; __4'__ ; ☒ solid; ☐ spaced ___ " o.c.; ☐ diagonal; ___
Siding __MASONITE HARD BD__ grade __A__ ; type __LAP__ ; size __8"__ exposure __7"__ ; fastening __8d NAILS__
Shingles ___ ; grade ___ ; type ___ ; size ___ ; exposure ___ ; fastening ___
Stucco ___ ; thickness ___ " ; Lath ___ ; weight ___ lb.
Masonry veneer ___   Sills ___   Lintels ___   Base flashing ___

Description of Material List.

Masonry: ☐ solid   ☐ faced   ☐ stuccoed;   total wall thickness _____ ; facing thickness _____ ; facing material _____

Backup material _____ ; thickness _____ ; bonding _____

Door sills _____ Windows sills _____ Lintels _____ Base flashing _____

Interior surfaces: dampproofing _____ coats of _____ ; furring _____

Additional information: _____

Exterior painting: material __EXTERIOR LATEX_____ ; number of coats __2_____

Gable wall construction: ☒ same as main walls; ☐ other construction _____

6. FLOOR FRAMING:

Joists: wood, grade, and species __SYP #3_____ ; other __SEE FOUND. PLAN__ bridging __SOLID_____ ; anchors _____

Concrete slab: ☐ basement floor; ☐ first floor; ☐ ground supported; ☐ self-supporting; mix _____ ; thickness _____

reinforcing _____ ; insulation _____ ; membrane _____

Fill under slab; material __#57 STONE GRAVEL_____ ; thickness __4"__ ; Additional information : _____

__6X6X 10/10 WWM IN GARAGE FLOOR SLAB_____

7. SUBFLOORING: (describe underflooring for special floors under item 21.)

Material: grade and species __WAFERBOARD_____ ; size __3/4" X 4'-8'_____ ; type __T & G____

Laid: ☒ first floor; ☒ second floor; ☒ attic _____ sq. ft.; ☐ diagonal; ☐ right angles. Additional information: __GLUE & NAIL__

8. FINISH FLOORING: (Wood only. Describe other finish flooring under item 21.)

Location	Rooms	Grade	Species	Thickness	Width	Bldg. Paper	Finish
First floor	Entry	#1	OAK	3/8"	2"		POLYURETHANE
Second floor							
Attic floor	sq. ft.						

Additional information:

9. PARTITION FRAMING:

Studs: wood, grade and species __DOUGLAS FIR CONSTRUCTION GRADE__ ; size and spacing __2X4 - 16" OC__ Other _____

Additional information: _1.   2"X4" BOTTOM PLATE_____   2. 2"X4" TOP PLATES_____

10. CEILING FRAMING:

Joists: wood, grade, and species __STANDARD GRADE DOUGLAS FIR__ Other _____ Bridging _____

Additional information: _2X8 - 16" OC_____

11. ROOF FRAMING:

Rafters: wood, grade, and species __STANDARD GRADE DOUGLAS FIR__ Roof trusses (see detail): grade and species _____

Additional information: _2X6 - 16" OC_____

12.. ROOFING:

Sheathing: wood, grade, and species __DFPA 3/16" WAFERBOARD_____ ☒ solid; ☐ spaced _____ " o.c.

Roofing __COMPOSITION SHINGLES_____ ; grade _____ ; size __12X36_____ ; type __CD 240_____

Underlay __#15 FELT_____ ; weight or thickness _____ ; size _____ ; fastening __NAILS__

Built-up roofing _____ ; number of plies _____ ; surfacing material _____

Flashing: material __ALUMINUM_____ ; gage or weight __26 6A65__ ; ☐ gravel stops; ☐ snow guards

Additional information: __ALUMINUM DRIP EDGES @ EAVES__

13. GUTTERS AND DOWNSPOUTS:

Gutters: material __ALUMINUM_____ ; gage or weight __26 G__ ; size __6"__ ; shape __OGEE_____

Downspouts: material __ALUMINUM_____ ; gage or weight __26 G__ ; size __4"__ ; shape __RECTANGULAR__ number __8__

Downspouts connected to: ☐ Storm sewer; ☐ sanitary sewer; ☐ dry-well. ☐ Splash blocks; material and size _____

Additional information: _____

14. LATH AND PLASTER

Lath ☐ walls, ☐ ceilings: material _____ ; weight or thickness _____ Plaster: coats _____ ; finish _____

Dry-wall ☒ walls, ☒ ceilings: material __GYPSUM BD_____ ; thickness 1/2" ; finish __SMOOTH__

Joint treatment __TAPE AND FINISH AS RECOMMENDED BY MANUFACTURER__

15. DECORATING: (Paint, wallpaper, etc.)

Rooms	Wall Finish Material and Application	Ceiling Finish Material and Application
Kitchen	LATEX PAINT	GYP. BOARD - SMOOTH
Bath	LATEX PAINT	GYP. BOARD - SMOOTH
Other ALL	LATEX PAINT	GYP.BOARD - SMOOTH

Description of Material List (continued).

Additional information: _____

16. INTERIOR DOORS AND TRIM:
Doors: type __HOLLOW CORE - PANEL__ ; material __MASONITE__ ; thickness __1 3/8"__
Door trim: type __MOLDED__ ; material __PINE__ ; base: type __MOLDED__ ; material __PINE__ ; size __2 3/4"__
Finish: doors __LATEX SEMI-GLOSS__ ; trim __LATEX SEMI-GLOSS__
Other trim (item, type and location) _____
Additional Information: _____

17. WINDOWS:
Windows: type __DOUBLE HUNG__ ; make __CAROLINA BUILDER__ ; material __VINYL CLAD WOOD-PINE__ ; sash thickness __1 1/2"__
Glass: grade __DOUBLE PANE__ ; ☐ sash weights; ☒ balances, type __SPRING__ ; head flashing __ALUMINUM DRIP__
Trim: type __MOLDED__ ; material __PINE__ ; Paint __LATEX SEMI-GLOSS__ ; number of coats __2__
Weatherstripping: type __INTEGRAL__ ; material __VINYL__ ; Storm sash, number _____
Screens: ☒ full; ☐ half; type _____ ; number __ALL__ ; screen cloth material __ALUMINUM__
Basement windows: type _____ ; material _____ ; screens, number _____ ; Storm sash, number _____
Special windows _____
Additional information: _____

18. ENTRANCES AND EXTERIOR DETAIL:
Main entrance door: material __WOOD__ ; width __3'-0"__ ; thickness __1 3/4"__ ; Frame: material __PINE__ ; thickness __3/4"__
Other entrance doors: material __METAL__ ; width __2'-8"__ ; thickness __1 3/4"__ ; Frame: material __PINE__ ; thickness __3/4"__
Head flashing __ALUMINUM DRIP CAP__ ; Weatherstripping: type __VINYL FOAM__ ; saddles _____
Screen doors: thickness ____" ; number ____ ; screen cloth material _____ ; Storm doors: thickness ____"; number _____
Combination storm and screen doors: thickness _____ ; number _____'; screen cloth material_____
Shutters: ☐ hinged; ☐ fixed. Railings __PORCH, PRESSURE TREATED WOOD__ Attic louvers __PER PLANS__
Exterior millwork; grade and species __C & B GRADE CLEAR PINE__ Paint __EXTERIOR LATEX SEMI-GLOSS__ number of coats __2__
Additional information: _____

19. CABINETS AND INTERIOR DETAIL
Kitchen cabinets, wall units; material __BASS PREFINISHED MED. DENSITY FIBERBD__; lineal feet of shelves __SEE PLANS__; shelf width __12"__
Base units: material __BASS PRE.FIN. MDF__ counter top __PLASTIC LAM__ ; edging __VINYL__
Back and end splash __PLASTIC LAM__ Finish of cabinets __PREFINISHED SPRAYED VINYL LACQUER__ ; number of coats _____
Medicine cabinets: __None__ ; model _____
Other cabinets and built-in furniture __P. LAM BENCH SEAT IN KITCHEN__
Additional information: _____

20. STAIRS:

	Treads		Risers		Strings		Handrail		Balusters	
	Material	Thickness	Material	Thickness	Material	Size	Material	Size	Material	Size
Basement										
Main	SYP #2	3/4"	WP #2	3/4"	SYP #2	2 X 112	RED OAK	1 ½" DIA	SYP	1 1/4" TURNED
Attic										

Disappearing: make and model number _____
Additional information: _____

21. SPECIAL FLOORS AND WAINSCOT: (Describe Carpet as listed in Certified Products Directory)

	Location	Material, Color, Border, Sizes, Gage, Etc.	Threshold Material	Wall Base Material	Underfloor Material
Floor	Kitchen	"SUCCESSOR" VINYL BY ARMSTRONG	METAL	WOOD	
	Bath				
		ALL CARPETING EITHER MEETS OR EXCEEDS HUD/FHA SPECIFICATIONS			

	Location	Material, Color, Border, Cap. Sizes, Gage, Etc.	Height	Height Over Tub	Height in Showers (From Floor)
Wainscott	Bath				

Description of Material List (continued).

Bathroom accessories: ☐ Recessed; material _____; number____; ☒ Attached; material __CAST ALUM.__; number __9__

Additional information: __5 SOAP DISHES - CERAMIC / 3 TOOTHBRUSH HOLDERS - CERAMIC__

22. PLUMBING:

Fixture	Number	Location	Make	MFR's Fixture Identification No.	Size	Color
Sink	1	KITCHEN	ELKAY	833-8	DOUBLE BOWL	POLISHED STAINLESS
Lavatory	5	SINKS	AMERICAN STANDARD	889-2CPS	OVAL 15"	WHITE
Water closet	3	M. BATH	AMERICAN STANDARD	"ANTIQUITY"	19" OVAL	WHITE
Bathtub	1	M. BATH	AMERICAN STANDARD	"AMERICAST"	60"	WHITE
Shower over tub	1	ATTIC BATH	AQUA -GLASS	6857-01	60"	WHITE
Stall shower	1	M. BATH		A69448	48"	WHITE
Laundry trays						
PEDISTAL SINK	2	1ST & 2ND FLOOR POWDER	AMERICAN STANDARD	"ANTIQUITY"	19" OVAL	WHITE

☒ Curtain rod ☐ Door Shower pan: material _____

Water supply: ☒ public; ☐ community system; ☐ individual (private) system.*

Sewage disposal: ☐ public; ☐ community system; ☒ individual (private) system.*

Show and describe individual system in complete detail in separate drawings and specifications according to requirements.

House drain (inside): ☐ cast iron; ☐ tile; ☒ other __PVC__ House sewer (outside): ☐ cast iron; ☐ tile; ☒ other __PVC__

Water piping: ☐ galvanized steel; ☒ copper tubing; ☐ other _____ Sill cocks, number _____

Domestic water heater: type _____GAS_____; make and model __RHEEM 21V50__; heating capacity __50 GAL__

_____ gph. 100° rise. Storage tank: material __GLASS LINED__; capacity __50 GAL__

Gas service: ☒ utility company; ☐ liq. pet gas; ☐ other _____ Gas piping: ☐ cooking; ☒ house heating.

Footing drains connected to: ☐ storm sewer; ☐ sanitary sewer; ☐ dry well. Sump pump; make and model _____

_____; capacity _____; discharges into _____

23. HEATING:

☐ Hot water. ☐ Steam. ☐ Vapor. ☐ One-pipe system. ☐ Two-pipe system.

☐ Radiators. ☐ Convectors. ☐ Baseboard radiation. Make and model _____

Radiant panel; ☐ floor; ☐ wall; ☐ ceiling. Panel coil: material _____

☐ Circulator. ☐ Return pump. Make and model _____; capacity _____ gpm.

Boiler: make and model _____ Output _____ Btuh.; net rating _____ Btuh.

Additional information: _____

Warm air: ☐ Gravity. ☒ Forced. Type of system __CENTRAL GAS FIRED - COMFORTMAKER__

Duct material: supply _____; return _____ Insulation _____, thickness _____ ☐ Outside air intake.

Furnace: make and model __COMFORTMAKER__ Input _____ Btuh.; output _____ Btuh.

Additional information _____

☐ Space heater; ☐ floor furnace; ☐ wall heater. Input _____ Btuh.; output _____ Btuh; number units _____

Make, model _____ Additional information: _____

Controls: make and types _____

Additional information: _____

Fuel: ☐ Coal; ☐ oil; ☐ gas; ☐ liq. Pet.gas; ☐ electric; ☐ other _____; storage capacity _____

Additional information: _____

Firing equipment furnished separately: ☐ Gas burner, conversion type. ☐ Stoker: hopper feed ☐ bin feed

Oil burner: ☐ pressure atomizing; ☐ vaporizing _____

Description of Material List (continued).

Make and model _____   Control _____

Additional information: _____

Electric heating system: type _____   Input _____ watts; @ _____ volts; output _____ Btuh.

Additional information: _____

Ventilating equipment: attic fan, make and model __1,200 sq. ft. CAPACITY_____ ; capacity _____cfm.

kitchen exhaust fan, make and model ___G.E.   RECIRCULATING_____

Other heating, ventilating, or cooling equipment _____COMFORTMAKER BRAND AIR CONDITIONING_____

__MODEL # 5BA030 - FBA018   2 ½ x 1 ½ TONS_____

24. ELECTRIC WIRING:

Service: ☐ overhead; ☒ underground. Panel: ☐ fuse box; ☒ circuit-breaker; make ____G.E____ AMP's __200__ No. Circuits __20__

Wiring: ☐ conduit; ☐ armored cable; ☒ nonmetallic cable; ☐ knob and tube; ☐ other _____

Special outlets: ☒ range; ☒ water heater; ☒ other ____Dryer_____

☒ Doorbell. ☐ Chimes. Push-button locations ____FRONT ENTRANCE____ Additional information _____

__SMOKE DETECTORS EACH FLOOR_____

25. LIGHTING FIXTURES:

Total number of fixtures _____PER PLANS____ Total allowance for fixtures, typical installation, $ __773.87__

Nontypical installation _____

Additional information: _____

26. INSULATION;

Location	Thickness	Material, Type, and Method in Installation	Vapor Barrier
Roof	8"	R-19 FIBERGLASS BATTS	
Ceiling	10"	R-30 FIBERGLASS BATTS	
Wall	5.5"	R-19 FIBERGLASS BATTS	
Floor	10"	R-30 FIBREGLASS BATTS	

27. MISCELLANEOUS: *(Describe any main dwelling materials, equipment, or construction items not shown elsewhere; or use to provide additional information where the space provided was inadequate. Always reference by item number to correspond to numbering used on this form.)*

_____
_____
_____
_____
_____
_____
_____

HARDWARE: *(make, material, and finish.)* _____POLISHED BRASS - SCHLAGE_____
_____

SPECIAL EQUIPMENT: *(State material or make, model and quantity. Include only equipment and appliances which are acceptable by local law, custom and applicable FHA standards. Do not include items which, by established custom, are supplied by occupant and removed when he vacates premises or chattles prohibited by law from becoming reality.)*

OVEN / STOVE   G.E. #JBP66GWWH
DISHWASHER   G.E. #GSD900XBA
REFRIGERATOR   G.E.#GSD900XBA
MICROWAVE / HOOD FAN   G.E. #JUM 1340 WW
_____
_____
_____

PORCHES:
PER PLANS
_____
_____

TERRACES:
_____
_____

Description of Material List (continued).

GARAGES:

2 CAR

WALKS AND DRIVEWAYS:

Driveway: width __10'__ ; base material _____ ; thickness _____"; surfacing material __GRAVEL #57__ ; thickness ___4"___

Front walk: width __3'__ ; material __BRICK__ ; thickness __2 1/4"__ . Service walk: width _____ ; material __BRICK__ ; thickness 2 1/4"

Steps: material _____ ; treads _____" ; risers _____". Cheek walls _____

OTHER ONSITE IMPROVEMENTS:

(Specify all exterior onsite improvements not described elsewhere, including items such as unusual grading, drainage structures, retaining walls, fence, railing and accessory structures.)

LANDSCAPING, PLANTING, AND FINISH GRADING:

Topsoil __4"__ thick: ☒front yard; ☐ side yards; ☐ rear yard to _____ feet behind main building

Lawns *(seeded, sodded, or sprigged):* ☒ front yard __SEEDED__ ; ☐ side yards _____ ; ☐ rear yard _____

Planting: ☐ as specified and shown on drawings; ☐ as follows:

_____	Shade trees, deciduous, _____" caliper.			_____	Evergreen trees. _____ ' to _____ ; B & B.
_____	Low flowering trees, deciduous,	_____ ' to _____ '		_____	Evergreen shrubs. _____ ' to _____ ; B & B.
_____	High-growing shrubs, deciduous,	_____ ' to _____ '		_____	Vines, 2-year _____
__2__	Medium-growing shrubs, deciduous,	__2__ ' to __3__ '			
_____	Low-growing shrubs, deciduous,	__1__ ' to __2__ '			

Identification--This exhibit shall be identified by the signature of the building, or sponsor, and/or the proposed mortgagor if the latter is known at the time of application.

Date ___1-7-97___

Signature X_____

Signature X_____

Description of Material List (continued).

Glossary

Acrylics. Transparent material that can be used for window glazing, roof domes, skylights, and translucent panels for ceiling lighting systems.

Active solar system. Consists of a collector, a thermal storage area, and pumps or fans to distribute the heat to the desired locations of the building. The two types of active solar methods are the warm-air and warm-water systems.

Actual lumber size. Thickness and width dimensions after the rough surface is removed from a piece of lumber and the lumber has seasoned (dried and shrunk).

Aggregate. Fine particles of sand and coarse gravel or crushed rock used in a concrete mix.

Air outlet. As part of an HVAC system, an opening at the end of a duct through which hot or cold air is supplied to a space. Also called *supply outlet*.

Air plenum heating. System in crawl-space foundation in which the air is heated in the crawl space and distributed through floor registers to the floor above.

Allowable span. The distance permitted between bearing points for a particular joist material, based on the strength and thickness of the material and the effect of the live and dead loads the joists must support.

American Institute of Architects (AIA). A national professional architect's association.

American Society for Testing and Materials (ASTM). A recognized industry-wide testing agency for building materials.

Ampere. The unit for measuring the amount of electrical flow (current) through an electric wire.

Appearance lumber. Lumber that presents a finished appearance. Also called *finish lumber*.

Appliance circuit. Usually serves the convenience outlets in rooms such as the kitchen, laundry room, pantry, dining area, and workshop.

Architect. Prepares plans for an individual client or for a building contractor; and must be licensed and registered by the state in which he or she practices.

Architect's scale. A rule used to scale plans drawn to English measurement.

Areaway. A sunken space next to the foundation wall that permits light or air through a basement window.

Ashlar masonry. Squared pieces of stone used to face masonry walls.

Asphalt-saturated rag felt. A weather-resistive barrier placed over the outside surface of a wall to prevent moisture infiltration and damage to the interior wall covering.

Atrium. A glazed, open area for collecting solar heat.

Awning window. A window hinged at the top and swinging out at the bottom.

Balusters. Slender vertical pieces running from the stair railing to a bottom rail or stair tread.

Barge board. A board or rafter placed at the projecting end of a gable overhang.

Batch plant. A plant where ready-mixed concrete is produced and delivered by truck to the job site.

Batt fiberglass insulation. Flexible paper-backed batts fastened between wall studs and between floor and ceiling joists.

Battered foundation. A foundation wall with one vertical and one sloped side.

Bay window. Window placed in a square, rectangular, or angled projection from an exterior wall.

Beam pocket. An opening prepared in a concrete wall to receive the end of a girder. Also called a *girder pocket*.

Benchmark (BM). A fixed and identified point at the job site such as a street curb; stake driven into the ground, mark on a power pole or some other immovable object.

Bid. As it applies to the building industry, an offer to contract the performance of work described in the construction documents at a specified cost.

Blanket fiberglass insulation. Flexible paper-backed blankets fastened between wall studs and between floor and ceiling joists.

Blind valley construction. A method of intersecting roof construction that eliminates the need for valley rafters. One roof section is framed and sheathed, then the intersecting roof section is framed over the sheathed area.

Blueprints. An older process of plan reproduction that created white lines against a blue background.

Boards. A piece of lumber less than 2″ in nominal thickness and 2″, or more in, nominal width.

BOCA International Plumbing Code. A model plumbing code that supersedes the BOCA National Plumbing Code.

Bolts. Fasteners used to connect heavier wood and metal materials.

Bond beam. A horizontal row of concrete masonry units placed toward the top of a stem wall containing a continuous row or rows of horizontal rebars.

Bonds. Financial guarantees by a surety company assuring payment for labor, material, and all other obligations related to the performance of work described in the contract.

Bottom cripple stud. Short vertical pieces placed between a rough window sill and the bottom plate.

Bottom chord. The lower part of a roof truss that acts as a ceiling joist.

Bottom plate. Horizontal piece nailed to the bottom of a wall stud. Also called *sole plate*.

Box beam. A beam fabricated with plywood sides nailed to top and bottom chords that are held together with vertical stiffeners.

Break line. A long line broken by zigzags. It is used where space does not permit the continuation of a

scaled drawing, or where the continuation of the drawing is not necessary to provide the needed information.

Building code. A set of legally binding regulations that cover every aspect of a building's construction.

Building contractor. An individual who is in overall charge of a construction project and directly employs office personnel, estimators, carpenters, and laborers. Also called a *general contractor*.

Building drain. A pipe connecting to the lowest points of the drainage system of a building and conveying the waste material to the municipal sewer system. Also called a *house sewer*.

Building height. The vertical distance from a grade or datum point to the finished surface of a flat roof or to the height of the highest point of a pitched roof.

Building Officials & Code Administrators National Building Code (BOCA). A model building code book commonly used in the eastern United States and some parts of the midwest.

Building paper. A weather-resistive barrier placed over the outside surface of a wall to prevent moisture infiltration and damage to the interior wall covering.

Building permit. An official document or certificate issued by a local building department to authorize the beginning of a construction project.

Building plans. A set of drawings that guide the construction of a building.

Building section drawings. Drawings in a set of building plans showing what would be seen if a vertical cut were made through the length or width of a building.

Building stone. A natural rock taken from the earth and broken down into sizes suitable for construction.

Cantilevered joist. A joist projecting past its supporting wall in order to hold up a balcony or another wall located past the wall below.

Cape Cod house. A one-and-a-half-story design featuring dormers built into the roof.

Capital. The topmost member, usually decorated, of a pilaster or column.

Carriage. The main support of a stairway. They are either cut out or dadoed to receive the finish treads and risers. Also called a *stair stringer*.

Carriage bolt. A bolt with an oval head and a square shoulder under the head. The shoulder is embedded in the wood and prevents the bolt from turning as it is being tightened with a nut at the threaded end.

Casement window. A window hinged on one side with the swing action of a door.

Caulking. A sealing material that usually comes in a cartridge and is applied with a caulking gun.

Cellulose insulation. Ground-up or shredded paper chemically treated to make it resistant to fire and vermin. It can be poured or blown into walls and ceilings.

Cement. The ingredient of a concrete mix that binds together the sand and gravel after water is added.

Center line. In building plans, a line consisting of a series of alternating long and short dashes extending at a right angle from the center of an object such as a wall, door or window opening, pier footing, column, etc.

Centimeter. A unit in metric measurement equal to 1/100 of a meter.

Ceramic. A material made of a clay mixture that is subjected to high temperature during manufacture.

Ceramic tile. Flat pieces of ceramic material used as finish floor and wall covering.

Certificate of occupancy. A document granted by a building department after the final inspection and approval of the building.

Chase wall. A wall between two rooms containing shared plumbing. Also called a *wet wall*.

Chemical anchor. A device consisting of a bolt placed in a pre-drilled hole and held in place by a synthetic resin and quartz aggregate filler.

Circuit breaker. A device that shuts down an electrical circuit if an overload occurs.

Clay masonry unit. A solid brick, hollow tile, or terra cotta unit used for structural and veneer walls.

Cleanouts. Openings found at the base of soil and waste stacks making it possible to eliminate congestion and backup in the stacks.

Coarse-grained soil. Soil that generally consists of a sand and gravel mixture.

Collar tie. A horizontal piece nailed to opposite rafters that helps to further stiffen the roof rafters.

Column. A cylindrical or rectangular supporting structure, usually placed beneath a beam or girder.

Common nail. A flat-head nail with a thicker shank, used most often for rough work such as wood framing.

Common rafter. A roof rafter that runs at a right angle from the plate line to the ridge board.

Component (building). An individual member or prefabricated unit of a structure.

Composite panels. A panel product consisting of plywood outside veneers glued to a core of reconstituted wood fibers.

Computer-aided drafting and design (CADD). A computer-generated program for creating and reproducing construction drawings.

Concrete. A mixture of cement, aggregate, and water.

Concrete block. A hollow block manufactured from a concrete mixture and used in the construction of foundations and walls. Also called a *concrete masonry unit (CMU)*.

Concrete masonry unit (CMU). See *Concrete block*.

Concrete mix. The proportions of cement, fine and coarse aggregates, and water in a batch of concrete.

Coniferous trees. Trees that do not lose their leaves during the winter. Also called *evergreen trees*.

Contour lines. Curved line connecting the same elevations on a lot.

Convection. Heat distributed in a circular motion caused by the constant rise of warm air toward the top of a heated space and its replacement by cold air at the bottom of the space.

Convenience outlets. Outlets that provide receptacles for plugging in lamps, clocks, television sets, radios, and portable appliance objects such as vacuum cleaners. Also used for ceiling and wall lighting outlets.

Conventions. Generally accepted practices for conveying information in construction drawings.

Corner post. Component constructed of studs and/or blocks and placed at the outside and inside corners of framed walls.

Cornice. The area under the overhang where the roof and sidewalls meet.

Cost-plus contract. An agreement in which the total cost of a building project is based on the labor and material plus a percentage of the labor cost. Also called a *Work-Plus-Fee contract.*

Counterflashing. A thin strip of flashing material bent over base flashing to prevent water leakage.

Crawl-space foundation. A foundation system providing an unfinished accessible space below the first floor, not high enough for a basement.

Cricket (roof). A small saddle-shaped structure placed on the upper side of a chimney projecting from a sloping roof. Its purpose is to divert water from around the chimney. Also called a *saddle.*

Customary measurement. Linear measurement based on yards, feet, and inches. Also called *English measurement.*

Cutting-plane line. A solid line with short, right-angle lines and arrows at each end. The cutting-plane line shows where a wall or object is cut, revealing a cross section.

Dado. A rectangular groove cut into a board so that a like-dimensioned piece can fit into it.

Datum point. A fixed elevation point used as a reference to establish other elevations.

Dead load. The total weight of the building including the foundation, floors, wall, roof, and stationary mechanical equipment.

Deciduous trees. Trees that lose their leaves during the fall and winter.

Decimeter. The largest division of a meter. A decimeter is 1/10 of a meter.

Detail drawing. An enlarged drawing of a small part of another drawing.

Diagonal brace. A piece running at an angle and required for a framed wall if structurally rated sheathing is not used. May consist of a 1 × 4 board notched into the studs, or a metal strap.

Diazo reproduction. Method of copying prints that produces blue or black lines against a white background.

Dimension line. In construction plans, a solid, thin line with terminators (arrows, dots, slash marks, or accent marks) at each end.

Dimension lumber. Considered a category of framing lumber. Dimension lumber are pieces 2″ to 4″ in nominal thickness and 2″ or more in nominal width.

Dimmer switch. A switch that can adjust a light to a desired brightness.

Dining room. A clearly defined space or room designed for eating meals.

Direct measurement. A measurement taken with a measuring tool such as a tape or ruler.

Disconnect switch. A switch placed in a separate box or in the service panel that can turn the electricity on or off for the entire building. Also called a *service switch.*

Disposal field. A network of trenches containing distribution pipes carrying effluent from a septic tank for distribution into the surrounding soil.

Distribution lines. The cold-water line from the water service to the fixture branch lines.

Distribution panel. Receives the electrical current from the outside power source and distributes the electricity through circuits in the building. Also called a *service panel.*

Door jamb. The finished frame for door openings.

Dormers. Pitched structures extending from the roof. They provide extra light and ventilation to an attic area.

Double-hung window. A window with two vertical sashes that slide past each other.

Double plates. The two plates nailed at the top of a framed wall. They strengthen the top of the wall and help tie together the corners of the adjoining walls.

Downdraft furnace. A type of furnace used in buildings over crawl-space foundations. The furnace is placed in the garage or elsewhere on the main floor level. Also called a *downflow furnace*.

Draftsperson. Individual who draws the building plans under the direction of an architect.

Drain lines. Clay, perforated plastic, or fiber pipe carrying waste from a septic tank and distributing the waste into the surrounding soil. Also called *leach lines*.

Drive pin. A specially hardened nail driven by a powder-actuated tool and used to fasten material to concrete or masonry. Also called a *stud*.

Duct. A round or rectangular pipe that transports the cool or hot air in a forced-air heating system.

Eaves flashing. An added layer of bituminous material placed over the roofing paper along the perimeter of a roof to help protect the sheathing from damage caused by ice dams formed in freezing weather.

Edge band. A thin piece of material applied to finish off the edges of a door or countertop.

Electrical engineer. A professional who designs more complicated power and electrical systems for a building.

Electrostatic reproduction. A machine method used to reproduce black or blue lines against a white background.

Elevation plan. A drawing that provides a vertical view of an exterior or interior wall surface.

Elevations. The ground levels above or below an established reference point at the corners of and within the property.

Engineer's scale. Scale ruler based on feet and tenths of a foot. It is used most often for dimensions covering a larger area such as site (plot) plans, survey plans, subdivision maps, and landscape plans.

English linear measurement. See *Customary measurement*.

Estimator. A professional who calculates the approximate cost of a proposed building project.

Evaporative cooler. A cooling unit in which warm air flows into the unit and is then propelled by a fan through special pads moistened by a continuous stream of cold water. It is a practical system for dry-climate areas with low humidity.

Existing elevation. An elevation point on a lot prior to beginning groundwork.

Expansion anchor. A fastening device used to secure lighter objects to concrete or masonry.

Expansion joint. A joint provided in masonry or concrete to allow for expansion and contraction in adjacent parts of the building caused by temperature changes.

Expansion shield. A medium- to heavy-duty anchoring device used to fasten materials to concrete. Various types can receive machine bolts or lag screws.

Extension line. On a building plan, a thin line projecting at a right angle from walls or other objects at which dimension lines terminate.

Exterior envelope. Pertains to the structural, protective, and insulating materials of the outside walls and roof of the building.

Family room. A room providing an area for family entertainment. Also called a *recreation room*.

Fascia. Finish boards nailed to the ends of the roof rafters.

Fiberglass. A mineral fiber insulating material produced from molten silica.

Fine-grained soil. Earth materials such as silt or clay. They provide less stable and sometimes inadequate load-bearing support.

Finish elevation. Shows the final elevation points after the groundwork has been completed.

Finish nail. Nail with a very small head and a thin shank. It is used to fasten finish woodwork such as molding and can be set below the surface with a nail set.

Fire door. A door constructed of materials that conform to the standards established by an approved testing agency such as the ASTM. A typical fire door is covered with metal and contains a solid core of fire-treated wood or other fire-resistant materials.

Fire-resistive rating. A rating determined by the number of hours a building or a building component sustains its capability to contain a fire and continue to function as a structural member.

Fixed-sash window. A stationary window that cannot be opened.

Fixture branch line. Water line that runs from the distribution line to the fixture supply line.

Fixture drain pipe. Connects to the fixture trap and carries the discharge to a horizontal branch or vertical stack.

Fixture supply line. Water line that connects to a fixture.

Flashing. Thin, waterproof material such as metal, plastic, or felt placed at areas susceptible to water leakage. Flashing is normally used at roof valleys, around door and window openings, and where a horizontal surface butts up against a vertical surface.

Floor area. As defined in all three model codes, it is the number of square feet within the surrounding exterior walls of a building, excluding vent shafts and courts.

Floor framing plan. On a building plan, provides a plan view guiding the construction of the floor unit.

Floor plenum. The part of a HVAC system where the air is heated or cooled in the crawl space beneath the main floor of the building.

Floor truss. Consists of a flat top and bottom chords and webs tied with metal fasteners or plywood cleats. It is used for the same purpose as floor joists.

Flush beam. A beam with a lower side that is flush with the bottom of the joists that butt up against the beam.

Footing. The bottom part of a foundation wall that spreads the load over a wider area of soil beneath the foundation.

Forced-air cooler. A cold-air system separate from or combined with a forced-air heating unit. In either case, the cold-air system normally uses the same ducts and openings as the heating system to direct cool air into the building.

Forced-air heating. A heating system in which the heated air originates from a furnace fueled by natural gas, liquid petroleum (LP) gas, or electricity.

Foundation plans. On a building plan, these plans provide the dimensions and drawings for a foundation system.

Foyer. A space inside the main entry of the house providing a buffer space between the front entry and the living room.

French door. A swinging door consisting mostly of glass panes framed by top and bottom rails and vertical stiles.

Frieze block. A flat piece placed at the top of an exterior wall where the siding meets the soffit of the roof overhang. Also called a *frieze board.*

Frost line. The depth to which the ground freezes below the surface during the winter season.

Full-basement foundation. A foundation system extending below the ground and providing an additional storage and work area beneath the first floor of the building.

Full bathroom. Contains a water closet (toilet), a lavatory (sink), and a bathtub and shower with the possible addition of a whirlpool and bidet.

Furring strips. Narrow wood pieces fastened to the wall and ceiling surfaces. They are used as a nailing base for finish materials.

Gable roof. A roof sloping in two directions with a ridge at the center.

Gambrel roof. A roof resembling a gable roof with an intermediate break in its slope.

General contractor. See *Building contractor.*

General-purpose circuit. Controls the electrical current going to convenience outlets. Also called a *lighting circuit.*

Girder pocket. See *Beam pocket.*

Glued-laminated timber. A timber made up of several pieces of lumber joined with a very strong adhesive. Often referred to as *glulam.*

Grade beam foundation. A concrete foundation system in which the walls are supported by piers that extend down to load-bearing soil.

Greenhouse. In solar construction, a glazed enclosure collecting solar heat that can be directed to other parts of the building. Also called a *sunspace or atrium.*

Ground Fault Interrupter (GFI). A safety device to protect against the possibility of electric shock, particularly in damp areas.

Grout. A thin concrete mix that can be placed in smaller spaces such as cracks and joints. It is also used to fill the hollow cores of concrete masonry units.

Half bathroom. A bathroom that contains only a water closet and lavatory.

Handrail. A member parallel to and running along the length of a stairway. It may be attached to a wall or the posts of a stairway system.

Hardboard. A panel product made of wood fiber compressed and bonded into sheets.

Hardwood. Lumber manufactured from broad-leafed (deciduous) trees, most of which shed their leaves at the end of the growing season.

Header (door and window). The horizontal member placed over door, window, and other types of wall openings.

Header joist. A continuous joist at the outside perimeter of a framed floor unit. The outside ends of the floor joists are spiked to the rim joists. Also called a *rim joist.*

Headroom (stair). The minimum vertical distance from the tread nosing to the ceiling above.

Heating pump. A very efficient year-round heating and cooling system for buildings in milder climates. Rather than producing the heat, the heat pump moves the heat.

Heating, ventilating, and air conditioning (HVAC). The climate-control system of a building's interior.

Hidden line. On building plans, a series of short dashes showing the edges of a part of a structure that would not be seen from the view shown in the drawing.

Hip rafter. Rafters that run at a 45-degree angle from the building corner to the ridge board.

Hip roof. A roof that has a ridge at its center and slopes in four directions.

Horizontal travel. The level plane of travel of a means of egress.

House sewer. See *Building drain.*

Hydration. A chemical process in which water is added to the cement and aggregates, and the cement acts as a paste binding the mixture together.

Hydronic heating. A heating system in which warm air is derived from heated water.

Hydronic loop. A hydronic heating system in which hot water is pumped from a boiler and carried by pipe to the heat outlets or distributors. As the water cools, it flows back to the boiler.

Illustrator. Prepares presentation drawings for a client based on the building plans.

Imperial measurement. See *English measurement.*

Indirect gain. A passive solar system in which the sun's rays heat a thermal mass placed between the glazed area and the interior space of the building.

Indirect measurement. A measurement derived by mathematical calculation.

Insulation. Building material controlling heat flow through the outer shell of the building.

Insurance. Job insurance obtained by the contractor for the protection of the owner and the contractor.

Isolated gain. A system that uses an area outside the living space to collect, store, and distribute solar heat.

Jack rafter. Roof framing members that fill the spaces between ridge boards and hip and valley rafters.

Joist. A plank placed on its edge that spans the distance between the walls and/or beams to support a floor or ceiling unit.

Joist hanger. A metal device used to support joists that butt up against a beam or another joist.

Keyway. A groove formed at the top of the foundation footing to help secure the wall to the footing.

Kilowatt-hour. The unit used by electric power companies to measure the consumption of electricity. A kilowatt equals 1000 watts.

Kitchen. A room with the primary purpose of food preparation. A section of a larger kitchen is often used as an eating area.

Knee wall. Short wall that provides intermediate support for roof rafters.

Lag bolt. A bolt with a square or hexagonal head and a section of coarse-pitched thread. Also called a *lag screw*.

Lally pipe column. A cylindrically shaped steel column used to support steel or wooden beams.

Laundry room. A room that may contain a washer, dryer, ironing board, and a counter for folding clothing. It may also provide storage for cleaning equipment and supplies. Also called a *utility room*.

Lavatory. A washbasin with a water supply and drainage. A room containing a washbasin.

Leach lines. See *Drain lines*.

Leader line. On building plans, a short line drawn at an angle to an object. It extends from a description, note, or measurement and points to the object being described.

Level. A horizontal line or plane that is parallel to the surface of still water. A tool used for leveling.

Light. A section of glass. Also spelled *lite*.

Lighting circuit. See *General-purpose circuit*.

Lighting fixture schedule. A schedule providing code numbers of lighting fixtures as they appear on the plan, along with other relevant information.

Lighting outlet. Provides a connection for light fixtures placed in the ceiling and upper wall sections.

Linear. A straight-line distance between two horizontal or vertical points.

Live load. The total anticipated weight of people in the building, along with the equipment and furniture. It also includes the weight exerted by snow, ice, and water on the roof.

Living area. The section of the building containing the living room, dining room, foyer, recreation room, and special-purpose room.

Living room. The main social center for the family and entertaining guests.

Longitudinal section. A cut along the length of a building shown in a building plan.

Loose-fill fiberglass. An insulating material blown into an area by pressurized hose.

Lot. A piece of land identified with established boundaries.

Machine bolt. A bolt with a hexagonal or square head. A nut tightened at the threaded end holds the bolted pieces together.

Mansard roof. Similar to a hip roof, with a double slope on all four sides.

Manufactured homes. Factory-produced homes constructed in sections. The sections are delivered by truck to the job site, where they are joined.

Mean sea level. The average sea level between high and low tides.

Means of egress. A continuous and unobstructed path of travel from any point in a building to the outside ground level.

Mechanical engineer. A professional who designs the more complex HVAC methods and other mechanical systems for a building.

Melamine. A plastic best known as the plastic laminate used for countertops and cabinet finishes.

Mercury switch. Electrical switch containing a sealed glass tube of mercury that provides a contact when the switch is turned on. It is silent and shockproof.

Meter. The basic linear unit in the metric system. Also spelled *metre*.

Metric scale. A scale ruler used when construction plans are drawn with metric linear measurements.

Metric system. A worldwide system of measurement based on the meter as a unit of length.

Millimeter. The smallest division of a meter. A millimeter is 1/1000 of a meter.

Model codes. National building codes written for the purpose of helping local building agencies develop workable local building codes.

Molding. A decorative piece of trim material (usually wood) placed over joints where a ceiling butts against the top of a wall; vertical inside and outside corners and joints created when different materials butt against each other.

Monolithic concrete. A system of placing concrete in a continuous pour, allowing only for construction joints.

Monument. A permanent marker such as a steel stake driven into the ground or an aluminum cap embedded in the street or sidewalk surface. Used to establish legal reference points for surveying purposes.

Mud room. A variation of a utility room, usually placed inside a rear entrance. In colder climates it serves as an area for taking off wet boots and clothing before entering the other rooms.

Mullion. The horizontal or vertical divisions between single windows in a multiple window unit.

Muntin. A short bar used to separate glass in a sash into vertical lights.

National Building Code of Canada. A national model code for Canada.

National Electric Code. A national model electrical code.

National Plumbing Code. A national model plumbing code.

Newel post. A post at the head or foot of a stairway supporting the end of a handrail.

Object line. On a building plan, a line used to identify the edges or outlines of objects such as walls, buildings, and walks.

Occupancy classification. Building code definition of the purpose for which a building, or portion thereof, is used.

One-and-one-half-story house. A building with a bottom floor and a large attic area that allows for added bedrooms, a bathroom, or other additional living space.

Orientation. The placement of a building on a lot. Proper orientation makes the best use of prevailing climate conditions and the surrounding physical environment.

Oriented strandboard. A panel product fabricated from long, strandlike particles compressed and glued together..

P-trap. A curved device found beneath each water fixture. Water remains in the bend of the trap, preventing the escape of gases through the drain opening of the fixture.

Panel products. Large sheets of lumber manufactured from thin veneers or reconstituted wood particles glued together.

Particleboard. A panel product made of smaller wood chips and particles.

Partition. An interior dividing wall.

Passive solar system. A heating and cooling system in which solar energy is collected, stored, and distributed mainly through the design of the building.

Penny. A term identifying the length of a nail. It is abbreviated as d.

Perimeter heating. A heating system in which all the warm-air outlets are placed against the outside walls and under the windows.

Phantom line. On building plans, phantom lines look the same as hidden lines. However, phantom lines identify an object that is not shown rather than one that is hidden from view. An example is the edge of a roof overhang on a floor plan.

Phenolic. A thermoset plastic commonly used in the manufacture of electrical parts such as sockets, switch boxes, and circuit breakers.

Pictorial drawing. A three-dimensional drawing.

Pier footing. A rectangular or round base resting on the soil and supporting a pier or column.

Pilaster. A column attached to and projecting from a wall.

Pitch (roof). The slope of a roof, expressed as the ratio of the vertical unit rise to the horizontal unit run.

Plank. Dimension lumber 2″ to 4″ thick and 6″ or more wide.

Plastic foams. A plastic that can be poured or sprayed into a wall or other types of cavities.

Plastics. Synthetically produced material made up of various chemical combinations that can be molded and formed into different solid shapes or very thin sheets.

Plat plans. Drawings available from government agencies defining the borders of land divisions in a given geographical area. Also called *subdivision maps.*

Plates. Horizontal pieces nailed to the tops and bottoms of the wall studs.

Platform framing. A framing system in which the wood framed floor unit provides a working platform for the wall construction at each story of the building.

Plot plan. See *Site plan.*

Plywood. A panel product made up of several wood veneer sheets placed at right angles to each other and glued together under high pressure.

Point of beginning. A fixed reference point on a property from which other dimensions and elevations are established.

Polyester. A thermoset plastic. Some examples of polyester products are translucent sheets for roofing, molded sinks, bathtubs, and shower stalls.

Polyethylene. A thermoplastic sheet material often used as a vapor barrier for floors, walls, and concrete slabs resting on the ground.

Polypropylene. A thermoplastic material well suited for insulation around piping and pipe fittings.

Polystyrene. A thermoplastic material used for the manufacture of lighting fixtures and molded pieces of hardware. Foamed (expanding) polystyrene is frequently used for wall, floor, and ceiling insulation.

Powder-actuated tool. A tool that uses an explosive charge to drive fastening devices into concrete.

Prefabricate. To construct building sections and components at a mill or plant not located at the job site.

Pressure-treated lumber. Lumber chemically injected with a preservative under pressure.

Prints. See *Building plans.*

Property line. On a construction drawing, a heavy solid line with one long and two alternating short dashes or dots. It identifies the boundaries of the property.

Purlin. A component that runs horizontally below the roof rafters and is nailed to the braces.

Pushbutton switch. A wall switch turned on and off by pressure on a button.

Rabbet. A longitudinal recessed groove cut along the edge of a piece of material.

Radiant heat. A heating system that transfers heat through space.

Radiant hydronic heating. A heating system in which heated water is pumped from a boiler to copper tubing placed in the floor or ceiling. The warm surfaces radiate heat to the enclosed room area.

Radiant wire heating. A grid pattern of heat-resistant wires placed in the floor or ceiling to produce heat that radiates into the room.

Radiation. The direct movement of heat through space from a warm source.

Radon. A gas produced by the radioactive decay of radium given off by some types of soils and rock.

Rafter. An inclined roof framing member that extends from the outside wall plate to a ridge.

Rail (door). The top and bottom horizontal section of a panel door.

Rail (stair). The top piece of a railing.

Railing. A combination of rails, posts, and balusters protecting the open ends of stairs and balconies.

Ranch-style house. A one-story house best suited for larger lots. Ranch-style houses usually feature low-pitched roofs.

Ready-mixed concrete. Concrete manufactured in a batch plant and delivered by truck to the job site.

Rebars. See *Reinforcing steel bars.*

Receptacle. A device attached to an electrical box that receives electricity from a power source. Plugs at the end of electrical cords are inserted into the receptacles.

Recessed wall heater. An individual heater recessed into a wall and used to heat smaller areas.

Reconstituted wood panels. A panel product consisting of wood particles, flakes, or strands bonded together and molded into full-size sheets.

Recreation room. See *Family room.*

Register. An adjustable grilled cover over a warm-air or cold-air outlet.

Registered design professional. Usually pertains to an architect or engineer licensed to design structures and systems that are part of a building structure.

Reinforced concrete. Concrete structural components reinforced with rebars or welded wire mesh.

Reinforcing steel bars. Steel bars placed horizontally and/or vertically in the walls to strengthen the lateral resistance of the walls. Also placed in floor slabs that must resist heavier vertical pressures. Commonly called rebars

Resistance value. The rating of a material's thermal resistance to heat flow. Commonly referred to as *R-value.*

Return air. Air returned to a heating or cooling source after it has circulated through a structure.

Rigid insulation. Panels made of insulating materials, often used as wall sheathing and also attached to concrete walls.

Rim joist. A continuous joist at the outside perimeter of a framed floor unit. The outside ends of the floor joists are spiked to the rim joists. Also called a *Header joist.*

Riser. A finish piece forming the vertical face of a stair step.

Roof truss. A prefabricated roof component consisting of top and bottom chords and webs tied together with metal fasteners or plywood cleats.

Roof window. A window built into the slope of the roof.

Room planning. Relates to the most efficient arrangement of rooms in a building.

Rough-in work. The concealed work within walls, ceilings, and floors of plumbing and electrical systems.

Rough-sawed. Pertains to lumber with a surface that has not been planed to a smooth finish.

Rough sill. Lower horizontal member of a framed window opening.

Rubble masonry. Random-shaped pieces of rock held together by mortar and used to construct solid or veneer walls.

Run (plumbing). The section of pipe continuing in a straight line in the direction of the water flow.

Sash. The frame of a window that holds the glass.

Scale drawing. A drawing in which the actual dimensions of a building and its components are represented by a division of an inch or inches. For example, 1/4″ = 1′-0″ means that 1/4″ on the drawing is equal to 1′-0″ of the object being drawn.

Schematic drawings. Elevations of a plumbing system showing piping and fixtures.

Section view. A view of an object as it would appear if a cut were made revealing its internal structure.

Septic system. A waste system consisting of a septic tank and disposal field used in areas that do not have a public sewer system.

Service area. Area of the building that includes the kitchen, utility and/or mud rooms, and a garage or carport.

Service drop. The point at which feed wires originating from a transformer connect to electrical wires leading to the service panel of the building.

Service entrance. Pertains to the point of entry of electrical equipment and wiring into the building.

Service panel. A panel that receives electrical current from a power source and then distributes the electrical current throughout the building through a series of branch circuits. Also called a *distribution panel.*

Service switch. A switch containing fuses or service breakers controlling the electrical current after it has passed through the meter. It can turn the electrical current to the service panel on or off.

Sheathing. The panel material nailed to the outside surface of an exterior wall.

Shed roof. A roof that slopes down in one direction.

Shut-off valve. A control valve placed in the piping of water and gas systems.

Sill plate. A horizontal wood piece bolted to the top of the foundation wall. Also called a *mudsill*.

Site plan. Provides information about the setbacks of buildings and any other structures on the lot. The site plan includes much of the information found on a survey plan. Also called a *plot plan*.

Slab-on-grade foundation. Consists of a concrete slab receiving its main support from the soil. Additional support is provided by footings around the perimeter and within the slab. Also called a *slab foundation*.

Sleeping area. The area of a residence that consists of the bedrooms, bathroom(s), and bedroom closets.

Soffit. The exposed underside of building components such as a balcony, roof cornice, arch, beam, and staircase.

Softwood. Lumber that comes from softwood (coniferous) trees that bear needlelike or scalelike leaves during all climate seasons.

Soil engineer. A professional employed to test for unstable ground conditions and soils requiring deep excavations.

Solar collector. A device that absorbs radiation from the sun and provides warm air or water within the building.

Sole plate. The bottom plate of a framed wall.

Span (roof). Total horizontal width of a building beneath the run of the roof rafters.

Special-purpose room. A room other than the living room that may be found in the living area, such as an atrium, home office, or den.

Special-service outlet. An outlet serving one piece of equipment, such as a washing machine or dishwasher. These outlets often have one receptacle wired for 220 volts.

Specifications. A legal document included with a set of construction plans containing important written information that is not noted or fully explained in the drawings. These documents are commonly called *specs* by persons working in the construction industry.

Specifications writer. A professional who writes specifications for a set of building plans.

Spike. A nail 10d or longer.

Split-level house. A type of house often found on steeply sloping lots. The floor levels are separated by half a story joined by short flights of stairs.

Split-wire outlet. A duplex receptacle in which the top receptacle is controlled by a wall switch and the bottom receptacle is connected to a hot circuit.

Spread foundation. A foundation design consisting of a stem wall placed over a concrete footing. Also called a *T-foundation*.

Stairway. A single flight of stairs or series of flights of stairs between two levels, including the supports and handrails. Also called a *staircase*.

Stairwell. The horizontal opening and vertical space for a stairway.

Standard building code. A model code used primarily in the southern states.

Staple. A U-shaped metal fastener with pointed ends, driven with a staple gun.

Steel I-beam. A beam made up of top and bottom flanges separated by a vertical web. The beam resembles a capital letter I when viewed from one end. Also called a *wide-flange beam*.

Stem wall. A vertical concrete or CMU wall supported by footings.

Stepped foundation. A foundation system consisting of a series of stepped-up walls and footings. Normally used for buildings erected on steeply sloping lots.

Stile (door). The side frame of a panel door.

Stipulated-sum contract. A contract that includes an agreed upon fixed price for a construction project.

Stock plans. A published set of building plans that can be purchased for a fee.

Storm sash. An extra window placed on the outside of an existing window to add additional protection against severely cold weather. Also called a *storm window*.

Storm sewer. A separate sewer to carry rainwater (not sewage) coming from downspouts connected to the roof gutters.

Story. The portion of a building between the upper surface of one floor and the upper surface of another floor or the ceiling above.

Stringer (stair). See *Carriage.*

Structural engineer. A professional who designs the main structural components of a building (walls, floors, beams, columns) to withstand the maximum stress and loads expected.

Structural lumber. Lumber material such as beams, stringers, joists, planks, posts, and timbers.

Stud (framing member). A vertical piece of a wood framed wall nailed between the top and bottom plates.

Stud (powder-activated). A specially hardened nail driven into concrete by a powder-actuated tool. Also called a *drive pin.*

Subcontractor. A licensed firm that contracts with the general contractor to perform work not done by the general contractor, such as plumbing, electrical, sheet metal, roofing, painting, and other specialized work.

Subdivision maps. See *Plat plan.*

Sunspace. See *Greenhouse.*

Supply outlet. See *Air outlet.*

Survey plan. Provides information about the shape, size, and topographical features of the building site.

Switch. A device to open or close an electric circuit. Used mainly to control lighting sources such as permanent overhead and wall light fixtures.

Système International d'Unités. Translated from French as International System of Units. The basic units of measurement for the metric system.

T-foundation. See *Spread foundation.*

Temper. A heating and cooling process used to harden and strengthen metal, glass, and other materials.

Thermal mass. Pertains to masonry, stone, or concrete wall and floor materials that can store heat.

Thermoplastic. A plastic material that can be softened by a heating process and then remolded to another shape that rehardens as it cools.

Thermoset. A plastic material that cannot be softened after it has been formed. Thermosets are generally harder and more heat-resistant than thermoplastic materials.

Thermosiphon. A passive solar method consisting of a glazed collector usually placed below a south-facing window area, with a rock bed serving as a heat-storage area.

Thermostat. A device that responds to temperature changes. It can be set to a desired heating level and will continue to control the heating system to maintain that level.

Three-quarter bathroom. A bathroom containing a toilet, sink, and shower.

Threshold. A finished strip placed beneath an interior door to cover the joint between two different floor materials. Also the term for a device placed below an exterior door to prevent air infiltration.

Timber. Piece of lumber 5" or more in nominal thickness and width. Timbers are commonly used for posts and beams.

Toe space. A recessed space at the front and bottom of a base cabinet.

Toggle switch. Switches operationg with a snap lever. They are used more frequently than other kinds.

Tongue and groove. A type of joint between panels or boards in which a tongue on one edge of a piece of material fits into a groove on the edge of an adjoining piece of material.

Top chord. The top member of a roof or floor truss.

Topography. Pertains to the physical and graphic description of land surface features.

Total rise (roof). The total height of a sloped roof.

Total rise (stairs). The vertical measurement from floor to floor of a stairway.

Total run (roof). A horizontal distance equal to one half of the span.

Total run (stairs). The horizontal measurement of the stairway.

Transverse section. A cross-sectional view along the width of a building.

Travel distance. The total length of a path that must be traveled to exit from any location within the occupied part of a building.

Tread. The horizontal surface of a stair step.

Tread nosing. A rounded horizontal projection of the stair tread extending past the the face of the riser below.

Trimmer studs. The studs supporting the ends of a window or door header.

Trombe wall. A passive solar system featuring an 8″ to 16″ darkly colored masonry wall for heat storage. It is placed very close behind the glazed area.

Two-story house. A residence consisting of two complete floor levels.

Uniform Building Code. A model code widely used in the western and some midwestern United States. It is published by the International Conference of Building Officials.

Uniform Heating and Comfort Cooling Code. A national HVAC model code.

Uniform Plumbing Code. A model code book published by the International Conference of Building Officials.

Uniform Zoning Code. A model zoning code book published by the International Conference of Building Officials.

Unit rise (roof). The vertical rise of a roof rafter for every foot of unit run.

Unit rise (stair). The height of a stair riser.

Unit run (roof). A horizontal measurement that is a unit of the total run. It will always be 12″.

Unit run (stairs). The width of a stair tread.

Upflow furnace. A furnace that is usually installed in a full basement.

Utility easement. A portion of land on a piece of property that can be used by the utility company to place water and gas lines as well as other utility devices.

Utility room. See *Laundry room*.

Valley rafter. A roof rafter placed at the valley meeting point of intersecting roofs.

Vapor retarder. A material usually placed against the warm side of studs or joists to help prevent moisture and condensation within a building. Also called a *vapor barrier*.

Veneer (masonry). An outside facing of brick or stone over a concrete or wood framed wall.

Veneer (wood). A thin layer of wood used as a finish surface over less attractive wood. A thin sheet used as one of several plies of plywood.

Veneered I-beam. A type of beam consisting of several veneer pieces glued together. The top and bottom flanges are notched to receive the edges of the veneered pieces.

Veneered joist. A joist made up of veneer pieces glued together.

Vent. A screened and louvered opening enabling air to circulate in an attic space or crawl-space foundation area.

Vinyl. A thermoplastic compound that is one of the more widely used plastics in construction work. Some examples of its use are sheet and tile floor finish, gutters, downspouts, exterior siding, and flashing.

Volt. The unit used for measuring the force that causes the electrical current to flow through a conductor such as an electrical wire.

Waferboard. A panel product made up of large wafer-like wood flakes compressed and bonded together.

Water main. The large pipes that convey the water from its source to the community where it is being used.

Water meter. A device for recording the volume of water passing through the pipes of the water supply system of a building.

Water storage wall. Used as part of a passive solar system, the storage wall consists of water-filled drums, fiberglass tubes, or large pipes placed directly behind the glazed area.

Water supply line. A pipe that taps into the water main and conveys water to the building. Also called a *water service line.*

Watt. A unit of electrical power.

Watt-meter. A device that records the amount of electrical energy used by a consumer.

Weatherstripping. A material attached to the edges of doors and windows to prevent cold-air infiltration.

Web (truss). A wood or metal member running between the top and bottom chords, tying them together.

Welded wire mesh. A grid of steel wires welded together and used to reinforce concrete slabs. Also called *welded wire fabric.*

Wet wall. See *Chase wall.*

Wide-flange beam. See Steel I-beam.

Work-plus-fee contract. See *Cost-plus contract.*

Working drawings. The actual building plans used to construct a building.

Zero-clearance fireplace. A prefabricated, insulated metal fireplace unit that can be placed close to the framework around the fireplace opening.

Zone-controlled heating. A heating and cooling system in which different areas of the building are controlled by separate thermostats.

Zoning codes. A local code written to control the design, location, land use, and type of occupancy of all buildings in the areas within the code's jurisdiction.

Abbreviations

A

above	ABV
above finish floor	AFF
access panel	AP
acoustic plaster	AC PL
acoustical	ACOUS
actual	ACT.
addition	ADD.
adhesive	ADH
adjustable	ADJ
aggregate	AGGR
air	A
air conditioning	AC
air vent	AV
alternate	ALT
alternating current	AC
aluminum	ALUM
American Concrete Institute	ACI
American Institute of Architects	AIA
American National Standards	ANS
American Plywood Association	APA
American Society for Testing and Materials	ASTM
American Standards Association	ASA
amount	AMT
ampere	AMP
anchor bolt	AB
angle	ANG
apartment	APT
approved	APPD
approximate	APPROX
architectural	ARCH.
area	A
asbestos	ASB
asphalt	ASPH
at	@
automatic	AUTO
avenue	AVE
average	AVG

B

balcony	BALC
base	B
basement	BSMT
bathroom	B
bathtub	BT
batten	BT
beam	BM
bearing	BRG
bedroom	BR
below	BLW
benchmark	BM
better	BTR
between	BET
bevel	BEV
block	BLK
blocking	BLKG
blower	BLO
blueprint	BP
board	BD
board feet	BD FT
boiler	BLR
both sides	BS
both ways	BW
bottom	BTM
bottom of footing	BF
bracket	BRKT
brass	BR
bread board	BRD BD
brick	BRK
British Thermal Unit	BTU
bronze	BRZ
broom closet	BCL
building	BLDG
building line	BL
Building Officials and Code Administrators	BOCA
built-in	BLTIN
buzzer	BUZ
by	X

C

cabinet	CAB.
cabinet light	CAB LT
cast iron	CI
catalog	CAT.
catch basin	CB
caulking	CLKG
ceiling	CLG

ceiling joist	CLG JST	dead load	DL
Celsius	C	decibel	DB
cement	CEM	deck	DK
center	CTR	decking	DKG
centerline	CL	degree	DEG
center to center	CC	depth	DP
centimeter	CM	design	DSGN
ceramic	CER	detail	DET
ceramic tile	CT	diagonal	DIAG
chamfer	CHAM	diagram	DIAG
channel	CHAN	diameter	DIA
check	CHK	dimension	DIMEN or DIM
chimney	CHM	dimmer	DIM. or DMR
cinder block	CIN BLK	dining room	DR
circle	CIR	direct current	DC
circuit	CKT	disconnect	DISC
circuit breaker	CKT BKR	dishwasher	DW
circumference	CIRC	disposal	DISP
cleanout	CO	ditto	DO
clear	CLR	division	DIV
closet	CL	door	DR
coated	CTD	double	DBL
cold	C	double blocking	DBL BLKG
cold water	CW	double hung	DH
cold water supply	CWS	Douglas fir	DF
column	COL	dowel	DWL
combination	COMB.	down	DN
common	COM	downspout	DS
composition	COMP	drain	DR
concrete	CONC	drawer	DRWR or DWR
concrete masonry unit	CMU	drawing	DWG
conduit	CND	drinking fountain	DF
construction	CONSTR or CONST	dryer	D
		drywall	DW
construction joint	CJ	duplicate	DUP
Construction Standards Institute	CSI		
continue	CONT		
continuous	CONT		
contractor	CONTR		
control joint	CJ	each	EA
copper	COP	each face	EF
corrugate	CORR	each way	EW
counter	CNTR	east	E
countersink	CSK	elbow	ELB
courses	C	electric	ELEC
cross section	X-SECT or CS	electric metallic tubing	EMT
cubic	CU	electric water cooler	EWC
cubic feet	CU FT	elevation	EL or ELEV
cubic feet per minute	CFM	elevator	ELEV
cubic inch	CU IN.	emergency	EMER
cubic yard	CU YD	enamel	ENAM
		enclosure	ENCL
		engineer	ENGR
		entrance	ENT
damper	DMPR	Environmental Protection Agency	EPA
dampproofing	DP	equal	EQ

D

E

equipment	EQUIP.	furred ceiling	FC
estimate	EST	furring	FUR.
excavate	EXC		
exhaust	EXH		
existing	EXIST. or EXST		
expansion	EXP		
expansion joint	EXP JT		
exposed	EXPO		
extension	EXTN		
exterior	EXT		

G

		gage, gauge	GA
		galvanized	GALV
		galvanized iron	GI
		galvanized sheet metal	GSM or GALV SH MTL

F

		garage	GAR
		gas	G
fabricate	FAB	girder	G
face of studs	FOS	glass	GL
face of wall	FOW	glazed	GL
Fahrenheit	F	glue laminated beam	GLB
family room	FAM RM	glued laminated	GLULAM
Federal Housing Administration	FHA	grade	GR
feet	FT	grade beam	GB or GR BM
feet board measure	FBM	grade line	GL
feet per minute	FPM	granular	GRAN
figure	FIG.	grating	GRTG
filter	FLTR	gravel	GVL
finish	FIN.	ground	GND
finish all over	FAO	ground fault interrupter	GFI
finished floor	FIN. FL	gypsum	GYP
finished grade	FIN. GR	gypsum wall board	GYP BD or GWB
firebrick	FBRK		
fire extinguisher	F EXT		
fireplace	FP		
fireproof	FPRF		
fitting	FTG		

H

fixed window	FX WDW		
fixture	FIX.		
flammable	FLAM		
flange	FLG	hall	H
flashing	FL	hardboard	HBD or HDBD
flexible	FLEX.	hardware	HDW
floor	FL or FLR	hardwood	HDWD
floor drain	FD	head	HD
floor joist	FL JST	header	HDR
flooring	FLG	heater	HTR
fluorescent	FLUOR	heating	HTG
folding	FLDG	heating-ventilating-air conditioning	HVAC
foot candle	FC	heat-vent-light	HVL
footing	FTG	height	HT
foundation	FDN	hemlock	HEM
frame	FR	hexagonal	HEX.
freeze proof hose bib	FPHB	high	H
freezer	FRZR	hollow core	HC
full size	FS	horizontal	HOR or HORIZ
furnace	FURN	horsepower	HP
		hose bibb	HB
		hot	H
		hot water	HW
		hot water heater	HWH
		hour	HR

house	HSE
hundred	HUN or C
hypotenuse	H

I

I-beam	I
illuminate	ILLUM
impregnate	IMPG
incandescent	INCAN
inch	IN.
incinerator	INCIN
inflammable	INFL
inside diameter	ID
install	INSTL
insulate	INS
insulation	INSUL
interior	INT
iron	I

J

jamb	JMB
joint	JT
joist	JST
junction	JCT

K

kick plate	KP
kiln dried	KD
kilometer	km
kilowatt	kW or KW
kilowatt hour	kWh
Kip	K
kitchen	KIT.
kitchen cabinet	KC
kitchen sink	KS
knockout	KO

L

laboratory	LAB
ladder	LAD.
laminate	LAM
laminated	LAM
laminated veneer lumber	LVL
landing	LDG
lateral	LAT
lath	LTH

laundry	LAU
lavatory	LAV
lazy susan	LS
left	L
length	LG
length overall	LOA
level	LEV
light	LT
linear	LIN
linear feet	LIN FT or LF
linen closet	L CL
linoleum	LINO
live load	LL
living room	LIV RM or LR
long	LG
louver	LV
lumber	LBR

M

machine bolt	MB
main	MN
manhole	MH
manual	MAN.
manufacture	MFR or MANU
marble	MRB
masonry	MAS
masonry opening	MO
masonry veneer	MAS VEN
master	MA or MSTR
master bathroom	MA B or MSTR B
master bedroom	MA BR or MSTR BR
material	MATL
maximum	MAX.
mean sea level	MSL
mechanical	MECH
medicine cabinet	MC or MED CAB
medium	MED
membrane	MEMB
metal	MET.
metal lath	ML
meter (metric)	m
mezzanine	MEZZ
millimeter	mm or MIL
microwave	MV
minimum	MIN
minute	MIN
mirror	MIR
miscellaneous	MISC
mixture	MIX.
model	MOD
modular	MOD

molding	MLDG	perpendicular	PERP
motor	MOT	phase	PH
motor operated	MO	piece	PC
mullion	MULL.	plaster	PLS
		plasterboard	PLS BD
		plastic	PLAS
		plate	P or PL
N		plate glass	PL GL
		plate height	PL HT
National Electric Code	NEC	platform	PLAT.
natural	NAT	plumbing	PLMB
natural grade	NG	plywood	PLWD
nominal	NOM	plywood end nail	PEN.
north	N	point	PT
nosing	NOS	point of beginning	POB
not applicable	NA	polyethylene	POLY
not in contract	NIC	polyvinyl chloride	PVC
not to scale	NTS	porcelain	PORC
number	NO.	pound	LB or #
		pounds per square foot	PSF
		pounds per square inch	PSI
O		power	PWR
		precast	PRCST
obscure	OB	prefabricated	PREFAB
obscure glass	OBSC GL	preferred	PFD
	or OGL	pressure	PRESS.
office	OFF.	pressure reducing valve	PRV
on center	OC	pressure treated	PT
opening	OPNG	primary	PRI
opposite	OPP	property	PROP.
optional	OPT	property line	PL
ounce	OZ	pull switch	PS
outlet	OUT.	pump	PMP
outside diameter	OD	pushbutton	PB
outside face	OF.		
over	O or O/		
overall	OA		
overhang	OH or OVHG	**Q**	
overhead	OVHD		
		quality	QUAL
		quantity	QTY
		quarry tile	QT
P		quarter	QTR
painted	PTD		
pair	PR		
panel	PNL	**R**	
pantry	PAN.		
parallel	PAR.		
part	PT	radiator	RAD
partition	PTN	radius	R
passage	PASS.	rain water leader	RWL
paving	PAV	range	R
penny (nail size)	d	receptacle	RCPT
perforate, perforated	PERF	recessed	REC
perimeter	PER or PERIM	redwood	RDWD or RWD
permanent	PERM	reference	REF
		refrigerator	REF

register	REG	sleeve	SL
reinforce	REINF	sliding	SL
reinforced concrete pipe	RCP	soil	S
reinforcing bar	REBAR	soil pipe	SP
reproduce	REPRO	solid	SOL
required	REQD	solid blocking	SOL BLK
return	RET	solid core	SC
return air	RA	solid grout	SG
revision	REV	south	S
ridge	RDG	speaker	SPKR
right hand	RH	specifications	SPECS
right of way	ROW.	sprinkler	SPR
riser	R or RS	square	SQ
roof	RF	square foot	SQ FT
roof drain	RD	square inch	SQ IN.
roofing	RFG	stainless steel	SST
room	RM	stairs	ST
rough	RGH	standard	STD
rough opening	RO	Standard Building Code	SBC
rough surfaced	RS	steam	STM
round	RD	steel	STL or ST
		stiffener	STIF
		stirrup	STIR.
		stock	STK

S

		stone	STN
saddle	SDL	storage	STG or STOR
safety	SAF	storm drain	SD
safety valve	SV	straight	STR
sanitary	SAN	street	ST
scale	SC	structural	STR
schedule	SCH	structural clay tile	SCT
screen	SCR or SCRN	sump pump	SP
screw	SCR	supply	SUP
scupper	SCUP	surface	SUR
second	SEC	surfaced four sides	S4S
section	SECT.	surfaced two sides	S2S
see structural drawings	SSD	suspended	SUSP
select	SEL	suspended ceiling	SUSP CLG
self-closing	SC	switch	SW
service	SERV	symbol	SYM
sewer	SEW.	system	SYS
sheathing	SHTG or SHTHG		
sheet	SH		
sheet metal	SM		
shelf	SH	**T**	
shelf and rod	SH & RD		
shiplap	SHLP	tangent	TAN.
shower	SH	tar and gravel	T & G
shutoff valve	SOV	tarpaulin	TARP
siding	SDG	tee	T
sill cock	SC	telephone	TEL
similar	SIM	television	TV
single	SGL	temperature	TEMP
sink	S or SK	tempered	TMPD
slate	SL	template	TEMPL
		temporary	TEMP

tensile	TNSL	vitreous clay tile	VCT
tensile strength	TS	vitreous pipe	VP
terminal	TERM.	volt	V
terra-cotta	TC	volume	VOL
terrazzo	TER		
thermostat	THERMO		
thick	THK		
thousand	THOUS or M		

W

wainscot	WSCT
wall	W
wall vent	WV
washing machine	WM
washroom	WR
waste	W
waste stack	WS
water	WTR
water closet	WC
water fountain	WF
water heater	WH
water meter	WM
waterproof	WP
water resistant	WR
watt	W
watt-hour	WHR
weather stripping	WS
weatherproof	WP
weep hole	WH
weight	WT
welded wire fabric	WWF
welded wire mesh	WWM
west	W
white pine	WP
wide	W
wide flange	WF
width	WD
window	WDW
wire glass	W GL
with	W/
without	W/O
wood	WD
wrought iron	WI

thousand board feet | MBF
threshold | THR
through | THRU
toilet | T
tongue and groove | T&G
top | T
top of | TO.
total | TOT.
transformer | TRANS
tread | TR
tubing | TUB.
typical | TYP

U

underground	UGND
Underwriters' Laboratories, Inc.	UL
unfinished	UNFIN
Uniform Building Code	UBC
unit heater	UH
unless otherwise note	UON
urinal	UR
utility	UTIL

V

vacuum	VAC
valve	V
vanity	VAN.
vapor barrier	VB
vaporproof	VAP PFR
varies	VAR
vent	V
vent pipe	VP
vent stack	VS
ventilate	VENT.
verify in field	VIF
vertical	VERT
vestibule	VEST.
vinyl	VIN
vinyl tile	VT
vitreous	VIT

Y

yard	YD
yellow pine	YP

Z

zinc	Z

Architectural Materials Symbols

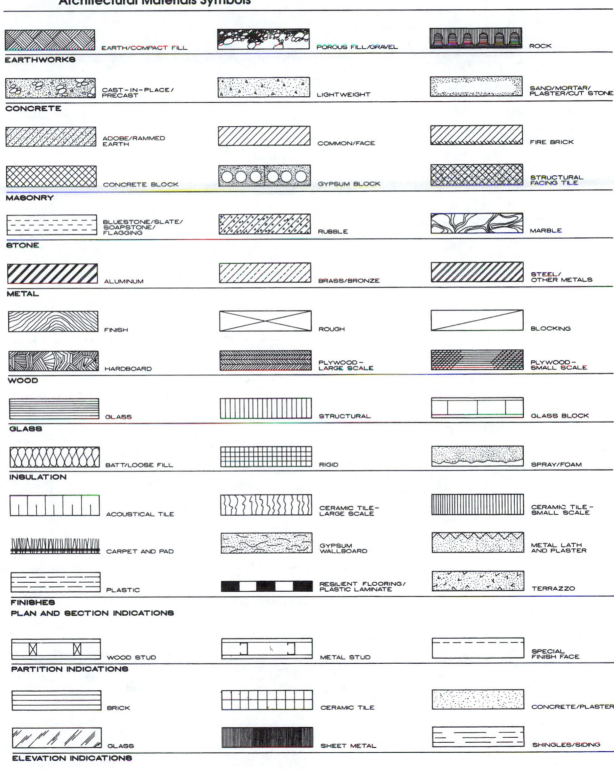

EARTH/COMPACT FILL   POROUS FILL/GRAVEL   ROCK

EARTHWORKS

CAST-IN-PLACE/PRECAST   LIGHTWEIGHT   SAND/MORTAR/PLASTER/CUT STONE

CONCRETE

ADOBE/RAMMED EARTH   COMMON/FACE   FIRE BRICK

CONCRETE BLOCK   GYPSUM BLOCK   STRUCTURAL FACING TILE

MASONRY

BLUESTONE/SLATE/SOAPSTONE/FLAGGING   RUBBLE   MARBLE

STONE

ALUMINUM   BRASS/BRONZE   STEEL/OTHER METALS

METAL

FINISH   ROUGH   BLOCKING

HARDBOARD   PLYWOOD – LARGE SCALE   PLYWOOD – SMALL SCALE

WOOD

GLASS   STRUCTURAL   GLASS BLOCK

GLASS

BATT/LOOSE FILL   RIGID   SPRAY/FOAM

INSULATION

ACOUSTICAL TILE   CERAMIC TILE – LARGE SCALE   CERAMIC TILE – SMALL SCALE

CARPET AND PAD   GYPSUM WALLBOARD   METAL LATH AND PLASTER

PLASTIC   RESILIENT FLOORING/PLASTIC LAMINATE   TERRAZZO

FINISHES
PLAN AND SECTION INDICATIONS

WOOD STUD   METAL STUD   SPECIAL FINISH FACE

PARTITION INDICATIONS

BRICK   CERAMIC TILE   CONCRETE/PLASTER

GLASS   SHEET METAL   SHINGLES/SIDING

ELEVATION INDICATIONS

John Ray Hoke, Jr., FAIA; Washington D.C.

(Courtesy of *Architectural Graphic Standards*, published by John Wiley and Sons, Inc.)

Drawing Conventions and Symbols

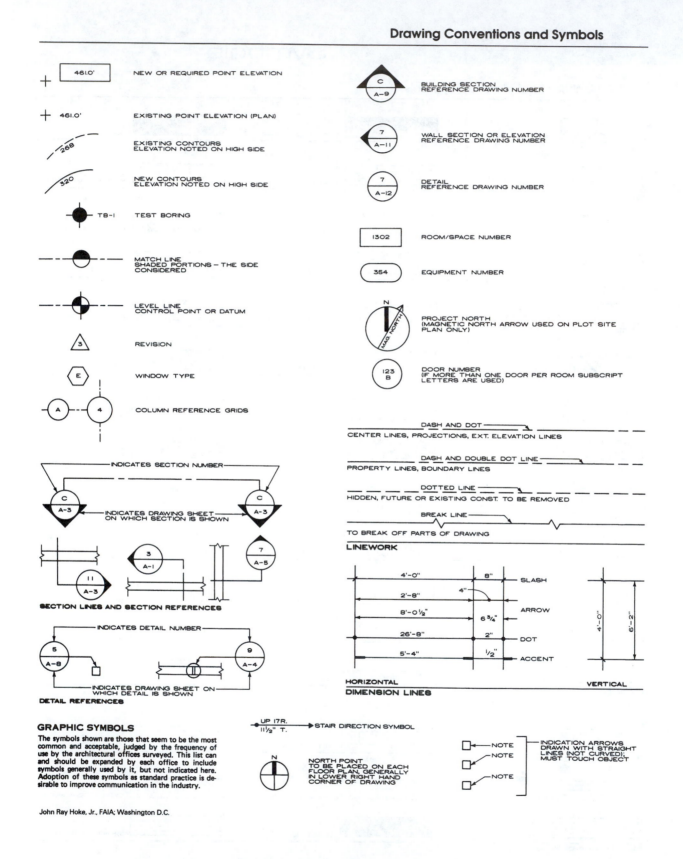

GRAPHIC SYMBOLS

The symbols shown are those that seem to be the most common and acceptable, judged by the frequency of use by the architectural offices surveyed. This list can and should be expanded by each office to include symbols generally used by it, but not indicated here. Adoption of these symbols as standard practice is desirable to improve communication in the industry.

John Ray Hoke, Jr., FAIA; Washington D.C.

(Courtesy of *Architectural Graphic Standards*, published by John Wiley and Sons, Inc.)

Methods of Dimensioning Doors and Windows

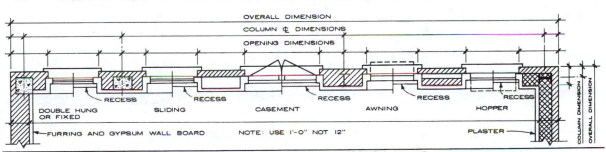

METHOD FOR DIMENSIONING EXTERIOR WINDOW OPENINGS IN MASONRY WALLS (DOORS SIMILAR)

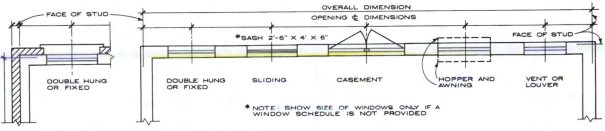

METHOD FOR DIMENSIONING EXTERIOR WINDOW OPENINGS IN FRAME WALLS (DOORS SIMILAR)

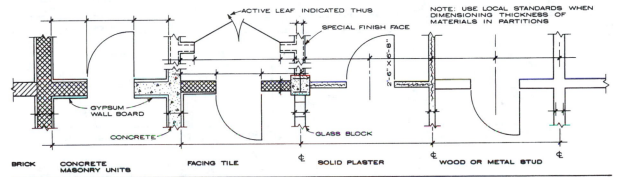

METHOD FOR DIMENSIONING AND INDICATIONS OF INTERIOR PARTITIONS AND DOORS

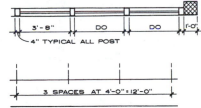

REPETITIVE DIMENSIONING

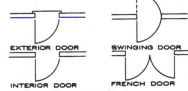

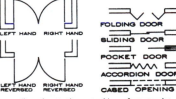

GENERAL NOTES

Dimensioning should start with critical dimensions and should be kept to a minimum. Consideration must be given to the trades using them and the sequencing adjusted to their respective work. It is also necessary to bear in mind that tolerances in actual construction will be varied. This means that as-built dimensions do not always coincide with design dimensions. Dimensioning from established grids or structural elements, such as columns and structural walls, assists the trades that must locate their work prior to that of others.

John Ray Hoke, Jr., FAIA; Washington D.C.

RECOMMENDATIONS

1. Dimensions under 1 ft shall be noted in inches. Dimensions 1 ft and over shall be expressed in feet.
2. Fractions under 1 in. shall NOT be preceded by a zero. Fractions must have a diagonal dividing line between numerator and denominator.
3. Dimension points to be noted with a short blunt 45° line. Dash to be oriented differently for vertical (⚊) and horizontal (⚋) runs of dimensions. Modular dimension points may be designated with an arrow or a dot.
4. Dimension all items from an established grid or reference point and do not close the string of dimensions to the next grid or reference point.
5. Dimension: to face of concrete or masonry work; to centerlines of columns or other grid points; to centerlines of partitions. In nonmodular wood construction dimension to critical face of studs. When a clear dimension is required, dimension to the finish faces and note as such. Do not use the word "clear."
6. Dimension as much as possible from structural elements.
7. Overall readability, conciseness, completeness, and accuracy must be foremost in any dimensional system. It takes experience to determine how to use dimensions to the best advantage.

(Courtesy of *Architectural Graphic Standards*, published by John Wiley and Sons, Inc.)

Plumbing Fixture and Miscellaneous Symbols

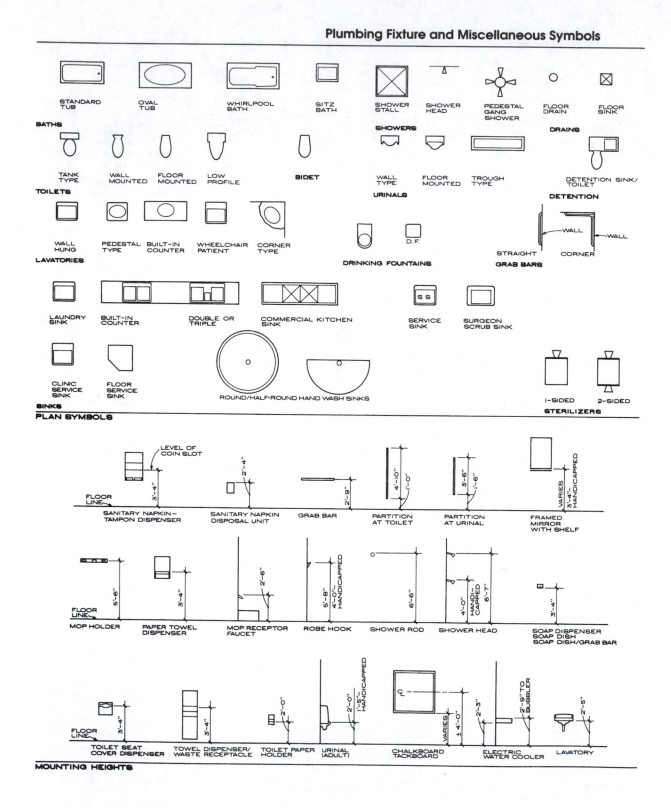

BATHS

STANDARD TUB | OVAL TUB | WHIRLPOOL BATH | SITZ BATH | SHOWER STALL | SHOWER HEAD | PEDESTAL GANG SHOWER | FLOOR DRAIN | FLOOR SINK

SHOWERS — **DRAINS**

TOILETS

TANK TYPE | WALL MOUNTED | FLOOR MOUNTED | LOW PROFILE | BIDET | WALL TYPE | FLOOR MOUNTED | TROUGH TYPE | DETENTION SINK/ TOILET

URINALS — **DETENTION**

LAVATORIES

WALL HUNG | PEDESTAL TYPE | BUILT-IN COUNTER | WHEELCHAIR PATIENT | CORNER TYPE | D.F. | STRAIGHT | CORNER

DRINKING FOUNTAINS — **GRAB BARS** — WALL / WALL

SINKS

LAUNDRY SINK | BUILT-IN COUNTER | DOUBLE OR TRIPLE | COMMERCIAL KITCHEN SINK | SERVICE SINK | SURGEON SCRUB SINK

CLINIC SERVICE SINK | FLOOR SERVICE SINK | ROUND/HALF-ROUND HAND WASH SINKS | 1-SIDED | 2-SIDED

STERILIZERS

PLAN SYMBOLS

SANITARY NAPKIN-TAMPON DISPENSER — LEVEL OF COIN SLOT — 3'-4" | SANITARY NAPKIN DISPOSAL UNIT — 2'-4" | GRAB BAR — 2'-9" | PARTITION AT TOILET — 4'-0" / 1'-0" | PARTITION AT URINAL — 3'-6" / 1'-6" | FRAMED MIRROR WITH SHELF — VARIES / 3'-4" HANDICAPPED

MOP HOLDER — 5'-6" | PAPER TOWEL DISPENSER — 3'-4" | MOP RECEPTOR FAUCET — 2'-6" | ROBE HOOK — 5'-8" / 4'-0" HANDICAPPED | SHOWER ROD — 6'-6" | SHOWER HEAD — 4'-0" HANDICAPPED / 6'-7" | SOAP DISPENSER SOAP DISH SOAP DISH/GRAB BAR — 3'-4"

TOILET SEAT COVER DISPENSER — 3'-4" | TOWEL DISPENSER/ WASTE RECEPTACLE — 3' | TOILET PAPER HOLDER — 2'-0" | URINAL (ADULT) — 2'-0" / 1'-5" HANDICAPPED | CHALKBOARD TACKBOARD — VARIES / ± 4'-0" | ELECTRIC WATER COOLER — 2'-3" / 2'-9" TO BUBBLER | LAVATORY — 2'-5"

FLOOR LINE

MOUNTING HEIGHTS

Dale Switzer, AIA; Hope Architects & Engineers; San Diego, California

(Courtesy of *Architectural Graphic Standards*, published by John Wiley and Sons, Inc.)

Electrical Symbols

INSTITUTIONAL COMMERCIAL AND INDUSTRIAL OCCUPANCIES

NURSES CALL SYSTEM DEVICES. (ANY TYPE)

PAGING SYSTEM DEVICES (ANY TYPE)

FIRE ALARM SYSTEM DEVICES (ANY TYPE)

STAFF REGISTER SYSTEM (ANY TYPE)

ELECTRICAL CLOCK SYSTEM DEVICES (ANY TYPE)

COMPUTER DATA SYSTEM DEVICES

PRIVATE TELEPHONE SYSTEM DEVICES

WATCHMAN SYSTEM DEVICES

SOUND SYSTEM

FACP FIRE ALARM CONTROL PANEL

SC SIGNAL CENTRAL STATION

CR CARD READER

AUXILIARY SYSTEM CIRCUITS

Any line without further designation indicates two-wire system. For a greater number of wires, designate with numerals in manner similar to: 12- no. 18W - ¾" C. Designate by numbers corresponding to listing in schedule.

A, B, C, ETC. SPECIAL AUXILIARY OUTLETS

Subscript lettering refers to notes on drawings or detailed description in specifications.

PANELBOARDS

FLUSH MOUNTED PANELBOARD AND CABINET

SURFACE - MOUNTED PANELBOARD AND CABINET

BUSDUCTS AND WIREWAYS

T T T TROLLEY DUCT

B B B BUSWAY (SERVICE, FEEDER OR PLUG-IN)

C C C CABLE THROUGH LADDER OR CHANNEL

W W W WIREWAY

SIGNALING SYSTEM OUTLETS RESIDENTIAL OCCUPANCIES

PUSH BUTTON

BUZZER

BELL

BELL AND BUZZER COMBINATION

ANNUNCIATOR

COMPUTER DATA OUTLET

INTERCONNECTING TELEPHONE

TELEPHONE SWITCHBOARD

BT BELL RINGING TRANSFORMER

D ELECTRIC DOOR OPENER

CH CHIME

TV TELEVISION OUTLET

T THERMOSTAT

UNDERGROUND ELECTRICAL DISTRIBUTION OR LIGHTING SYSTEM

M MANHOLE

H HANDHOLE

TM TRANSFORMER- MANHOLE OR VAULT

TP TRANSFORMER PAD

UNDERGROUND DIRECT BURIAL CABLE

UNDERGROUND DUCT LINE

STREET LIGHT STANDARD FED FROM UNDERGROUND CIRCUIT

ELECTRICAL DISTRIBUTION OR LIGHTING SYSTEM, AERIAL

POLE

STREET LIGHT AND BRACKET

TRANSFORMER

PRIMARY CIRCUIT

SECONDARY CIRCUIT

DOWN GUY

HEAD GUY

SIDEWALK GUY

SERVICE WEATHER

PANELS CIRCUITS AND MISCELLANEOUS

LIGHTING PANEL

POWER PANEL

WIRING, CONCEALED IN CEILING OR WALL

WIRING, CONCEALED IN FLOOR

WIRING EXPOSED

HOME RUN TO PANEL BOARD.

Indicate number of circuits by number of arrows. Any circuit without such designation indicates a two-wire circuit. For a greater number of wires indicate as follows: ─╫╫╫─ (3 wires) ─╫╫╫╫─ (4 wires), etc.

FEEDERS

Use heavy lines and designate by number corresponding to listing in feeder schedule.

WIRING TURNED UP

WIRING TURNED DOWN

G GENERATOR

M MOTOR

I INSTRUMENT (SPECIFY)

T TRANSFORMER (OR DRAW TO SCALE)

CONTROLLER

EXTERNALLY OPERATED DISCONNECT SWITCH

PULL BOX

Frederick R. Brown, PE; Ayres, Cohen and Hayakawa, Consulting Engineers; Los Angeles/San Francisco, California
Richard F. Humenn, PE; Joseph R. Loring & Associates, Inc., Consulting Engineers; New York, New York

(Courtesy of *Architectural Graphic Standards*, published by John Wiley and Sons, Inc.)

Electrical Symbols

LIGHTING OUTLETS

CEILING, WALL

○ —○ OUTLET BOX AND INCANDESCENT LIGHTING FIXTURE. SLASH INDICATES FIXTURE ON EMERGENCY SERVICE

⊡ ⊡ INCANDESCENT LIGHTING TRACK

Ⓑ —Ⓑ BLANKED OUTLET

Ⓓ DROP CORD

⊗ —⊗ EXIT LIGHT AND OUTLET BOX, DIRECTIONAL ARROWS AS INDICATED. SHADED AREAS DENOTE FACES

○—●—○ OUTDOOR POLE ARM MOUNTED FIXTURES

Ⓙ —Ⓙ JUNCTION BOX

Ⓛps —Ⓛps LAMP HOLDER WITH PULL SWITCH

▽▽▽ MULTIPLE FLOODLIGHT ASSEMBLY

EMERGENCY BATTERY PACK WITH CHARGER AND SEALED BEAM HEADS

D —D REMOTE EMERGENCY SEALED BEAM HEAD WITH OUTLET BOX

Ⓛ —Ⓛ OUTLET CONTROLLED BY LOW VOLTAGE SWITCHING WHEN RELAY IS INSTALLED IN OUTLET BOX

INDIVIDUAL FLUORESCENT FIXTURE. SLASH INDICATES FIXTURE ON EMERGENCY SERVICE

⊢ OUTLET BOX AND FLUORESCENT LIGHTING STRIP FIXTURE

⊡⊡⊡ CONTINUOUS ROW FLUORESCENT FIXTURE

⊡ ⊡ SURFACE-MOUNTED FLUORESCENT

RECEPTACLE OUTLETS

—⊖ SINGLE RECEPTACLE OUTLET

⇒ DUPLEX RECEPTACLE OUTLET

⇛ TRIPLEX RECEPTACLE OUTLET

⊕ QUADRUPLEX RECEPTACLE OUTLET

⇒ DUPLEX RECEPTACLE OUTLET—SPLIT WIRED

⇛ TRIPLEX RECEPTACLE OUTLET—SPLIT WIRED

—△ SINGLE SPECIAL PURPOSE RECEPTACLE OUTLET

⇒△ DUPLEX SPECIAL PURPOSE RECEPTACLE OUTLET

⇒R RANGE OUTLET

—●DW SPECIAL PURPOSE CONNECTION

▭◁ CLOSED CIRCUIT TELEVISION CAMERA

—Ⓒ CLOCK HANGER RECEPTACLE

Ⓕ FAN HANGER RECEPTACLE

⊟ FLOOR SINGLE RECEPTACLE OUTLET

⊟ FLOOR DUPLEX RECEPTACLE OUTLET

△ FLOOR SPECIAL PURPOSE OUTLET

◫ DATA OUTLET IN FLOOR

◫ FLOOR TELEPHONE OUTLET—PRIVATE

UNDERFLOOR DUCT AND JUNCTION BOX FOR TRIPLE, DOUBLE, OR SINGLE DUCT SYSTEM AS INDICATED BY NUMBER OF PARALLEL LINES

CELLULAR FLOOR HEADER DUCT

SWITCH OUTLETS

S SINGLE POLE SWITCH

S_2 DOUBLE POLE SWITCH

S_3 THREE-WAY SWITCH

S_4 FOUR-WAY SWITCH

S_D AUTOMATIC DOOR SWITCH

S_K KEY OPERATED SWITCH

S_P SWITCH AND PILOT LAMP

S_{CB} CIRCUIT BREAKER

S_{WCB} WEATHERPROOF CIRCUIT BREAKER

S_{DM} DIMMER

S_{RC} REMOTE CONTROL SWITCH

S_{WP} WEATHERPROOF SWITCH

S_F FUSED SWITCH

S_{WF} WEATHERPROOF FUSED SWITCH

S_L SWITCH FOR LOW VOLTAGE SWITCHING SYSTEM

S_{LM} MASTER SWITCH FOR LOW VOLTAGE SWITCHING SYSTEM

S_T TIME SWITCH

Ⓢ CEILING PULL SWITCH

—○s SWITCH AND SINGLE RECEPTACLE

⇒s SWITCH AND DOUBLE RECEPTACLE

○ A,B,C ETC. ⎫
⇒ A,B,C ETC. ⎬ SPECIAL OUTLETS
S A,B,C ETC. ⎭

Any standard symbol given above with the addition of lowercase subscript lettering may be used to designate some special variation of standard equipment of particular interest in a specific set of architectural plans.

When used they must be listed in the schedule of symbols on each drawing and if necessary further described in the specifications.

Frederick R. Brown, PE; Ayres, Cohen and Hayakawa, Consulting Engineers; Los Angeles/San Francisco, California
Richard F. Humenn, PE; Joseph R. Loring & Associates, Inc., Consulting Engineers; New York, New York

(Courtesy of *Architectural Graphic Standards*, published by John Wiley and Sons, Inc.)

HVAC Ductwork Symbols

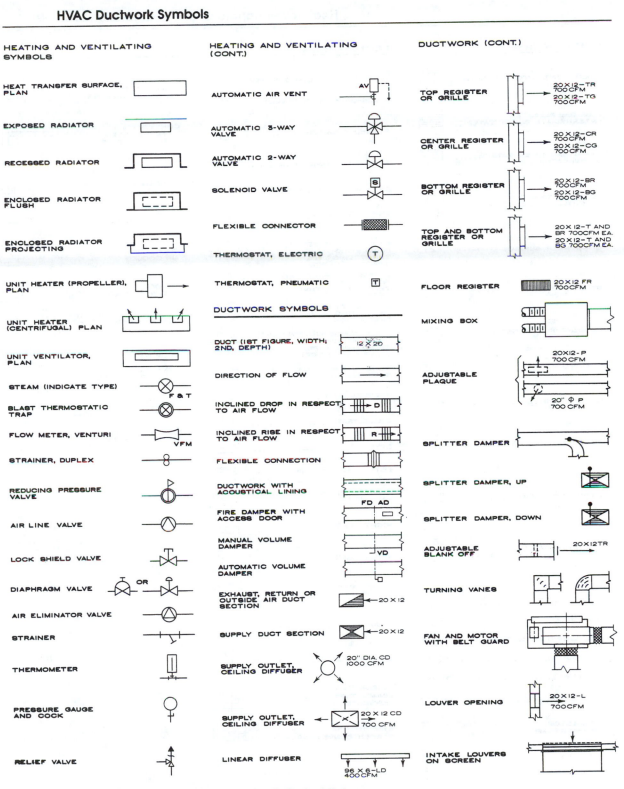

HEATING AND VENTILATING SYMBOLS

- HEAT TRANSFER SURFACE, PLAN
- EXPOSED RADIATOR
- RECESSED RADIATOR
- ENCLOSED RADIATOR FLUSH
- ENCLOSED RADIATOR PROJECTING
- UNIT HEATER (PROPELLER), PLAN
- UNIT HEATER (CENTRIFUGAL) PLAN
- UNIT VENTILATOR, PLAN
- STEAM (INDICATE TYPE)
- BLAST THERMOSTATIC TRAP
- FLOW METER, VENTURI
- STRAINER, DUPLEX
- REDUCING PRESSURE VALVE
- AIR LINE VALVE
- LOCK SHIELD VALVE
- DIAPHRAGM VALVE
- AIR ELIMINATOR VALVE
- STRAINER
- THERMOMETER
- PRESSURE GAUGE AND COCK
- RELIEF VALVE

HEATING AND VENTILATING (CONT.)

- AUTOMATIC AIR VENT
- AUTOMATIC 3-WAY VALVE
- AUTOMATIC 2-WAY VALVE
- SOLENOID VALVE
- FLEXIBLE CONNECTOR
- THERMOSTAT, ELECTRIC
- THERMOSTAT, PNEUMATIC

DUCTWORK SYMBOLS

- DUCT (1ST FIGURE, WIDTH; 2ND, DEPTH)
- DIRECTION OF FLOW
- INCLINED DROP IN RESPECT TO AIR FLOW
- INCLINED RISE IN RESPECT TO AIR FLOW
- FLEXIBLE CONNECTION
- DUCTWORK WITH ACOUSTICAL LINING
- FIRE DAMPER WITH ACCESS DOOR
- MANUAL VOLUME DAMPER
- AUTOMATIC VOLUME DAMPER
- EXHAUST, RETURN OR OUTSIDE AIR DUCT SECTION
- SUPPLY DUCT SECTION
- SUPPLY OUTLET, CEILING DIFFUSER
- SUPPLY OUTLET, CEILING DIFFUSER
- LINEAR DIFFUSER

DUCTWORK (CONT.)

- TOP REGISTER OR GRILLE
- CENTER REGISTER OR GRILLE
- BOTTOM REGISTER OR GRILLE
- TOP AND BOTTOM REGISTER OR GRILLE
- FLOOR REGISTER
- MIXING BOX
- ADJUSTABLE PLAQUE
- SPLITTER DAMPER
- SPLITTER DAMPER, UP
- SPLITTER DAMPER, DOWN
- ADJUSTABLE BLANK OFF
- TURNING VANES
- FAN AND MOTOR WITH BELT GUARD
- LOUVER OPENING
- INTAKE LOUVERS ON SCREEN

Amor Halperin, PE; Ayres, Cohen and Hayakawa, Consulting Engineers; Los Angeles/San Francisco, California
Joseph R. Loring & Associates, Inc., Consulting Engineers; New York, New York

(Courtesy of *Architectural Graphic Standards*, published by John Wiley and Sons, Inc.)

Heat-Power Apparatus and Refrigeration Symbols

HEAT-POWER APPARATUS

STEAM GENERATOR (BOILER) .....

FLUE GAS REHEATER
(INTERMEDIATE SUPERHEATER) ..

LIVE STEAM SUPERHEATER ......
OR REHEATER

FEED HEATER WITH
AIR OUTLET ....

CONDENSER, SURFACE ....

STEAM TURBINE ....

CONDENSING TURBINE ....

OPEN TANK ....

CLOSED TANK ....

AUTOMATIC REDUCING VALVE ....

AUTOMATIC BYPASS VALVE .....

AUTOMATIC VALVE
OPERATED BY GOVERNOR ......

BOILER FEED PUMP ....

SERVICE PUMP ....

CONDENSATE PUMP ....

CIRCULATING WATER PUMP .....

AIR PUMP ....

OIL PUMP ....

RECIPROCATING PUMP ....

AIR EJECTOR
(DYNAMIC PUMP)

VACUUM TRAP ....

REFRIGERATION

THERMOSTAT, SELF-CONTAINED

THERMOSTAT, REMOTE BULB ...

PRESSURE SWITCH ....

EXPANSION VALVE, HAND ....

EXPANSION VALVE, AUTOMATIC .

EXPANSION VALVE,
THERMOSTATIC ....

EVAPORATOR PRESSURE
REGULATING VALVE,
THROTTLING TYPE
(EVAPORATOR SIDE)

EVAPORATOR PRESSURE
REGULATING VALVE,
THERMOSTATIC, THROTTLING
TYPE

EVAPORATOR PRESSURE
REGULATING VALVE
SNAP-ACTION

COMPRESSOR SUCTION VALVE,
PRESSURE LIMITING,
THROTTLING TYPE
(COMPRESSOR SIDE)

CONSTANT PRESSURE VALVE,
SUCTION

THERMAL BULB ....

SCALE TRAP ....

DRYER ....

FILTER AND STRAINER   ....

COMBINATION STRAINER
AND DRYER

SIGHT GLASS ....

FLOAT VALVE
HIGH SIDE

FLOAT VALVE
LOW SIDE

GAUGE ....

COOLING TOWER ....

EVAPORATOR,
FINNED TYPE, NATURAL
CONVECTION

EVAPORATOR,
FORCED CONVECTION ....

IMMERSION   COOLING UNIT ....

CONDENSER,
AIR-COOLED,
FINNED, FORCED AIR

CONDENSER,
WATER-COOLED,
SHELL AND TUBE

CONDENSER
EVAPORATIVE

HEAT EXCHANGER

CONDENSING UNIT
AIR COOLED

CONDENSING UNIT
WATER COOLED

PRESSURE SWITCH WITH
HIGH PRESSURE CUT-OUT

COMPRESSOR

COMPRESSOR
OPEN CRANKCASE
RECIPROCATING, DIRECT
DRIVE

COMPRESSOR
OPEN CRANKCASE
RECIPROCATING BELTED

COMPRESSOR
ENCLOSED CRANKCASE,
ROTARY, BELTED

Amor Halperin, PE; Ayres, Cohen and Hayakawa, Consulting Engineers; Los Angeles/San Francisco, California

(Courtesy of *Architectural Graphic Standards*, published by John Wiley and Sons, Inc.)

Piping Symbols: Plumbing, Heating, and Air Conditioning

PLUMBING PIPING

SOIL, WASTE OR LEADER (ABOVE GRADE)	—————
SOIL, WASTE OR LEADER (BELOW GRADE)	— — — —
VENT	- - - - - -
COMBINATION WASTE AND VENT	——SV——
ACID WASTE	——AW——
ACID VENT	— — AV — —
INDIRECT DRAIN	——IW——
STORM DRAIN	——S——
COLD WATER	— — - —
SOFT COLD WATER	——SW——
INDUSTRIALIZED COLD WATER	——ICW——
CHILLED DRINKING WATER SUPPLY	——DWS——
CHILLED DRINKING WATER RETURN	——DWR——
HOT WATER	— - - — - - —
HOT WATER RETURN	— - - - — - - - —
SANITIZING HOT WATER SUPPLY (180° F.)	—/— - —/— -/
SANITIZING HOT WATER RETURN (180° F.)	—/— - - —/— - -
INDUSTRIALIZED HOT WATER SUPPLY	——IHW——
INDUSTRIALIZED HOT WATER RETURN	——IHR——
TEMPERED WATER SUPPLY	——TWS——
TEMPERED WATER RETURN	——TWR——
FIRE LINE	—F——F—
WET STANDPIPE	——WSP——

DRY STANDPIPE	—DSP—
COMBINATION STANDPIPE	— CSP —
MAIN SUPPLIES SPRINKLER	—S—
BRANCH AND HEAD SPRINKLER	—o——o—
GAS - LOW PRESSURE	—G——G—
GAS - MEDIUM PRESSURE	—MG—
GAS - HIGH PRESSURE	—HG—
COMPRESSED AIR	—A—
VACUUM	—V—
VACUUM CLEANING	—VC—
OXYGEN	—O—
LIQUID OXYGEN	—LOX—
NITROGEN	—N—
LIQUID NITROGEN	—LN—
NITROUS OXIDE	—NO—
HYDROGEN	—H—
HELIUM	—HE—
ARGON	—AR—
LIQUID PETROLEUM GAS	—LPG—
INDUSTRIAL WASTE	—INW—
PNEUMATIC TUBES TUBE RUNS	—PN—
CAST IRON	—CI—
CULVERT PIPE	—CP—
CLAY TILE	—CT—
DUCTILE IRON	—DI—
REINFORCED CONCRETE	—RCP—
DRAIN - OPEN TILE OR AGRICULTURAL TILE	══ ══ ══

HEATING PIPING

HIGH PRESSURE STEAM	—HPS—
MEDIUM PRESSURE STEAM	—MPS—
LOW PRESSURE STEAM	—LPS—
HIGH PRESSURE RETURN	—HPR—
MEDIUM PRESSURE RETURN	—MPR—
LOW PRESSURE RETURN	—LPR—
BOILER BLOW OFF	—BD—
CONDENSATE OR VACUUM PUMP DISCHARGE	—VPD—
FEEDWATER PUMP DISCHARGE	—PPD—
MAKE UP WATER	—MU—
AIR RELIEF LINE	—V—
FUEL OIL SUCTION	—FOS—
FUEL OIL RETURN	—FOR—
FUEL OIL VENT	—FOV—
COMPRESSED AIR	—A—
HOT WATER HEATING SUPPLY	—HW—
HOT WATER HEATING RETURN	—HWR—

AIR CONDITIONING PIPING

REFRIGERANT LIQUID	—RL—
REFRIGERANT DISCHARGE	—RD—
REFRIGERANT SUCTION	—RS—
CONDENSER WATER SUPPLY	—CWS—
CONDENSER WATER RETURN	—CWR—
CHILLED WATER SUPPLY	—CHWS—
CHILLED WATER RETURN	—CHWR—
MAKE UP WATER	—MU—
HUMIDIFICATION LINE	—H—
DRAIN	—D—
BRINE SUPPLY	—B—
BRINE RETURN	—BR—

Amor Halperin, PE; Ayres, Cohen and Hayakawa, Consulting Engineers; Los Angeles/San Francisco, California
Joseph R. Loring & Associates, Inc., Consulting Engineers; New York, New York

(Courtesy of *Architectural Graphic Standards*, published by John Wiley and Sons, Inc.)

Appendix A

SECTION 1
UNIT 1 CONSTRUCTION DRAWINGS

1. working
2. pictorial
3. working
4. photographs
5. orthographic
6. two
7. one
8. two
9. orthographic
10. specifications
11. schedules

UNIT 2 LINEAR MEASUREMENT

1. English
2. customary
3. direct
4. inches
5. 14'-2"
6. decimal
7. ten
8. meter
9. millimeter
10. 304.8

UNIT 3 LINES

1. property
2. dimension
3. extension
4. short
5. hidden
6. leader
7. cutting
8. arrows
9. break
10. center
11. property
12. dimension
13. short dimension
14. hidden
15. leader
16. cutting plane
17. object
18. break
19. center

UNIT 4 SCALE

1. proportion
2. site
3. metric
4. architect's
5. 0
6. 3/4' = 1'-0"
7. decimal
8. 60
9. millimeters
10. 75

UNIT 5 SYMBOLS & ABBREVIATIONS

1. topographic
2. architectural
3. electrical
4. abbreviations
5. uppercase
6. DMPR
7. circuit breaker
8. C
9. ridge
10. COMB.

SECTION 2
UNIT 6 SURVEY AND SITE PLANS

1. 24
2. sections
3. plat
4. site
5. survey
6. surveyor
7. elevations
8. bench
9. finish
10. contour
11. minutes

12. building
13. distance
14. natural
15. setback
16. F
17. L
18. G
19. D
20. A
21. J
22. H
23. B
24. E
25. C
26. I
27. K
28. BM ELEV
29. CATV
30. E
31. EXIST
32. N
33. PL
34. PUE
35. R

Unit 7 Foundation Plans

1. basement
2. 18″
3. 7 to 8 feet
4. steel
5. slab-on-grade
6. spread
7. footing
8. twice
9. grade beam
10. piers
11. dead
12. coarse
13. away
14. frost line
15. cement
16. reinforced
17. 7/8″
18. welded wire fabric
    welded wire mesh
19. section
20. dimension
21. E
22. J
23. M
24. C
25. L
26. A
27. H

28. R
29. D
30. U
31. K
32. B
33. G
34. O
35. T
36. S
37. P
38. F
39. Q
40. N
41. L
42. @
43. BET.
44. BSMT
45. BTM
46. CL
47. CONC
48. CONT
49. CMU
50. DIA
51. FOW
52. FDN
53. FL JSTS
54. FTG
55. GALV
56. INSUL
57. JTS
58. MIN
59. OC
60. PERF
61. PERIM
62. PVC
63. PT
64. REBAR
65. SG
66. SIM
67. ST
68. S4S
69. SQ FT
70. TYP
71. W/
72. WD

Unit 8 Floor Frame Plans

1. platform
2. joists
3. 16 or 24 in.
4. allowable span
5. stairway
6. rim or header

7. beams
8. blocking
9. joist hangers
10. panels
11. doubled
12. 8′
13. blanket or batt
14. lap
15. tail joists
16. H
17. F
18. D
19. G
20. A
21. C
22. F
23. E
24. B
25. BLDG
26. DKG
27. DBL BLKG
28. EA
29. EXT
30. FL
31. GYP BD
32. HT
33. JSTS
34. MIL
35. OC
36. PLYWD
37. POLY
38. REDWD

Unit 9 Floor Plan

1. wall studs
2. 2 × 4 & 2 × 6
3. bottom
4. double plate
5. outside
6. 45 to 60 degrees
7. trimmer
8. rough sill
9. cripple
10. floor
11. code
12. frame
13. lines
14. center
15. 16″ and 24″
16. U
17. L
18. H
19. Q
20. I

21. A
22. R
23. O
24. C
25. M
26. B
27. E
28. T
29. J
30. K
31. S
32. G
33. P
34. X
35. W
36. V
37. D
38. F
39. Y
40. N
41. @
42. ABV
43. B
44. BR
45. BLW
46. CL
47. CNTR
48. DBL
49. ELEC
50. EQ
51. GYP BD
52. HR
53. INSUL
54. KIT.
55. LAU
56. LR
57. LS
58. MA
59. MECH
60. MIN
61. RMS
62. PAN.
63. SH
64. SQ FT
65. WDW
66. WH

Unit 10 Ceiling and Roof Framing Plans

1. span
2. rafters
3. attic
4. toenailed
5. third
6. two

7. hip
8. live
9. roof pitch
10. unit rise
11. 12″
12. rise
13. ridge
14. 45 degrees
15. trusses
16. C
17. P
18. B
19. A
20. BM
21. CHM
22. CLG JSTS
23. DN
24. EA
25. GALV SH METAL
26. GYP BD
27. GLU-LAM
28. HDR
29. HT
30. OH
31. PL
32. LB
33. RS
34. SQ FT
35. W/

UNIT 11 SECTIONS AND DETAIL DRAWINGS

1. building
2. detail
3. circle
4. longitudinal
5. width
6. stringers
7. 30 to 45 degrees
8. 7″ to 7 1/2″
9. total rise
10. vertical
11. horizontal
12. fireplace
13. hearth
14. side
15. center
16. J
17. F
18. H
19. E
20. K
21. A
22. D
23. I

24. C
25. G
26. B
27. @
28. BM
29. BLKG
30. BTM
31. CLG
32. DBL
33. DIA
34. DR
35. EA
36. EXT
37. FL JSTS
38. GA
39. GALV
40. GYP BD
41. HDR
42. HT
43. INSUL
44. JSTS
45. LR
46. MA B
47. MA BR
48. MANU
49. MIL
50. PL
51. PLYWD
52. POLY
53. S4S
54. SIM
55. TYP
56. W/
57. WDW

UNIT 12 ELEVATION PLANS

1. elevation
2. projecting
3. floor
4. side
5. roof
6. vents
7. downspouts
8. compass
9. interior
10. symbol
11. M
12. G
13. E
14. L
15. Q
16. J
17. A
18. H

19. R
20. N
21. K
22. B
23. O
24. D
25. I
26. C
27. P
28. @
29. ABV
30. ADJ
31. BRD BD
32. CNTR
33. CMU
34. D
35. DP
36. DW
37. CAB.
38. DRWR
39. ELEC
40. FLR
41. GA
42. GALV
43. LAM
44. LR
45. MAS VEN
46. MC
47. MSTR BR
48. MW
49. PL HT
50. PREFAB
51. RDWD
52. S4S
53. W/
54. WD
55. WM

Unit 13 Electrical, Plumbing, Heating Plans

1. service
2. circuits
3. power poles
4. fuses - circuit breakers
5. floor
6. code
7. waste
8. cost
9. gypsum
10. warm water
11. W
12. Q
13. I
14. U
15. X

16. aa
17. K
18. E
19. A
20. V
21. R
22. H
23. S
24. N
25. F
26. T
27. D
28. Y
29. C
30. Z
31. O
32. G
33. P
34. B
35. L
36. J
37. K
38. C
39. CAB LT
40. CLG
41. DIM.
42. D
43. ELEC
44. FIN FL
45. FPHB
46. GFI
47. GYP
48. H
49. HVL
50. MW
51. PS
52. RM
53. TV
54. WM
55. WP

Unit 14 Specifications

1. CSI Master Format
2. divisions
3. General Conditions
4. contractor
5. site work
6. earth formed
7. Concrete & Masonry
8. mix
9. carpentry
10. general contractor
11. anchors
12. millwork

13. appearance
14. exterior
15. name
16. cabinet
17. walls-floor-ceiling
18. X-5/8" gypsum
19. valley - ridge
20. manufacturers'
21. owner
22. finish hardware
23. thermal - sound
24. heating - hot water
25. electrical
26. UL
27. plumbing
28. fixtures

SECTION 3
UNIT 15 ELECTRICAL PLANS

1. NEC
2. service entrance
3. volt
4. watts
5. transformer
6. service drop
7. watt meter
8. circuit breaker
9. branch circuits
10. general purpose
11. special service
12. ends
13. 6'
14. toggle
15. 3 way
16. mercury
17. wiring or power
18. branch circuits
19. lighting
20. fixture

UNIT 16 PLUMBING PLANS

1. model
2. water mains
3. water supply, water service
4. water meter
5. distribution
6. fixture supply
7. stacks
8. 1/4"
9. fixture trap

10. 12"
11. vent stacks
12. cleanout
13. company
14. septic
15. back to back
16. chase
17. plan view
18. schematic
19. isometric
20. schedule

UNIT 17 HVAC PLANS

1. 70 to 75 degrees
2. conduction, convection, radiation
3. zoned
4. furnace
5. return air
6. full basement
7. perimeter
8. hydronic
9. radiators or convectors
10. radiant
11. heat pump
12. plenum chamber
13. crawl space
14. ducts - registers
15. evaporative
16. window
17. room plenum
18. primary distribution
19. ducts
20. WA & RA

SECTION 4
UNIT 18 DRAWING THE PLANS

1. architect
2. civil (structural) engineer
3. AIA
4. custom
5. speculative
6. stock
7. blueprint
8. diazo - electrostatic
9. CADD
10. pen plotter

UNIT 19 THE CONSTRUCTION SITE AND TOPOGRAPHY

1. contours
2. ranch
3. zoning

4. deciduous
5. orientation
6. lower
7. split level
8. north
9. insulating
10. double paned

UNIT 20 BASIC HOUSE DESIGN

1. ranch
2. one-and-one-half
3. dormers
4. space
5. split -level
6. square
7. half bathroom
8. foyer
9. service
10. utility rooms

UNIT 21 ENERGY AND BUILDING DESIGN

1. 70%
2. reflect
3. overhangs
4. louvers and vents
5. storm
6. weatherstripping
7. fiberglass, plastic foam, cellulose
8. condensation
9. R-value
10. south

SECTION 5
UNIT 22 ZONING CODES AND REGULATIONS

1. agriculture
2. R
3. dwelling uints
4. C
5. M
6. C
7. manufacturing
8. R
9. 20'-0"
10. 30'-0"

UNIT 23 BUILDING CODES

1. model
2. Standard Building Code

3. BOCA
4. Uniform Building Code
5. provincial
6. six
7. eighteen
8. A-1
9. 101.3.2
10. square feet
11. ASTM
12. fire
13. means - egress
14. horizontal
15. guardrails
16. 6"
17. 2"
18. 12"
19. 16"
20. 7'-4"

UNIT 24 BUILDING PERMITS AND INSPECTIONS

1. building permit
2. alter
3. change
4. 9'-5"
5. retaining wall
6. location
7. two
8. value
9. 6 months
10. site-planning-structural
11. structural
12. rough-in
13. footing
14. final
15. Certificate of Occupancy

UNIT 25 CONTRACTS AND BIDS

1. general contractor
2. AIA
3. fixed
4. contractor
5. architect
6. limits
7. percentage
8. general contractor
9. bidding
10. bond

SECTION 6
UNIT 26 CONVERTING LINEAR MEASUREMENTS

Quiz 26-1 Convert Inches to Feet
1. 4'-0"
2. 6'-0"
3. 5'-0"
4. 8'-0"
5. 2'-5"
6. 6'-9"
7. 4'-4"
8. 1'-4"
9. 9'-7"
10. 14'-5"

Quit 26-2 Convert Feet to Inches
1. 36"
2. 84"
3. 132"
4. 252"
5. 20"
6. 53"
7. 111"
8. 184"
9. 218"
10. 295"

Quiz 26-3 Convert Fractions to Decimals
1. .25
2. .625
3. .33
4. .438
5. .8333
6. 6.75
7. 9.6
8. 3.88
9. 5.0625
10. 7.34375

Quiz- 26-4 Convert Decimals to Fractions
1. 1/4
2. 7/16
3. 9/16
4. 1/8
5. 15/16
6. 3/16
7. 3/8
8. 13/16
9. 7/16
10. 11/16

Quiz 26-5 Convert Inch to Decimal Fractions of a Foot
1. .5'
2. .5833'
3. .75'
4. .9167'
5. .4167'
6. .3438'
7. .7292'
8. .4583'
9. .2760'
10. .6563'

Quiz 26-6 Decimals of Feet to Inches
1. 3"
2. 6"
3. 6 1/2"
4. 4 11/16"
5. 9 15/16"
6. 8 7/16"
7. 5 7/8"
8. 6 13/16"
9. 9 15/16"
10. 8 1/2"

UNIT 27 CALCULATING LINEAR ENGLISH MEASUREMENT

Quiz 27-1 Calculating English Measurement
1. 7'-9"
2. 10'-7"
3. 34'-9"
4. 21'-5"
5. 25'-4"
6. 20'-7"
7. 66'-9"
8. 79'-2"
9. 34'-2"
10. 686'-1"

Quiz 27-2 Add Separate and Combined Feet and Inches
1. 67"
2. 296"
3. 174"
4. 321"
5. 241"
6. 8'-11"
7. 22'-3"
8. 26'-2"
9. 39'-10"
10. 23'-11"

Quiz 27-3 Subtracting Feet and Inches
1. 6'-6"
2. 8'-2"
3. 17'-9"
4. 7'-8"
5. 12'-8"
6. 13'-11"

7. 28'-10"
8. 2'-10"
9. 30'-8"
10. 63'-11"

Quiz 27-4 Subtracting Feet and Inches from Feet and Inches
1. 5"
2. 29"
3. 7"
4. 34"
5. 280"
6. 1'-4"
7. 1'-2"
8. 1'-10"
9. 3'-10"
10. 25'-4"

Quiz 27-5 Multiplying Feet and Inches
1. 33'-9"
2. 4'-10"
3. 49'-8"
4. 43'-2"
5. 55'-6"
6. 108'-8"
7. 147'-4"
8. 250'-0"
9. 247'-6"
10. 908'-3"

Quit 27-6 Dividing Combined Feet and Inches
1. 9'-2"
2. 8'-2"
3. 3'-6"
4. 4'-2"
5. 12'-3"
6. 2'-1"
7. 5'-2"
8. 6'-5 1/8"
9. 9'-3 7/16"
10. 7'-7 1/2"

UNIT 28 COMPUTING SCALE

Quiz 28-1 Finding Scale Using Rule
1. 4 1/2"
2. 18 3/4"
3. 2 7/16"
4. 3/8"
5. 5 5/8"
6. 13 1/2"
7. 5 7/8"
8. 12 3/4"
9. 9 15/16"
10. 7 1/4"

11. 3 3/16"
12. 5 3/16"
13. 13 1/4"
14. 2 3/16"
15. 5 3/8"
16. 10 7/8"
17. 3 3/16"
18. 3 3/8"
19. 16 7/16"
20. 27 1/8"

UNIT 29 RISE AND RUN OF ROOFS

Quiz 29-1 Total Rise of Roofs
1. 6'-8"
2. 16'-0"
3. 13'-0"
4. 4'-6"
5. 16'-6"
6. 18'-8"
7. 15'-0"
8. 14'-0"
9. 3'-2"
10. 5'-5"
11. 3'-0"
12. 8'-2"
13. 5'-1"
14. 12'-0"
15. 4'-7 1/2"
16. 7'-2"
17. 25'-3"
18. 14'-11 11/16"
19. 23'-3 9/16"
20. 44'11 "

UNIT 30 STAIR DIMENSIONS

Quiz 30-1 Riser and Treads

	Riser	Tread
1.	7 9/16"	10 1/2"
2.	7"	9 5/8"
3.	7 1/4"	11"
4.	7 1/8"	10 3/8"
5.	7 5/16"	10"
6.	7 9/16"	10 1/2"
7.	7 5/16"	10 7/16"
8.	7 5/8"	10 1/8"
9.	7 1/8"	10 7/8"
10.	7"	11 7/8"
11.	7 1/4"	9 1/2"
12.	7 5/16"	11 3/8"
13.	7 1/8"	12 1/4"
14.	7 11/16"	10 3/8"
15.	7 3/8"	9 1/4"
16.	7 1/2"	12 3/16"

17. 7 1/16″   11″
18. 7 1/8″   9 1/2″
19. 7 1/8″   9 1/2″
20. 7 1/4″   11 5/16″

UNIT 31 METRIC CONVERSION

Quiz 31-1 Convert Inches to Millimeters

1. 431.8 mm
2. 635 mm
3. 876.3 mm
4. 2216.15 mm
5. 6256.78 mm

Quiz 31-2 Convert Millimeters to Inches

1. 36.24″
2. 57.92″
3. 24.90″
4. 149.88″
5. 644.23″

Quiz 31-3 Convert Inches to Centimeters

1. 35.56 cm
2. 78.74 cm
3. 49.10 cm
4. 54.61 cm
5. 108.25 cm

Quiz 31-4 Convert Centimeters to Inches

1. 12.61″
2. 27.19″
3. 31.91″
4. 40.19″
5. 28.76″

Quiz 31-5 Convert Inches to Meters

1. .99 m
2. 1.60 m
3. .53 m
4. 4.55 m
5. .36 m

Quiz 31-6 Convert Meters to Inches

1. 157.48″
2. 118.11″
3. 68.90″
4. 209.84″
5. 364.17″

Quiz 31-7 Convert Feet to Meters

1. 10.68 m
2. 4.27 m
3. 22.25 m
4. .38 m
5. 14.78 m

Quiz 31-8 Convert Meters to Feet

1. 42.65′
2. 82.02′
3. 143.70′
4. .82′
5. 305.11′

Quiz 31-9 Convert Feet and Inches to Millimeters

1. 1396 mm
2. 2819 mm
3. 4165 mm
4. 1943 mm
5. 2800 mm
6. 3534 mm
7. 5654 mm
8. 6623 mm
9. 9864 mm
10. 14017 mm

SECTION 7
UNIT 32 CONCRETE AND MASONRY

1. cement-aggregate-water
2. sand-gravel
3. rebars
4. WMM or WWF
5. CMU
6. grout
7. clay
8. hollow tiles
9. rubble - ashlar
10. uncoursed

UNIT 33 LUMBER

1. softwood
2. broad leaf or deciduous
3. structural
4. actual
5. 1 1/2 × 3 1/2
6. dimension
7. timbers
8. plywood
9. grade
10. veneer sheets
11. reconstituted
12. composite
13. chords - webs
14. glulam
15. veneered I-beams

UNIT 34 METALS

1. I-beams or wide flange beams
2. lally

3. common & finish
4. 1" to 6"
5. machine
6. self driving
7. machine
8. lag bolts
9. powder activated
10. machine bolts & lag screws
11. chemical anchors
12. metal connectors
13. wood
14. steel clad
15. ductwork

Unit 35 Plastics

1. thermoplastic
2. acrylic
3. polyethylene
4. polystyrene
5. vinyl
6. thermoset
7. melamines, phenolics, polyesters
8. plastic laminate
9. phenolics
10. fiberglass

Appendix B

DIMENSION LUMBER GRADES   Table 2.1

Product	Grades	WWPA Western Lumber Grading Rules Section Reference	Uses
Structural Light Framing (SLF) 2″ to 4″ thick 2″ to 4″ wide	SELECT STRUCTURAL NO.1 NO.2 NO.3	(42.10) (42.11) (42.12) (42.13)	Structural applications where highest design values are needed in light framing sizes.
Light Framing (LF) 2″ to 4″ thick 2″ to 4″ wide	CONSTRUCTION STANDARD UTILITY	(40.11) (40.12) (40.13)	Where high-strength values are not required, such as wall framing, plates, sills, cripples, blocking, etc.
Stud 2″ to 4″ thick 2″ and wider	STUD	(41.13)	An optional all-purpose grade designed primarily for stud uses, including bearing walls.
Structural Joists and Planks (SJ&P)	SELECT STRUCTURAL NO.1 NO.2 NO.3	(62.10) (62.11) (62.12) (62.13)	Intended to fit engineering applications for lumber 5″ and wider, such as joists, rafters, headers, beams, trusses, and general framing.

STRUCTURAL DECKING GRADES   Table 2.2

Product	Grades	WWPA Western Lumber Grading Rules Section Reference	Uses
Structural Decking 2″ to 4″ thick 4″ to 12″ wide	SELECTED DECKING	(55.11)	Used where the appearance of the best face is of primary importance.
	COMMERCIAL DECKING	(55.12)	Customarily used when appearance is not of primary importance.

TIMBER GRADES   Table 2.3

Product	Grades	WWPA Western Lumber Grading Rules Section Reference	End Uses
Beams and Stringers 5″ and thicker, width more than 2″ greater than thickness	DENSE SELECT STRUCTURAL* DENSE NO. 1* DENSE NO. 2* SELECT STRUCTURAL NO.1 NO.2	(53.00 & 170.00) (53.00 & 170.00) (53.00 & 170.00) (70.10) (70.11) (70.12)	Grades are designed for beam and stringer type uses when sizes larger than 4″ nominal thickness are required.
Post and Timbers 5″ × 5″ and larger, width not more than 2″ greater than thickness	DENSE SELECT STRUCTURAL* DENSE NO. 1* DENSE NO. 2* SELECT STRUCTURAL NO.1 NO.2	(53.00 & 170.00) (53.00 & 170.00) (53.00 & 170.00) (80.10) (80.11) (80.12)	Grades are designed for vertically loaded applications where sizes larger than 4″ nominal thickness are required.

*Douglas Fir or Douglas Fir-Larch only.

(Courtesy of Western Wood Products Association.)

APPEARANCE LUMBER GRADES   Table **2.5**

	Product	Grades[1]	Equivalent Grades in Idaho White Pine	WWPA Grading Rules Section Number
Highest Quality Appearance Grades	Selects *(all species)*	B & BTR SELECT C SELECT D SELECT	SUPREME CHOICE QUALITY	10.11 10.12 10.13
	Finish *(usually available only in Doug Fir and Hem-Fir)*	SUPERIOR PRIME E		10.51 10.52 10.53
	Special Western Red Cedar Pattern Grades	CLEAR HEART A GRADE B GRADE		20.11 20.12 20.13
General Purpose Grades	Common Boards (WWPA Rules) *(primarily in pines, spruces, and cedars)*	1 COMMON 2 COMMON 3 COMMON 4 COMMON 5 COMMON	COLONIAL STERLING STANDARD UTILITY INDUSTRIAL	30.11 30.12 30.13 30.14 30.15
	Alternate Boards (WCLIB Rules) *(primarily in Doug Fir and Hem-Fir)*	SELECT MERCHANTABLE CONSTRUCTION STANDARD UTILITY ECONOMY		**WCLIB[3]** 118-a 118-b 118-c 118-d 118-e
	Special Western Red Cedar Pattern[2] Grades	SELECT KNOTTY QUALITY KNOTTY		**WCLIB[3]** 111-e 111-f

[1]Refer to WWPA's *Vol. 2, Western Wood Species* book for full-color photography and to WWPA's *Natural Wood Siding* for complete information on siding grades, specification, and installation.

[2]"PATTERN" includes Finish, Paneling, Ceiling, and Siding grades.

[3]West Coast Lumber Inspection Bureau's *West Coast Lumber Standard Grading Rules.*

(Courtesy of Western Wood Products Association.)

Appendix C

Southern Pine Grade Descriptions

Based on 1994 SPIB Grading Rules

Product	Grade	Grade Characteristics and Typical Uses
Dimension Lumber: 2″ to 4″ thick, 2″ and wider *See table 1 for design values*		
	*Dense Select Structural Select Structural *NonDense Select Structural	High quality, relatively free of characteristics which impair strength or stiffness. Recommended for uses where high strength, stiffness and good appearance are desired.
	*No. 1 Dense No. 1 *No. 1 NonDense	Recommended for general utility and construction where high strength, stiffness and good appearance are desired.
	*No. 2 Dense No. 2 *No. 2 NonDense	Recommended for most general construction uses where moderately high design values are required. Allows well-spaced knots of any quality.
	No. 3	Assigned design values meet a wide range of design requirements. Recommended for general construction purposes where appearance is not a controlling factor. Many pieces included in this grade would qualify as No. 2 except for a single limiting characteristic.
	Stud	Suitable for stud uses including use in load-bearing walls. Composite of No. 3 strength and No. 1 nailing edge characteristics.
	*Construction (2″ to 4″ wide only)	Recommended for general framing purposes. Good appearance, but graded primarily for strength and serviceability.
	*Standard (2″ to 4″ wide only)	Recommended for same purposes as Construction grade. Characteristics are limited to provide good strength and excellent serviceability.
	*Utility (2″ to 4″ wide only)	Recommended where a combination of economical construction and good strength is desired. Used for such purposes as studding, blocking, plates, bracing and rafters.
	Design values are not assigned Economy	Usable lengths suitable for bracing, blocking, bulk heading and other utility purposes where strength and appearance are not controlling factors.
*Timbers: 5″ x 5″ and larger *See table 2 for design values*		
	Dense Select Structural Select Structural	Recommended where high strength, stiffness and good appearance are desired.
	No. 1 Dense No. 1 No. 2 Dense No. 2	No. 1 and No. 2 are similar in appearance to corresponding grades of 2″ thick Dimension Lumber. Recommended for general construction uses.
	Design values are not assigned No. 3	Non-stress rated, but economical for general construction purposes such as blocking, fillers, etc.

*Most mills do not manufacture all products and make all grade separations. Those products and grades not manufactured by most mills are noted with an asterisk.

(Courtesy of The Southern Forest Products Association.)

SOUTHERN PINE GRADE DESCRIPTIONS (cont'd)

Based on 1994 SPIB Grading Rules

Product	Grade	Grade Characteristics and Typical Uses
*Mechanically Graded Lumber — Machine Stress Rated (MSR) Lumber: 2″ and less in thickness, 2″ and wider See table 3 for design values		
	1650f – 1.5E thru 3000f – 2.4E	Machine Stress Rated (MSR) lumber is lumber that has been evaluated by mechanical stress rating equipment. MSR lumber is distinguished from visually stress graded lumber in that each piece is non-destructively tested. MSR lumber is also required to meet certain visual grading requirements.
*Mechanically Graded Lumber — Machine Evaluated Lumber (MEL): 2″ and less in thickness, 2″ and wider See SPIB Grading Rules for design values		
	M – 5 thru M – 28	Well-manufactured material evaluated by calibrated mechanical grading equipment which measures certain properties and sorts the lumber into various strength classifications. Machine Evaluated Lumber is also required to meet certain visual requirements.
*Scaffold Plank: 2″ and 3″ thick, 8″ and wider See table 4 for design values		
	Dense Industrial 72 Dense Industrial 65	All Scaffold Plank design values are calculated using ASTM Standards D245 and D2555. These values are modified using procedures shown in "Calculating Apparent Reliability of Wood Scaffold Planks," as published by the Journal on Structural Safety, 2 (1984) 47-57.
	MSR: 2400f – 2.0E MSR: 2200f – 1.8E	Dressed to standard dry size prior to machine stress rating, and visually graded to assure that characteristics affecting strength are no more serious than the limiting characteristics for each grade. MSR Scaffold Plank is available 2″ thick only.
*Stadium Grade: 2″ thick, 4″ to 12″ wide See table 1 for design values		
	No. 1 Dense No. 1	For outdoor seating. Free of pitch pockets, pitch streaks and medium pitch on one wide face, but otherwise conforms to No. 1 Dense or No. 1 Dimension Lumber.
*Prime Dimension: 2″ to 4″ thick, 2″ to 12″ wide See table 1 for design values		
	No. 1 Prime	Grade based on No. 1 Dimension Lumber characteristics except that holes, skip and wane are closely limited to provide a high-quality product.
	No. 2 Prime	Grade based on No. 2 Dimension Lumber characteristics except that holes, skip and wane are closely limited to provide a high-quality product.

*Most mills do not manufacture all products and make all grade separations. Those products and grades not manufactured by most mills are noted with an asterisk.

(Courtesy of The Southern Forest Products Association.)

SOUTHERN PINE GRADE DESCRIPTIONS (cont'd)

Based on 1994 SPIB Grading Rules

Product	Grade	Grade Characteristics and Typical Uses
*Marine Grades: 1″ to 20″ thick, 2″ to 20″ wide *See tables 1 and 2 for design values*		
	Any grade of Dimension Lumber or Timbers	All four longitudinal faces must be free of pith and/or heartwood. Application of the product requires pressure treatment by an approved treating process and preservative for marine usage.
*Decking, Heavy Roofing and Heavy Shiplap: 2″ to 4″ thick, 2″and wider *See SPIB Grading Rules for design values*		
	Dense Standard Decking	High-quality product, suitable for plank floor where face serves as finish floor. Has a better appearance than No. 1 Dense Dimension Lumber because of additional restrictions on pitch, knots, pith and wane.
	Dense Select Decking Select Decking	An excellent decking grade that can be used face side down for roof decking or face side up for floor decking.
	Dense Commercial Decking Commercial Decking	An economical roof decking which conforms to No. 2 Dimension Lumber characteristics.
Boards: 1″ to 1-1/2″ thick, 2″ and wider *See SPIB Grading Rules for design values*		
	Industrial 55	Graded as per No. 1 Dimension.
	Industrial 45	Graded as per No. 2 Dimension.
	Industrial 26	Graded as per No. 3 Dimension.
Design values are not assigned	No. 1	High quality with good appearance characteristics. Generally sound and tight-knotted. Largest hole permitted is 1/16″. Superior product suitable for a wide range of uses including shelving, crating, and form lumber.
	No. 2	Good-quality sheathing, fencing, shelving and other general purpose uses.
	No. 3	Good, serviceable sheathing; usable for many economical applications without waste.
	No. 4	Admits pieces below a No. 3 grade which can be used without waste, or which contain less than 25% waste by cutting.
*Industrial Lumber: 4″ and less in thickness, 12″ and less in width *See SPIB Special Product Rules for design values*		
	Industrial 86	Appearance is same as B&B Finish. Larger sizes conform to Dense Structural 86 Structural Lumber except for dense grain requirement.
	Industrial 72	Appearance is same as C Finish. Larger sizes conform to Dense Structural 72 Structural Lumber except for dense grain requirement.
	Industrial 65	Appearance is same as D Finish. Larger sizes conform to Dense Structural 65 Structural Lumber except for dense grain requirement.

*Most mills do not manufacture all products and make all grade separations. Those products and grades not manufactured by most mills are noted with an asterisk.

(Courtesy of The Southern Forest Products Association.)

SOUTHERN PINE GRADE DESCRIPTIONS (cont'd)

Based on 1994 SPIB Grading Rules

Product	Grade	Grade Characteristics and Typical Uses
*Structural Lumber: 2″ and thicker, 2″ and wider *See SPIB Special Product Rules for design values*		
	Dense Structural 86 Dense Structural 72 Dense Structural 65	Premier structural grades. Provides good appearance with some of the highest design values available in any softwood species.
Radius Edge Decking: 1-1/4″ thick, 4″ to 6″ wide *See SPIB Special Product Rules for recommended spans*		
	Premium	High-quality product, recommended where smallest knots are desired and appearance is of utmost importance. Excellent for painting or staining.
	Standard	Slightly less restrictive than premium grade. A very good product to use where appearance is not the major factor. Excellent for painting or staining.
Finish: 3/8″ to 4″ thick, 2″ and wider *Design values are not assigned*		
	*B&B	Highest recognized grade of Finish. Generally clear, although a limited number of pin knots are permitted. Finest quality for natural or stain finish.
	C	Excellent for painting or natural finish where requirements are less exacting. Reasonably clear, but permits limited number of surface checks and small tight knots.
	C&Btr	Combination for B&B and C grades; satisfies requirements for high-quality finish.
	D	Economical, serviceable grade for natural or painted finish.
Flooring, Drop Siding, Paneling, Ceiling and Partition, OG Batts, Bevel Siding, Miscellaneous Millwork *Design values are not assigned*		
	*B&B, C C&Btr, D	See Finish grades for face side; reverse side wane limitations are lower.
	No. 1	No. 1 Flooring and Paneling not provided under SPIB Grading Rules as a separate grade, but if specified, will be designated and graded as D; No. 1 Drop Siding is graded as No. 1 Boards.
	No. 2	Graded as No. 2 Boards. High utility value where appearance is not a factor.
	No. 3	More manufacturing imperfections allowed than in No. 2, but suitable for economical use.
*Shop and Moulding *Design values are not assigned*		
	No. 1 No. 2 No. 3	Recommended for moulding and millwork applications. Currently graded according to rules developed by the Western Wood Products Association.

*Most mills do not manufacture all products and make all grade separations. Those products and grades not manufactured by most mills are noted with an asterisk.

(Courtesy of The Southern Forest Products Association.)

Appendix D

Standard Sizes of Framing and Appearance Lumber

STANDARD SIZES—FRAMING LUMBER
Nominal & Dressed (Based on *Western Lumber Grading Rules*)

Table **2.4**

Product	Description	Nominal Size Thickness (inches)	Nominal Size Width (inches)	Dressed Dimensions Thicknesses & Widths (inches) Surfaced Dry	Dressed Dimensions Thicknesses & Widths (inches) Surfaced Unseasoned	Length (feet)
DIMENSION	S4S	2 3 4	2 3 4 5 6 8 10 12 over 12	1½ 2½ 3½ 4½ 5½ 7¼ 9¼ 11¼ off ¾	1⁹⁄₁₆ 2⁹⁄₁₆ 3⁹⁄₁₆ 4⅝ 5⅝ 7½ 9½ 11½ off ½	6' and longer, generally shipped in multiples of 2'
TIMBERS	Rough or S4S (shipped unseasoned)	5 and larger	5 and larger	**Thickness (unseasoned)** 1/2 off nominal (S4S). See 3.20 of WWPA Grading Rules for Rough.	**Width (unseasoned)** 1/2 off nominal (S4S). See 3.20 of WWPA Grading Rules for Rough.	6' and longer, generally shipped in multiples of 2'
DECKING	2" (Single T&G)	2	5 6 8 10 12	**Thickness (dry)** 1½	**Width (dry)** 4 5 6¾ 8¾ 10¾	6' and longer, generally shipped in multiples of 2'
DECKING	3" and 4" (Double T&G)	3 4	6	2½ 3½	5¼	

Abbreviations: FOHC—Free of Heart Center   T&G—Tongued and grooved   Rough Full Sawn—Unsurfaced lumber cut to full specified size
S4S—Surfaced four sides

(Courtesy of Western Wood Products Association.)

STANDARD SIZES—APPEARANCE LUMBER
Nominal & Dressed (Based on *Western Lumber Grading Rules*)

Table **2.6**

Product	Description	Nominal Size		Dry Dressed Dimensions		
		Thickness (inches)	Width (inches)	Thickness (inches)	Width (inches)	Lengths (feet)
SELECTS AND COMMONS	S1S, S2S, S4S, S1S1E, S1S2E	4/4 5/4 6/4 7/4 8/4 9/4 10/4 11/4 12/4 16/4	2 3 4 5 6 7 8 & wider	3/4 1 5/32 1 13/32 1 19/32 1 13/16 2 3/32 2 3/8 2 9/16 2 3/4 3 3/4	1 1/2 2 1/2 3 1/2 4 1/2 5 1/2 6 1/2 3/4 off nominal	6' and longer in multiples of 1', except Douglas Fir and Larch Selects shall be 4' and longer with 3% of 4' and 5' permitted.
FINISH AND ALTERNATE BOARD GRADES	S1S, S2S, S4S, S1S1E, S1S2E	3/8 1/2 5/8 3/4* 1* 1 1/4* 1 1/2* 1 3/4 2 2 1/2 3 3 1/2 4	2 3 4 5 6 7 8 & wider	5/16 7/16 9/16 5/8 3/4 1 1 1/4 1 3/8 1 1/2 2 2 1/2 3 3 1/2	1 1/2 2 1/2 3 1/2 4 1/2 5 1/2 6 1/2 3/4 off nominal	3' and longer. In SUPERIOR grade, 3% of 3' and 4' and 7% of 5' and 6' are permitted. In PRIME grade 20% of 3' to 6' is permitted.

*These sizes apply only to Alternate Board grades.

Abbreviations:
S1S—Surfaced one side
S2S—Surfaced two sides
S4S—Surfaced four sides
S1S1E—Surfaced one side, one edge
S1S2E—Surfaced one side, two edges

(Courtesy of Western Wood Products Association.)

METRIC EQUIVALENTS FOR STANDARD NOMINAL AND DRESSED LUMBER SIZES   Table **A.3**
Based on WWPA Grading Rules

Product	Description	Nominal Size (U.S.)		Dry Dressed Dimensions		
		Thickness (in.)	Width (in.)	Thickness (mm)	Width (mm)	Lengths (meters)
Selects & Commons	S1S, S2S, S4S, S1S1E, S1S2E	4/4 5/4 6/4 7/4 8/4 9/4 10/4 11/4 12/4 16/4	2 3 4 5 6 7 8 & wider	19 29 36 40 46 53 60 65 70 95	38 64 89 114 140 165 19 Off nominal	1.8 and longer in multiples of 0.3 except Douglas Fir and Larch Selects shall be 1.2 and longer with 3% of 1.2 and 1.5 permitted.
Finish & Boards	S1S, S2S, S4S, S1S1E, S1S2E These sizes apply only to Alternate Board Grades.	3/8 3/4 1 1 1/4 1 1/2 1 3/4 2 2 1/2 3 3 1/2 4	2 5 6 7 8 & wider	8 16 19 25 32 35 38 51 64 76 89	38 114 140 165 19 Off nominal	0.9 and longer. In Superior Grade, 3% of 0.9 and 1.2 and 7% of 1.5 and 1.8 are permitted. In Prime Grade, 20% of 0.9 to 1.8 is permitted.
Paneling & Siding	T&G or Shiplap	1	6 8 10 12	19	138 181 232 283	1.2 and longer in multiples of 0.3
Factory & Shop Lumber	S2S	1 (4/4) 1 1/4 (5/4) 1 1/2 (6/4) 1 3/4 (7/4) 2 (8/4) 2 1/2 (10/4) 3 (12/4) 4 (16/4)	See individual grade descriptions in WWPA's *Western Lumber Grading Rules*	19 29 36 40 46 60 70 95	Usually sold random width	1.8 and longer in multiples of 0.3

Product	Description	Nominal Size (U.S.)		Dressed Dimensions		
				Thicknesses & Widths (mm)		
		Thickness (in.)	Width (in.)	Surfaced Dry	Surfaced Unseasoned	Lengths (meters)
Dimension	S4S Other surface combinations are available.*	2 3 4	2 3 4 5 6 8 10 12 Over 12	38 64 89 114 140 184 235 286 Off 19	40 65 90 117 143 191 241 292 Off 13	1.8m and longer in multiples of 0.3m
Scaffold Plank	Rough Full Sawn or S4S (Usually shipped unseasoned)	1 1/4 & Thicker	8 & Wider	If Dressed, refer to "DIMENSION" sizes.		1.8m and longer in multiples of 0.3m
Timbers	Rough or S4S (Shipped unseasoned)	5 & Larger		Thicknesses and Width 13mm Off Nominal (S4S) See 3.20 of WWPA Grading Rules for Rough.		1.8m and longer in multiples of 0.3m

*Surfacing Abbreviations: S1S–Surfaced one side   S1S1E–Surfaced one side, one edge   (Courtesy of Western Wood Products Association.)
S2S–Surfaced two sides   S1S2E–Surfaced one side, two edges
S4S–Surfaced four sides

Appendix F

Panel Grades

TABLE 1

VENEER GRADES

 Smooth, paintable. Not more than 18 neatly made repairs, boat, sled, or router type, and parallel to grain, permitted. Wood or synthetic repairs permitted. May be used for natural finish in less demanding applications.

 Solid surface. Shims, sled or router repairs, and tight knots to 1 inch across grain permitted. Wood or synthetic repairs permitted. Some minor splits permitted.

 Improved C veneer with splits limited to 1/8-inch width and knotholes or other open defects limited to 1/4 x 1/2 inch. Wood or synthetic repairs permitted. Admits some broken grain.
Plugged

C Tight knots to 1-1/2 inch. Knotholes to 1 inch across grain and some to 1-1/2 inch if total width of knots and knotholes is within specified limits. Synthetic or wood repairs. Discoloration and sanding defects that do not impair strength permitted. Limited splits allowed. Stitching permitted.

 Knots and knotholes to 2-1/2-inch width across grain and 1/2 inch larger within specified limits. Limited splits are permitted. Stitching permitted. Limited to Exposure 1 or Interior panels.

Veneer grades describe the face and back surface veneers of plywood panels.

(Courtesy of APA - The Engineered Wood Association.)

TABLE 2

GUIDE TO APA PERFORMANCE RATED PANELS[a][b]
FOR APPLICATION RECOMMENDATIONS, SEE FOLLOWING PAGES.

APA RATED SHEATHING
Typical Trademark

Specially designed for subflooring and wall and roof sheathing. Also good for a broad range of other construction and industrial applications. Can be manufactured as plywood, as a composite, or as OSB. EXPOSURE DURABILITY CLASSIFICATIONS: Exterior, Exposure 1, Exposure 2. COMMON THICKNESSES: 5/16, 3/8, 7/16, 15/32, 1/2, 19/32, 5/8, 23/32, 3/4.

**APA STRUCTURAL I
RATED SHEATHING**[c]
Typical Trademark

Unsanded grade for use where shear and cross-panel strength properties are of maximum importance, such as panelized roofs and diaphragms. Can be manufactured as plywood, as a composite, or as OSB. EXPOSURE DURABILITY CLASSIFICATIONS: Exterior, Exposure 1. COMMON THICKNESSES: 5/16, 3/8, 7/16, 15/32, 1/2, 19/32, 5/8, 23/32, 3/4.

APA RATED STURD-I-FLOOR
Typical Trademark

Specially designed as combination subfloor-underlayment. Provides smooth surface for application of carpet and pad and possesses high concentrated and impact load resistance. Can be manufactured as plywood, as a composite, or as OSB. Available square edge or tongue-and-groove. EXPOSURE DURABILITY CLASSIFICATIONS: Exterior, Exposure 1, Exposure 2. COMMON THICKNESSES: 19/32, 5/8, 23/32, 3/4, 1, 1-1/8.

APA RATED SIDING
Typical Trademark

For exterior siding, fencing, etc. Can be manufactured as plywood, as a composite or as an overlaid OSB. Both panel and lap siding available. Special surface treatment such as V-groove, channel groove, deep groove (such as APA Texture 1-11), brushed, rough sawn and overlaid (MDO) with smooth- or texture-embossed face. Span Rating (stud spacing for siding qualified for APA Sturd-I-Wall applications) and face grade classification (for veneer-faced siding) indicated in trademark. EXPOSURE DURABILITY CLASSIFICATION: Exterior. COMMON THICKNESSES: 11/32, 3/8, 7/16, 15/32, 1/2, 19/32, 5/8.

Example:
Grade APA Rates Sturd-I-Floor panels are designed for combination subfloor and underlayment.

Panels are 23/32" thick.
Maximum span between joists below is 24".
Exposure 1 indicated the highest weather exposure durability.

(Courtesy of APA - The Engineered Wood Association.)

TABLE 3

GUIDE TO APA SANDED & TOUCH-SANDED PLYWOOD PANELS[a][b][c]
FOR APPLICATION RECOMMENDATIONS, SEE FOLLOWING PAGES.

APA A-A
Typical Trademark

| A-A • G-1 • EXPOSURE 1-APA • 000 • PS1-95 |

Use where appearance of both sides is important for interior applications such as built-ins, cabinets, furniture, partitions; and exterior applications such as fences, signs, boats, shipping containers, tanks, ducts, etc. Smooth surfaces suitable for painting. EXPOSURE DURABILITY CLASSIFICATIONS: Interior, Exposure 1, Exterior. COMMON THICKNESSES: 1/4, 11/32, 3/8, 15/32, 1/2, 19/32, 5/8, 23/32, 3/4.

APA A-B
Typical Trademark

| A-B • G-1 • EXPOSURE 1-APA • 000 • PS1-95 |

For use where appearance of one side is less important but where two solid surfaces are necessary. EXPOSURE DURABILITY CLASSIFICATIONS: Interior, Exposure 1, Exterior. COMMON THICKNESSES: 1/4, 11/32, 3/8, 15/32, 1/2, 19/32, 5/8, 23/32, 3/4.

APA A-C
Typical Trademark

APA
THE ENGINEERED
WOOD ASSOCIATION
A-C GROUP 1
EXTERIOR
000
PS 1-95

For use where appearance of only one side is important in exterior or interior applications, such as soffits, fences, farm buildings, etc.[f] EXPOSURE DURABILITY CLASSIFICATION: Exterior. COMMON THICKNESSES: 1/4, 11/32, 3/8, 15/32, 1/2, 19/32, 5/8, 23/32, 3/4.

APA A-D
Typical Trademark

APA
THE ENGINEERED
WOOD ASSOCIATION
A-D GROUP 1
EXPOSURE 1
000
PS 1-95

For use where appearance of only one side is important in interior applications, such as paneling, built-ins, shelving, partitions, flow racks, etc.[f] EXPOSURE DURABILITY CLASSIFICATIONS: Interior, Exposure 1. COMMON THICKNESSES: 1/4, 11/32, 3/8, 15/32, 1/2, 19/32, 5/8, 23/32, 3/4.

APA B-B
Typical Trademark

| B-B • G-2 • EXPOSURE 1-APA • 000 • PS1-95 |

Utility panels with two solid sides. EXPOSURE DURABILITY CLASSIFICATIONS: Interior, Exposure 1, Exterior. COMMON THICKNESSES: 1/4, 11/32, 3/8, 15/32, 1/2, 19/32, 5/8, 23/32, 3/4.

APA B-C
Typical Trademark

APA
THE ENGINEERED
WOOD ASSOCIATION
B-C GROUP 1
EXTERIOR
000
PS 1-95

Utility panel for farm service and work buildings, boxcar and truck linings, containers, tanks, agricultural equipment, as a base for exterior coatings and other exterior uses or applications subject to high or continuous moisture.[f] EXPOSURE DURABILITY CLASSIFICATION: Exterior. COMMON THICKNESSES: 1/4, 11/32, 3/8, 15/32, 1/2, 19/32, 5/8, 23/32, 3/4.

APA B-D
Typical Trademark

APA
THE ENGINEERED
WOOD ASSOCIATION
B-D GROUP 2
EXPOSURE 1
000
PS 1-95

Utility panel for backing, sides of built-ins, industry shelving, slip sheets, separator boards, bins and other interior or protected applications.[f] EXPOSURE DURABILITY CLASSIFICATIONS: Interior, Exposure 1. COMMON THICKNESSES: 1/4, 11/32, 3/8, 15/32, 1/2, 19/32, 5/8, 23/32, 3/4.

Continued on next page

(Courtesy of APA - The Engineered Wood Association.)

TABLE 3

CONTINUED

APA UNDERLAYMENT Typical Trademark		For application over structural subfloor. Provides smooth surface for application of carpet and pad and possesses high concentrated and impact load resistance. For areas to be covered with resilient flooring, specify panels with "sanded face."[e] EXPOSURE DURABILITY CLASSIFICATIONS: Interior, Exposure 1. COMMON THICKNESSES[d]: 1/4, 11/32, 3/8, 15/32, 1/2, 19/32, 5/8, 23/32, 3/4.
APA C-C PLUGGED[g] Typical Trademark		For use as an underlayment over structural subfloor, refrigerated or controlled atmosphere storage rooms, pallet fruit bins, tanks, boxcar and truck floors and linings, open soffits, and other similar applications where continuous or severe moisture may be present. Provides smooth surface for application of carpet and pad and possesses high concentrated and impact load resistance. For areas to be covered with resilient flooring, specify panels with "sanded face."[e] EXPOSURE DURABILITY CLASSIFICATION: Exterior. COMMON THICKNESSES[d] : 11/32, 3/8, 15/32, 1/2, 19/32, 5/8, 23/32, 3/4.
APA C-D PLUGGED Typical Trademark		For open soffits, built-ins, cable reels, separator boards and other interior or protected applications. Not a substitute for Underlayment or APA Rated Sturd-I-Floor as it lacks their puncture resistance. EXPOSURE DURABILITY CLASSIFICATIONS: Interior, Exposure 1. COMMON THICKNESSES: 3/8, 15/32, 1/2, 19/32, 5/8, 23/32, 3/4.

Example 1:
Grade APA A-C panels are used where appearance of only one side is important.

    See Table 1 for veneer descriptions.
    Group 1 identifies the category of species used to manufacture plywood.
    Exterior grade plywood is used.

Example 2:
Grade APA underlayment is placed over structural subfloors to provide a smooth surface for a carpet and pad or resilient flooring.

    Panels range from 1/4" to 3/4" in thickness.
    Group 1 identifies the category of species used in the manufacture of the panels.
    Exposure 1 indicates the highest weather exposure durability.

(Courtesy of APA - The Engineered Wood Association.)

TABLE 4

GUIDE TO APA SPECIALTY PLYWOOD PANELS[a]
FOR APPLICATION RECOMMENDATIONS, SEE FOLLOWING PAGES.

APA DECORATIVE
Typical Trademark

> **APA**
> THE ENGINEERED
> WOOD ASSOCIATION
> DECORATIVE
> GROUP 2
> EXPOSURE 1
> 000
> PS 1-95

Rough-sawn, brushed, grooved, or striated faces. For paneling, interior accent walls, built-ins, counter facing, exhibit displays. Can also be made by some manufacturers in Exterior for exterior siding, gable ends, fences and other exterior applications. Use recommendations for Exterior panels vary with the particular product. Check with the manufacturer. EXPOSURE DURABILITY CLASSIFICATIONS: Interior, Exposure 1, Exterior. COMMON THICKNESSES: 5/16, 3/8, 1/2, 5/8.

APA HIGH DENSITY OVERLAY (HDO)[b]
Typical Trademark

> HDO • A-A • G-1 • EXT-APA • 000 • PS 1-95

Has a hard semi-opaque resin-fiber overlay on both faces. Abrasion resistant. For concrete forms, cabinets, countertops, signs, tanks. Also available with skid-resistant screen-grid surface. EXPOSURE DURABILITY CLASSIFICATION: Exterior. COMMON THICKNESSES: 3/8, 1/2, 5/8, 3/4.

APA MEDIUM DENSITY OVERLAY (MDO)[b]
Typical Trademark

> **APA**
> THE ENGINEERED
> WOOD ASSOCIATION
> M. D. OVERLAY
> GROUP 1
> EXTERIOR
> 000
> PS 1-95

Smooth, opaque, resin-fiber overlay on one or both faces. Ideal base for paint, both indoors and outdoors. For exterior siding, paneling, shelving, exhibit displays, cabinets, signs. EXPOSURE DURABILITY CLASSIFICATION: Exterior. COMMON THICKNESSES: 11/32, 3/8, 15/32, 1/2, 19/32, 5/8, 23/32, 3/4.

APA MARINE
Typical Trademark

> MARINE • A-A • EXT-APA • PS 1-95

Ideal for boat hulls. Made only with Douglas-fir or western larch. Subject to special limitations on core gaps and face repairs. Also available with HDO or MDO faces. EXPOSURE DURABILITY CLASSIFICATION: Exterior. COMMON THICKNESSES: 1/4, 3/8, 1/2, 5/8, 3/4.

APA B-B PLYFORM CLASS I[b]
Typical Trademark

> **APA**
> THE ENGINEERED
> WOOD ASSOCIATION
> PLYFORM
> B-B CLASS 1
> EXTERIOR
> 000
> PS 1-95

Concrete form grades with high reuse factor. Sanded both faces and mill-oiled unless otherwise specified. Special restrictions on species. Also available in HDO for very smooth concrete finish, and with special overlays. EXPOSURE DURABILITY CLASSIFICATION: Exterior. COMMON THICKNESSES: 19/32, 5/8, 23/32, 3/4.

APA PLYRON
Typical Trademark

> PLYRON • EXPOSURE 1-APA • 000

Hardboard face on both sides. Faces tempered, untempered, smooth or screened. For countertops, shelving, cabinet doors, flooring. EXPOSURE DURABILITY CLASSIFICATIONS: Interior, Exposure 1, Exterior. COMMON THICKNESSES: 1/2, 5/8, 3/4.

Example:

Grade APA B-B Plyform Class 1 used to construct concrete forms with a high reuse factor.

> Veneer grades B-B indicate sanded surfaces to provide smooth finishes for concrete.
> Thicknesses range from 19/32" to 3/4".
> The plywood is of exterior design.

(Courtesy of APA - The Engineered Wood Association.)

Appendix G

Joist and Rafter Span Tables

FLOOR JOISTS

Residential occupancies include private dwelling, private apartment and hotel guest rooms. Deck under CABO and Standard Codes.[1]

40# Live Load, 10# Dead Load, $\ell/360$

Table FJ-1

Species or Group	Grade	2 x 6 12" oc	2 x 6 16" oc	2 x 6 24" oc	2 x 8 12" oc	2 x 8 16" oc	2 x 8 24" oc	2 x 10 12" oc	2 x 10 16" oc	2 x 10 24" oc	2 x 12 12" oc	2 x 12 16" oc	2 x 12 24" oc
Douglas Fir-Larch	Sel. Struc.	11-4	10-4	9-0	15-0	13-7	11-11	19-1	17-4	15-2	23-3	21-1	18-5
	No. 1 & Btr.	11-2	10-2	8-10	14-8	13-4	11-8	18-9	17-0	14-5	22-10	20-5	16-8
	No. 1	10-11	9-11	8-8	14-5	13-1	11-0	18-5	16-5	13-5	22-0	19-1	15-7
	No. 2	10-9	9-9	8-1	14-2	12-7	10-3	17-9	15-5	12-7	20-7	17-10	14-7
	No. 3	8-8	7-6	6-2	11-0	9-6	7-9	13-5	11-8	9-6	15-7	13-6	11-0
Douglas Fir-South	Sel. Struc.	10-3	9-4	8-2	13-6	12-3	10-9	17-3	15-8	13-8	21-0	19-1	16-8
	No. 1	10-0	9-1	7-11	13-2	12-0	10-5	16-10	15-3	12-9	20-6	18-1	14-9
	No. 2	9-9	8-10	7-9	12-10	11-8	10-0	16-5	14-11	12-2	19-11	17-4	14-2
	No. 3	8-6	7-4	6-0	10-9	9-3	7-7	13-1	11-4	9-3	15-2	13-2	10-9
Hem-Fir	Sel. Struc.	10-9	9-9	8-6	14-2	12-10	11-3	18-0	16-5	14-4	21-11	19-11	17-5
	No. 1 & Btr.	10-6	9-6	8-4	13-10	12-7	11-0	17-8	16-0	13-9	21-6	19-6	16-0
	No. 1	10-6	9-6	8-4	13-10	12-7	10-9	17-8	16-0	13-1	21-6	18-7	15-2
	No. 2	10-0	9-1	7-11	13-2	12-0	10-2	16-10	15-2	12-5	20-4	17-7	14-4
	No. 3	8-8	7-6	6-2	11-0	9-6	7-9	13-5	11-8	9-6	15-7	13-6	11-0
Spruce-Pine-Fir (South)	Sel. Struc.	10-0	9-1	7-11	13-2	12-0	10-6	16-10	15-3	13-4	20-6	18-7	16-3
	No. 1	9-9	8-10	7-9	12-10	11-8	10-2	16-5	14-11	12-5	19-11	17-7	14-4
	No. 2	9-6	8-7	7-6	12-6	11-4	9-6	15-11	14-3	11-8	19-1	16-6	13-6
	No. 3	8-0	6-11	5-8	10-2	8-9	7-2	12-5	10-9	8-9	14-4	12-5	10-2
Western Woods	Sel. Struc.	9-9	8-10	7-9	12-10	11-8	10-2	16-5	14-11	12-7	19-11	17-10	14-7
	No. 1	9-6	8-7	7-0	12-6	10-10	8-10	15-4	13-3	10-10	17-9	15-5	12-7
	No. 2	9-2	8-4	7-0	12-1	10-10	8-10	15-4	13-3	10-10	17-9	15-5	12-7
	No. 3	7-6	6-6	5-4	9-6	8-3	6-9	11-8	10-1	8-3	13-6	11-8	9-6

Example 1:

Select structural Douglas Fir or Larch lumber species is to be used for floor joists.

Joist size is 2 × 10 spaced 16" OC.
The allowable span is 17'-4".

Example 2:

#2 grade Hemlock-Fir lumber species is to be used for the floor joists.

Joist size is 2 × 6 spaced 12" OC.
The allowable span is 10'-0".

The floor joist span table shown above is designed for a live load of 40 pounds and dead load of 10 pounds per square foot (psf). The Western Wood Products Association provides additional floor joist span tables for a wide range of live and dead loads.

(Courtesy of Western Wood Products Association.)

477

CEILING JOISTS

Use these loading conditions for the following: Limited attic storage where development of future rooms is not possible. Ceilings where the roof slope is steeper than 3 in 12. Where the clear height in the attic is greater than 30 inches. Plaster ceilings.

Species or Group	Grade	2 × 4			2 × 6			2 × 8			2 × 10		
		12" oc	16" oc	24" oc	12" oc	16" oc	24" oc	12" oc	16" oc	24" oc	12" oc	16" oc	24" oc
Douglas Fir-Larch	Sel. Struc.	9-1	8-3	7-3	14-4	13-0	11-4	18-10	17-2	15-0	24-1	21-10	19-1
	No. 1 & Btr.	8-11	8-1	7-1	14-1	12-9	11-2	18-6	16-10	14-8	23-8	21-6	18-7
	No. 1	8-9	8-0	7-0	13-9	12-6	10-11	18-2	16-6	14-2	23-2	21-1	17-4
	No. 2	8-7	7-10	6-10	13-6	12-3	10-6	17-10	16-2	13-3	22-9	19-10	16-3
	No. 3	7-8	6-8	5-5	11-2	9-8	7-11	14-2	12-4	10-0	17-4	15-0	12-3
Douglas Fir-South	Sel. Struc.	8-3	7-6	6-6	12-11	11-9	10-3	17-0	15-6	13-6	21-9	19-9	17-3
	No. 1	8-0	7-3	6-4	12-7	11-5	10-0	16-7	15-1	13-2	21-2	19-3	16-5
	No. 2	7-10	7-1	6-2	12-3	11-2	9-9	16-2	14-8	12-10	20-8	18-9	15-9
	No. 3	7-6	6-6	5-3	10-11	9-6	7-9	13-10	12-0	9-9	16-11	14-8	11-11
Hem-Fir	Sel. Struc.	8-7	7-10	6-10	13-6	12-3	10-9	17-10	16-2	14-2	22-9	20-8	18-0
	No. 1 & Btr.	8-5	7-8	6-8	13-3	12-0	10-6	17-5	15-10	13-10	22-3	20-2	17-8
	No. 1	8-5	7-8	6-8	13-3	12-0	10-6	17-5	15-10	13-10	22-3	20-2	16-11
	No. 2	8-0	7-3	6-4	12-7	11-5	10-0	16-7	15-1	13-1	21-2	19-3	16-0
	No. 3	7-8	6-8	5-5	11-2	9-8	7-11	14-2	12-4	10-0	17-4	15-0	12-3
Spruce-Pine-Fir (South)	Sel. Struc.	8-0	7-3	6-4	12-7	11-5	10-0	16-7	15-1	13-2	21-2	19-3	16-10
	No. 1	7-10	7-1	6-2	12-3	11-2	9-9	16-2	14-8	12-10	20-8	18-9	16-0
	No. 2	7-7	6-11	6-0	11-11	10-10	9-6	15-9	14-3	12-4	20-1	18-3	15-0
	No. 3	7-1	6-1	5-0	10-4	8-11	7-4	13-1	11-4	9-3	16-0	13-10	11-4
Western Woods	Sel. Struc.	7-10	7-1	6-2	12-3	11-2	9-9	16-2	14-8	12-10	20-8	18-9	16-3
	No. 1	7-7	6-11	6-0	11-11	10-10	9-0	15-9	14-0	11-5	19-9	17-1	14-0
	No. 2	7-4	6-8	5-10	11-7	10-6	9-0	15-3	13-10	11-5	19-5	17-1	14-0
	No. 3	6-8	5-9	4-8	9-8	8-5	6-10	12-4	10-8	8-8	15-0	13-0	10-7

Example 1:

#3 grade Douglas Fir (South) lumber species is to be used for the ceiling joists.

The joist size is 2 × 4 spaced 24″ OC.
The allowable span is 5′-3″.

Example 2:

#1 grade Spruce-Pine-Fir (South) is to be used for the ceiling joists.

The joist size is 2 × 8 spaced 12″ OC.
The allowable span is 16′-2″.

The ceiling joist span table shown above is designed for a live load of 20 pounds and dead load of 10 pounds per square foot (psf). The Western Wood Products Association provides additional ceiling joist span tables for a wide range of live and dead loads.

(Courtesy of Western Wood Products Association.)

ROOF RAFTERS

20# Live Load, 7# Dead Load, ℓ/180

Table RR-1

No snow load.[1]
Roof slope greater than 3 in 12.
Light roof covering. No ceiling finish.

Species or Group	Grade	2 × 6 12" oc	2 × 6 16" oc	2 × 6 24" oc	2 × 8 12" oc	2 × 8 16" oc	2 × 8 24" oc	2 × 10 12" oc	2 × 10 16" oc	2 × 10 24" oc	2 × 12 12" oc	2 × 12 16" oc	2 × 12 24" oc
Douglas Fir-Larch	Sel. Struc.	18-0	16-4	14-4	23-9	21-7	18-10	30-4	27-6	24-1	36-10	33-6	28-6
	No. 1 & Btr.	17-8	16-1	14-1	23-4	21-2	17-11	29-9	26-10	21-11	35-11	31-1	25-5
	No. 1	17-4	15-9	13-3	22-11	20-6	16-9	28-11	25-0	20-5	33-6	29-0	23-8
	No. 2	17-0	15-2	12-4	22-2	19-2	15-8	27-0	23-5	19-1	31-4	27-2	22-2
	No. 3	13-3	11-5	9-4	16-9	14-6	11-10	20-5	17-8	14-5	23-8	20-6	16-9
Douglas Fir-South	Sel. Struc.	16-3	14-9	12-11	21-5	19-6	17-0	27-5	24-10	21-9	33-4	30-3	26-5
	No. 1	15-11	14-5	12-6	20-11	19-0	15-10	26-9	23-9	19-5	31-10	27-6	22-6
	No. 2	15-6	14-1	12-0	20-5	18-6	15-2	26-0	22-9	18-7	30-5	26-4	21-6
	No. 3	12-10	11-2	9-1	16-4	14-1	11-6	19-11	17-3	14-1	23-1	20-0	16-4
Hem-Fir	Sel. Struc.	17-0	15-6	13-6	22-5	20-5	17-10	28-7	26-0	22-9	34-10	31-8	27-8
	No. 1 & Btr.	16-8	15-2	13-3	21-11	19-11	17-2	28-0	25-5	20-11	34-1	29-9	24-3
	No. 1	16-8	15-2	12-10	21-11	19-11	16-4	28-0	24-5	19-11	32-8	28-3	23-1
	No. 2	15-11	14-5	12-2	20-11	18-11	15-5	26-8	23-1	18-10	30-11	26-9	21-10
	No. 3	13-3	11-5	9-4	16-9	14-6	11-10	20-5	17-8	14-5	23-8	20-6	16-9
Spruce-Pine-Fir (South)	Sel. Struc.	15-11	14-5	12-7	20-11	19-0	16-7	26-9	24-3	21-2	32-6	29-6	25-9
	No. 1	15-6	14-1	12-2	20-5	18-6	15-5	26-0	23-1	18-10	30-11	26-9	21-10
	No. 2	15-0	13-8	11-5	19-10	17-9	14-6	25-0	21-8	17-8	29-0	25-2	20-6
	No. 3	12-2	10-7	8-7	15-5	13-4	10-11	18-10	16-4	13-4	21-10	18-11	15-5
Western Woods	Sel. Struc.	15-6	14-1	12-3	20-5	18-6	15-8	26-0	23-5	19-1	31-4	27-2	22-2
	No. 1	15-0	13-1	10-8	19-1	16-6	13-6	23-4	20-2	16-6	27-0	23-5	19-1
	No. 2	14-7	13-1	10-8	19-1	16-6	13-6	23-4	20-2	16-6	27-0	23-5	19-1
	No. 3	11-5	9-11	8-1	14-6	12-7	10-3	17-8	15-4	12-6	20-6	17-9	14-6

[1] A 1.25 Duration of Load adjustment has been applied. See page 5.

Example 1:

#2 grade Douglas Fir - Larch lumber species is to be used for roof rafters.

The rafter size is 2 × 12 spaced 24" OC.
The allowable span is 22'-2".

Example 2:

Select structural grade Hemlock-Fir species to be used for the roof rafters.

The rafter size is 2 × 6 spaced 16" OC.
The allowable span is 15'-6".

The rafter span table shown above is designed for a live load of 20 pounds and dead load of 7 pounds per square foot (psf). The Western Wood Products Association provides additional roof rafter span tables for a wide range of live and dead loads.

(Courtesy of Western Wood Products Association.)

Appendix H

Truss Designs

FRENCHWOOD® GLIDING PATIO DOORS

A N D E R S E N®

Table of Basic Unit Sizes—6' 8" Height Scale 1/8" = 1'-0" (1:96)

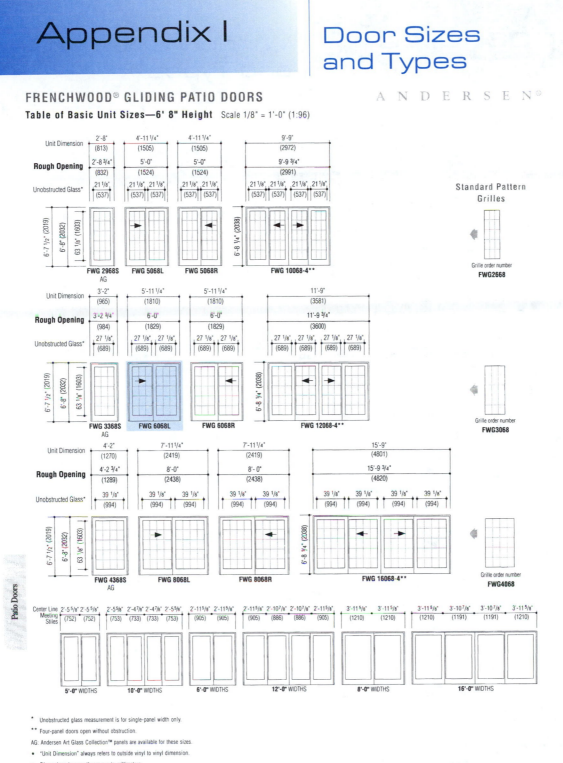

*   Unobstructed glass measurement is for single-panel width only.
** Four-panel doors open without obstruction.
AG: Andersen Art Glass Collection™ panels are available for these sizes.
*   "Unit Dimension" always refers to outside vinyl to vinyl dimension.
*   Dimensions in parentheses are in millimeters.
*   Combinations of window units joined together may require horizontal or vertical reinforcement. This reinforcement should be specified by a professional engineer.
*   When ordering, be sure to specify color desired: white, Sandtone, or Terratone®.

Example: A unit identification number FWG 6068L two-door sliding door unit is 5'-11 1/4" (1810 mm) wide and 6'-7 1/2" (2019 mm) high. The rough opening is 6'-0" (1829 mm) wide and 6'-8" (2032 mm) high. The unobstructed glass for each door is 27 1/8" (689 mm) wide and 63 1/8" (1603 mm) high.

This table of standard gliding patio door sizes gives the unit identification number of each door unit, width and height of the door unit, and the rough opening into which the door is placed. Also provided are the dimensions of the unobstructed glass, which is the actual glass area minus the door rails, mullions, and muntins. The dimensions are given in both English and metric measurement. *(Courtesy of Andersen Windows, Inc.)*

NARROLINE® DOUBLE-HUNG

Table of Basic Unit Sizes Scale 1/8" = 1'-0" (1:96)

A N D E R S E N®

* These units have restricted sash travel.
• Unobstructed glass measurement is for single sash only.
• "Unit Dimension" always refers to outside vinyl to vinyl dimension.
• Dimensions in parentheses are in millimeters.

Narroline® units are available in **white color only.**

Example: A unit identification number 24310 window unit is 2'-5 5/8" (752 mm) wide and 4'-1 1/4" (1251 mm) high. The rough opening is 2'-6 1/8" (765 mm) wide and 4'-1 1/4" (1251 mm) high. The unobstructed glass is 24 7/16" (621 mm) wide and 19 15/16" (506 mm) for each (top and bottom) window.

This table of standard double-hung window sizes gives the unit identification number of each window, its width and height, and the dimensions of the rough opening into which the window is placed. Also provided are the dimensions of the unobstructed glass, which is the actual glass area minus the window, sash, mullions, and muntins. The dimensions are given in both English and metric measurement. *(Courtesy of Andersen Windows, Inc.)*

Appendix K

CSI Masterformat

DIVISION 0—BIDDING AND CONTRACT REQUIREMENTS

Reference
Number
00010 Pre-bid Information
00100 Instructions to Bidders
00200 Information Available to Bidders
00300 Bid/Tender Forms
00400 Supplements to Bid/Tender Forms
00500 Agreement Forms
00600 Bonds and Certificates
00700 General Conditions of the Contract
00800 Supplementary Conditions
00900 Addenda and Modifications
00950 Drawings Index

DIVISION 1—GENERAL REQUIREMENTS
01010 General Contractors
01020 Allowances
01030 Special Project Procedures
01040 Coordination
01050 Field Engineering
01060 Regulatory Requirements
01070 Abbreviations and Symbols
01080 Identification Systems
01100 Alternates/Alternatives
01150 Measurement and Payment
01200 Project Meetings
01300 Submittals
01400 Quality Control
01500 Construction Facilities and Temporary Controls
01600 Material and Equipment
01650 Starting of Systems
01660 Testing, Adjusting, and Balancing of Systems
01700 Contract Closeout
01800 Maintenance Materials

DIVISION 2—SITEWORK
02010 Subsurface Investigation
02050 Demolition
02100 Site Preparation
02150 Underpinning
02200 Earthwork
02300 Tunneling
02350 Piles, Caissons, and Cofferdams
02400 Drainage
02440 Site Improvements
02480 Landscaping
02500 Paving and Surfacing
02580 Bridges
02590 Ponds and Reservoirs
02700 Piped Utilities
02800 Power and Commmunication Utilities
02850 Railroad Work
02880 Marine Work

DIVISION 3—CONCRETE
03010 Concrete Materials
03050 Concreting Procedures
03100 Concrete Formwork
03150 Forms

03180 Form Ties and Accessories
03200 Concrete Reinforcement
03250 Concrete Accessories
03300 Cast-in-place Concrete
03350 Special Concrete Finishes
03360 Specially Placed Concrete
03370 Concrete Curing
03400 Precast Concrete
03500 Cementitious Decks
03700 Concrete Restoration and Cleaning

DIVISION 4—MASONRY
04050 Masonry Procedures
04100 Mortar
04150 Masonry Accessories
04200 Unit Masonry
04400 Stone
04500 Masonry Restoration and Cleaning
04550 Refractories
04600 Corrosion-resistant Masonry

DIVISION 5—METALS
05010 Metal Materials and Methods
05050 Metal Fastening
05100 Structural Metal Framing
05200 Metal Joists
05300 Metal Decking
05400 Cold-formed Metal Framing
05500 Metal Fabrications
05700 Ornamental Metal
05800 Expansion Control
05900 Metal Finishes

DIVISION 6—WOOD AND PLASTICS
06050 Fasteners and Supports
06100 Rough Carpentry
06130 Heavy Timber Construction
06150 Wood-Metal Systems
06170 Prefabricated Structural Wood
06200 Finish Carpentry
06300 Wood Treatment
06400 Architectural Woodwork
06500 Prefabricated Structural Plastics
06600 Plastic Fabrications

DIVISION 7—THERMAL AND MOISTURE PROTECTION
07100 Waterproofing
07150 Dampproofing
07200 Insulation
07250 Fireproofing
07300 Shingles and Roofing Tiles
07400 Preformed Roofing and Siding
07500 Membrane Roofing
07570 Traffic Topping
07600 Flashing and Sheet Metal
07800 Roof Accessories
07900 Sealants

(Courtesy of Construction Specification Institute and Construction Specifications Canada.)

DIVISION 8—DOORS AND WINDOWS
08100   Metal Doors and Frames
08200   Wood and Plastic Doors
08250   Door Opening Assemblies
08300   Special Doors
08400   Entrances and Storefronts
08500   Metal Windows
08600   Wood and Plastic Windows
08650   Window Covering
08700   Hardware
08800   Glazing
08900   Glazed Curtain Walls

DIVISION 9—FINISHES
09100   Metal Support Systems
09200   Lath and Plaster
09230   Aggregate Coatings
09300   Tile
09400   Terrazzo
09500   Acoustical Treatment
09550   Wood Flooring
09600   Stone and Brick Flooring
09650   Resilient Flooring
09680   Carpeting
09700   Special Flooring
09760   Floor Treatment
09800   Special Coatings
09900   Painting
09950   Wall Covering

DIVISION 10—SPECIALITIES
10100   Chalkboards and Tackboards
10150   Compartments and Cubicles
10200   Louvers and Vents
10240   Grills and Screens
10230   Service Wall Systems
10260   Wall and Corner Guards
10270   Access Flooring
10280   Speciality Modules
10290   Pest Control
10300   Fireplaces and Stoves
10340   Prefabricated Steeples, Spires, and Cupolas
10350   Flagpoles
10400   Identifying Devices
10450   Pedestrian Control Devices
10500   Lockers
10520   Fire Extinguishers, Cabinets, and Accessories
10530   Protective Covers ●
10550   Postal Specialities
10600   Partitions
10650   Scales
10670   Storage Shelving
10700   Exterior Sun Control Devices
10750   Telephone Enclosures
10800   Toilet and Bath Accessories
10900   Wardrobe Specialities

DIVISION 11—EQUIPMENT
11010   Maintenance Equipment
11020   Security and Vault Equipment
11030   Checkroom Equipment
11040   Ecclesiastical Equipment
11050   Library Equipment

11060   Theater and Stage Equipment
11070   Musical Equipment
11080   Registration Equipment
11100   Mercantile Equipment
11110   Commercial Laundry and Dry Cleaning Equipment
11120   Vending Equipment
11130   Audio-Visual Equipment
11140   Service Station Equipment
11150   Parking Equipment
11160   Loading Dock Equipment
11170   Waste Handling Equipment
11190   Detention Equipment
11200   Water Supply and Treatment Equipment
11300   Fluid Waste Disposal and Treatment Equipment
11400   Food Service Equipment
11450   Residential Equipment
11460   Unit Kitchens
11470   Darkroom Equipment
11480   Athletic, Recreational, and Therapeutic Equipment
11500   Industrial and Process Equipment
11600   Laboratory Equipment
11650   Planetarium and Observatory Equipment
11700   Medical Equipment
11780   Mortuary Equipment
11800   Telecommunication Equipment
11850   Navigation Equipment

DIVISION 12—FURNISHINGS
12100   Artwork
12300   Manufactured Cabinets and Casework
12500   Window Treatment
12550   Fabrics
12600   Furniture and Accessories
12670   Rugs and Mats
12700   Multiple Seating
12800   Interior Plants and Plantings

DIVISION 13—SPECIAL CONSTRUCTION
13010   Air-supported Structures
13020   Integrated Assemblies
13030   Audiometric Rooms
13040   Clean Rooms
13050   Hyperbaric Rooms
13060   Insulated Rooms
13070   Integrated Ceilings
13080   Sound, Vibration, and Seismic Control
13090   Radiation Protection
13100   Nuclear Reactors
13110   Observatories
13120   Pre-engineered Structures
13130   Special-purpose Rooms and Buildings
13140   Vaults
13150   Pools
13160   Ice Rinks
13170   Kennels and Animal Shelters
13200   Seismographic Instrumentation
13210   Stress Recording Instrumentation
13220   Solar and Wind Instrumentation
13410   Liquid and Gas Storage Tanks
13510   Restoration of Underground Pipelines
13520   Filter Underdrains and Media
13530   Digestion Tank Covers and Appurtenances

(Courtesy of Construction Specification Institute and Construction Specifications Canada.)

3540 Oxygenation Systems
13550 Thermal Sludge Conditioning Systems
13560 Site Constructed Incinerators
13600 Utility Control Systems
13700 Industrial and Process Control Systems
13800 Oil and Gas Refining Installations and Control Systems
13900 Transportation Instrumentation
13940 Building Automation Systems
13970 Fire Suppression and Supervisory Systems
13980 Solar Energy Systems
13990 Wind Energy Systems

DIVISION 14—CONVEYING SYSTEMS
14100 Dumbwaiters
14200 Elevators
14300 Hoists and Cranes
14400 Lifts
14500 Material Handling Systems
14600 Turntables
14700 Moving Stairs and Walks
14800 Powered Scaffolding
14900 Transportation Systems

DIVISION 15—MECHANICAL
15050 HVAC and Piping Contractors
15200 Noise, Vibration, and Seismic Control
15250 Insulation
15300 Special Piping Systems
15400 Plumbing Systems
15450 Plumbing Fixtures and Trim
15500 Fire Protection
15600 Power or Heat Generation
15650 Refrigeration
15700 Liquid Heat Transfer
15800 Air Distribution
15900 Controls and Instrumentation

DIVISION 16—ELECTRICAL
16050 Electrical Contractors
16200 Power Generation
16300 Power Transmission
16400 Service and Distribution
16500 Lighting
16600 Special Systems
16700 Communications
16850 Heating and Cooling
16900 Controls and Instrumentation

(Courtesy of Construction Specification Institute and Construction Specifications Canada.)

Appendix L

There are hundreds of industry-related organizations. They are a valuable source of information regarding types and appplication of materials, industry standards, and construction information. Following is a partial list of the addresses of some of the more prominent organizations.*

INDUSTRY ASSOCIATIONS

ASA Acoustical Society of America
500 Sunnyside Blvd.
Woodberryl, NY 11797

ASC Adhesive and Sealant Council, Inc.
1627 K St. NW, Ste. 1000
Washington, DC 20006-1707

ARI Air-Conditioning and Refrigeration Institute
4301 N. Fairfax Drive, Ste. 425
Arlington, VA 22203

AA Aluminum Association
900 19th St. NW
Washington, DC 20006

ACI American Concrete Institute
PO Box 9094
Farmington Hills, MI 48333-9090

ACPA American Concrete Pavement Association
5420 Old Orchard Rd, No. A100
Skokie, IL 60077

ACPA American Concrete Pipe Association
222 W. Las Colinas Blvd, Ste 641
Irving, TX 75039

ACEC American Consulting Engineers Council
1015 15th St. NW, Suite 802
Washington, DC 20005

AFA American Fibreboard Association
1210 Northwest Highway
Palatine, IL 60007

AGA American Gas Association, Inc.
1515 Wilson Blvd.
Arlington, VA 22209

AHA American Hardboard Association
1210 W. Northwest Hwy.
Palatine, IL 60067

AHMA American Hardware Manufacturers Association
801 N. Plaza Drive Rd.
Schaumburg, IL 60173

AIA American Institute of Architects
1735 New York Ave., NW
Washington, DC   20006

AIC American Institute of Constructors
466 94th Ave.
St. Petersburg, FL 33702

AISC American Institute of Steel Construction, Inc.
1 E Wacker Dr., Ste. 3100
Chicago, IL   60611-2001

ASID American Society of Interior Designers
608 Massachusetts Ave. NE
Washington DC, 20002

AITCH American Institute of Timber Construction
7012 S Revere Pky, Ste. 140
Englewood, CO 80112

AISI American Iron and Steel Institute
1101 17th St. NW
Washington, DC 20036-4700

AAMA American Architectural Manufacturers Association
1827 Walden Office Sq., Suite 104
Schaumberg, IL 60173

APFA American Pipe Fitting Association
111 Park Place
Falls Church, VA 22046-4513

ARTBA American Road and Transportation Builders Association
The ARTBA Bldg.
1010 Massachusetts Ave. NW, 6th flr
Washington, DC 20001

ASCE American Society of Civil Engineers
1015 15th St. NW, Ste 600
Washington, DC 20005

ASHRAE American Society of Heating, Refrigerating and Air-Conditioning Engineers
1791 Tullie Circle, NE
Atlanta, GA 30329

ASME American Society of Mechanical Engineers
United Engineering Center
345 E. 47th St.
New York, NY 10017

ASPE American Society of Professional Estimators
11141 Georgia Ave., Ste 412
Wheaton, MD 20902

ASSE American Society of Sanitary Engineers
28901 Clemens Rd., Ste 100
Cleveland, OH 44145-1166

AWS American Welding Institute
10628 Dutchtown Rd.
Knoxville, TN 37932.

AAMA Architectural Aluminum Manufacturers Association
2700 River Rd., Suite 118
Des Plaines, IL 60018

APA Architectural Precast Association
P.O. Box 08669
Ft. Myers, FL 33908-0669

AWI Architectural Woodwork Institute
1952 Isaac Newton Square
Reston, VA 20190

AIA/NA Asbestos Information Association/North America
1745 Jefferson Davis Hwy, Ste 406
Arlington, VA 22202

AI Asphalt Institute
Research Park Drive
P.O. Box 14052
Lexington, Ky 40512-4052

ASC Associated Specialty Contractors
3 Bethesda Metro Center, Ste 1100
Bethesda, MD 20814

AWCI Association of the Wall and Ceiling Industries International
307E. Anadale Rd., Suite 200
Falls Church, VA 22042

BIA Brick Institute of America
11490 Commerce Park Dr., Suite 300
Reston, VA 22091

BHMA Builder's Hardware Manufacturers Association, Inc.
355 Lexington Ave., 17th flr.
New York, NY 10017

BSI Building Stone Institute
P.O. Box 507
Purdys, NY 10578

BSC Building Systems Council of NAHB
1201 15th St. NW
Washington, DC 20005

CPA Composite Panel Association
18928 Premiere Ct.
Gaithersburg, MD 20879

CRI Carpet and Rug Institute
310 Holiday Ave.
Dalton, GA 30722

CISPI Cast Iron Soil Pipe Institute
5959 Shallowford Rd., Ste 419
Chattanooga, TN 37421

CSSB Cedar Shake & Shingle Bureau
515 116th Ave., NE, Suite 275
Bellevue, WA 98004-5294

CISCA Ceilings and Interior Systems Construction Association
500 Lincoln Hwy, No. 202
St. Charles, IL 60174

CIMA Construction Industry Manufacturers Association
111 E. Wisconsin Ave.
Milwaukee WI 53202

CABO Council of American Building Officials
5203 Leesburg Pike, Suite 708
Falls Church, VA 22041

DFI Deep Foundation Institute
120 Charlotte Pl., 3rd flr
Englewood Cliffs, NJ 07632

DHI Door and Hardware Institute
14170 Newbrook Dr.
Chantilly, VA 20151-2223

DIPRA Ductile Iron Pipe Research Association
245 Riverchase Parkway E., Suite 0
Birmingham, AL 35244

EEBA Energy Efficient Building Association
2950 Metro Drive, Suite 108
Minneapolis, MN 55425

EIA Environmental Information Association
3050 Presidential Drive
Atlanta, GA 30340

ESCSI Expanded Shale, Clay and Slate Institute
2225 E. Murray Holladay Rd. Ste 102
Salt Lake City, UT 84117

FTI Facing Tile Institute
P.O. Box 8880
Canton, OH 44418

GA Gypsum Association
810 1st St. NE, No. 510
Washington, DC 20002

HPMA Hardwood Plywood & Veneer Manufacturers
Association
P.O. Box 2789
Reston, VA 22090

IESNA Illuminating Engineering Society of North
America
120 Wall St., 17th Flr.
New York, NY 10005-4001

IFI Industrial Fasteners Institute
1717 E. 9th St.. Ste. 1105
Cleveland, OH 44114-2879

IHEA Industrial Heating Equipment Association
1901 N. Moore St.
Arlington, VA 22209

IEEE Institute of Electrical and Electronics Engineers
345 E. 47th St.
New York, NY 10017

ICAA Insulation Contractors Association of America
P.O. Box 26237
Alexandria, VA 22313

IALD International Association of Lighting Designers
1133 Broadway, Ste. 520
New York, NY 10010-7903

IAPMO International Association of Plumbing and
Mechanical Officials
20001 Walnut Dr., S
Walnut, CA 91789-2825

ICBO International Council of Building Officials
5360 S. Workman Mill Rd.
Whittier, CA 90601

IILP International Institute for Lath and Plaster
820 Transfer Rd.
St. Paul, MN 55114-1406

IMI International Masonry Institute
823 15th St., NW
Washington, DC 20005

IRF International Road Federation
2600 Virginai Ave. NW, Ste 208
Washington, DC 20037

KCMA Kitchen Cabinet Mfg. Association
1899 Preston White Drive
Reston, VA 20191-5435

MSSVFI Manufacturers Standardization Society of the
Valve and Fittings Industry
127 Park St., NE
Vienna, VA 22180

MFMA Maple Flooring Manufacturers Association
60 Revere Dr., Suite 500
Northbrook, Il 60062

MIA Marble Institute of America
33505 State St.
Framington, MI 48024

MCAA Mechanical Contractors Association of
America
1385 Piccard Dr.
Rockville, MD 30850

MBSA Metal Building Manufacturers Association
1300 Summer Ave.
Cleveland, OH 44115

MFMA Metal Framing Manufacturers Association
401 N. Michigan
Chicago, IL 60611

MLSFA Metal Lath/Steel Framing Association
600 S. Federal St., Ste. 400
Chicago, IL 60605

NAPA National Asphalt Pavement Association
NAPA Bldg.
5100 Forbes Blvd.
Lanham, MD 20706-4413

NAFCD National Association of Floor Covering
Distributors
401 N. Michigan Ave.
Chicago, IL 60611
Washington, DC 20005

NAHRO National Association of Housing
Redevelopment Officials
630 Eye St. NW
Washington, DC 20001

NAPHCC National Association of Plumbing-Heating-
Cooling Contractors
303 E. Wacker Dr., Suite 711
Chicago, IL 60601

NARI National Association of the Remodeling
Industry
4900 Seminary Rd., Suite 320
Alexandria, VA 22311

NAWC National Association of Waterproofing
Contractors
25550 Shagrin St., Suite 403
Cleveland, OH 44122

NAWIC National Association of Women in
Construction
327 S. Adams St.
Fort Worth, TX 76104

NBMA National Building Manufacturers Association
142 Lexington Ave.
New York, NY 10016

NCMA National Concrete Masonry Association
2302 Horse Pen Rd.
Herndon, VA 20171-3499

NCSBCS National Conference of States on Building
Codes and Standards
505 Huntmar Park Drive, Ste 210
Herndon, VA 22070

NCA National Constructors Association
1730 M St. NW, Ste 503
Washington, DC 20036

NCRP National Council on Radiation Protection and
Measurement
7910 Woodmont Ave., Suite 800
Bethesda, MD 20814

NEMA National Electrical Manufacturers Association
1300 N. 17th St., Ste 1847
Rosslyn, VA 22209

NFPA National Fire Protection Association
Batterymarch Park
P.O. Box 9101
Quincy, MA 02269-9101

NGA National Glass Association
8200 Greensboro Dr., 3rd floor
McLean, VA 22102

NHRA National Housing Rehabilitation Association
1726 18th St., NW
Washington, DC 20009

NLBMDA National Lumber and Building Material
Dealers Association
40 Ivy St., SE
Washington, DC 20003

NOFMA National Oak Flooring Manufacturers
Association
P.O. Box 3009
Memphis, TN 38173-0009

NPCA National Paint and Coatings Association
1500 Rhode Island Ave., NW
Washington, DC 20005

NPCA National Precast Concrete Association
10333 N. Meridian St., Ste 272
Indianapolis, IN 46290-1081

NRMCA National Ready Mixed Concrete Association
900 Spring Street
Silver Spring, MD 20910

NRCA National Roofing Contractors Association
10255 W. Higgins Rd., Ste 600
Rosemont, IL 60018-5607

NSDJA National Sash and Door Jobbers Association
10225 Robert Trent Jones Pkwy.
New Point Richey, FL 34655

NSPE National Society of Professional Engineers
1420 King St.
Alexandria, VA 22314

NSA National Stone Association
1415 Elliot Pl. NW
Washington, DC 20007

NTMA National Terrazzo and Mosaic Association
3166 Des Plaines Ave., Suite 121
Des Plaines, IL 60018

NTRMA National Tile Roofing Manufacturers
Association
P.O. Box 40337
Eugene, OR 97404

NWWDA National Wood Window and Door
Association
1400 E. Touhy Ave., No. 470
Des Plaines, IL 60018

NWMA National Woodwork Manufacturers
Association
400 W. Madison St.
Chicago, IL 60606

NABMDA North American Building Material
Distributors Association
401 N. Michigan Ave.
Chicago, IL 60611

NAIMA North American Insulation Manufacturers
Association
44 Canal Center Plaza, Suite 310
Alexandria, VA 22314

PLCA Pipe Line Contractors Association
1700 Pacific Ave., Ste 4100
Dallas, TX 75201

PPI Plastic Pipe Institute
1801 K St. NW, Ste 600K
Washington, DC 20006

PDI Plumbing and Drainage Institute
45 Bristol Dr., Ste 101
South Easton, MA 02375-1916

PMI Plumbing Manufacturers Institute
Bldg. C, Ste 20
800 Roosevelt Rd.
Glen Ellyn, IL 60137

PCA Portland Cement Association
5420 Old Orchard Rd.
Skokie, IL 60077

PTI Post Tensioning Institute
1717 W. Northern Ave. Suite 114
Phoenix, AZ 85021

PCI Prestressed Concrete Institute
175 W. Jackson Blvd., Suite 1859
Chicago, IL 60604

RCRC Reinforced Concrete Research Council
Texas A & M University
Dept of Civil Engineering
College Station, TX 77843-3136

RFCI Resilient Floor Covering Institute
966 Hungerford Drive, Ste 12-B
Rockville, MD 20850

SSFI Scaffolding, Shoring, and Forming Institute, Inc.
c/o Thomas Assoc., Inc.
1300 Summer Ave.
Cleveland, OH 44115

SMA Screen Manufacturers Association
2850 S. Ocean Blvd., No. 114
Palm Beach, FL 33480-5535

SWRI Sealant Waterproofing & Restoration Institute
2841 Main
Kansas City, MO 64108

SIGMA Sealed Insulating Glass Manufacturers
Association
401 N. Michigan Ave.
Chicago, IL 60611

SDI Steel Deck Institute
P.O. Box 9506
Canton, OH 44711

SDI Steel Door Institute
30200 Detroit Rd.
Cleveland, OH 44145

SSPC Steel Structures Painting Council
40 24th St., 6th flr.
Pittsburgh, PA 15222-4643

SWI Steel Window Institute
1300 Sumner Ave.
Cleveland, OH 44115

SMA Stucco Manufacturers Association
507 Evergreen Ave.
Pacific Grove, CA 93590

SBA System Builders Association
28 Lowry Dr.
P.O. Box 117
West Milton, OH 45383

TCA Tile Council of America
P.O. Box 1787
Clemson, SC 29633

TCA Tilt-up Concrete Association
121 1/2 1st St. W.
P.O. Box 204
Mt. Vernon, IA 52314

TPI Truss Plate Institute
583 D'Onofrio Dr., Suite 200
Madison, WI 53719

WA Wallcoverings Association
401 N. Michigan Ave.
Chicago, IL 60611-4267

WRI Wire Reinforcement Institute
P .O. Box 450
Findlay, OH 45839

WMMPA Wood Moulding and Millwork Producers
Association
507 1st St.
Woodland, CA 95695

EMPLOYER ORGANIZATIONS

ACCA Air Conditioning Contractors of America
1712 New Hampshire Ave. NW
Washington, DC 20009

ABC Association of Bituminous Contractors
1747 Pennsylvania Ave. NW
Washington, DC 20006

ABCA American Building Contractors Association
P.O. Box 39277
Downey, CA 90239-0277

ASA American Subcontractors Association
1004 Duke St.
Alexandria, VA 22314

ABC Associated Builders and Contractors, Inc.
1300 N. 17th St.
Rosslyn, VA 22209

AGC Associated General Contractors of America
1957 East St. NW
Washington, DC 20006

GBCA General Building Contractors Association
36 S. 18th St.
P.O. Box 15959
Philadelphia, PA 19103

MCAA Mason Contractors Association of America
1910 S. Highland Ave., Ste 101
Lombard, IL 60148

MCAA Mechanical Contractors Association of
America
5410 Grosvener Ave., Suite 120
Bethesda, MD 20814

NADC National Association of Demolition
Contractors
16 N. Franklin St.
Doylestown, PA 18901

NAEC National Association of Elevator Contractors
1298 Wellbrook Cir.
Conyers, GA 30207

NAHB National Association of Home Builders
15th and M St., NW
Washington, DC 20005

NAPHCC National Association of Plumbing, Heating,
and Cooling Contractors
180 S. Washington St.
P.O. Box 6808
Falls Church, VA 22040

NARSC National Association of Reinforcing Steel
Contractors
10382 Main St., Ste. 300
P.O. Box 280
Fairfax, VA 22030

NECA National Electrical Contractors Association
3 Bethesda Metro Center, Ste 1100
Bethesda, MD 20814

NRCA National Roofing Contractors Association
10255 W. Higgins Rd., Ste 600
Rosemont, IL 60018-5607

PDCA Painting and Decorating Contractors of America
3913 Old Lee Hwy, Ste 33B
Fairfax, VA 22030

SMACNA Sheet Metal & Air Conditioning Contractors National Association, Inc.
4201 LaFayette Center Dr.
Chantilly, VA 20151-1209

LABOR UNIONS

BCTD Building and Construction Trades Department - AFL CIO
1155 15th St. NW, 4th Floor
Washington, DC 20005

IABSOIW International Association of Bridge, Structural and Ornamental Iron Workers
1750 New York Ave. NW
Washington, DC 20006

IBEW International Brotherhood of Electrical Workers
1125 15th Street, NW
Washington, DC 20005

IBPAT International Brotherhood of Painters and Allied Trades
1750 New York Ave, NW
Washington, DC 20006

IUBAC International Union of Bricklayers and Allied Craftsmen
Bowen Building
815 15th St. NW
Washington, DC 20005

IUEC International Union of Elevator Constructors
Clark Building
5565 Sterrett Place, Suite 310
Columbia, MD 21044

IUOE International Union of Operating Engineers
1125 17th Street, NW
Washington, DC 20036

LIUNA Laborers' International Union of North America
16th St. NW
Washington, DC 20006

OPCMIA Operative Plasterers and Cement Masons International Association of the United States and Canada
1440 S. Laurel Place, Suite 300
Laurel, MD, 20707

SMWIA Sheet Metal Workers International Association
1750 New York Ave. NW
Washington, DC 20006

UAPPS United Association of Journeymen & Apprentices of the Plumbing & Pipe Fitting Industry of the United States & Canada
P.O. Box 37800
Washington, DC 20013

UBCJA United Brotherhood of Carpenters and Joiners of America
101 Constitution Ave. NW
Washington, DC 20001

UURWAW United Union of Roofers, Waterproofers and Allied Workers
1125 17th St. NW, 5th Floor
Washington, DC 20036

GRADE, STANDARDS, AND RESEARCH INSTITUTES

AF&PA American Forest & Paper Association
1250 Connecticut Ave. NW, Suite 200
Washington, DC 20036

ALSC American Lumber Standard Committee
P.O. Box 210
Germantown, MD 20875

ANSI American National Standards Institute
11 W. 42nd St., 13th flr.
New York, NY 10036

APA American Plywood Association—An Engineered Wood Research Association
P.O. Box 11700
Tacoma, WA 98411

ASTM American Society for Testing and Materials
100 Barr Harbor Dr.
W. Conshohocken, PA 19428

BMRI Building Materials Research Institute, Inc.
501 5th Ave., #1402
New York, NY 10007

BRC Building Research Council
U of Illinois at Urbana-Champaign
1E St. Mary's Road
Champaign, IL 61820

CRA California Redwood Association
405 Enfrente Dr., Suite 200
Novato, CA 94949

CSI Construction Specifications Institute
601 Madison St.
Alexandria, VA 22314

FPS Forest Products Society
2801 Marshall Ct.
Madison, WI 53705-2295

MMSA Material and Methods Standards Association
P.O. Box 350
Grand Haven, MI 49417

UL Underwriters' Laboratories, Inc
333 Pfingsten Rd.
Northbrook, IL 60062

WWPA Western Wood Products Association
Yeon Building
522 S.W. 5th Ave.
Portland, OR 97204-2122

MODEL BUILDING CODES

ASME American Society of Mechanical Engineers -
National Plumbing Code
345 E. 47th St.
New York, NY 10017

BOCA Building Officials and Code Aministrators
International - BOCA National Building Code
4051 Flossmoor Rd.
Country Club Hills, IL 60478

ICBO International Conference of Building Officials -
Uniform Building Code
5360 Workman Mill Rd.
Whittier, CA 90601

NFPA National Fire Protection Association - National
Electrical Code
1 Batterymarch Park
P.O. Box 9101
Quincy, MA 02269-9101

SBCCI Southern Building Code Congress
International, Inc. - Standard Building Code
900 Montclair Rd.
Birmingham, AL 35213

GOVERNMENT AGENCIES

EPA Environmental Protection Agency
401 M St. SW
Washington, DC 20460

FHA Federal Housing Administration
451 7th St. SW, Rm 3158
Washington, DC 20410

OSHA U.S. Department of Labor/Occupational Safety
and Health Administration
200 Constitution Ave. NW
Washington, DC 20210

*All the above addresses are as they appear in the
Encyclopedia of Associations 1998.

Index